HARMONIC FUNCTIONS AND RANDOM WALKS ON GROUPS

Research in recent years has highlighted the deep connections between the algebraic, geometric, and analytic structures of a discrete group. New methods and ideas have resulted in an exciting field, with many opportunities for new researchers. This book is an introduction to the area from a modern vantage point. It incorporates the main basics, such as Kesten's amenability criterion, the Coulhon and Saloff-Coste inequality, random walk entropy and bounded harmonic functions, the Choquet–Deny theorem, the Milnor–Wolf theorem, and a complete proof of Gromov's theorem on polynomial growth groups.

The book is especially appropriate for young researchers, and those new to the field, accessible even to graduate students. An abundance of examples, exercises, and solutions encourage self-reflection and the internalization of the concepts introduced. The author also points to open problems and possibilities for further research.

Ariel Yadin is Professor in the Department of Mathematics at Ben-Gurion University of the Negev, Israel. His research is focused on the interplay between random walks and the geometry of groups. He has taught a variety of courses on the subject and has been part of a new wave of investigation into the structure of spaces of unbounded harmonic functions on groups.

T0332204

CAMBRIDGE STUDIES IN ADVANCED MATHEMATICS

All the titles listed below can be obtained from good booksellers or from Cambridge University Press.
For a complete series listing, visit www.cambridge.org/mathematics.

Harmonic Functions and Random Walks on Groups

ARIEL YADIN

Ben-Gurion University of the Negev

CAMBRIDGE
UNIVERSITY PRESS

Shaftesbury Road, Cambridge CB2 8EA, United Kingdom

One Liberty Plaza, 20th Floor, New York, NY 10006, USA

477 Williamstown Road, Port Melbourne, VIC 3207, Australia

314–321, 3rd Floor, Plot 3, Splendor Forum, Jasola District Centre, New Delhi – 110025, India

103 Penang Road, #05–06/07, Visioncrest Commercial, Singapore 238467

Cambridge University Press is part of Cambridge University Press & Assessment, a department of the University of Cambridge.

We share the University's mission to contribute to society through the pursuit of education, learning and research at the highest international levels of excellence.

www.cambridge.org
Information on this title: www.cambridge.org/9781009123181

DOI: 10.1017/9781009128391

First published 2024

A catalogue record for this publication is available from the British Library

A Cataloging-in-Publication data record for this book is available from the Library of Congress

ISBN 978-1-009-12318-1 Hardback

Contents

Preface

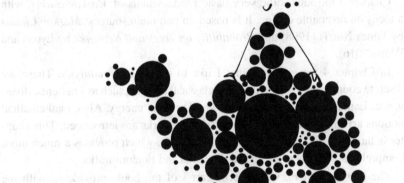

What This Book Is About

In recent years I have become more and more interested in the connections between the different possible mathematical structures on a set, given by algebra, by geometry, and by analysis. A group, by definition, has an obvious algebraic structure – its multiplication table. Finitely generated groups come equipped with a metric (inherited from the graph structure of the *Cayley graph*). Also, since Polyá, Feller, and Kesten it has been noted that the stochastic process known as a *random walk* constrains and is constrained by these structures. One may say that this book is about the interplay between the above structures.

Chapter 1 introduces the basic concepts we will work with throughout the book: random walks, harmonic functions, group properties, and basic examples.

In Chapter 2 we review an important probabilistic object: the *martingale*. This chapter is largely based on the highly influential book *Probability: Theory and Examples* by Rick Durrett (2019).

Chapter 3 introduces the very basic fundamentals of *Markov chains*, with a focus on reversible chains. It is based on two main sources: *Markov Chains* by James Norris (1998) and *Probability on Trees and Networks* by Lyons and Peres (2016).

In Chapter 4 we get into what I like to call *discrete analysis*. These are discrete counterparts to objects from classical analysis such as gradients, divergence, Laplacian, Green's function, and Dirichlet energy. Also, mathematical notions arising from models of electrical networks are introduced. This chapter is largely based on Lyons and Peres (2016), which provides a much more comprehensive study of electrical networks and random walks.

The above chapters, which form Part I of the book, provide us with the necessary tools to study harmonic functions and random walks on groups. Part II of the book deals with the basic questions regarding these objects and how to apply them to the study of the geometric and algebraic properties of the group.

We start with the relationship of isoperimetry and amenability with the random walk, and specifically *heat kernel decay*. This is done in Chapter 5. The main source for this chapter is *Random Walks on Infinite Graphs and Groups* by Woess (2000). In this chapter, Kesten's thesis is presented, which relates the algebraic property of *amenability* to the analytic property of a spectral gap for the *random walk Laplacian*.

In Chapter 6 we study the space of bounded harmonic functions. This is a well-studied topic, and has a huge body of literature expanding into many other areas of mathematics not mentioned here. This chapter only presents the tip of the iceberg. The main result here is the well-known *entropic criterion* for the existence of nonconstant bounded harmonic functions (the *Liouville property*).

Still related to bounded harmonic functions, in Chapter 7 we discuss the Choquet–Deny phenomena, sometimes known as the *strong Liouville property*, where a group never has any nonconstant bounded harmonic functions (with respect to any random walk). This chapter includes some very recent work by Frisch, Hartman, Tamuz, and Vahidi-Ferdowsi (2019), which shows that the (seemingly analytic) *Choquet–Deny property* is equivalent to a group being *virtually nilpotent* (an algebraic property).

The notion of *growth* is studied further in Chapter 8. The main result of the chapter is a dichotomy, originally due to results of Milnor and Wolf, for *solvable groups*. The algebra of these groups constrains their geometry, and the Milnor–Wolf theorem states that they can only have *growth* that is either polynomial or exponential, but not in between. The relationship between algebra and geometry is strengthened here. This chapter is based on material from Woess (2000) and *Geometric Group Theory* by Druțu and Kapovich (2018).

Finally, we culminate with Chapter 9, in which Gromov's theorem is proved. The theorem relates the geometric property of *polynomial growth* to the algebraic property of *virtual nilpotence*, telling us that these properties are actually equivalent. The proof presented for Gromov's theorem is elementary, and utilizes the tools previously developed. This proof is based on ideas from Shalom and Tao (2010) and from Tao's blog *What's New*, as well as the paper by Ozawa (2018) proving Gromov's theorem.

This book has been written from the perspective of someone trained in probability. Many times probabilistic proofs are used, even if other proofs are available. Also, an effort has been made to keep everything as elementary as possible, and facilitate reading for students and beginners. This sometimes leads to repetition of ideas and more "hands-on" proofs.

The chapters give only a taste of the material mentioned. More depth can be achieved by going to one of the relevant sources mentioned in this section.

An advantage of this book is that it brings together topics that usually appear in separate texts in a unified and self-contained fashion. Another advantage is that it is elementary in the sense that it is suitable for newcomers. For example, I have avoided the use of more advanced or less known topics such as the *Tits Alternative, Zariski topology, property (T)*, and *ultraproducts*. However, these are all extremely interesting and important notions: I do not wish anyone to be discouraged from reading about these in other sources!

How to Read This Book

This book grew out of lecture notes written for courses I have taught over the years about random walks, harmonic functions, and their relationship to the geometry of finitely generated groups. The main objective in writing the book is to enable a beginner to enter into the field of random walks on groups. The book provides the main examples and basic results in the field. An effort has been made to keep everything elementary, in the sense that a good third-year undergraduate or first-year graduate student can read the book in a self-contained fashion.

The reader should be familiar with the basics of mathematics as taught in most undergraduate studies at universities around the world. Specifically, one requires linear algebra, the basics of probability and measure theory, as well as a very basic introduction to abstract group theory. A beginner's familiarity with Hilbert spaces, Shannon entropy, and couplings is an advantage; the main definitions and results required are in the appendices. Part I of the book contains all the necessary tools for the bigger theorems presented in Part II. Some of the topics covered in Part I may be well known to some readers; I have opted to make this book accessible to more people and to keep the book self contained.

For many of the theorems in the book, the proofs are broken down into steps that are spelt out through lemmas and various exercises. I encourage the reader to attempt to solve the exercises along the way themselves, which would assist not only in internalizing the notions but also in grasping the main ideas. Many technical steps from proofs are also delegated to exercises, so that hopefully in the actual proofs one can see the big picture without getting lost in the details.

After the table of contents there is a list of notation and definitions. This is to facilitate finding the original place where some notion or notation is defined.

Acknowledgments

As mentioned, this book evolved through lecture notes for courses. Naturally, many of the proofs have not been preserved in the original form in which they appeared for the first time. The typical evolutionary path a proof would have taken could be described generically as follows:

- A proof was published in a paper.
- Someone gave a course or wrote a book including the proof.
- I read the proof or learned it in some book or course as a graduate student.
- I explained the proof to some peers, so had to recall what I read or heard in a past lecture.
- The proof was eventually written down by me in some lecture notes I handed out.
- The proof was edited in subsequent lecture notes, most likely influenced by remarks from colleagues and students.

At this point in time, it is very hard for me to trace back exactly where the different proofs in this book came from, the intermediate "fossils" having been lost to time (sitting somewhere on an old lost hard drive).

However, I can list the sources that have definitely influenced and taught me over the years, which probably cover the vast majority of the material in this book.

I was introduced to random walks in my graduate studies. Definitely the most influential resource I had then for random walks was *Probability on Trees and Networks* by Russell Lyons and Yuval Peres. At the time it was yet unpublished, and I had different versions throughout the years.

I was seduced into the world of random walks on groups and harmonic functions by a course I took in 2011 at Tel Aviv University by Yehuda Shalom, where he presented an elementary proof of Gromov's theorem based on his

xv

work with Terence Tao (Shalom and Tao, 2010). I took notes during that course, which I supplemented with ideas from Tao's blog *What's new*.

Gabor Pete's book in preparation, *Probability and Geometry on Groups*, is one of my go-to resources for anything under that topic. It is not easy for me to isolate Gabor's specific fingerprint from my notes. Most likely his influence is diffused throughout everything I have written.

Another source I often use is *Random Walks on Infinite Graphs and Groups* by Wolfgang Woess, especially regarding Varopolous' theorem, recurrence, and transience.

Anything appearing here is most likely some kind of logical descendant of the above, if not even more directly related. I am greatly in debt to the authors mentioned.

Finally, many people have offered me their comments and assistance in different stages of writing, teaching, and presenting the material included. I will make an attempt to list them all, but most likely there will be someone left out and I apologize in advance to anyone for which this is the case.

The material present has benefitted from many discussions I had with many people. Among them are Itai Benjamini, Hugo Duminil-Copin, Yair Glasner, Yair Hartman, Gady Kozma, Tom Meyerovitch, Yuval Peres, Idan Perl, Gabor Pete, Yehuda Shalom, Maud Szusterman, Matthew Tointon, Wolfgang Woess, Amir Yehudayoff, and Ofer Zeitouni.

At different stages, many people have contributed remarks on versions of the manuscript. These include Adam Dor-On, Guy Drori, Dor Elimelech, Yair Glasner, Yair Hartman, Gady Kozma, Tom Meyerovitch, Christophe Pittet, Liran Ron-George, Guy Salomon, Matthew Tointon, Vasiliki Velona-Anastasiou, Yeari Vigder, and Jiayan Ye.

I also thank the team of editors and typesetters at Cambridge University Press, who gave me many comments and corrections to typos. Some worked in the background and are unknown to me, and with some I have had direct correspondence. Among these, let me especially mention Eleanor Bolton, Clare Dennison, Jasintha Jacob Srinivasan, Anna Scriven, and David Tranah.

Itai Benjamini has graciously decorated the book with his wonderful illustrations, and I thank him for this nice touch.

A very special thanks goes to my partner Tovi, for all her support and encouragement. "The book won't write itself."

Notation

PART I

Tools and Theory

1

Background

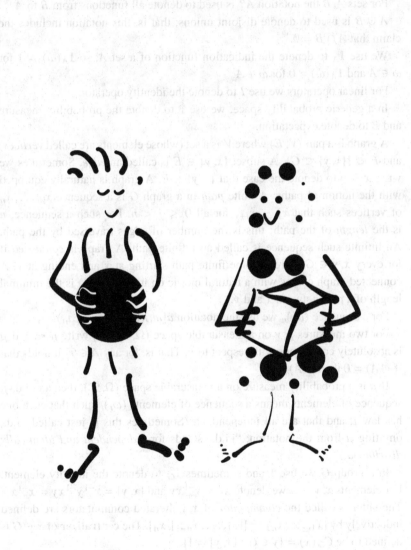

1.1 Basic Notation

The sets $\mathbb{N}, \mathbb{Z}, \mathbb{Q}, \mathbb{R}$, and $\mathbb{C}$ denote the sets of natural, whole, rational, real, and complex numbers, respectively. We assume that $\mathbb{N}$ contains the number 0.

For real numbers $a, b \in \mathbb{R}$ we use $a \wedge b = \min\{a, b\}$ and $a \vee b = \max\{a, b\}$.

For sets A, B the notation A^B is used to denote all functions from B to A.

$A \uplus B$ is used to denote disjoint unions; that is, this notation includes the claim that $A \cap B = \emptyset$.

We use $\mathbf{1}_A$ to denote the indication function of a set A; so $\mathbf{1}_A(\omega) = 1$ for $\omega \in A$ and $\mathbf{1}_A(\omega) = 0$ for $\omega \notin A$.

For linear operators we use I to denote the identity operator.

In a generic probability space, we use $\mathbb{P}$ to denote the probability measure and $\mathbb{E}$ to denote expectation.

A graph is a pair (V, E) where V is a set (whose elements are called *vertices*) and $E \subset \{\{x, y\} \subset G\}$. A subset $\{x, y\} \in E$ is called an *edge*. Sometimes we write $x \sim y$ to denote the case that $\{x, y\} \in E$. A graph is naturally equipped with the notion of paths: A finite *path* in a graph G is a sequence $x_0, \ldots, x_n$ of vertices such that $x_j \sim x_{j+1}$ for all $0 \le j < n$. For such a sequence, n is the *length* of the path; this is the number of edges traversed by the path. An infinite such sequence is called an infinite path. A graph is *connected* if for every $x, y \in G$ there is some finite path starting at x and ending at y. A connected graph comes with a natural metric on it: $\mathrm{dist}_G(x, y)$ is the minimal length of a path between x and y.

For a sequence $(a_n)_n$ we use the notation $a[m, n] = (a_m, \ldots, a_n)$.

For two measures μ, ν on a measurable space $(\Omega, \mathcal{F})$, we write $\mu \ll \nu$ if μ is absolutely continuous with respect to ν. That is, for any $A \in \mathcal{F}$ it holds that if $\nu(A) = 0$ then $\mu(A) = 0$.

If μ is a probability measure on a measurable space $(\Omega, \mathcal{F})$, then an i.i.d.-μ sequence of elements means a sequence of elements $(\omega_t)_t$ such that each one has law μ and that are all independent. (Sometimes this is just called i.i.d., omitting μ from the notation; "i.i.d." stands for *independent and identically distributed*.)

In a group G we use 1 and sometimes 1_G to denote the identity element. For elements $x, y \in G$ we denote $x^y = y^{-1}xy$ and $[x, y] = x^{-1}y^{-1}xy = x^{-1}x^y$. The latter is called the *commutator* of x, y. Iterated commutators are defined inductively by $[x_1, \ldots, x_n] := [[x_1, \ldots, x_{n-1}], x_n]$. The **centralizer** of $x \in G$ is defined to be $C_G(x) = \{y \in G : [x, y] = 1\}$.

For $A \subset G$ we write $A^x = \{a^x : a \in A\}$ and $A^{-1} = \{a^{-1} : a \in A\}$. A is called **symmetric** if $A = A^{-1}$. A group G is **generated** by a subset $S \subset G$ if every element of G can be written a product of finitely many elements from

$S \cup S^{-1}$. Also, $\langle A \rangle$ denotes the subgroup generated by the elements of A; that is, all elements that can be written as a product of finitely many elements from $A \cup A^{-1}$. For two subsets $A, B \subset G$ we write $[A, B] = \langle [a, b] : a \in A, \ b \in B \rangle$ (note that this is the group generated by all commutators, not just the set of commutators). We also denote $AB = \{ab : a \in A, \ b \in B\}$.

1.2 Spaces of Sequences

Let G be a countable set. Let us briefly review the formal setup of the canonical probability spaces on $G^{\mathbb{N}}$. This is the space of sequences $(\omega_n)_{n=0}^{\infty}$ where $\omega_n \in G$ for all $n \in \mathbb{N}$. A cylinder set is a set of the form

$$C(J, \omega) = \left\{ \eta \in G^{\mathbb{N}} \mid \forall j \in J, \ \eta_j = \omega_j \right\}, \qquad J \subset \mathbb{N}, 0 < |J| < \infty, \quad \omega \in G^{\mathbb{N}}.$$

It is also natural to define $C(\emptyset, \omega) = G^{\mathbb{N}}$. Let $X_j : G^{\mathbb{N}} \to G$ be the map $X_j(\omega) = \omega_j$ projecting onto the jth coordinate. For times $t > s$ we also use the notation $X[s, t] = (X_s, X_{s+1}, \ldots, X_t)$.

Define the **cylinder σ-algebra**

$$\mathcal{F} = \sigma(X_0, X_1, X_2, \ldots) = \sigma\left(X_n^{-1}(g) \mid n \in \mathbb{N}, \ g \in G \right).$$

Exercise 1.1 Show that

$$\mathcal{F} = \sigma(X_0, X_1, X_2, \ldots) = \sigma\left(C(J, \omega) \mid 0 < |J| < \infty, \ J \subset \mathbb{N}, \quad \omega \in G^{\mathbb{N}} \right)$$
$$= \sigma\left(C(\{0, \ldots, n\}, \omega) \mid n \in \mathbb{N} \quad \omega \in G^{\mathbb{N}} \right).$$

Show that $\eta \in C(J, \omega)$ if and only if $C(J, \omega) = C(J, \eta)$.

For $t \geq 0$ we denote

$$\mathcal{F}_t = \sigma(X_0, \ldots, X_t).$$

Exercise 1.2 Show that $\mathcal{F}_t \subset \mathcal{F}_{t+1} \subset \mathcal{F}$. (A sequence of σ-algebras with this property is called a *filtration*.) Conclude that

$$\mathcal{F} = \sigma\left(\bigcup_t \mathcal{F}_t \right).$$

Theorems of Carathéodory and Kolmogorov tell us that the probability measure $\mathbb{P}$ on $\left(G^{\mathbb{N}}, \mathcal{F} \right)$ is completely determined by knowing the marginal probabilities $\mathbb{P}[X_0 = g_0, \ldots, X_n = g_n]$ for all $n \in \mathbb{N}, g_0, \ldots, g_n \in G$. That is, when G is countable, Kolmogorov's extension theorem implies the following:

Theorem 1.2.1 *Let $(P_t)_t$ be a sequence of probability measures, where each P_t is defined on $\mathcal{F}_t$. Assume that these measures are consistent in the sense that for all t,*

$$P_{t+1}(\{(X_0, \ldots, X_t) = (g_0, \ldots, g_t)\}) = P_t(\{(X_0, \ldots, X_t) = (g_0, \ldots, g_t)\})$$
(1.1)

for any $g_0, \ldots, g_t \in G$. Then, there exists a unique probability measure $\mathbb{P}$ on $\left(G^{\mathbb{N}}, \mathcal{F}\right)$ such that for any $A \in \mathcal{F}_t$ we have $\mathbb{P}(A) = P_t(A)$.

Details can be found in Durrett (2019, appendix A).

Exercise 1.3 Let $(P_t)_t$ be a sequence of probability measures, where each P_t is defined on $\mathcal{F}_t$. Show that (1.1) holds if and only if for any $t < s$ and any $A \in \mathcal{F}_t$, we have $P_s(A) = P_t(A)$.

The space $G^{\mathbb{N}}$ comes equipped with a natural *shift operator*: $\theta \colon G^{\mathbb{N}} \to G^{\mathbb{N}}$ given by $\theta(\omega)_t = \omega_{t+1}$ for all $t \in \mathbb{N}$.

Exercise 1.4 Show that $\theta^t(A) \in \mathcal{F}$ for any $A \in \mathcal{F}$.   ▷ solution ◁

Exercise 1.5 Let $\mathcal{K} \subset \mathcal{F}$ be a collection of events. Show that if $\mathcal{G} = \sigma(\mathcal{K})$ is the σ-algebra generated by $\mathcal{K}$, then $\theta^{-t}\mathcal{G} := \{\theta^{-t}(A) : A \in \mathcal{G}\}$ is a σ-algebra, and in fact $\theta^{-t}\mathcal{G} = \sigma\left(\theta^{-t}(K) : K \in \mathcal{K}\right)$.   ▷ solution ◁

Exercise 1.6 Show that $\theta^{-1}(A) \in \mathcal{F}$ for any $A \in \mathcal{F}$.   ▷ solution ◁

Exercise 1.7 Define

$$\sigma(X_t, X_{t+1}, \ldots) = \sigma\left(X_{t+j}^{-1}(g) : g \in G, \, j \geq 0\right).$$

Show that $\sigma(X_t, X_{t+1}, \ldots) = \theta^{-t}\mathcal{F} = \{\theta^{-t}(A) : A \in \mathcal{F}\}$.   ▷ solution ◁

Exercise 1.8 If $\sim$ is an equivalence relation on Ω, we say that a subset $A \subset \Omega$ **respects** $\sim$ if for any $\omega \sim \eta \in \Omega$ we have $\omega \in A \iff \eta \in A$.

Show that the collection of subsets A that respect the equivalence relation $\sim$ forms a σ-algebra on Ω.

Exercise 1.9 Define an equivalence relation on $G^{\mathbb{N}}$ by $\omega \sim_t \omega'$ if $\omega_j = \omega'_j$ for all $j = 0, 1, \ldots, t$.

Show that this is indeed an equivalence relation.

Show that $\sigma(X_0, X_1, \ldots, X_t) = \{A : A \text{ respects } \sim_t\}$.

1.3 Group Actions

A (left) group action $G \curvearrowright X$ is a function from $G \times X$ to X, $(\gamma, x) \mapsto \gamma.x$, that is *compatible* in the sense that $(\gamma\eta).x = \gamma.(\eta.x)$, and such that $1.x = x$ for all $x \in X$. We usually denote $\gamma.x$ or γx for the action of $\gamma \in G$ on $x \in X$.

A *right* action is analogously defined for $(x, \gamma) \mapsto x.\gamma$ and compatibility is $x.(\gamma\eta) = (x.\gamma).\eta$ (and $x.1 = x$ for all $x \in X$).

Exercise 1.10 Let $G \curvearrowright X$ be a left group action. For any $\gamma \in G$ and $x \in X$ define $x.\gamma := \gamma^{-1}.x$. Show that this defines a right action of G on X.

Conversely, show that if G acts on X from the right, then defining $\gamma.x = x.\gamma^{-1}$ is a left action.

The bijections on a set X form a group with the group operation given by composition of functions. A (left) group action $G \curvearrowright X$ can be thought of as a homomorphism from the group into the group of bijections on X.

Sometimes, we wish to restrict to some subgroup of bijections on X when X has some additional structure. For example, if X is a topological space, we say that G acts on X by homeomorphisms if every element of G is a homeomorphism of X, when thinking of elements of G as identified with their corresponding bijection of X. That is, an action by homeomorphisms is a group homomorphism from G into the set of homeomorphisms of X.

Similarly, if $\mathbb{H}$ is some Hilbert space, then a group G acts on $\mathbb{H}$ by unitary operators if every element of G is mapped to a unitary operator of $\mathbb{H}$. This is just a group homomorphism from G into the group of unitary operators on $\mathbb{H}$.

Exercise 1.11 Show that any group acts on itself by left multiplication; that is, $G \curvearrowright G$ by $x.y := xy$.

Exercise 1.12 Let $\mathbb{C}^G$ be the set of all functions from $G \to \mathbb{C}$. Show that $G \curvearrowright \mathbb{C}^G$ by $(x.f)(y) := f\left(x^{-1}y\right)$.

Show that $f^x(y) := f\left(yx^{-1}\right)$ defines a *right* action.

Exercise 1.13 Generalize the previous exercise as follows:

Suppose that $G \curvearrowright X$. Consider $\mathbb{C}^X$, all functions from $X \to \mathbb{C}$. Show that $G \curvearrowright \mathbb{C}^X$ by $\gamma.f(x) := f\left(\gamma^{-1}.x\right)$, for all $f \in \mathbb{C}^X, \gamma \in G, x \in X$.

Show that the action $G \curvearrowright \mathbb{C}^X$ is *linear*; that is, $\gamma.(\zeta f + h) = \zeta(\gamma.f) + h$, for all $f, h: X \to \mathbb{C}, \zeta \in \mathbb{C}$, and $\gamma \in G$.

Show that $f^\gamma(x) := f(\gamma.x)$ defines a right action of G on $\mathbb{C}^X$. Show that this right action is linear as well.

Exercise 1.14 Let $G \curvearrowright X$. Let $\mathcal{M}_1(X)$ be the set of all probability measures on $(X, \mathcal{F})$, where $\mathcal{F}$ is some σ-algebra on X. Suppose that for any $g \in G$ the function $g \colon X \to X$ given by $g(x) = g.x$ is a measurable function. (In this case we say that G acts on X by measurable functions.)

Show that $G \curvearrowright \mathcal{M}_1(X)$ by

$$\forall A \in \mathcal{F} \qquad g.\mu(A) := \mu\left(g^{-1}A\right),$$

where $g^{-1}A := \left\{g^{-1}.x \colon x \in A\right\}$.

Exercise 1.15 Let $F = \{f \colon G \to \mathbb{C} \colon f(1) = 0\}$. Show that

$$(x.f)(y) := f\left(x^{-1}y\right) - f\left(x^{-1}\right)$$

defines a left action of G on F.

Notation Throughout this book, unless specified otherwise, we will always use the left action $\gamma.f(x) = f\left(\gamma^{-1}x\right)$ for $G \curvearrowright X$ and $f \colon X \to \mathbb{C}$.

Definition 1.3.1 Let $G \curvearrowright X$ be a (left) action. For $A \subset X$ and $\gamma \in G$ define $\gamma.A = \{\gamma.x \colon x \in A\}$. For $F \subset G$ denote $F.A = \{\gamma.x \colon \gamma \in F, \ x \in A\}$.

A subset $A \subset X$ is called G-**invariant** if $\gamma.A \subset A$ for all $\gamma \in G$; equivalently, $G.A = A$.

Definition 1.3.2 For a group action $G \curvearrowright X$ and some $x \in X$, the set $G.x := \{g.x \colon g \in G\}$ is called the **orbit** of x under G. The **stabilizer** of x is the subgroup $\text{stab}(x) = \{g \in G \colon g.x = x\}$.

Exercise 1.16 Show that for $G \curvearrowright X$ any stabilizer $\text{stab}(x)$ is indeed a subgroup.

▷ solution ◁

Exercise 1.17 (Orbit-Stabilizer theorem) Let $G \curvearrowright X$.

Show that $|G.x| = [G : \text{stab}(x)]$.

▷ solution ◁

One nice consequence of the orbit-stabilizer theorem is that intersections of finite-index subgroups have finite index.

Proposition 1.3.3 *Let G be a group and $H, N \leq G$ be subgroups.*

Then, $[G : H \cap N] \leq [G : H] \cdot [G : N]$.

Proof If either $[G : H] = \infty$ or $[G : N] = \infty$ there is nothing to prove because the right-hand side is infinite. So assume that $[G : H] < \infty$ and $[G : N] < \infty$.

Let $X = G/H \times G/N$. That is, elements of X are pairs of cosets $(\alpha H, \beta N)$. Therefore, X is finite, since $|X| = |G/H| \cdot |G/N|$.

The group G acts on X by $g.(\alpha H, \beta N) = (g\alpha H, g\beta N)$. The stabilizer of (H, N) may easily be computed: $\mathrm{stab}(H, N) = H \cap N$. Thus, $[G : H \cap N] \leq |X| \leq [G : H] \cdot [G : N]$. □

1.4 Discrete Group Convolutions

Throughout this book we will almost exclusively deal with countable groups. Given a countable group G, one may define the **convolution** of functions $f, g \colon G \to \mathbb{C}$ as follows.

Definition 1.4.1 Let G be a countable group. Let $f, g \colon G \to \mathbb{C}$. The **convolution** of f and g is the function $f * g \colon G \to \mathbb{C}$ defined by

$$(f * g)(x) := \sum_y f(y)g\left(y^{-1}x\right) = \sum_y f(y)(y.g)(x),$$

as long as the above sum converges absolutely.

This is the analogue of the usual convolution of functions on the group $\mathbb{R}$:

$$(f * g)(x) = \int f(y)g(x - y)dy.$$

However, the convolution is not necessarily commutative, as is the case for Abelian groups.

Exercise 1.18 Show that

$$(f * g)(x) = \sum_y f\left(xy^{-1}\right)g(y).$$

Give an example for which $f * g \neq g * f$.

Exercise 1.19 (Left action and convolutions) Show that $x.(f * g) = (x.f * g)$ for the canonical left action $x.f(y) = f\left(x^{-1}y\right)$.

When G is countable, a probability measure μ on G may be thought of as a function $\mu \colon G \to [0, 1]$ so that $\mu(A) = \sum_{a \in A} \mu(a)$.

Exercise 1.20 Let μ be a probability measure on a countable group G, and let X be a random element of G with law μ. Show that

$$\mathbb{E}\left[f\left(x \cdot X^{-1}\right)\right] = (f * \mu)(x)$$

whenever the above quantities are well defined.

Definition 1.4.2 Let μ be a probability measure on G. We will use the notation μ^t to denote the t-fold convolution of μ with itself. Specifically, $\mu^1 = \mu$ and $\mu^{t+1} = \mu * \mu^t = \mu^t * \mu$.

Exercise 1.21 Let G be a countable group. Let μ, ν be probability measures on G. Let X, Y be independent random elements in G such that X has law μ, and Y has law ν.

Show that the law of $X \cdot Y$ is $\mu * \nu$.   ▷ solution ◁

Exercise 1.22 Show that for any $p \geq 1$ we have $\|x.f\|_p = \|f\|_p$. Here, $\|f\|_p^p = \sum_x |f(x)|^p$ and $\|f\|_\infty = \sup_x |f(x)|$.

Show that $\|\check{f}\|_p = \|f\|_p$, where $\check{f}(x) = f\left(x^{-1}\right)$.

Exercise 1.23 Prove *Young's inequality for products*: For all $a, b \geq 0$ and any $p, q > 0$ such that $p + q = 1$, we have $ab \leq pa^{1/p} + qb^{1/q}$.   ▷ solution ◁

Exercise 1.24 Prove the *generalized Hölder inequality*: for all $p_1, \ldots, p_n \in [1, \infty]$ such that $\sum_{j=1}^n \frac{1}{p_j} = 1$, we have

$$\|f_1 \cdots f_n\|_1 \leq \prod_{j=1}^n \|f_j\|_{p_j}.$$ ▷ solution ◁

Exercise 1.25 Prove *Young's inequality for convolutions*: For any $p, q \geq 1$ and $1 \leq r \leq \infty$ such that $\frac{1}{p} + \frac{1}{q} = \frac{1}{r} + 1$, we have

$$\|f * g\|_r \leq \|f\|_p \cdot \|g\|_q.$$ ▷ solution ◁

1.5 Basic Group Notions

Here we briefly recall some basic notions and examples from group theory. Further depth on any of these notions can be found in any basic textbook on group theory.

1.5.1 Basic Linear Groups

If R is a ring we use the notation $M_n(R)$ to denote the set of $n \times n$ matrices with entries in R. For example, $M_n(\mathbb{Z})$ is the set of all $n \times n$ matrices with integer entries. These do not necessarily form a group. By $GL_n(\mathbb{R})$ we denote the group of $n \times n$ invertible matrices with real entries. The group operation here is matrix multiplication. In more generality, $GL_n(\mathbb{C})$ is the group of $n \times n$ matrices with complex entries, so that $GL_n(\mathbb{R}) \leq GL_n(\mathbb{C})$.

Exercise 1.26 Show that $GL_2(\mathbb{R}) \cap M_2(\mathbb{Z})$ is not a group with matrix multiplication.

▷ solution ◁

A nontrivial fact is that if we restrict to integer entries with determinant ± 1, we do have a group. We denote

$$GL_n(\mathbb{Z}) = \{M \in M_n(\mathbb{Z}) : |\det(M)| = 1\}.$$

Proposition 1.5.1 $GL_n(\mathbb{Z})$ *is a group with matrix multiplication.*

Proof The main property we will use is that for any $M \in GL_n(\mathbb{Z})$ the number $\det(M)$ is invertible in the ring $\mathbb{Z}$. (This proof generalizes to matrices over a commutative ring with unit such that the determinants are invertible in the ring; see Exercise 1.102.)

Recall Cramer's Rule: For $b \in \mathbb{R}^d$ and $A \in GL_n(\mathbb{R})$, we may compute the solution to $Ax = b$ by $x_i = \frac{\det(A(i, b))}{\det(A)}$ for each $i = 1, \ldots, n$, where $A(i, b)$ is the matrix A with ith column replaced by the vector b.

Let e_i denote the standard basis for $\mathbb{R}^n$. So e_i is a vector with 1 in the ith position, and 0 everywhere else.

Now let $A \in GL_n(\mathbb{Z})$. We want to compute A^{-1} and show that it has integer entries. Let x_i be the ith column of A^{-1}. Then $Ax_i = e_i$. Consequently,

$$\left(A^{-1}\right)_{i,j} = (x_i)_j = \frac{\det(A(j, e_i))}{\det(A)}.$$

Note that since A, e_i have integer entries, then so does $A(j, e_i)$. Since $\det(A) \in \{-1, 1\}$, we conclude that A^{-1} has integer entries.

Thus, if $A \in GL_n(\mathbb{Z})$ then $A^{-1} \in GL_n(\mathbb{Z})$.

The fact that $GL_n(\mathbb{Z})$ is closed under matrix multiplication is easy to prove, and is left to the reader. □

The following notation is also standard. Define:

$$SL_n(\mathbb{Z}) = \{A \in GL_n(\mathbb{Z}) \mid \det(A) = 1\}.$$

Exercise 1.27 Show that $\mathrm{SL}_n(\mathbb{Z}) \lhd \mathrm{GL}_n(\mathbb{Z})$ and that $[\mathrm{GL}_n(\mathbb{Z}) : \mathrm{SL}_n(\mathbb{Z})] = 2$.

▷ solution ◁

Exercise 1.28 For $1 \leq i, j \leq n$ let $E_{i,j}$ denote the $n \times n$ matrix with 1 only in the (i, j) entry, and 0 in all other entries.

Show that $\mathrm{SL}_n(\mathbb{Z}) = \langle I + E_{i,j} \mid 1 \leq i \neq j \leq n \rangle$, where I is the $n \times n$ identity matrix.

▷ solution ◁

1.5.2 Abelian Groups

A group G is called **Abelian**, or **commutative**, if $xy = yx$ for all $x, y \in G$.

A group G is called **finitely generated** if there exists a finite generating set for G. That is, if there exists a finite set $S \subset G$, $|S| < \infty$, such that for any $x \in G$ there are $s_1, \ldots, s_n \in S \cup S^{-1}$ such that $x = s_1 \cdots s_n$. We will come back to finitely generated groups in Section 1.5.7.

Exercise 1.29 Show that the group $\mathbb{Z}^d$ (with vector addition as the group operation) is a finitely generated Abelian group, with the standard basis serving as a finite generating set.

Finitely generated Abelian groups have a special structure. The classification of these groups is given by the so-called *fundamental theorem of finitely generated Abelian groups*. We will prove a simplified version of this theorem.

Theorem 1.5.2 *Let G be a finitely generated Abelian group. Then there exists a finite Abelian group F and some integer $d \geq 0$ such that $G \cong \mathbb{Z}^d \times F$. Also, $d > 0$ if and only if $|G| = \infty$.*

Proof Let $U = \{u_1, \ldots, u_n\}$ be a finite generating set for G. Consider the vector space $V = \mathbb{Q}^n$. Define a map $\psi : \mathbb{Z}^n \to G$ by

$$\psi(z_1, \ldots, z_n) = (u_1)^{z_1} \cdots (u_n)^{z_n}.$$

Note that since G is Abelian and since U generates G, the map ψ is surjective. Also, it is simple to check that because G is Abelian we have that ψ is a homomorphism.

Let $K = \mathrm{Ker}\psi = \{\vec{z} \in \mathbb{Z}^n : \psi(\vec{z}) = 1\}$. Let $W = \mathrm{span}(K)$, which is a subspace of V. The quotient vector space V/W has dimension $d \leq n$, so we can choose $\vec{b}_1, \ldots, \vec{b}_d \in V$ such that $\{b_j + W : 1 \leq j \leq d\}$ forms a (linear) basis for V/W. Let $\vec{w}_1, \ldots, \vec{w}_k$ be a basis for W. By multiplying by a large enough integer, we can assume without loss of generality that $\vec{b}_j \in \mathbb{Z}^n$ for all $1 \leq j \leq d$ and $\vec{w}_j \in \mathbb{Z}^n$ for all $1 \leq j \leq k$.

Define $s_j = \psi(\vec{b}_j)$ for all $1 \leq j \leq d$ and $f_j = \psi(\vec{w}_j)$ for all $1 \leq j \leq k$.

We claim that the map $(z_1, \ldots, z_d) \mapsto (s_1)^{z_1} \cdots (s_d)^{z_d}$ is an isomorphism from $\mathbb{Z}^d$ onto $Z = \langle s_1, \ldots, s_d \rangle$. It is immediate to verify that this is a surjective homomorphism. To show it is injective, assume that $(s_1)^{z_1} \cdots (s_d)^{z_d} = 1$. Then,

$$\psi\left(z_1\vec{b}_1 + \cdots + z_d\vec{b}_d\right) = (s_1)^{z_1} \cdots (s_d)^{z_d} = 1,$$

implying that $z_1\vec{b}_1 + \cdots + z_d\vec{b}_d \in K \subset W$. Since $\vec{b}_1 + W, \ldots, \vec{b}_d + W$ are linearly independent, it must be that $z_1 = \cdots = z_d = 0$. This proves injectivity, showing that $Z \cong \mathbb{Z}^d$.

Now fix some $\vec{z} \in \mathbb{Z}^n \cap W$. Then since $W = \text{span}(K)$, there exist some $q_1, \ldots, q_m \in \mathbb{Q}$ and $\vec{z}_1, \ldots, \vec{z}_m \in K$ such that $\vec{z} = q_1\vec{z}_1 + \cdots + q_m\vec{z}_m$. So there exists a large enough integer $r \neq 0$ such that $r\vec{z} \in K$, implying that $\psi(\vec{z})^r = \psi(r\vec{z}) = 1$. This implies that any element of $F = \langle f_1, \ldots, f_k \rangle$ is *torsion*; that is, for any $x \in F$ there exists an integer $r \neq 0$ such that $x^r = 1$. (One can check that in fact F is exactly the subgroup of all torsion elements of G.) So we may take $r > 0$ large enough so that $(f_j)^r = 1$ for all $1 \leq j \leq k$. Since F is generated by $f_1, \ldots, f_k$, and since F is Abelian, we have that the map $\{0, \ldots, r-1\}^k \to F$ given by $(a_1, \ldots, a_k) \mapsto (f_1)^{a_1} \cdots (f_k)^{a_k}$ is surjective, and thus F is a finite group.

Let $x \in Z \cap F$. So $x^r = 1$ for some integer $r > 0$. Then

$$\psi\left(rz_1\vec{b}_1 + \cdots + rz_d\vec{b}_d\right) = ((s_1)^{z_1} \cdots (s_d)^{z_d})^r = x^r = 1,$$

for some integers $z_1, \ldots, z_d \in \mathbb{Z}$. This implies that $rz_1\vec{b}_1 + \cdots + rz_d\vec{b}_d \in K \subset W$, and as before we get that $z_1 = \cdots = z_d = 0$, so that $x = 1$. That is, we have shown that $Z \cap F = \{1\}$.

Finally, recall that the map $\psi : \mathbb{Z}^n \to G$ is surjective. Any $\vec{z} \in \mathbb{Z}^n$ can be written as $\vec{z} = \vec{v} + \vec{w}$ where $\vec{v} = z_1\vec{b}_1 + \cdots + z_d\vec{b}_d$ and $\vec{w} = a_1\vec{w}_1 + \cdots + a_k\vec{w}_k$ for integers $z_1, \ldots, z_d, a_1, \ldots, a_k$. Thus, for any $x \in G$ there exist $\psi(\vec{v}) \in Z$ and $\psi(\vec{w}) \in F$ such that $x = \psi(\vec{v}) \cdot \psi(\vec{w})$.

To conclude, we have $Z \lhd G$ with $Z \cong \mathbb{Z}^d$ and $F \lhd G$ with $|F| < \infty$, and these have the following properties:

- $G = ZF = \{zf : z \in Z, f \in F\}$,
- $Z \cap F = \{1\}$,
- and for any $z \in Z, f \in F$ we have $zf = fz$.

It is an exercise to show that this implies that $G \cong Z \times F$. $\qquad\square$

Exercise 1.30 Let G be a group and let Z, F be subgroups such that $G = ZF$, $Z \cap F = \{1\}$, and $zf = fz$ for all $z \in Z$, $f \in F$.

Show that $G \cong Z \times F$.

1.5.3 Virtual Properties

A **group property** is a class of groups $\mathcal{P}$ such that if $G \cong H$ and $G \in \mathcal{P}$, then also $H \in \mathcal{P}$. For $G \in \mathcal{P}$ we sometimes say that G **is** $\mathcal{P}$.

For a group property $\mathcal{P}$, we may define the property **virtually** $\mathcal{P}$. A group G is **virtually** $\mathcal{P}$ if there exists a finite index subgroup $[G : H] < \infty$ such that H is $\mathcal{P}$.

Example 1.5.3 A group G is virtually finitely generated if there exists a finite index subgroup $H \leq G$, $[G : H] < \infty$ such that H is finitely generated. ◁ ▽ ▷

Exercise 1.31 Show that if G is virtually finitely generated then G is finitely generated. ▷ solution ◁

Example 1.5.4 A group G is called **indicable** if there exists a surjective homomorphism from G onto $\mathbb{Z}$.

A group G is therefore **virtually indicable** if there exists a finite index subgroup $[G : H] < \infty$ and a surjective homomorphism $\varphi \colon H \to \mathbb{Z}$. ◁ ▽ ▷

Exercise 1.32 Show that if G is finitely generated and there exists a homomorphism $\varphi \colon G \to A$ where A is an Abelian group and $|\varphi(G)| = \infty$, then G is indicable. ▷ solution ◁

Exercise 1.33 Let G be a finitely generated group. Show that $|G/[G, G]| = \infty$ if and only if there exists a surjective homomorphism $\varphi \colon G \to \mathbb{Z}$. ▷ solution ◁

Every group $G \in \mathcal{P}$ is also virtually $\mathcal{P}$, as it has index 1 in itself. But not every property $\mathcal{P}$ is the same as virtually $\mathcal{P}$.

For example: the *infinite dihedral group* D_∞, see Exercise 1.72, is virtually $\mathbb{Z}$ (i.e. contains a finite index subgroup isomorphic to $\mathbb{Z}$) but is not Abelian, and so definitely not isomorphic to $\mathbb{Z}$.

1.5.4 Nilpotent Groups

Definition 1.5.5 For a group G, we define the **lower central series** inductively as follows: $\gamma_0 = \gamma_0(G) = G$ and $\gamma_{n+1} = \gamma_{n+1}(G) = [\gamma_n(G), G]$ for all $n \geq 0$.
We define the **upper central series** as: $Z_0 = \{1\}$ and for all $n \geq 0$,

$$Z_{n+1} = Z_{n+1}(G) := \{x \in G : \forall\, y \in G\, [x, y] \in Z_n\}.$$

$Z_1 = Z_1(G)$ is called the **center** of G, and is sometimes denoted just $Z(G)$.

Exercise 1.34 Assume that $G = \langle S \rangle$, for some set of elements $S \subset G$.
Show that

$$\gamma_n(G) = \langle [s_0, \ldots, s_n]^x : s_0, \ldots, s_n \in S,\ x \in G \rangle.$$ ▷ solution ◁

Exercise 1.35 Let φ be an automorphism of a group G. Show that $\varphi(\gamma_n(G)) = \gamma_n(G)$ and that $\varphi(Z_n(G)) = Z_n(G)$.
Conclude that $\gamma_n(G), Z_n(G)$ are normal subgroups of G. ▷ solution ◁

Exercise 1.36 Show that for $k \leq n$ we have $Z_k(G) \triangleleft Z_n(G)$.
Show that $Z_n(G)/Z_k(G) = Z_{n-k}(G/Z_k(G))$. ▷ solution ◁

Exercise 1.37 Show that if $\gamma_n(G) = \{1\}$ then $Z_n(G) = G$. ▷ solution ◁

Exercise 1.38 Show that if $Z_n(G) = G$ then $\gamma_n(G) = \{1\}$. ▷ solution ◁

Exercise 1.39 Show that if G is finitely generated, then γ_k/γ_{k+1} is also finitely generated for any $k \geq 0$. ▷ solution ◁

Definition 1.5.6 A group G is called n-step nilpotent if $\gamma_n(G) = \{1\}$ and $\gamma_{n-1}(G) \neq \{1\}$. (By convention, 0-step nilpotent is just the trivial group.)
A group is called **nilpotent** if it is n-step nilpotent for some $n \geq 0$.

Note that 0-step nilpotent is the trivial group $\{1\}$. Note too that 1-step nilpotent is just Abelian.

Exercise 1.40 Show that a group is n-step nilpotent if and only if $Z_n(G) = G$ and $Z_{n-1}(G) \neq G$.
Show that G is $(n + 1)$-step nilpotent if and only if $G/Z_1(G)$ is n-step nilpotent. ▷ solution ◁

Exercise 1.41 Show that $G/\gamma_n(G)$ is at most n-step nilpotent. ▷ solution ◁

Exercise 1.42 Show that if G is a nilpotent group and $H \leq G$, then H is nilpotent as well. ▷ solution ◁

Exercise 1.43 Show that if G is nilpotent and $N \lhd G$, then G/N is also nilpotent. ▷ solution ◁

Let us go through some basic examples of nilpotent groups.

Some readers may have seen the following definition: An $n \times n$ matrix $M \in M_n(\mathbb{R})$ is called k-**step nilpotent** if $M^{k-1} \neq 0$ and $M^k = 0$. This is related to nilpotence of groups, as the following exercises show.

Exercise 1.44 Let $T_n(\mathbb{R})$ denote all $n \times n$ upper triangular matrices with real entries. For $1 \leq k \leq n$ define

$$D_k = \{M \in T_n(\mathbb{R}) : \forall j \leq i + k - 1, \ M_{i,j} = 0\}.$$

That is, all the first k diagonals of M are 0. (So e.g. $D_0 = T_n(\mathbb{R})$.)
Show that if $M \in D_k, N \in D_\ell$ then $MN \in D_{k+\ell}$. ▷ solution ◁

Exercise 1.45 Fix $n > 1$. Let $T_n(\mathbb{R})$ denote all $n \times n$ upper triangular matrices. For $1 \leq k \leq n$ define

$$D_k = \{M \in T_n(\mathbb{R}) : \forall j \leq i + k - 1, \ M_{i,j} = 0\},$$

and define $D_k(\mathbb{Z}) = D_k \cap M_n(\mathbb{Z})$ (recall that $M_n(\mathbb{Z})$ is the set of $n \times n$ matrices with integer entries).
  Set

$$Q_{n,k} = \{I + N : N \in D_k(\mathbb{Z})\}.$$

Show that $Q_{n,k}$ is a group (with the usual matrix multiplication). ▷ solution ◁

Exercise 1.46 Let $n > 1$. Let $H_n(\mathbb{Z})$ be the collection of all upper triangular $n \times n$ matrices, with 1 on the diagonal, and only integer entries.
  Show that $H_n(\mathbb{Z})$ is a group (with the usual matrix multiplication).
  Show that for $0 \leq k \leq n - 1$ we have

$$\gamma_k(H_n(\mathbb{Z})) \subset Q_{n,k+1} \subset Z_{n-k-1}(H_n(\mathbb{Z})).$$ ▷ solution ◁

1.5.5 Solvable Groups

Definition 1.5.7 Let G be a group. The **derived series** is defined inductively as follows: $G^{(0)} = G$, and $G^{(n+1)} = \left[G^{(n)}, G^{(n)} \right]$.

Definition 1.5.8 A group G is n-**step solvable** if $G^{(n)} = \{1\}$ and $G^{(n-1)} \neq \{1\}$. (By convention, 0-step solvable is the trivial group.)

A group is **solvable** if it is solvable for some $n \geq 0$.

Note that the properties of 1-step solvable, 1-step nilpotent, and Abelian all coincide.

Exercise 1.47 Show that any nilpotent group is solvable.   ▷ solution ◁

Exercise 1.48 Show that if G is 2-step solvable, then $G^{(1)}$ is Abelian.   ▷ solution ◁

Exercise 1.49 Show that the following are equivalent:

- G is a solvable group.
- $G^{(n)}$ is solvable for all $n \geq 0$.
- $G^{(n)}$ is solvable for some $n \geq 0$.   ▷ solution ◁

Exercise 1.50 Show that if G is solvable and infinite then $[G : [G, G]] = \infty$.

▷ solution ◁

Exercise 1.51 Show that if G is a solvable group and $H \leq G$ then H is solvable.

▷ solution ◁

Exercise 1.52 Let Δ_n^+ denote the collection of all $n \times n$ diagonal matrices with real entries and only positive values on the diagonal.

Show that Δ_n^+ is an Abelian group (with the usual matrix multiplication).

▷ solution ◁

Exercise 1.53 Fix $n > 1$, and recall D_k, the collection of all $n \times n$ upper triangular matrices, with first k diagonals equal to 0 (from Exercise 1.44).

Recall also Δ_n^+, the collection of all $n \times n$ diagonal matrices with only positive values on the diagonal.

For $k \geq 1$ define

$$P_{n,k} = P_{n,k} := \{T + M : T \in \Delta_n^+, \, M \in D_k\}.$$

Show that $P_{n,k}$ is a group (with the usual matrix multiplication).

Show that $[P_{n,k}, P_{n,k}] \subset \{I + M : M \in D_k\}$.

Show that $P_{n,k}$ is solvable, of step at most $\lceil \log_2(n) \rceil + 1$.

Show that $P_{n,k}$ is not nilpotent when $k < n$. ▷ solution ◁

Exercise 1.54 Let $r > 1$ and consider $\omega = e^{2\pi i/r}$, the rth root of unity. Define

$$D = \left\{ \sum_{k=0}^{r-1} a_k \omega^k : a_k \in \mathbb{Z} \right\}$$

and

$$G = \left\{ \begin{bmatrix} \omega^z & d \\ 0 & 1 \end{bmatrix} : z \in \mathbb{Z}, d \in D \right\}.$$

Show that G is a finitely generated virtually Abelian group that is not nilpotent.

▷ solution ◁

1.5.6 Free Groups

Let S be a finite set. For each element $s \in S$, consider a new element $\bar{s}$, and define $\bar{S} = \{\bar{s} : s \in S\}$. Consider all possible finite words in the letters $S \cup \bar{S}$, including the empty word $\varnothing$, and denote this set by Ω_S. That is,

$$\Omega_S := \{a_1 \cdots a_n : n \in \mathbb{N}, a_j \in S \cup \bar{S}\} \cup \{\varnothing\}.$$

Define the *reduction operation* $R : \Omega_S \to \Omega_S$ as follows: Call a word $a_1 \cdots a_n \in \Omega_S$ **reduced** if for all $1 \leq j < n$ we have that $(a_j, a_{j+1}) \notin \{(s, \bar{s}), (\bar{s}, s) : s \in S\}$. The empty word $\varnothing$ is reduced by convention. Let $\mathbb{F}_S$ denote the collection of all reduced words. Now, for a word $\omega \in \mathbb{F}_S$, define $R(\omega) = \omega$. For a word $a_1 \cdots a_n \notin \mathbb{F}_S$, let j be the smallest index for which $(a_j, a_{j+1}) \in \{(s, \bar{s}), (\bar{s}, s) : s \in S\}$, and define $R(a_1 \cdots a_n) = a_1 \cdots a_{j-1} a_{j+2} \cdots a_n$ (if $j = 1$ this means $R(a_1 \cdots a_n) = a_3 \cdots a_n$).

It is easy to see that for any word $a_1 \cdots a_n \in \Omega_S$, at most n applications of R will result in a reduced word. Let $R^\infty(a_1 \cdots a_n)$ denote this reduced word. So $R^\infty : \Omega_S \to \mathbb{F}_S$, which fixes any word in $\mathbb{F}_S$.

Define a product structure on $\mathbb{F}_S$: For two reduced words $a_1 \cdots a_n$ and $b_1 \cdots b_m$ define

$$\varnothing a_1 \cdots a_n = a_1 \cdots a_n \varnothing = a_1 \cdots a_n$$

and

$$a_1 \cdots a_n \cdot b_1 \cdots b_m = R^\infty(a_1 \cdots a_n b_1 \cdots b_m).$$

It is easily verified that this turns $\mathbb{F}_S$ into a group with identity element $\varnothing$.

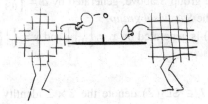

Definition 1.5.9 $\mathbb{F}_S$ is called the **free group** on generators S.

Since the actual letters generating the free group are not important, we will usually write $\mathbb{F}_d$ for the free group generated by d elements.

If G is a finitely generated group, generated by a finite set S, then consider $\mathbb{F}_S$ and define a map $\varphi \colon \mathbb{F}_S \to G$ by $\varphi(\varnothing) = 1$, for $s \in S$ we set $\varphi(s) = s$ and $\varphi(\bar{s}) = s^{-1}$, and finally for general reduced words set $\varphi(a_1 \cdots a_n) = \varphi(a_1) \cdots \varphi(a_n)$. This is easily seen to be a surjective homomorphism, so $G \cong \mathbb{F}_S/\mathrm{Ker}\varphi$.

Remark 1.5.10 Let G be a group generated by a finite set S. We have seen that there exists a normal subgroup $R \lhd \mathbb{F}_S$ such that $\mathbb{F}_S/R \cong G$. In this case we write $G = \langle S \mid R \rangle$.

Moreover, suppose there exist $(r_n)_n \subset R$ such that R is the smallest normal subgroup containing all $(r_n)_n$. Then we write $G = \langle S \mid (r_n)_n \rangle$.

We will come back to this *presentation* in Section 1.5.8.

There is a classical method of proving that certain groups (or subgroups) are isomorphic to a free group. We will not require it but include it for the educational value.

Exercise 1.55 (Ping-pong lemma) Let G be a group acting on some set X. Let $a, b \in G$.

Suppose that there exist disjoint non-empty subsets $A, B \subset X$, $A \cap B = \varnothing$ such that for all $0 \neq z \in \mathbb{Z}$ we have $a^z(B) \subset A$ and $b^z(A) \subset B$. (This is known as: *a, b play ping-pong*.)

Then $H = \langle a, b \rangle \leq G$ is isomorphic to $\mathbb{F}_2$.

▷ solution ◁

Exercise 1.56 Consider $a = \begin{bmatrix} 1 & 2 \\ 0 & 1 \end{bmatrix}$ and $b = \begin{bmatrix} 1 & 0 \\ 2 & 1 \end{bmatrix}$ in $SL_2(\mathbb{Z})$.
Show that $S = \langle a, b \rangle$ is a free group generated by 2 elements.   ▷ solution ◁

Remark 1.5.11 The group S above, generated by $a = \begin{bmatrix} 1 & 2 \\ 0 & 1 \end{bmatrix}$ and $b = \begin{bmatrix} 1 & 0 \\ 2 & 1 \end{bmatrix}$, is
sometimes called the *Sanov subgroup*.
   Note that $SL_2(\mathbb{Z})$ is generated by $x = \begin{bmatrix} 1 & 1 \\ 0 & 1 \end{bmatrix}$ and $y = \begin{bmatrix} 1 & 0 \\ 1 & 1 \end{bmatrix}$, and that $a = x^2$
and $b = y^2$.

Exercise 1.57 Let $I \in SL_2(\mathbb{Z})$ denote the 2×2 identity matrix. Show that
$\{-I, I\} \triangleleft SL_2(\mathbb{Z})$.

   Denote $PSL_2(\mathbb{Z}) = SL_2(\mathbb{Z})/\{-I, I\}$.

Exercise 1.58 Let $x = \begin{bmatrix} 1 & 1 \\ 0 & 1 \end{bmatrix}$ and $y = \begin{bmatrix} 1 & 0 \\ 1 & 1 \end{bmatrix}$ and let $a = x^2, b = y^2$. Set
$t = \begin{bmatrix} 0 & -1 \\ 1 & 0 \end{bmatrix}$ and $s = xt$.
   Show that $t^2 = s^3 = -I$, where I is the 2×2 identity matrix.
   Show that $x = -st$ and $y = -s^2t$.
   Let $\pi \colon SL_2(\mathbb{Z}) \to PSL_2(\mathbb{Z})$ be the canonical homomorphism. Show that
$PSL_2(\mathbb{Z}) = \langle \pi(t), \pi(s) \rangle$.
   Show that for any $z \in SL_2(\mathbb{Z})$ there exist $\varepsilon_1, \ldots, \varepsilon_n \in \{-1, 1\}$ and $\alpha, \beta \in$
$\{0, 1\}$ such that $z \equiv t^\alpha s^{\varepsilon_1} t s^{\varepsilon_2} \cdots t s^{\varepsilon_n} t^\beta \pmod{\{-I, I\}}$.

Exercise 1.59 Let x, y, a, b, s, t be as in Exercise 1.58.
   Let $S = \langle a, b \rangle \leq SL_2(\mathbb{Z})$ be the Sanov subgroup (from Exercise 1.56).
   Show that $a = stst$ and $b = s^2ts^2t$.
   Let $\pi \colon SL_2(\mathbb{Z}) \to PSL_2(\mathbb{Z})$ be the canonical projection.
   Show that for any $z \in SL_2(\mathbb{Z})$ there exist $w \in S$ and $p \in \{1, s, s^2, t, st, s^2t\}$
such that $\pi(z) = \pi(w)\pi(p)$.
   Show that $[PSL_2(\mathbb{Z}) : \pi(S)] \leq 6$.
   Conclude that $[SL_2(\mathbb{Z}) : S] \leq 12$.   ▷ solution ◁

1.5.7 Finitely Generated Groups

Exercise 1.60 Let $H \leq G$ and let S be a finite generating set for G. Let T be a
right-traversal of H in G; that is, a set of representatives for the right-cosets of
H containing $1 \in T$. So $G = \uplus_{t \in T} Ht$.
   Show that H is generated by $TST^{-1} \cap H$.   ▷ solution ◁

Exercise 1.61 Show that a finite index subgroup of a finitely generated group is also finitely generated.

Exercise 1.62 Let $H \lhd G$ and let $\pi \colon G \to G/H$ be the canonical projection. Assume that H is generated by U and G/H is generated by $\tilde{S}$.
Show that if $S \subset G$ is such that $\pi(S) = \tilde{S}$, then $U \cup S$ generates G.
Conclude that if H and G/H are finitely generated, then so is G.   ▷ solution ◁

A nice property of finitely generated groups is that there cannot be too many finite index subgroups of a given index.

Theorem 1.5.12 *Let G be a finitely generated group, generated by d elements. Then for any n, the set $\{H \leq G \mid [G : H] = n\}$ has size at most $(n!)^d$.*

Proof Assume that $S \subset G$ is a finite generating set for G of size $|S| = d$.
Let Π_n be the group of permutations of the set $\{1, 2, \ldots, n\}$.
Let $X = \{H \leq G : [G : H] = n\}$. If $X = \emptyset$ then it is of course finite. So assume that $X \neq \emptyset$.
Consider $H \in X$. Write $G/H = \{xH : x \in G\} = \{x_1 H, x_2 H, \ldots, x_n H\}$, where $x_1 = 1$. G acts on G/H by $x(yH) = xyH$. Define $\psi_H \colon G \to \Pi_n$ by $x \mapsto \pi_x \in \Pi_n$, where π_x is the permutation for which $\pi_x(i) = j$ for the unique $1 \leq i, j \leq n$ such that $xx_i H = x_j H$. Note that $\pi_x(1) = 1$ if and only if $x \in H$.
It is easy to see that ψ_H is a homomorphism from G into Π_n.
We claim that $H \mapsto \psi_H$ is an injective map from X into $\mathsf{Hom}(G, \Pi_n)$. Indeed, if $H \neq K \in X$, then without loss of generality we may take $x \in H \backslash K$ (otherwise $x \in K \backslash H$, and reverse the roles of H and K in what follows). Let $\pi = \psi_H(x)$ and $\sigma = \psi_K(x)$. Since $x \in H$ we have that $\pi(1) = 1$. Since $x \notin K$ we have that $\sigma(1) \neq 1$. So $\psi_H(x) \neq \psi_K(x)$, implying that $\psi_H \neq \psi_K$.
We conclude that $|X| \leq |\mathsf{Hom}(G, \Pi_n)|$, so we only need to bound the size of this last quantity.
Any homomorphism $\psi \in \mathsf{Hom}(G, \Pi_n)$ is completely determined by the values $\{\psi(s) : s \in S\}$. Thus,

$$|\mathsf{Hom}(G, \Pi_n)| \leq \left|(\Pi_n)^S\right| = (n!)^d. \qquad \square$$

1.5.8 Finitely Presented Groups

Definition 1.5.13 Let G be a group generated by a finite set S. Consider the free group on the generators S, $\mathbb{F}_S$. If it is possible to find a normal subgroup $R \lhd \mathbb{F}_S$ and finitely many $r_1, \ldots, r_k \in R$ such that R is the smallest normal subgroup containing $r_1, \ldots, r_k$, we write $G = \langle S \mid r_1, \ldots, r_k \rangle$, and in this special case we say that G is **finitely presented**.

The elements of S are called **generators** of G, and the elements of R are called **relations** of G.

The next lemma shows how the property of finite presentation can be moved from quotients by finitely presented groups to the mother group.

Lemma 1.5.14 *Let G be a group and $N \lhd G$. Assume that both N and G/N are finitely presented. Then, G is finitely presented as well.*

Proof Assume that

$$N = \langle s_1, \ldots, s_k \mid r_1, \ldots, r_\ell \rangle \quad \text{and} \quad G/N = \langle a_1, \ldots, a_d \mid p_1, \ldots, p_m \rangle.$$

Let $\mathbb{F} = \mathbb{F}_{d+k}$ be the free group on $d + k$ generators. Denote the generators of $\mathbb{F}$ by $\{f_1, \ldots, f_d, t_1, \ldots, t_k\}$. Let $F = \langle f_1, \ldots, f_d \rangle \leq \mathbb{F}$ and $T = \langle t_1, \ldots, t_k \rangle \leq \mathbb{F}$. So F is a free group on d generators, and T is a free group on k generators.

For any $1 \leq j \leq d$ choose an element $g_j \in G$ such that g_j is mapped to a_j under the canonical projection $G \to G/N$ (i.e. $a_j = Ng_j$).

Let $\varphi \colon \mathbb{F} \to G$ be the homomorphism defined by $\varphi(f_j) = g_j$ for $1 \leq j \leq d$ and $\varphi(t_j) = s_j$ for $1 \leq j \leq k$. By our assumptions on the presentation for N, there exist words $r_1, \ldots, r_\ell \in T$ such that if R is the smallest normal subgroup of T containing $r_1, \ldots, r_k$, then $\varphi|_T \colon T \to N$ with $\mathrm{Ker}(\varphi|_T) = \mathrm{Ker}\varphi \cap T = R$.

Also, by our assumptions on the presentation of G/N, there exist words $p_1, \ldots, p_m \in F$ such that if P is the smallest normal subgroup of F containing $p_1, \ldots, p_m$, then $\varphi^{-1}(N) \cap F = P$

For any $1 \leq i \leq k$ and $1 \leq j \leq d$, we have that $\varphi\left((t_i)^{f_j}\right) = (n_i)^{g_j} \in N$. So there exists $u_{i,j} \in T$ such that $\varphi\left((t_i)^{f_j}\right) = \varphi(u_{i,j})$. Define $q_{i,j} = (t_i)^{f_j}(u_{i,j})^{-1}$. Observe that $q_{i,j} \in \mathrm{Ker}\varphi$ for all i, j.

For any $1 \leq j \leq m$ we have that $\varphi(p_j) \in N$, by our assumptions on the presentation of G/N. So there exists $w_j \in T$ such that $\varphi(p_j) = \varphi(w_j)$. Define $z_j = p_j(w_j)^{-1}$. Observe that $z_j \in \mathrm{Ker}\varphi$ for all j.

Denote $K := \mathrm{Ker}\varphi$. Let Q be the smallest normal subgroup of $\mathbb{F}$ containing $\{q_{i,j} \colon 1 \leq i \leq k, \ 1 \leq j \leq d\}$. Let Z be the smallest normal subgroup of $\mathbb{F}$ containing $z_1, \ldots, z_m$.

Let $M \lhd \mathbb{F}$ be any normal subgroup containing

$$\{r_1, \ldots, r_\ell, z_1, \ldots, z_m\} \bigcup \{q_{i,j} \colon 1 \leq i \leq k, \ 1 \leq j \leq d\} \subset M.$$

Since M is an arbitrary normal subgroup containing the above relations, we only need to show that $K \lhd M$ for all such M, which will prove that G is finitely presented, since $G \cong \mathbb{F}/K$.

To this end, we will prove that

$$K \subset RQZ := \{rqz : r \in R, q \in Q, z \in Z\} \subset M. \tag{1.2}$$

We move to prove (1.2). It will be convenient to use the notations

$$AB = \{ab : a \in A, b \in B\} \quad \text{and} \quad A^B = \{a^b : a \in A, b \in B\}$$

for subsets $A, B \subset \mathbb{F}$.

Step I. Let $t \in T$ and $f \in F$. Then replacing $(t_i)^{f_j} = u_{i,j} q_{i,j}$, since $Q \lhd \mathbb{F}$, we have that $t^f = uq$ for some $u \in T$ and $q \in Q$. That is, $T^F \subset TQ$.

Step II. For any $1 \le j \le m$, and any $f \in F$, we have that $(p_j)^f = (w_j z_j)^f$. Since $Z \lhd \mathbb{F}$ and since $P = \langle (p_j)^f : 1 \le j \le m, f \in F \rangle$, we have that $P \subset T^F Z \subset TQZ$.

Step III. For any $x \in \mathbb{F}$ we can write $x = h_1 v_1 \cdots h_n v_n$ for some $h_1, \ldots, h_n \in F$ and $v_1, \ldots, v_n \in T$. By conjugating the v_j, we have that $x = (u_1)^{d_1} \cdot (u_n)^{d_n} f$ for some $u_1, \ldots, u_n \in T$ and $d_1, \ldots, d_n, f \in F$. Since $Q \lhd \mathbb{F}$, we conclude that $\mathbb{F} \subset TQF$.

Step IV. Let $x \in K$. Write $x = tqf$ for $t \in T$, $q \in Q$, and $f \in F$. So $\varphi(tf) = 1$, implying that $f \in \varphi^{-1}(N) \cap F = P$. This implies that

$$K \subset TQP \subset TQTQZ \subset TQZ.$$

Hence, for any $x \in K$ we can write $x = tqz$ for some $t \in T$, $q \in Q$, and $z \in Z$. Since $Q, Z \subset K$, we have that $t \in T \cap K = R$. So we have shown that $K \subset RQZ$, which is (1.2). $\quad\square$

Theorem 1.5.15 *Suppose G is a group, and suppose that there exists a sequence of subgroups $G = H_0 \rhd H_1 \rhd \cdots \rhd H_n = \{1\}$, with the property that every quotient H_j / H_{j+1} is finitely presented.*

Then G is finitely presented.

Proof This is proved by induction on n. If $n = 1$, then $G = H_0$ is finitely presented by assumption.

For $n > 1$, let $H = H_1$. By induction, considering the sequence $H = H_1 \rhd \cdots \rhd H_n = \{1\}$ we have that H is finitely presented. Also, by assumption G/H_1 is finitely presented. So G is finitely presented by Lemma 1.5.14, completing the induction. $\quad\square$

Exercise 1.63 Show that if G is a finite group then it is finitely presented.

▷ solution ◁

Exercise 1.64 Assume that a group G is virtually-$\mathbb{Z}$; that is, there exists a finite index normal subgroup $H \lhd G$, $[G : H] < \infty$ such that $H \cong \mathbb{Z}$.

Show that G is finitely presented.   ▷ solution ◁

Exercise 1.65 Let G be a n-step solvable group. Assume that $G^{(k)}/G^{(k+1)}$ is virtually-$\mathbb{Z}$ for every $0 \le k < n$.

Show that G is finitely presented.   ▷ solution ◁

Exercise 1.66 Show that $\mathbb{Z}^d$ is finitely presented.

Show that any finitely generated virtually Abelian group is finitely presented.

▷ solution ◁

Exercise 1.67 Show that if G is a finitely generated nilpotent group, then G is finitely presented.   ▷ solution ◁

1.5.9 Semi-direct Products

In this exercise, we introduce the notion of *semi-direct products*.

Recall that a *direct product* of groups G, H is the group whose elements are the pairs $G \times H$ and the group operation is given by $(g, h)(g', h') = (gg', hh')$ for all $g, g' \in G$ and $h, h' \in H$.

Exercise 1.68 Let G, H be groups. Assume that G acts on H by automorphisms. That is, each $g \in G$ can be thought of as an automorphism of H. A different way of thinking of this is that there is a homomorphism $\rho \colon G \to \mathrm{Aut}(H)$; that is, $g.h = (\rho(g))(h)$ for any $g \in G$ and $h \in H$.

Define the **semi-direct product** of G acting on H (with respect to ρ) as the group $G \ltimes H$ (also sometimes denoted $H \rtimes_\rho G$), whose elements are $G \times H = \{(g, h) \mid g \in G, \ h \in H\}$ and where multiplication is defined by

$$(g, h)(g', h') = (gg', h \cdot g.h').$$

Show that this defines a group structure. Determine the identity element in $G \ltimes H$ and the inverse of (g, h).

Show that the set $\{1_G\} \times H$ is an isomorphic copy of H sitting as a normal subgroup inside $G \ltimes H$. Show that $G \ltimes H/(\{1_G\} \times H) \cong G$.   ▷ solution ◁

A useful (but not completely precise) way to think about semi-direct product $G \ltimes H$ is to think of matrices of the form $\left[\begin{smallmatrix} g & h \\ 0 & 1 \end{smallmatrix}\right]$, $g \in G, h \in H$. This is especially aesthetic when H is Abelian, so that multiplication in H can be written additively. Indeed, when multiplying two such matrices we have

$$\begin{bmatrix} g & h \\ 0 & 1 \end{bmatrix} \cdot \begin{bmatrix} g' & h' \\ 0 & 1 \end{bmatrix} = \begin{bmatrix} gg' & gh'+h \\ 0 & 1 \end{bmatrix},$$

which is reminiscent of $(g, h)(g', h') = (gg', h + gh')$. In the non-Abelian case matrix multiplication must be interpreted properly:

$$\begin{bmatrix} g & h \\ 0 & 1 \end{bmatrix} \cdot \begin{bmatrix} g' & h' \\ 0 & 1 \end{bmatrix} = \begin{bmatrix} gg' & h \cdot gh' \\ 0 & 1 \end{bmatrix}.$$

Also, it may be worth pointing out that $G \ltimes H$ hints at which group is acting on which: $\ltimes$ has a small triangle, similar to the symbol $\rhd$, which reminds us that $H \cong \{1_G\} \times H \lhd G$.

Exercise 1.69 Let G, H be groups. Define an action $\rho \colon G \to \mathrm{Aut}(H)$ of G on H by $\rho(g).h = h$ for all $h \in H$ and $g \in G$.
Show that $G \ltimes H = G \times H$.

So a semi-direct product generalizes the notion of a direct product of groups.

Exercise 1.70 Recall from Sections 1.5.4 and 1.5.5 the following groups of $n \times n$ matrices: For $1 \le k \le n$, the group D_k is the additive group of all upper-triangular $n \times n$ real matrices A such that $A_{i,j} = 0$ for all $j \le i + k - 1$ (so the first k diagonals are 0). Here, Δ_n^+ is the multiplicative group of diagonal matrices with only strictly positive entries on the diagonal.
Show that Δ_n^+ acts on D_k by left multiplication.
Show that $\Delta_n^+ \ltimes D_k$ is 2-step solvable.
Show that if $\Delta_n^+ \ltimes D_k$ is nilpotent, then $k \ge n$. ▷ solution ◁

Exercise 1.71 Let V be a vector space over $\mathbb{C}$. $\varphi \colon V \to V$ is an **affine transformation** if $\varphi(v) = \alpha v + u$ for some fixed scalar $0 \ne \alpha \in \mathbb{C}$ and fixed vector u (α is called the *dilation* and u the *translation*).
Let A be the collection of all affine transformations on V. Show that A is a group with multiplication given by composition.
Show that $A \cong \mathbb{C}^* \ltimes V$ where $\mathbb{C}^*$ is the multiplicative group $\mathbb{C}\backslash\{0\}$ and V is considered as an additive group.
Is A Abelian? Nilpotent? Solvable? ▷ solution ◁

Exercise 1.72 The **infinite dihedral group** is $D_\infty = \big\langle a, b \mid baba, b^2 \big\rangle$.
Let $\varphi \in \mathrm{Aut}(\mathbb{Z})$ be given by $\varphi(x) = -x$. Let $\mathbb{Z}_2 = \{-1, 1\}$ be the group on 2 elements (the group operation given by multiplication). Show that $D_\infty \cong \mathbb{Z}_2 \ltimes \mathbb{Z}$ where $\mathbb{Z}_2$ acts on $\mathbb{Z}$ via $\varepsilon.x = \varepsilon \cdot x$ for $\varepsilon \in \{-1, 1\}$.

Show that D_∞ is not nilpotent.

Show that D_∞ is 2-step solvable.

Show that D_∞ is virtually $\mathbb{Z}$.   ▷ solution ◁

Exercise 1.73 Consider the following group: Let S_d be the group of permutations on d elements. Let S_d act on $\mathbb{Z}^d$ by permuting the coordinates; that is, $\sigma(z_1, \ldots, z_d) = \left(z_{\sigma^{-1}(1)}, \ldots, z_{\sigma^{-1}(d)} \right).$

Show that this is indeed a left action.

Consider the group $G = S_d \ltimes \mathbb{Z}^d$. Show that there exist $H \lhd G$ such that $G/H \cong S_d$ and $H \cong \mathbb{Z}^d$. (Specifically H is Abelian.)

Show that G is not Abelian for $d > 2$.   ▷ solution ◁

1.6 Measures on Groups and Harmonic Functions

1.6.1 Metric and Measure Structures on a Group

Definition 1.6.1 (Cayley graph) Let G be a finitely generated group. Let $S \subset G$ be a finite generating set. Assume that S is **symmetric**; that is, $S = S^{-1} := \{s^{-1} : s \in S\}$. The **Cayley graph** of G with respect to S is the graph with vertex set G and edges defined by the relations $x \sim y \iff x^{-1}y \in S$.

The distance in this Cayley graph is denoted by dist_S.

Exercise 1.74 Show that $\mathrm{dist}_S(x, y)$ is invariant under the diagonal G-action. That is, $\mathrm{dist}_S(gx, gy) = \mathrm{dist}_S(x, y)$ for any $g \in G$.

Due to this fact, we may denote $|x| = |x|_S := \mathrm{dist}_S(1, x)$. So that $\mathrm{dist}_S(x, y) = |x^{-1}y|$. Balls of radius r in this metric are denoted

$$B(x, r) = B_S(x, r) = \{y : \mathrm{dist}_S(x, y) \leq r\}.$$

Throughout the book, the underlying generating set will be implicit, and we will not specify it explicitly in the notation. If we wish to stress a specific generating set (or, sometimes, a specific group), we will use the notation $\mathrm{dist}_{G,S}(x, y) = \mathrm{dist}_S(x, y) = \mathrm{dist}_G(x, y)$ and $B_{G,S}(x, r) = B_S(x, r) = B_G(x, r)$.

Exercise 1.75 Let S, T be two finite symmetric generating sets of G. Show that there exists a constant $\kappa = \kappa_{S,T} > 0$ such that for all $x, y \in G$,

$$\kappa^{-1} \cdot \mathrm{dist}_T(x, y) \leq \mathrm{dist}_S(x, y) \leq \kappa \cdot \mathrm{dist}_T(x, y).$$ ▷ solution ◁

Definition 1.6.2 Let μ be a probability measure on G.

- We say that μ is **adapted** (to G) if any element $x \in G$ can be written as a product $x = s_1 \cdots s_k$ where $s_1, \ldots, s_k \in \text{supp}(\mu)$.
- μ is **symmetric** if $\mu(x) = \mu\left(x^{-1}\right)$ for all $x \in G$.
- μ has an **exponential tail** if for some $\varepsilon > 0$,

$$\mathbb{E}_\mu \left[e^{\varepsilon |X|} \right] = \sum_x \mu(x) e^{\varepsilon |x|} < \infty.$$

- We say that μ has k**th moment** if

$$\mathbb{E}_\mu \left[|X|^k \right] = \sum_x \mu(x) |x|^k < \infty.$$

By $\text{SA}(G, k)$ we denote the collection of symmetric, adapted measures on G with kth moment. By $\text{SA}(G, \infty)$ we denote the collection of symmetric, adapted, exponential tail measures on G.

Exercise 1.76 Show that if μ has kth moment with respect to a finite symmetric generating set S, then μ has kth moment with respect to *any* finite symmetric generating set.

Show that if μ has an exponential tail with respect to a finite symmetric generating set S, then μ has an exponential tail with respect to *any* finite symmetric generating set.

The most basic example of $\mu \in \text{SA}(G, \infty)$ is when μ is the uniform measure on some finite symmetric generating set S of a finitely generated group G.

Exercise 1.77 Show that if μ is a symmetric, adapted measure on G with finite support, then $\mu \in \text{SA}(G, \infty)$.

Exercise 1.78 Show that if μ, ν are symmetric probability measures on G, then $p\mu + (1 - p)\nu$ is also symmetric for $p \in (0, 1)$.

Exercise 1.79 Show that if μ is an adapted probability measure on G and ν is any probability measure on G, then for any $p \in (0, 1]$ we have that $p\mu + (1 - p)\nu$ is also adapted.

Exercise 1.80 Let $p \in (0, 1)$. Show that if $\mu \in \text{SA}(G, k)$ then $\nu = p\delta_1 + (1 - p)\mu \in \text{SA}(G, k)$. (Such a measure ν is called a *lazy version* of μ.)   ▷ solution ◁

1.6.2 Random Walks

Given a group G with a probability measure μ define the μ-**random walk** on G started at $x \in G$ as the sequence

$$X_t = xU_1U_2\cdots U_t,$$

where $(U_j)_j$ are i.i.d. with law μ.

The probability measure and expectation on $G^{\mathbb{N}}$ (with the canonical cylinder-set σ-algebra) are denoted $\mathbb{P}_x, \mathbb{E}_x$. When we omit the subscript x we refer to $\mathbb{P} = \mathbb{P}_1, \mathbb{E} = \mathbb{E}_1$. Note that the law of $(X_t)_t$ under $\mathbb{P}_x$ is the same as the law of $(xX_t)_t$ under $\mathbb{P}$. For a probability measure ν on G we denote $\mathbb{P}_\nu = \sum_x \nu(x)\,\mathbb{P}_x$ and similarly for $\mathbb{E}_\nu = \sum_x \nu(x)\,\mathbb{E}_x$. More precisely, given some probability measure ν on G, we define $\mathbb{P}_\nu$ to be the measure obtained by Kolmogorov's extension theorem, via the sequence of measures

$$P_t(\{(X_0,\ldots,X_t) = (g_0,\ldots,g_t)\}) = \nu(g_0) \cdot \prod_{j=1}^{t} \mu\left(g_{j-1}^{-1}g_j\right).$$

Exercise 1.81 Show that P_t above indeed defines a probability measure on $\mathcal{F}_t = \sigma(X_0,\ldots,X_t)$.

Exercise 1.82 Show that the μ-random walk on G is a Markov chain with transition matrix $P(x,y) = \mu\left(x^{-1}y\right)$. (Markov chains will be defined and studied in Chapter 3. For the unfamiliar reader, this exercise may be skipped in the meantime.)

Show that the corresponding Laplacian operator, usually defined $\Delta := I - P$, and the averaging operator P are given by

$$Pf(x) = f * \check{\mu}(x), \qquad \Delta f(x) = f * (\delta_1 - \check{\mu})(x),$$

where $\check{\mu}(y) = \mu\left(y^{-1}\right)$.

Exercise 1.83 Consider the matrix $P(x,y) = \mu\left(x^{-1}y\right)$ from the previous exercise. Show that if P^t is the tth matrix power of P then

$$\mathbb{E}_x[f(X_t)] = \left(P^t f\right)(x).$$

Exercise 1.84 Let μ be a probability measure on G, and let $P(x,y) = \mu\left(x^{-1}y\right)$.

- Show that $P^t(1,x) = \check{\mu}^{*t}(x)$, where $\check{\mu}^t$ is convolution of $\check{\mu}$ with itself t times. $\left(\check{\mu}(y) = \mu(y^{-1})\right)$.

- Show that μ is adapted if and only if for every $x, y \in G$ there exists $t \geq 0$ such that $P^t(x, y) > 0$. (This property is also called *irreducible*.)
- Show that μ is symmetric if and only if P is a symmetric matrix (if and only if $\check{\mu} = \mu$).

We will investigate random walks in more depth in Chapter 3.

1.6.3 Harmonic Functions

In classical analysis, a function $f : \mathbb{R}^n \to \mathbb{R}$ is harmonic at x if for any small enough ball around x, $B(x, r)$, it satisfies the mean value property: $\frac{1}{|\partial B(x,r)|} \int_{\partial B(x,r)} f(y) dy = f(x)$. Another definition is that $\Delta f(x) = 0$ where $\Delta = \sum_j \frac{\partial^2}{\partial x_j^2}$ is the Laplace operator. (Why these two definitions should coincide is a deep fact, outside the scope of our current discussion.)

Definition 1.6.3 Let G be a finitely generated group and μ a probability measure on G. A function $f : G \to \mathbb{C}$ is μ-**harmonic** (or simply, *harmonic*) at $x \in G$ if

$$\sum_y \mu(y) f(xy) = f(x)$$

and the above sum converges absolutely.

A function is harmonic if it is harmonic at every $x \in G$.

Exercise 1.85 Show that f is μ-harmonic at x if and only if $\mathbb{E}_\mu[f(xU)] = f(x)$, if and only if $\Delta f(x) = 0$. (Here $\mathbb{E}_\mu$ is expectation with respect to μ, and U is a random element of G with law μ.)

Exercise 1.86 Prove the *maximum principle* for harmonic functions:

Consider an adapted probability measure μ on G. If f is harmonic, and there exists x such that $f(x) = \sup_y f(y)$, then f is constant.

Exercise 1.87 (L^2 **harmonic functions**) Consider the space $\ell^2(G)$ of functions $f : G \to \mathbb{C}$ such that $\sum_y |f(y)|^2 < \infty$. This space is a Hilbert space with the inner product $\langle f, g \rangle = \sum_y f(y) \overline{g(y)}$.

Prove the following "integration by parts" identity: for any $f, g \in \ell^2(G)$,

$$\sum_{x,y} P(x, y)(f(x) - f(y))(\bar{g}(x) - \bar{g}(y)) = 2 \langle \Delta f, g \rangle.$$

(The left-hand side above is $\langle \nabla f, \nabla g \rangle$, appropriately interpreted, hence the name

"integration by parts". This is also sometimes understood as Green's identity.)
Here as usual, $P(x, y) = \mu\left(x^{-1}y\right)$ for a symmetric measure μ.

Show that any $f \in \ell^2(G)$ that is harmonic must be constant.   ▷ solution ◁

Example 1.6.4 Consider the group $\mathbb{Z}$ and the measure $\mu = \frac{1}{2}\delta_1 + \frac{1}{2}\delta_{-1}$. Suppose that f is a μ-harmonic function. Then, for any $z \in \mathbb{Z}$, $f(z-1)+f(z+1) = 2f(z)$, which implies that

$$f(z + 1) = 2f(z) - f(z - 1),$$
$$f(z - 1) = 2f(z) - f(z + 1).$$

So the values of f are determined by the two numbers $f(0), f(1)$. This implies that the space $\mathrm{HF}(\mathbb{Z}, \mu) = \{f : \mathbb{Z} \to \mathbb{C} : \Delta f \equiv 0\}$ of all harmonic functions has dimension at most 2.

Moreover, any function $f(z) = \alpha z + \beta$, is a μ-harmonic function (check this!).

Thus, we conclude that $\mathrm{HF}(\mathbb{Z}, \mu)$ is the (2-dimensional) space of all linear maps $z \mapsto \alpha z + \beta$ for $\alpha, \beta \in \mathbb{C}$.   △▽△

Exercise 1.88 Show that if $G = \mathbb{Z}$ and μ is uniform measure on $\{-1, 1, -2, 2\}$ then the space of all μ-harmonic functions has dimension at least 2.

Is this dimension finite?   ▷ solution ◁

Exercise 1.89 Consider the group $G = \mathbb{Z}^2$ and the measure μ, which is uniform on the standard generators $\{(\pm 1, 0), (0, \pm 1)\}$.

Show that the functions $f(x, y) = x$, $h(x, y) = y$ and $g(x, y) = x^2 - y^2$ and $k(x, y) = xy$ are all μ-harmonic.

Consider a different measure ν, which is uniform on $\{(\pm 1, 0), (0, \pm 1), \pm(1, 1)\}$. Which of the above functions is harmonic with respect to ν?   ▷ solution ◁

Exercise 1.90 Let G be a finitely generated group. Let $\mu \in \mathrm{SA}(G, 1)$. Show that any homomorphism from G to the additive group $(\mathbb{C}, +)$ is a μ-harmonic function.   ▷ solution ◁

Exercise 1.91 Let μ be a symmetric and adapted probability measure on a finitely generated group G. Let $p \in (0, 1)$ and let $\nu = p\delta_1 + (1 - p)\mu$ be a lazy version of μ. Show that any function $f : G \to \mathbb{C}$ is μ-harmonic if and only if it is ν-harmonic.   ▷ solution ◁

1.7 Bounded, Lipschitz, and Polynomial Growth Functions

1.7.1 Bounded Functions

Recall for $f: G \to \mathbb{C}$ and $p > 0$ we have

$$\|f\|_p^p = \sum_x |f(x)|^p,$$

$$\|f\|_\infty = \sup_x |f(x)|.$$

Recall that $\|x.f\|_p = \|f\|_p$ for all $p \in (0, \infty]$.

Exercise 1.92 Show that $\|f\|_\infty \leq \|f\|_p$ for any $p > 0$.

For a finitely generated group G and a probability measure μ on G, we use BHF(G, μ) to denote the set of bounded μ-harmonic functions on G; that is,

$$\text{BHF}(G, \mu) = \{f: G \to \mathbb{C} : \|f\|_\infty < \infty, \ \Delta f \equiv 0\}.$$

Exercise 1.93 Show that BHF(G, μ) is a vector space over $\mathbb{C}$. Show that it is a G-invariant subspace; that is, $G.\text{BHF}(G, \mu) \subset \text{BHF}(G, \mu)$.

Any constant function is in BHF(G, μ), so $\dim \text{BHF}(G, \mu) \geq 1$. The question of whether BHF(G, μ) consists of more than just constant functions is an important one, and we will dedicate Chapter 6 to this investigation.

1.7.2 Lipschitz Functions

For a group G and a function $f: G \to \mathbb{C}$, define the right-derivative at y

$$\partial^y f: G \to \mathbb{C} \qquad \text{by} \qquad \partial^y f(x) = f\left(xy^{-1}\right) - f(x).$$

Given a finite symmetric generating set S, define the gradient $\nabla f = \nabla_S f: G \to \mathbb{C}^S$ by $(\nabla f(x))_s = \partial^s f(x)$. We define the Lipschitz semi-norm by

$$\|\nabla_S f\|_\infty := \sup_{s \in S} \sup_{x \in G} |\partial^s f(x)|.$$

Definition 1.7.1 A function $f: G \to \mathbb{C}$ is called **Lipschitz** if $\|\nabla_S f\|_\infty < \infty$.

Exercise 1.94 Show that for any two symmetric generating sets S_1, S_2, there exists $C > 0$ such that

$$||\nabla_{S_1} f||_\infty \le C \cdot ||\nabla_{S_2} f||_\infty.$$

Conclude that the definition of *Lipschitz function* does not depend on the choice of specific generating set.

Exercise 1.95 What is the set $\left\{ f \in \mathbb{C}^G : ||\nabla_S f||_\infty = 0 \right\}$?

We use LHF(G, μ) to denote the set of Lipschitz μ-harmonic functions; that is,

$$\text{LHF}(G, \mu) = \{f : G \to \mathbb{C} : ||\nabla_S f||_\infty < \infty, \ \Delta f \equiv 0\}.$$

Exercise 1.96 Show that LHF(G, μ) is a G-invariant vector space, by showing that

$$\forall x \in G \qquad ||\nabla_S x.f||_\infty = ||\nabla_S f||_\infty.$$

Exercise 1.97 (Horofunctions) Let G be a finitely generated group with a metric given by some fixed finite symmetric generating set S.

Consider the space

$$L = \{h : G \to \mathbb{C} : ||\nabla_S h||_\infty \le 1, \ h(1) = 0\}.$$

Show that L is compact under the topology of pointwise convergence.

Show that $x.h(y) = h\left(x^{-1}y\right) - h\left(x^{-1}\right)$ defines a left action of G on L.

Show that if h is fixed under the G-action (i.e. $x.h = h$ for all $x \in G$) then h is a homomorphism from G into the group $(\mathbb{C}, +)$.

Show that if h is a homomorphism from G into $(\mathbb{C}, +)$, then there exists $\alpha > 0$ such that $\alpha h \in L$.

For every $x \in G$ let $b_x(y) = \text{dist}_S(x, y) - \text{dist}_S(x, 1) = \left|x^{-1}y\right| - |x|$. Show that $b_x \in L$ for any $x \in G$. Prove that the map $x \mapsto b_x$ from G into L is an injective map.

▷ solution ◁

1.7.3 Polynomially Growing Functions

Let S be a finite, symmetric generating set for a group G. For $f : G \to \mathbb{C}$ and $k \ge 0$, define the kth degree **polynomial semi-norm** by

$$||f||_{S,k} := \limsup_{r \to \infty} r^{-k} \cdot \sup_{|x| \le r} |f(x)|.$$

Let

$$HF_k(G, \mu) = \left\{ f \in \mathbb{C}^G : f \text{ is } \mu\text{-harmonic}, \ \|f\|_{S,k} < \infty \right\}.$$

Exercise 1.98 Show that $\| \cdot \|_{S,k}$ is indeed a semi-norm.

Show that $\|x.f\|_{S,k} = \|f\|_{S,k}$.

Show that $HF_k(G, \mu)$ is a G-invariant vector space. ▷ solution ◁

Exercise 1.99 Show that if S, T are two finite symmetric generating sets for G then there exists some constant $C = C(S, T, k) > 0$ such that for any $f : G \to \mathbb{C}$ we have $\|f\|_{S,k} \leq C \cdot \|f\|_{T,k}$.

Specifically, the space $HF_k(G, \mu)$ does not depend on the specific choice of generating set. ▷ solution ◁

Exercise 1.100 Show that if $\|f\|_{S,k} < \infty$ then there exists $C > 0$ such that for all $x \in G$ we have $|f(x)| \leq C \left(|x|^k + 1 \right)$.

Exercise 1.101 Show that

$$\mathbb{C} \leq \mathsf{BHF}(G, \mu) \leq \mathsf{LHF}(G, \mu) \leq HF_1(G, \mu) \leq HF_k(G, \mu) \leq HF_{k+1}(G, \mu),$$

for all $k \geq 1$.

1.8 Additional Exercises

Exercise 1.102 Let R be a commutative ring. Define $GL_n(R)$ to be the collection of all $n \times n$ matrices M with entries in R such that $\det(M)$ is an invertible element in R.

Show that $GL_n(R)$ is a group. ▷ solution ◁

Exercise 1.103 Let I be the $n \times n$ identity matrix. Show that $\{I, -I\} \triangleleft GL_n(\mathbb{Z})$.

Define $PGL_n(\mathbb{Z}) = GL_n(\mathbb{Z})/\{-I, I\}$.

Show that $GL_{2n+1}(\mathbb{Z}) \cong \{-1, 1\} \times PGL_{2n+1}(\mathbb{Z})$.

Show that $SL_{2n+1}(\mathbb{Z}) \cong PGL_{2n+1}(\mathbb{Z})$. ▷ solution ◁

Exercise 1.104 Let $S \leq \mathrm{SL}_2(\mathbb{Z})$ be the Sanov subgroup (see Exercise 1.56 and Remark 1.5.11). Show that if $A \in S$ then

$$A = \begin{bmatrix} 4k+1 & 2n \\ 2m & 4\ell+1 \end{bmatrix}$$

for some integers $n, m, k, \ell \in \mathbb{Z}$.

⊳ solution ⊲

Exercise 1.105 Show that the Sanov subgroup is exactly

$$S = \left\{ A = \begin{bmatrix} 4k+1 & 2n \\ 2m & 4\ell+1 \end{bmatrix} \mid \det(A) = 1, \ k, \ell, n, m \in \mathbb{Z} \right\}.$$

⊳ solution ⊲

Exercise 1.106 Show that the Sanov subgroup S has finite index in $\mathrm{SL}_2(\mathbb{Z})$. (Hint: use the map taking the matrix entries modulo 4.)

⊳ solution ⊲

1.9 Solutions to Exercises

Solution to Exercise 1.4 :(
Let $\mathcal{G} = \{A : \theta^t(A) \in \mathcal{F}\}$. $\mathcal{G}$ is easily seen to be a σ-algebra. For any $t \leq n \in \mathbb{N}$ and $g \in G$, we have that

$$\theta^t\left(X_n^{-1}(g)\right) = \{\theta^t(\omega) : \omega_n = g\} = X_{n-t}^{-1}(g) \in \mathcal{F},$$

and if $t > n \in \mathbb{N}$ then $\theta^t\left(X_n^{-1}(g)\right) = G^{\mathbb{N}} \in \mathcal{F}$.
So $X_n^{-1}(g) \in \mathcal{G}$ for all $n \in \mathbb{N}$ and $g \in G$. This implies that $\mathcal{F} \subset \mathcal{G}$, which completes the proof.   :) ✓

Solution to Exercise 1.5 :(
$\theta^{-t}\mathcal{G}$ is a σ algebra because $\theta^{-t}\left(G^{\mathbb{N}}\right) = G^{\mathbb{N}}$ and $\theta^{-t}(\cup_n A_n) = \cup_n \theta^{-t}(A_n)$ and $\theta^{-t}(A^c) = (\theta^{-t}(A))^c$.
For any $k \in \mathcal{K}$ we have that $\theta^{-t}(K) \in \theta^{-t}\mathcal{G}$ by definition. So let $\mathcal{H}$ be any σ-algebra containing $\{\theta^{-t}(K) : K \in \mathcal{K}\}$. Define $\mathcal{G}' = \{A : \theta^{-t}(A) \in \mathcal{H}\}$. Then, similarly to the above, it is easy to see that $\mathcal{G}'$ is a σ-algebra. Moreover, $\mathcal{K} \subset \mathcal{G}'$, so it must be that $\mathcal{G} \subset \mathcal{G}'$. But then, $\theta^{-t}\mathcal{G} \subset \theta^{-t}\mathcal{G}' \subset \mathcal{H}$. Since $\mathcal{H}$ was any σ-algebra containing $\{\theta^{-t}(K) : K \in \mathcal{K}\}$, this implies that $\theta^{-t}\mathcal{G} = \sigma(\theta^{-t}(K) : K \in \mathcal{K})$.   :) ✓

Solution to Exercise 1.6 :(
This is immediate from

$$\theta^{-1}\mathcal{F} = \sigma\left(\theta^{-1}(X_n(g)) : n \in \mathbb{N}, \ g \in G\right) = \sigma(X_{n+1}(g) : n \in \mathbb{N}, \ g \in G) \subset \mathcal{F}.$$:) ✓

Solution to Exercise 1.7 :(
Note that

$$\theta^{-t}\left(X_n^{-1}(g)\right) = \{\omega : \theta^t(\omega) \in X_n^{-1}(g)\} = \{\omega : \omega_{t+n} = g\} = X_{t+n}^{-1}(g).$$

Since $\mathcal{F} = \sigma\left(X_n^{-1}(g) : n \in \mathbb{N}, \ g \in G\right)$ we have that

$$\theta^{-t}\mathcal{F} = \sigma\left(\theta^{-t}X_n^{-1}(g) : n \in \mathbb{N}, \ g \in G\right) = \sigma\left(X_{n+t}^{-1}(g) : n \in \mathbb{N}, \ g \in G\right) = \sigma(X_t, X_{t+1}, \ldots).$$:) ✓

Solution to Exercise 1.16 :(
If $g, \gamma \in \mathrm{stab}(x)$, then $\gamma g.x = \gamma.x = x$, and also $g^{-1}.x = g^{-1}g.x = x$.   :) ✓

Solution to Exercise 1.17 :(

Let $x \in X$ and $S = \text{stab}(x)$. The map of cosets of S into $G.x$ given by $gS \mapsto g.x$ is a well-defined bijection. Indeed, if $gS = \gamma S$ then $g = \gamma s$ for some $s \in S$. So $g.x = \gamma s.x = \gamma.x$, and the map is well defined.

It is obviously surjective, and if $g.x = \gamma.x$ then $\gamma^{-1}g \in S$, so $gS = \gamma S$, implying that the map is injective as well.
:) ✓

Solution to Exercise 1.21 :(

Compute:

$$\mathbb{P}[X \cdot Y = z] = \sum_x \mathbb{P}\left[X = x,\ Y = x^{-1}z\right] = \sum_x \mu(x)\nu\left(x^{-1}z\right) = (\mu * \nu)(z).$$
:) ✓

Solution to Exercise 1.23 :(

If $a = 0$ or $b = 0$ there is nothing to prove. So assume that $a, b > 0$. Consider the random variable X that satisfies $\mathbb{P}\left[X = a^{1/p}\right] = p$, $\mathbb{P}[X = b^{1/q}] = q$. Then $\mathbb{E}[\log X] = \log a + \log b$. Also, $\mathbb{E}[X] = pa^{1/p} + qb^{1/q}$. Jensen's inequality tells us that $\mathbb{E}[\log X] \leq \log \mathbb{E}[X]$, which results in $\log(ab) = \log a + \log b \leq \log\left(pa^{1/p} + qb^{1/q}\right)$.
:) ✓

Solution to Exercise 1.24 :(

The proof is by induction on n. For $n = 1$ there is nothing to prove. For $n = 2$, this is the "usual" Hölder inequality, which is proved as follows: denote $f = f_1$, $g = f_2$, $p = p_1$, $q = p_2$ and $\tilde{f} = \frac{f}{\|f\|_p}$, $\tilde{g} = \frac{g}{\|g\|_q}$. Then,

$$\|fg\|_1 = \|f\|_p \cdot \|g\|_q \cdot \sum_x |\tilde{f}(x)| \cdot |\tilde{g}(x)| \leq \|f\|_p \cdot \|g\|_q \cdot \sum_x \frac{1}{p}|\tilde{f}(x)|^p + \frac{1}{q}|\tilde{g}(x)|^q$$

$$= \|f\|_p \cdot \|g\|_q \cdot \left(\frac{1}{p}\|\tilde{f}\|_p^p + \frac{1}{q}\|\tilde{g}\|_q^q\right) = \|f\|_p \cdot \|g\|_q,$$

where the inequality is just Young's inequality for products: $ab \leq pa^{1/p} + qb^{1/q}$. A similar (and simpler) argument proves the case where $p = 1$, $q = \infty$.

Now for the induction step, $n > 2$. Let $q_n = \frac{p_n}{p_n - 1}$ and $q_j = p_j \cdot \left(1 - \frac{1}{p_n}\right) = \frac{p_j}{q_n}$ for $1 \leq j < n$. Then, $\frac{1}{p_n} + \frac{1}{q_n} = 1$ and

$$\sum_{j=1}^{n-1} \frac{1}{q_j} = \left(1 - \frac{1}{p_n}\right)^{-1} \cdot \sum_{j=1}^{n-1} \frac{1}{p_j} = 1.$$

By the induction hypothesis (for $n = 2$ and $n - 1$),

$$\|f_1 \cdots f_n\|_1 \leq \|f_n\|_{p_n} \cdot \|f_1 \cdots f_{n-1}\|_{q_n} = \|f_n\|_{p_n} \cdot (\||f_1|^{q_n} \cdots |f_{n-1}|^{q_n}\|_1)^{1/q_n}$$

$$\leq \|f_n\|_{p_n} \cdot \left(\prod_{j=1}^{n-1} \||f_j|^{q_n}\|_{q_j}\right)^{1/q_n} = \|f_n\|_{p_n} \cdot \prod_{j=1}^{n-1} \|f_j\|_{p_j}.$$
:) ✓

Solution to Exercise 1.25 :(

For any x, since $\frac{1}{r} + \frac{r-p}{pr} + \frac{r-q}{qr} = \frac{1}{p} + \frac{1}{q} - \frac{1}{r} = 1$,

$$|f * g(x)| \leq \sum_y \left|f(y)g\left(y^{-1}x\right)\right| = \sum_y \left(|f(y)|^p \left|g\left(y^{-1}x\right)\right|^q\right)^{1/r} \cdot |f(y)|^{(r-p)/r} \left|g\left(y^{-1}x\right)\right|^{(r-q)/r}$$

$$= \|f_1 \cdot f_2 \cdot f_3\|_1 \leq \|f_1\|_r \cdot \|f_2\|_{pr/(r-p)} \cdot \|f_3\|_{qr/(r-q)},$$

where the second inequality is the generalized Hölder inequality with

$$f_1(y) = \left(|f(y)|^p \left|g\left(y^{-1}x\right)\right|^q\right)^{1/r},$$

$$f_2(y) = |f(y)|^{(r-p)/r},$$

$$f_3(y) = \left|g\left(y^{-1}x\right)\right|^{(r-q)/r}.$$

Now,

$$||f_1||_r = \left(\sum_y |f(y)|^p \left| g\left(y^{-1}x\right) \right|^q \right)^{1/r},$$

$$||f_2||_{pr/(r-p)} = \left(\sum_y |f(y)|^p \right)^{(r-p)/pr} = ||f||_p^{(r-p)/r},$$

$$||f_3||_{qr/(r-q)} = \left(\sum_y |g(y^{-1}x)|^q \right)^{(r-q)/qr} = ||x.\check{g}||_q^{(r-q)/r} = ||g||_q^{(r-q)/r},$$

recalling that $\check{g}(z) = g\left(z^{-1}\right)$ and that $||x.\check{g}||_q = ||g||_q$. Combining all the above,

$$||f * g||_r^r = \sum_x |f * g(x)|^r \le \sum_{x,y} |f(y)|^p \left| g\left(y^{-1}x\right) \right|^q \cdot ||f||_p^{r-p} \cdot ||g||_q^{r-q}$$

$$= ||f||_p^{r-p} \cdot ||g||_q^{r-q} \cdot \sum_y |f(y)|^p \cdot ||y.g||_q^q$$

$$= ||g||_q^r \cdot ||f||_p^{r-p} \cdot \sum_y |f(y)|^p = ||g||_q^r \cdot ||f||_p^r. \qquad :) \checkmark$$

Solution to Exercise 1.26 :(
Inverses of invertible matrices with integer entries do not necessarily have to have integer entries. For example, take $M = \begin{bmatrix} 1 & 2 \\ 2 & 1 \end{bmatrix}$. The inverse is $M^{-1} = \frac{1}{3} \begin{bmatrix} -1 & 2 \\ 2 & -1 \end{bmatrix}$. $\qquad :) \checkmark$

Solution to Exercise 1.27 :(
The map $A \mapsto \det(A)$ is a homomorphism from $GL_n(\mathbb{Z})$ onto $\{-1, 1\}$. $SL_n(\mathbb{Z})$ is the kernel of this map. :) $\checkmark$

Solution to Exercise 1.28 :(
We use $e_1, \ldots, e_n$ to denote the standard basis of $\mathbb{R}^n$.
For a matrix A we write $c_j(A)$ for the jth column of A, and $r_j(A)$ for the jth row of A.
It is easy to see that $AE_{i,j}$ is a matrix with $c_k(AE_{i,j}) = 0$ for $k \ne j$ and $c_j(AE_{i,j}) = c_i(A)$. Thus, multiplying A on the right by $I + E_{i,j}$ results in adding $c_i(A)$ to $c_j(A)$. That is,

$$c_k(A(I + E_{i,j})) = \begin{cases} c_k(A) & \text{for } k \ne j, \\ c_j(A) + c_i(A) & \text{for } k = j. \end{cases}$$

Specifically, $(I + E_{i,j})^{-1} = I - E_{i,j}$. Applying $(I + E_{i,j})^z$ we see that we can add a z-multiple of column i to column j.
By transposing the matrices, we see that

$$r_k((I + E_{i,j})A) = \begin{cases} r_k(A) & \text{for } k \ne j, \\ r_i(A) + r_j(A) & \text{for } k = i. \end{cases}$$

Thus, we can add a multiple of some row i to another row j.
Also, for $i \ne j$, set $S_{i,j} = (I + E_{i,j})(I - E_{j,i})(I + E_{i,j})$. One may compute that

$$c_k(AS_{i,j}) = \begin{cases} c_k(A) & \text{for } k \notin \{i, j\}, \\ -c_j(A) & \text{for } k = i, \\ c_i(A) & \text{for } k = j. \end{cases}$$

That is, we can swap columns at the price of changing the sign of one of them. Multiplying by $S_{i,j}$ on the left we can also swap rows, changing the sign of one.
Denote $G_k = \left\langle I + E_{i,j} \mid 1 \le i \ne j \le k \right\rangle$.
We claim by induction on k that for any $A \in GL_k(\mathbb{Z})$ there exist $M, N \in G_k$ such that for any diagonal $(n - k) \times (n - k)$ matrix D with integer entries, if we consider the $n \times n$ matrix $A' = \begin{bmatrix} A & 0 \\ 0 & D \end{bmatrix}$, we find that

$MA'N$ is a diagonal matrix. The base case, where $k = 1$, is just the case where A' is already diagonal, so we may choose $M = N = I$.

So assume $1 < k \leq n$, and let $A \in \mathsf{GL}_k(\mathbb{Z})$. Let D be any diagonal $(n - k) \times (n - k)$ matrix D with integer entries, and define $A' = \begin{bmatrix} A & 0 \\ 0 & D \end{bmatrix}$. By swapping columns and/or rows, we may assume without loss of generality that $A'_{k,k} \neq 0$. Now, suppose that $A'_{i,k} \neq 0$ for some $1 \leq i < k$. Adding appropriate multiples of $r_k(A')$ to $r_i(A')$ and appropriate multiples of $r_i(A')$ to $r_k(A')$ sequentially, we arrive at a matrix $M \in G_k$ for which $(MA')_{k,k} \neq 0$ and $(MA')_{i,k} = 0$. Continuing this way for all $1 \leq i < k$, we find that there exists $M \in G_k$ such that $(MA')_{k,k} \neq 0$ and $(MA')_{i,k} = 0$ for all $1 \leq i < n$. The same procedure with columns instead of rows yields a matrix $N \in G_k$ such that $(MA'N)_{k,k} \neq 0$ and $(MA'N)_{i,k} = (MA'N)_{k,i} = 0$ for all $1 \leq i < n$.

Let B be the $(k - 1) \times (k - 1)$ matrix given by $B_{i,j} = (MA'N)_{i,j}$ for all $1 \leq i, j \leq k - 1$. Let D' be the $(n-k+1) \times (n-k+1)$ diagonal matrix given by $D'_{1,1} = (MA'N)_{k,k}$ and $D'_{1+i,1+i} = (MA'N)_{k+i,k+i} = D_{i,i}$ for all $1 \leq i \leq n - k$. We find that

$$MA'N = \begin{bmatrix} B & 0 \\ 0 & D' \end{bmatrix}.$$

Moreover,

$$\det(A) \cdot \det(D) = \det(A') = \det(MA'N) = \det(B) \cdot (MA'N)_{k,k} \cdot \det(D),$$

which implies that $\det(B) \cdot (MA'N)_{k,k} = \det(A)$. As these are all integers, and $|\det(A)| = 1$, we also find that $|\det(B)| = 1$, so that $B \in \mathsf{GL}_{k-1}(\mathbb{Z})$. By induction, there exist $M', N' \in G_{k-1}$ such that $M'MA'NN'$ is a diagonal matrix. Since $G_{k-1} \leq G_k$, we have that $M'M, MN' \in G_k$, completing the induction step.

Taking $k = n$ from the above induction claim, we see that for any $A \in \mathsf{GL}_n(\mathbb{Z})$ there exist $M, N \in G_n$ such that MAN is a diagonal matrix. Since $\det(A) = \det(MAN)$, and since MAN has integer entries, we find that $a_i := (MAN)_{i,i} \in \{-1, 1\}$ for all $1 \leq i \leq n$. Also, $\det(A) = \prod_{i=1}^{n} a_i$.

Now, if $A \in \mathsf{SL}_n(\mathbb{Z})$, then $\prod_{i=1}^{n} a_i = 1$. Let $J = \{1 \leq i \leq n : a_i = -1\}$.

If $J \neq \emptyset$, then since $(-1)^{|J|} = \prod_{j \in J} a_j = 1$, it must be that $|J| \geq 2$. Take any $i \neq j \in J$ and consider the matrix $B = S_{j,i} MANS_{i,j}$. B is a diagonal matrix, with $B_{j,j} = -a_i = 1$ and $B_{i,i} = -a_j = 1$ and $B_{k,k} = a_k$ for all $k \notin \{i, j\}$. Continuing this way, we find some matrices $S, T \in G_n$ such that $TMANS = I$. So $A = M^{-1}T^{-1}S^{-1}N^{-1} \in G_n$, and we are done.   :) ✓

Solution to Exercise 1.31 :(

Let G be virtually finitely generated. So there exists $H \leq G$, $[G : H] < \infty$ such that H is finitely generated.

Let $R \subset G$ be a set of representatives for the cosets of H in G; that is $G = \biguplus_{r \in R} Hr$, and $|R| = [G : H]$. Let S be a finite symmetric generating set for H.

Let $x \in G$. There are unique $y \in H$ and $r \in R$ such that $x = yr$. Since S generates H, there are $s_1, \ldots, s_n \in S$ such that $y = s_1 \cdots s_n$. Thus, $x = s_1 \cdots s_n \cdot r$.

This implies that $S \cup R$ is a finite generating set for G.   :) ✓

Solution to Exercise 1.32 :(

Since G is finitely generated, the image $\varphi(G)$ is a finitely generated Abelian group. By Theorem 1.5.2, $\varphi(G) \cong \mathbb{Z}^d \times F$ for a finite Abelian group F. If $d = 0$ then $|\varphi(G)| < \infty$. So under our assumptions, $d > 0$.

Since $|\varphi(G)| = \infty$, there must exist $0 \neq z \in \mathbb{Z}^d$ and $f \in F$ such that $(z, f) \in \varphi(G) \leq \mathbb{Z}^d \times F$. Since $z \neq 0$, there must exist $1 \leq j \leq d$ such that $z_j \neq 0$. Let $\pi \colon \mathbb{Z}^d \times F \to \mathbb{Z}$ be the homomorphism given by $\pi(w, f) = w_j$ for all $w \in \mathbb{Z}^d$ and $f \in F$. Then, $\psi = \pi \circ \varphi$ is a homomorphism from G into $\mathbb{Z}$. Since $0 \neq z_j \in \psi(G)$, we obtain that $z_j \mathbb{Z} \leq \psi(G)$, implying that $|\psi(G)| = \infty$. Since $\psi(G) \leq \mathbb{Z}$ it can only be trivial, or isomorphic to $\mathbb{Z}$. Thus, ψ maps G onto the group $\psi(G) \cong \mathbb{Z}$.   :) ✓

Solution to Exercise 1.33 :(

Let $\pi \colon G \to G/[G, G]$ be the canonical projection. If $G/[G, G]$ is infinite, then $\pi(G)$ is an infinite Abelian group, so Exercise 1.32 provides a surjective homomorphism onto $\mathbb{Z}$.

If on the other hand there exists a surjective homomorphism $\varphi \colon G \to \mathbb{Z}$, then $[G, G] \lhd \mathrm{Ker}\varphi$. Thus, $[G : [G, G]] \geq [G : \mathrm{Ker}\varphi] = \infty$.   :) ✓

Solution to Exercise 1.34 :(

We prove this by induction on n.

Note that

$$[xy, z] = y^{-1}x^{-1}z^{-1}xyz = ([x, z])^y \cdot [y, z],$$

so if $x = s_1 \cdots s_m$ then for any $y \in G$, there exist $z_1, \ldots, z_m$ such that

$$[x, y] = ([s_1, y])^{z_1} \cdots ([s_m, y])^{z_m}.$$

Expanding out y in a similar fashion shows that

$$\gamma_1(G) = \langle [s, s']^x : s, s' \in S,\ x \in G \rangle,$$

proving the claim for $n = 1$.

Assume now that $n > 1$. Recall that

$$\gamma_n(G) = \langle [x, z] : x \in \gamma_{n-1}(G),\ z \in G \rangle.$$

By induction on n, any $x \in \gamma_{n-1}(G)$ can be written as

$$x = [s_{1,1}, \ldots, s_{n-1,1}]^{z_1} \cdots [s_{1,m}, \ldots, s_{n-1,m}]^{z_m}$$

for $s_{i,j} \in S$ and $z_j \in G$. Thus, for any $s \in S$ there exist $w_1, \ldots, w_m \in G$ such that

$$[x, s] = [s_{1,1}, \ldots, s_{n-1,1}, s]^{w_1} \cdots [s_{1,m}, \ldots, s_{n-1,m}, s]^{w_m}.$$

Also, for any $y = r_1 \cdots r_\ell$ with $r_j \in S$ there exist $u_1, \ldots, u_\ell$ such that

$$[x, y]^{-1} = [y, x] = [r_1, x]^{u_1} \cdots [r_\ell, x]^{u_\ell}.$$

All this implies that for any $x \in \gamma_{n-1}(G)$ and any $y \in G$ we can write $[x, y]$ as a finite product of elements of the form $[s_1, \ldots, s_n]^z$ where $s_j \in S$ and $z \in G$. In other words, this proves the induction step. :) ✓

Solution to Exercise 1.35 :(

This is shown by induction on n.

For $n = 0$ it is immediate that we have $\varphi(\gamma_0(G)) = \varphi(G) = G$ and $\varphi(Z_0(G)) = \varphi(\{1\}) = \{1\}$.

For $n > 0$, note that $\varphi([x, y]) = [\varphi(x), \varphi(y)]$ for all $x, y \in G$. So by induction

$$\varphi([\gamma_{n-1}(G), G]) = [\varphi(\gamma_{n-1}(G)), \varphi(G)] = [\gamma_{n-1}(G), G] = \gamma_n(G).$$

Also by induction, $[\varphi(x), y] \in Z_{n-1}(G)$ for all $y \in G$ if and only if $\varphi([x, y]) \in Z_{n-1}(G) = \varphi(Z_{n-1}(G))$ for all $y \in G$, which is if and only if $[x, y] \in Z_{n-1}(G)$ for all $y \in G$. So $\varphi(Z_n(G)) = Z_n(G)$.

This completes the proof by induction.

Finally, for any $y \in G$, the map $\varphi_y(x) = x^y$ is an automorphism of G, so that $\gamma_n(G)^y = \gamma_n(G)$ and $Z_n(G)^y = Z_n(G)$ for all $y \in G$; that is, these are normal subgroups. :) ✓

Solution to Exercise 1.36 :(

Since $Z_k(G)$ is a normal subgroup, for any $x \in Z_k(G)$ and any $y \in G$ we have that $[x, y] = x^{-1}x^y \in Z_k(G)$. So $Z_k(G) \lhd Z_{k+1}(G)$. This proves the first assertion.

Now, the second assertion we prove by induction on $m := n - k$. Fix $k \geq 0$. The base step is $m = 0$, which is just $Z_k(G)/Z_k(G) = \{1\} = Z_0(G/Z_k(G))$.

For the induction step, let $m > 0$. Let $H = G/Z_k(G)$ and let $\pi : G \to H$ be the canonical projection. Since $Z_k(G) \lhd Z_{k+m}(G)$, it suffices to prove that $\pi(Z_{k+m}(G)) = Z_m(H)$. Indeed, we have by induction that for $x, y \in G$,

$$[x, y] \in Z_{k+m-1}(G) \iff [\pi(x), \pi(y)] = \pi([x, y]) \in Z_{k+m-1}(G)/Z_k(G) = Z_{m-1}(H),$$

so

$$\pi(Z_{k+m}(G)) = \{\pi(x) : \forall\, y \in G\ [x, y] \in Z_{k+m-1}(G)\}$$
$$= \{\pi(x) : \forall\, z \in H\ [\pi(x), z] \in Z_{m-1}(H)\} = Z_m(H),$$

completing the induction step. :) ✓

Solution to Exercise 1.37 :(

We do this by induction on n. For $n = 0$ this is obvious.

For $n > 0$, assume that $\gamma_n = \{1\}$. Then, $[\gamma_{n-1}, G] = \{1\}$ implies that $\gamma_{n-1} \lhd Z_1$. Let $H = G/Z_1$, and let

$\pi : G \to H$ be the canonical projection. It is easy to verify that for all $k \geq 1$, we have

$$\pi(\gamma_k) = [\pi(\gamma_{k-1}), \pi(G)] = [\gamma_{k-1}(H), H] = \gamma_k(H),$$

so $\gamma_{n-1}(H) = \{1\}$. By induction and a previous exercise,

$$G/Z_1 = H = Z_{n-1}(H) = Z_{n-1}(G/Z_1) = Z_n/Z_1.$$

As $Z_1 \lhd Z_n$, this can only happen if $G = Z_n$. :) ✓

Solution to Exercise 1.38 :(
Again this is by induction, where the base step $n = 0$ is obvious.
   Assume for $n > 0$ that $Z_n = G$. Set $H = G/Z_1$. Since $Z_{n-1}(H) = Z_n/Z_1 = H$, we have by induction that $\gamma_{n-1}(H) = \{1\}$. As before, if $\pi : G \to H$ is the canonical projection, then $\pi(\gamma_{n-1}) = \gamma_{n-1}(H) = \{1\}$, so $\gamma_{n-1} \lhd Z_1$. Thus,

$$\gamma_n = [\gamma_{n-1}, G] \lhd [Z_1, G] = \{1\}. \qquad \text{:) ✓}$$

Solution to Exercise 1.39 :(
Let $k \geq 1$. We know that $\gamma_k = \langle [x, y] : x \in \gamma_{k-1}, y \in G \rangle$. Consider γ_k/γ_{k+1} as a subgroup of G/γ_{k+1}. Note that since $[\gamma_k, G] = \gamma_{k+1}$, we have that $\gamma_k/\gamma_{k+1} \leq Z(G/\gamma_{k+1})$. Thus, for any $x \in \gamma_{k-1}, y, z \in G$ we get that

$$[x, yz] = x^{-1}z^{-1}y^{-1}xyz = [x, z]z^{-1}[x, y]z \equiv [x, z] \cdot [x, y] \pmod{\gamma_{k+1}}.$$

Also, if $x, y \in \gamma_{k-1}$ and $z \in G$ then

$$[xy, z] = y^{-1}x^{-1}z^{-1}xyz = y^{-1}[x, z]z^{-1}yz \equiv [x, z] \cdot [y, z] \pmod{\gamma_{k+1}}.$$

We conclude that if $\gamma_{k-1} = \langle X \rangle$ and $G = \langle S \rangle$ then

$$\gamma_k/\gamma_{k+1} = \langle [\gamma_{k+1}x, \gamma_{k+1}s] : x \in X, \ s \in S \rangle.$$

By induction on k, this proves that as long as G is finitely generated, the group γ_k/γ_{k+1} is finitely generated for all k. :) ✓

Solution to Exercise 1.40 :(
G is n-step nilpotent if and only if $\gamma_n(G) = \{1\}$ and $\gamma_{n-1}(G) \neq \{1\}$, which, by Exercises 1.37 and 1.38, is if and only if $Z_n = G$ and $Z_{n-1} \neq G$.
   The second assertion follows from the fact that $Z_{n+1}(G) = Z_n(G/Z_1)$ and $Z_n(G) = Z_{n-1}(G/Z_1)$. :) ✓

Solution to Exercise 1.41 :(
One verifies that $\gamma_k(G/\gamma_n) \leq \gamma_k/\gamma_n$, so $\gamma_n(G/\gamma_n) = \{1\}$. :) ✓

Solution to Exercise 1.42 :(
This follows from $\gamma_n(H) \leq \gamma_n(G)$ for all n, which is easily shown by induction, since for any subgroups $A \leq B \leq G$ and $C \leq D \leq G$ we have $[A, C] \leq [B, D]$. :) ✓

Solution to Exercise 1.43 :(
Let $\pi : G \to G/N$ be the canonical projection. Note that $\gamma_k(G/N) \leq \pi(\gamma_k(G))$. So if $\gamma_n(G) = \{1\} \leq N$, then $\gamma_n(G/N) = \{1\}$. :) ✓

Solution to Exercise 1.44 :(
Let $1 \leq j \leq i + k + \ell - 1 \leq n$. Compute for $M \in D_k$, $N \in D_\ell$:

$$(MN)_{i,j} = \sum_{t=1}^n M_{i,t} N_{t,j} 1_{\{t \geq i+k\}} 1_{\{j \geq t+\ell\}} = 0,$$

because $j - \ell < i + k$. :) ✓

Solution to Exercise 1.45 :(
If $M, N \in D_k(\mathbb{Z})$ then

$$(I + M)(I + N) = I + M + N + MN \in Q_{n,k}$$

because $MN \in D_{2k}(\mathbb{Z}) \subset D_k(\mathbb{Z})$.

Moreover, since D_n contains only the 0 matrix, we have that for any $N \in D_k(\mathbb{Z})$ we may choose $M = \sum_{j=1}^{n-1}(-N)^j \in D_k(\mathbb{Z})$, and we have that

$$(I + N)(I + M) = (I + N) \cdot \sum_{j=0}^{n-1}(-N)^j = I,$$

implying that $(I + N)^{-1} = I + M$ for this choice of M.

This proves that $Q_{n,k}$ is a group. :) ✓

Solution to Exercise 1.46 :(

Let $H = H_n(\mathbb{Z})$. Note that $H = Q_{n,1}$ from the previous exercise, so it is indeed a group.

We now show that for $0 \le k \le n - 1$ we have $\gamma_k(H) \subset Q_{n,k+1} \subset Z_{n-k-1}(H)$.

The case $k = 0$ is exactly what was shown above. For $k > 0$, if $I + M \in H$, $N \in D_k(\mathbb{Z})$ then

$$[(I + M), (I + N)] = (I + M)^{-1}(I + N)^{-1}(I + N + M(I + N))$$
$$= (I + M)^{-1}\left(I + ((I + N)^{-1} - I)M(I + N) + M(I + N)\right)$$
$$= (I + M)^{-1}((I + M) + L + MN) = I + (I + M)^{-1}(L + MN),$$

where

$$L = \left((I + N)^{-1} - I\right)M(I + N) = \sum_{j=1}^{n}(-N)^j M(I + N) \in D_{k+1}(\mathbb{Z}).$$

Since $MN \in D_{k+1}(\mathbb{Z})$ as well, we conclude inductively that $\gamma_k(H) \subset Q_{n,k+1}$.

Also, since D_n only contains the 0 matrix, it is immediate that $Q_{n,k+1} \subset Z_{n-k-1}(H)$ holds when $k = n-1$. For $k < n - 1$ and $N \in D_{k+1}(\mathbb{Z})$, for any $I + M \in H$, we have seen that $[(I + M), (I + N)] \in Q_{n,k+2} \subset Z_{n-k-2}(H)$ (inductively). Thus, $I + N \in Z_{n-k-1}(H)$ for any $N \in D_{k+1}(\mathbb{Z})$, as required. :) ✓

Solution to Exercise 1.47 :(

This follows since if $H \le G$ then $[G, H] \le [G, G]$.

So for any group G we have that $G^{(n)} \le \gamma_n(G)$, inductively. :) ✓

Solution to Exercise 1.48 :(

If G is 2-step solvable then $\left[G^{(1)}, G^{(1)}\right] = G^{(2)} = \{1\}$. :) ✓

Solution to Exercise 1.49 :(

This follows since $\left(G^{(n)}\right)^{(k)} = G^{(n+k)}$. :) ✓

Solution to Exercise 1.50 :(

There exists n such that $G^{(n)} \ne \{1\} = G^{(n+1)}$.

We prove this by induction on n. If $n = 0$ then G is infinite Abelian, in which case $[G, G] = \{1\}$. For $n > 0$, let $H = G/G^{(n)}$. We have that $H^{(n)} = \{1\}$, so by induction $[H : [H, H]] = \infty$. Also, $[H, H] = G^{(1)}/G^{(n)}$, so $[G : [G, G]] = [H : [H, H]] = \infty$, completing the induction. :) ✓

Solution to Exercise 1.51 :(

This follows from $H^{(n)} \le G^{(n)}$, which can be easily shown inductively. :) ✓

Solution to Exercise 1.52 :(

For $A, B \in \Delta_n^+$ we have that

$$(AB)_{i,j} = \sum_{\ell=1}^{n} A_{i,\ell}B_{\ell,j} = \mathbb{1}_{\{i=j\}}A_{i,i}B_{i,i}.$$

This immediately shows that $AB = BA$.

Also, since $A_{i,i} > 0$ for all i, we can choose $B_{i,i} = \frac{1}{A_{i,i}}$, to get $AB = I$, so $B = A^{-1}$. :) ✓

Solution to Exercise 1.53 :(

For $T + M, S + N \in P_{n,k}$ we have $(T + M)(S + N) = TS + TN + MS + MN$. Since $TN, MS, MN \in D_k$ we get that $P_{n,k}$ is closed under matrix multiplication.

Choosing $M = \sum_{j=1}^{n} \left(-S^{-1}N\right)^j \cdot S^{-1}$ and $T = S^{-1}$ will give us that

$$(T + M)(S + N) = I + S^{-1}N + M\left(I + S^{-1}N\right) = I,$$

so $(S + N)^{-1} = S^{-1} + M$ for this choice of M.

Now consider the map $\varphi\colon P_{n,k} \to \Delta_n^+$ given by $\varphi(T + M) = T$. One easily check that this is a surjective homomorphism, and that $\mathrm{Ker}\varphi = \{I + N : N \in D_k\}$. Since Δ_n^+ is an Abelian group, it must be that $[P_{n,k}, P_{n,k}] \lhd \mathrm{Ker}\varphi$.

As in Exercise 1.46, we compute commutators: for any $M \in D_\ell$, $N \in D_k$ we have $[(I + M), (I + N)] = I + (I + M)^{-1}(L + MN)$, where

$$L = \sum_{j=1}^{n} (-N)^j M(I + N) \in D_{\ell+k}$$

and also $MN \in D_{\ell+k}$ (by Exercise 1.44).

This implies inductively that

$$(P_{n,k})^{(\ell+1)} = ([P_{n,k}, P_{n,k}])^{(\ell)} \lhd \{I + N : N \in D_{2^\ell k}\}$$

for all $\ell \geq 0$. Since D_n contains only the 0 matrix, $P_{n,k}$ is solvable of step at most $\lceil \log_2(n/k)\rceil + 1$.

Finally, to show that $P_{n,k}$ is not nilpotent, we will show that $Z_1(P_{n,k}) = \{1\}$, which implies that $Z_\ell(P_{n,k}) = \{1\}$ for all $\ell \geq 0$. Indeed,

$$(T + M)(S + N) - (S + N)(T + M) = TN - NT + MS - SM + MN - NM.$$

If $S + N \in Z_1(P_{n,k})$, then by choosing $M \in D_{n-1}$, we have that $NM = MN = 0$. Also, an easy computation gives

$$MS - SM = (S_{n,n} - S_{1,1}) \cdot M.$$

Also, there exist t, s such that $N_{t,s} \neq 0$. Necessarily $s > t$. We choose $M_{i,j} = 1_{\{i=j=n\}}$ and $T_{i,j} = \alpha \cdot 1_{\{i=j=t\}}$ for some $\alpha > 0$. Then

$$(TN - NT)_{i,j} = \alpha \left(1_{\{i=t\}} - 1_{\{j=t\}}\right) N_{i,j}.$$

Hence

$$((T + M)(S + N) - (S + N)(T + M))_{t,s} = \alpha N_{t,s} + S_{n,n} - S_{1,1}.$$

Since we can choose $\alpha > 0$ such that this is nonzero, we find that $S + N$ does not commute with $T + M$ in this case.   :) ✓

Solution to Exercise 1.54 :(

It is easy to compute that

$$\begin{bmatrix} \omega^z & d \\ 0 & 1 \end{bmatrix} \cdot \begin{bmatrix} \omega^w & c \\ 0 & 1 \end{bmatrix} = \begin{bmatrix} \omega^{z+w} & \omega^z c + d \\ 0 & 1 \end{bmatrix},$$

so that $\begin{bmatrix} \omega^z & d \\ 0 & 1 \end{bmatrix}^{-1} = \begin{bmatrix} \omega^{-z} & -\omega^{-z}d \\ 0 & 1 \end{bmatrix}$ showing that G is a group.

For $d = \sum_{k=0}^{r-1} a_k \omega^k$ where $a_0, \ldots, a_{r-1} \in \mathbb{Z}$, and $z \in \mathbb{Z}$, we have that

$$\begin{bmatrix} \omega^z & d \\ 0 & 1 \end{bmatrix} = \begin{bmatrix} 1 & d \\ 0 & 1 \end{bmatrix} \cdot \left(\begin{bmatrix} \omega & 0 \\ 0 & 1 \end{bmatrix}\right)^z = \prod_{k=0}^{r-1} \begin{bmatrix} 1 & a_k\omega^k \\ 0 & 1 \end{bmatrix} \cdot \left(\begin{bmatrix} \omega & 0 \\ 0 & 1 \end{bmatrix}\right)^z = \prod_{k=0}^{r-1} \left(\begin{bmatrix} 1 & \omega^k \\ 0 & 1 \end{bmatrix}\right)^{a_k} \cdot \left(\begin{bmatrix} \omega & 0 \\ 0 & 1 \end{bmatrix}\right)^z,$$

implying that G is generated by the finite set

$$G = \left\langle \begin{bmatrix} \omega & 0 \\ 0 & 1 \end{bmatrix}, \begin{bmatrix} 1 & \omega^k \\ 0 & 1 \end{bmatrix} : 0 \leq k \leq r-1 \right\rangle.$$

Computing commutators we see that

$$\left[\begin{bmatrix} \omega^z & d \\ 0 & 1 \end{bmatrix}, \begin{bmatrix} \omega^w & c \\ 0 & 1 \end{bmatrix}\right] = \begin{bmatrix} \omega^{-z-w} & -\omega-z-wc-\omega^{-z}d \\ 0 & 1 \end{bmatrix}\begin{bmatrix} \omega^{z+w} & \omega^z c + d \\ 0 & 1 \end{bmatrix} = \begin{bmatrix} 1 & \omega^{-z-w}((\omega^z-1)c-(\omega^w-1)d) \\ 0 & 1 \end{bmatrix}.$$

As above, this shows that G is 2-step solvable, but not nilpotent, since $Z_1(G) = \{1\}$.

However, consider the map $\varphi \colon G \to \{0, 1, \ldots, r-1\}$ given by $\varphi\left(\begin{bmatrix} \omega^z & d \\ 0 & 1 \end{bmatrix}\right) = z \pmod{r}$. This is easily seen to be a well-defined surjective homomorphism, so $[G : \mathrm{Ker}\varphi] = r$. Moreover, $\begin{bmatrix} \omega^z & d \\ 0 & 1 \end{bmatrix} \in \mathrm{Ker}\varphi$ if and only if $z = 0 \pmod{r}$. Thus

$$\mathrm{Ker}\varphi = \left\{\begin{bmatrix} 1 & d \\ 0 & 1 \end{bmatrix} : d \in D\right\},$$

which is an Abelian group of finite index in G. :) ✓

Solution to Exercise 1.55 :(

Let $S = \{a, b\}$. Define $\varphi \colon \mathbb{F}_S \to H$ by $\varphi(a) = a$ and $\varphi(b) = b$ and extending in the canonical way to words in $\mathbb{F}_S$. This is a surjective homomorphism, and we want to show that it is injective as well.

Step I. Let $h = a^{z_1} b^{w_1} \cdots a^{z_n}$ be an element in H such that $z_n, z_k, w_k \in \mathbb{Z} \setminus \{0\}$ for all $1 \le k \le n-1$. For any $x \in B$,

$$h.x = a^{z_1} b^{w_1} \cdots a^{z_n}.x = a^{z_1} b^{w_1} \cdots a^{z_{n-1}} b^{z_{n-1}}(A) \subset A,$$

so it is impossible that $h.x = x$, implying that $h \ne 1$.

Step II. Now, for a general element $h = a^{z_1} b^{w_1} \cdots a^{z_n} b^{w_n}$, where $w_1, z_n, z_k, w_k \in \mathbb{Z} \setminus \{0\}$ for $2 \le k \le n-1$, but possibly $z_1 = 0$ or $w_n = 0$. In this case we can define:

$$g = \begin{cases} a^{-z_n} h a^{z_n} & \text{if } z_1 = w_n = 0, \\ a^{-1} h a & \text{if } z_1 = 0 \ne w_n, \\ h & \text{if } z_1 \ne 0 = w_n. \end{cases}$$

We see that in each of the above cases, the element g falls into the conditions of Step I. So $g \ne 1$. Since every time g is a conjugate of h, also $h \ne 1$. :) ✓

Solution to Exercise 1.56 :(

$\mathrm{SL}_2(\mathbb{Z})$ acts on $\mathbb{Z}^2$. Let $A = \left\{(x, y) \in \mathbb{Z}^2 : |y| < |x|\right\}$ and $B = \left\{(x, y) \in \mathbb{Z}^2 : |x| < |y|\right\}$.

Note that $a^z = \begin{bmatrix} 1 & 2z \\ 0 & 1 \end{bmatrix}$ and $b^z = \begin{bmatrix} 1 & 0 \\ 2z & 1 \end{bmatrix}$ for any $z \in \mathbb{Z}$.

We have that $a^z(x, y) = (x + 2zy, y)$. So if $(x, y) \in B$, since $|y| > |x|$ we get that

$$|x + 2zy| \ge 2|z||y| - |x| > (2|z| - 1)|y| \ge |y|$$

if $z \ne 0$. So $a^z(x, y) \in A$ for all $z \ne 0$ and $(x, y) \in B$.

Similarly, if $(x, y) \in A$ then

$$|2zx + y| \ge 2|z||x| - |y| > (2|z| - 1)|x| \ge |x|,$$

so $b^z(x, y) \in B$ for all $z \ne 0$ and $(x, y) \in A$.

This implies that $\langle a, b \rangle$ is isomorphic to $\mathbb{F}_2$ by the Ping-Pong Lemma. :) ✓

Solution to Exercise 1.59 :(

It was shown in Exercise 1.58 that $a = x^2 = (-st)^2 = stst$ and $b = y^2 = (-s^2 t)^2 = s^2 t s^2 t$.

Now, let $z \in \mathrm{SL}_2(\mathbb{Z})$. By Exercise 1.58, there exist $n \ge 0$ and $\varepsilon_1, \ldots, \varepsilon_n \in \{-1, 1\}$ and $\alpha, \beta \in \{0, 1\}$ such that $z \equiv t^\alpha s^{\varepsilon_1} t s^{\varepsilon_2} \cdots t s^{\varepsilon_n} t^\beta \pmod{\{-I, I\}}$. Choose a minimal $n = n(z)$ as above. We prove the assertion that there exist $w \in S$ and $p \in \left\{1, s, s^2, t, ts, ts^2\right\}$ such that $z \equiv wp \pmod{\{-I, I\}}$ by induction on n.

The base case is $n(z) = 0$, for which $z \equiv t^{\alpha+\beta} \pmod{\{-I, I\}}$ for some $\alpha, \beta \in \{0, 1\}$. In all cases one sees that the assertion holds with $w = 1$ and $p \in \{1, t\}$.

For the induction step, we have that $z \equiv t^\alpha s^{\varepsilon_1} t s^{\varepsilon_2} \cdots t s^{\varepsilon_n} t^\beta \pmod{\{-I, I\}}$ and $n \ge 1$. Set $\tilde{z} = t^\alpha s^{\varepsilon_1} t s^{\varepsilon_2} \cdots t s^{\varepsilon_{n-1}} t$. By induction, there exist $\tilde{w} \in S$ and $\tilde{p} \in \left\{1, s, s^2, t, ts, ts^2\right\}$ such that $\tilde{z} \equiv \tilde{w}\tilde{p} \pmod{\{-I, I\}}$. Note that modulo $\{-I, I\}$,

$$\tilde{p}s^{-1}t \equiv \begin{cases} s^{-1}t \equiv s^2 t & \tilde{p} = 1, \\ t & \tilde{p} = s, \\ st & \tilde{p} = s^2, \\ ts^{-1}t \equiv a^{-1}s & \tilde{p} = t, \\ sts^{-1}t \equiv ab^{-1}s^2 & \tilde{p} = st, \\ s^2 ts^{-1}t \equiv b & \tilde{p} = s^2 t, \end{cases} \qquad \tilde{p}st \equiv \begin{cases} st & \tilde{p} = 1, \\ s^2 t & \tilde{p} = s, \\ t & \tilde{p} = s^2, \\ tst \equiv b^{-1}s^2 & \tilde{p} = t, \\ stst = a & \tilde{p} = st, \\ s^2 tst \equiv ba^{-1}s & \tilde{p} = s^2 t, \end{cases}$$

which completes the induction step.

This immediately shows that the number of cosets of $\pi(S)$ is at most 6, that is, $[\text{PSL}_2(\mathbb{Z}) : \pi(S)] \leq 6$.

Finally, we also have that for any $z \in \text{SL}_2(\mathbb{Z})$ there exist $w \in S$ and $p \in \{1, s, s^2, t, st, s^2t\}$ such that $\pi(z) = \pi(wp)$. This implies that for some $\varepsilon \in \{-1, 1\}$ we have that $z = \varepsilon wp$. Hence, there are at most 12 cosets for S in $\text{SL}_2(\mathbb{Z})$; that is, $[\text{SL}_2(\mathbb{Z}) : S] \leq 12$. :) ✓

Solution to Exercise 1.60 :(
For every $x \in G$ there are unique elements $y_x \in H$ and $t_x \in T$ such that $x = y_x t_x$. For any $u \in T$ and $s \in S$ one has that $us(t_{us})^{-1} = y_{us} \in TST^{-1} \cap H$.

We will show that $\{y_{us} : u \in T, s \in S\}$ generate H. To this end, fix some $x \in G$ and write $x = s_1 \cdots s_n$ for $s_j \in S$. Define inductively $u_1 = s_1$ and $u_{k+1} = t_{u_k} s_{k+1}$. Then,

$$x = y_{s_1} t_{s_1} s_2 \cdots s_n = y_{u_1} y_{u_2} t_{u_2} s_3 \cdots s_n = \cdots = y_{u_1} y_{u_2} \cdots y_{u_n} t_{u_n}.$$

Note that $u_j \in TS$ so $y_{u_j} \in TST^{-1} \cap H$. Specifically, if $x \in H$ then it must be that $t_{u_n} = 1$ and $x = y_{u_1} \cdots y_{u_n}$. :) ✓

Solution to Exercise 1.62 :(
For any $x \in G$, we can write $\pi(x) = \pi(s_1) \cdots \pi(s_n)$ for some $s_j \in S$. Thus, there exists $h \in H$ such that $x = hs_1 \cdots s_n$. Writing $h = u_1 \cdots u_m$ for $u_i \in U$, we have that $U \cup S$ generates G. :) ✓

Solution to Exercise 1.63 :(
Let $G = \{g_1, \ldots, g_n\}$. Let $\mathbb{F} = \mathbb{F}_n$ be the free group on n generators, and denote the generators by $\{s_1, \ldots, s_n\}$. Consider the homomorphism $\varphi : \mathbb{F} \to G$ defined by setting $\varphi(s_j) = g_j$.

For every $1 \leq i, j \leq n$ there exists $1 \leq k = k(i, j) \leq n$ such that $g_i g_j = g_k$. Define the relation $r_{i,j} = s_i s_j (s_k)^{-1}$ for $k = k(i, j)$.

Let $K = \text{Ker}\varphi$ and let $R \triangleleft \mathbb{F}$ be the smallest normal subgroup containing $\{r_{i,j} : 1 \leq i, j \leq n\}$. Note that $R \triangleleft K$.

Let $\pi : \mathbb{F}/R \to G$ be the homomorphism defined by $\pi(Rx) = \varphi(x)$. This is well defined because $R \triangleleft K$. So $\mathbb{F}/R$ and K/R are finite groups. Since $(\mathbb{F}/R)/(K/R) \cong \mathbb{F}/K \cong G$, we have that $|G| \leq \frac{|G|}{|K/R|}$, which can only mean that $K = R$. Hence $G = \langle s_1, \ldots, s_n \mid r_{i,j} \ 1 \leq i, j \leq n \rangle$ is a finitely presented group. :) ✓

Solution to Exercise 1.64 :(
$\mathbb{Z}$ is finitely presented, as it is just the free group on 1 generator. Since G/H is finite, it is also finitely presented. Thus, G is finitely presented by Lemma 1.5.14. :) ✓

Solution to Exercise 1.65 :(
This follows directly from Theorem 1.5.15 and the fact that virtually-$\mathbb{Z}$ groups are finitely presented. :) ✓

Solution to Exercise 1.66 :(
If $e_1, \ldots e_d$ are the standard basis vectors spanning $\mathbb{Z}^d$, then defining $H_k = \langle e_1, \ldots, e_{d-k} \rangle$ for $0 \leq k < d$, and $H_d = \{1\}$, we have that $H_{k+1} \triangleleft H_k$ and $H_k/H_{k+1} \cong \mathbb{Z}$ for all $0 \leq k < d$. Thus $\mathbb{Z}^d$ is finitely presented by Theorem 1.5.15.

If G is a finitely generated virtually Abelian group, then $G \cong \mathbb{Z}^d \times F$ for some d and some finite group F (by Theorem 1.5.2). Thus, there exists a normal subgroup $N \triangleleft G$ such that $N \cong \mathbb{Z}^d$ and $G/N \cong F$. Since both N and F are finitely presented, so is G by Lemma 1.5.14. :) ✓

Solution to Exercise 1.67 :(
Assume that G is n-step nilpotent. We prove the claim by induction on n.

If $n = 1$ then G is Abelian, and since it was assumed to be finitely generated, G is finitely presented, completing the induction base.

For $n > 1$, consider the lower central series $G = \gamma_0 \triangleright \gamma_1 \triangleright \cdots \triangleright \gamma_n = \{1\}$. Consider the group $H = \gamma_{n-1}$. Since $[G, H] = \{1\}$, we have that H is Abelian. By Exercise 1.39, $H \cong \gamma_{n-1}/\gamma_n$ is finitely generated. Thus, H is finitely presented. Also, G/H is at most $(n-1)$-step nilpotent and finitely generated, so G/H is finitely presented by induction. Thus, G is also finitely presented, completing the induction step. :) ✓

Solution to Exercise 1.68 :(

Multiplication is associative since

$$((g,h)(g',h'))(g'',h'') = (gg',hg(h'))(g'',h'') = (gg'g'',hg(h')gg'(h'')),$$

$$(g,h)((g',h')(g'',h'')) = (g,h)(g'g'',h'g'(h'')) = (gg'g'',hg(h')gg'(h'')).$$

The identity is easily seen to be $(1_G, 1_H)$. Inverses are given by $(g,h)^{-1} = \left(g^{-1}, g^{-1}(h^{-1})\right)$.

The map $(g,h) \mapsto g$ is a homomorphism onto G with kernel $\{(1,h) \mid h \in H\}$, which is isomorphic to H.

:) ✓

Solution to Exercise 1.70 :(

Since $TM \in D_k$ for all $T \in \Delta_n^+$ and $M \in D_k$, it is obvious that Δ_n^+ acts on the set D_k. Also, $T(M+N) = TM + TN$ so this action is indeed a group automorphism (recall that D_k has an additive operation).

For $T \in \Delta_n^+$, $M \in D_k$ define a $2n \times 2n$ matrix by $\Psi(T,M) = \begin{bmatrix} T & M \\ 0 & I \end{bmatrix}$. Multiplying two such matrices by blocks gives $\Psi(T,M)\Psi(S,N) = \Psi(TS, TN + M)$. This immediately leads to the conclusion that $\Delta_n^+ \ltimes D_k \cong G := \{\Psi(T,M) : T \in \Delta_n^+, \ M \in D_k\}$.

Now, since $\Psi(T,M)^{-1} = \Psi\left(T^{-1}, -T^{-1}M\right)$, we have that

$$[\Psi(T,M), \Psi(S,N)] = \Psi\left(T^{-1}S^{-1}, -T^{-1}S^{-1}N - T^{-1}M\right)\Psi(TS, TN+M)$$

$$= \Psi\left(I, T^{-1}S^{-1}(TN+M) - T^{-1}S^{-1}N - T^{-1}M\right)$$

$$= \Psi\left(I, (I-T^{-1})S^{-1}N - (I-S^{-1})T^{-1}M\right).$$

Thus, $G^{(1)} \subset \{\Psi(I,M) : M \in D_k\}$. However, computing the commutator again (when $S = T = I$) we get that $G^{(2)} = \{I\}$, so G is 2-step solvable.

If $k \geq n$ then D_k is just the 0 matrix, so $\Delta_n^+ \ltimes D_k \cong \Delta_n^+$, which is Abelian.

To show that G is not nilpotent when $k < n$, we first compute the center $Z = Z_1(G)$. If $\Psi(S,N) \in Z$, then the commutator computation above implies that $(T-I)N = (S-I)M$ for all $T \in \Delta_n^+$, $M \in D_k$. Choosing $T = I$ and $M_{i,j} = \mathbf{1}_{(j \geq i+k)}$, we get that $S_{j,j} = 1$ for all $j \leq n-k$. Thus, $(S-I)M = 0$ for any $M \in D_k$. This leads to $(T-I)N = 0$ for all $T \in \Delta_n^+$, which cannot hold unless $N = 0$. We conclude that

$$Z = \{\Psi(S,0) : \forall j \leq n-k, \ S_{j,j} = 1\}.$$

Now, we compute the second center $Z_2 = Z_2(G) = \{x \in G : \forall y \in G \ [x,y] \in Z_1(G)\}$. Using the commutator formula above, we see that if $\Psi(S,N) \in Z_2$, then again $(T-I)N = (S-I)M$ for all $T \in \Delta_n^+$, $M \in D_k$, which leads to $N = 0$ and $S_{j,j} = 1$ for all $j \leq n-k$, as before. But then we get that $Z_2 = Z$, so the upper-central series stabilizes at Z, and G cannot be nilpotent.

:) ✓

Solution to Exercise 1.71 :(

For $\alpha \neq 0$ and $u \in V$, denote the transformation $v \mapsto \alpha v + u$ by the "matrix" $\begin{bmatrix} \alpha & u \\ 0 & 1 \end{bmatrix}$. (If V is finite dimensional, then this is an actual $(\dim V + 1) \times (\dim V + 1)$ matrix.)

One sees that the usual matrix multiplication provides us with composition of transformations:

$$\begin{bmatrix} \alpha & u \\ 0 & 1 \end{bmatrix} \cdot \begin{bmatrix} \beta & v \\ 0 & 1 \end{bmatrix} = \begin{bmatrix} \alpha\beta & \alpha v + u \\ 0 & 1 \end{bmatrix}.$$

The inverse transformation is given by

$$\begin{bmatrix} \alpha & u \\ 0 & 1 \end{bmatrix}^{-1} = \begin{bmatrix} \alpha^{-1} & -\alpha^{-1}u \\ 0 & 1 \end{bmatrix}.$$

This provides the group structure for the affine transformations of V.

In fact, note that the multiplicative group $\mathbb{C}^* = \mathbb{C} \setminus \{0\}$ acts on the additive group V, so the collection of affine transformations is just $\mathbb{C}^* \ltimes V$.

It is now straightforward to compute commutators:

$$\left[\begin{bmatrix} \alpha & u \\ 0 & 1 \end{bmatrix}, \begin{bmatrix} \beta & v \\ 0 & 1 \end{bmatrix}\right] = \begin{bmatrix} \alpha^{-1}\beta^{-1} & -\alpha^{-1}\beta^{-1}v - \alpha^{-1}u \\ 0 & 1 \end{bmatrix} \begin{bmatrix} \alpha\beta & \alpha v + u \\ 0 & 1 \end{bmatrix} = \begin{bmatrix} 1 & \alpha^{-1}\beta^{-1}((\alpha-1)v - (\beta-1)u) \\ 0 & 1 \end{bmatrix}.$$

Just as before, one sees that $\mathbb{C}^* \ltimes V$ is 2-step solvable, if $V \neq \{0\}$.

Also, if $\begin{bmatrix} \alpha & u \\ 0 & 1 \end{bmatrix} \in Z = Z_1(\mathbb{C}^* \ltimes V)$, then $(\alpha-1)v = (\beta-1)u$ for all $\beta \in \mathbb{C}^*$, $v \in V$. If $V \neq \{0\}$, this is only possible if $u = 0$ and $\alpha = 1$. Hence $Z = \{1\}$. That is, the only case where $\mathbb{C}^* \ltimes V$ is nilpotent is when $V = \{0\}$ and $\mathbb{C}^* \ltimes V \cong \mathbb{C}^*$, which is Abelian.

:) ✓

Solution to Exercise 1.72 :(
First, note that the collection $\{a^x b^\eta : \eta \in \{0, 1\}, \ x \in \mathbb{Z}\}$ forms a subgroup of D_∞. Since the generators a, b of D_∞ are contained in this subgroup, we get that any element of D_∞ is of the form $a^x b^\eta$ for some $\eta \in \{0, 1\}$ and $x \in \mathbb{Z}$.

It is not difficult to verify that the map $(\varepsilon, x) \mapsto a^x b^{(1+\varepsilon)/2}$ is a surjective homomorphism with trivial kernel.

Now,

$$\begin{aligned}
[(\varepsilon, x), (\delta, y)] &= (\varepsilon, -\varepsilon x)(\delta, -\delta y)(\varepsilon, x)(\delta, y) \\
&= (\varepsilon\delta, -\varepsilon x - \varepsilon\delta y)(\varepsilon\delta, x + \varepsilon y) \\
&= \left(\varepsilon^2\delta^2, -\varepsilon x - \varepsilon\delta y + \varepsilon\delta x + \varepsilon^2\delta y\right) \\
&= (1, \varepsilon(\delta - 1)x + \delta(1 - \varepsilon)y).
\end{aligned}$$

Thus $[(1, x), (-1, 0)] = (1, -2x)$ and $[(-1, x), (1, 1)] = (1, 2)$. Hence $Z(\mathbb{Z}_2 \ltimes \mathbb{Z}) = \{(1, 0)\}$, so $D_\infty \cong \mathbb{Z}_2 \ltimes \mathbb{Z}$ is not nilpotent.

Also, the above commutator calculation shows that $[(1, x), (1, y)] = (1, 0)$, so $D_\infty \cong \mathbb{Z}_2 \ltimes \mathbb{Z}$ is 2-step solvable.

Finally, the surjective homomorphism $(\varepsilon, x) \mapsto \varepsilon$ shows that $H = \{(1, x) : x \in \mathbb{Z}\}$ is a normal subgroup of $\mathbb{Z}_2 \ltimes \mathbb{Z}$ isomorphic to $\mathbb{Z}$, and of index 2 because $\mathbb{Z}_2 \ltimes \mathbb{Z}/H \cong \mathbb{Z}_2$. :) ✓

Solution to Exercise 1.73 :(
The group structure is easy to verify. The identity in G is $1_G = (1_{S_d}, 0)$ and $(\sigma, z)^{-1} = \left(\sigma^{-1}, -\sigma^{-1}z\right)$.

Now, note that

$$(\sigma, z)^{(\tau, w)} = \left(\tau^{-1}, -\tau^{-1}w\right)(\sigma\tau, z + \sigma w) = \left(\sigma^\tau, \tau^{-1}z + \sigma^\tau\tau^{-1}w - \tau^{-1}w\right).$$

Let $H = \{(1_{S_d}, z) : z \in \mathbb{Z}^d\}$. Then it is immediate that $H \cong \mathbb{Z}^d$ and from the above, $\left(1_{S_d}, z\right)^{(\tau, w)} = \left(1_{S_d}, \tau^{-1}z\right)$, so $H \lhd G$. Also, the map $\pi : G \to S_d$ given by $(\sigma, z) \mapsto \sigma$ is a homomorphism with kernel H. So $G/H \cong S_d$.

Finally,

$$[(\sigma, z), (\tau, w)] = \left(\sigma^{-1}, -\sigma^{-1}z\right)\left(\sigma^\tau, \tau^{-1}z + \sigma^\tau\tau^{-1}w - \tau^{-1}w\right) = \left([\sigma, \tau], -\sigma^{-1}z + \tau^\sigma w - \sigma^{-1}\tau^{-1}w\right).$$

Since S_d is non-Abelian (for $d > 2$) we may find $\sigma, \tau \in S_d$ such that $[\sigma, \tau] \neq 1_{S_d}$. :) ✓

Solution to Exercise 1.75 :(
It suffices to show only one inequality, as the other will follow by reversing the roles of S, T.

For any $t \in T$ let $s_{t,1}, \ldots, s_{t,n(t)} \in S$ be such that $s_{t,1} \cdots s_{t,n(t)} = t$ and $n(t) = |t|_S = \mathrm{dist}_S(1, t)$. Let $\kappa = \max_{t \in T} n(t)$.

Now, for any $x \in G$ let $t_1, \ldots, t_m \in T$ be such that $t_1 \cdots t_m = x$ and $m = |x|_T = \mathrm{dist}_T(1, x)$. Then,

$$x = t_1 \cdots t_m = s_{t_1,1} \cdots s_{t_1,n(t_1)} \cdot s_{t_2,1} \cdots s_{t_2,n(t_2)} \cdots s_{t_m,1} \cdots s_{t_m,n(t_m)},$$

so

$$|x|_S \leq \sum_{j=1}^{m} \sum_{k=1}^{n(t_j)} |s_{t_j,k}| \leq \kappa \cdot m = \kappa \cdot |x|_T.$$

Hence, for general $x, y \in G$ we have that

$$\mathrm{dist}_S(x, y) = \left|x^{-1}y\right|_S \leq \kappa \cdot \left|x^{-1}y\right|_T = \mathrm{dist}_T(x, y). \qquad \text{:) ✓}$$

Solution to Exercise 1.80 :(
Symmetry and adaptedness of ν follow from the previous exercises. Let U be a random element of law μ, and let $V = U$ with probability $1 - p$, and $V = 1$ with probability p. Then,

$$\mathbb{E}\left[|V|^k\right] = (1 - p)\,\mathbb{E}\left[|U|^k\right] < \infty,$$

implying that $\nu \in \mathrm{SA}(G, k)$. :) ✓

Solution to Exercise 1.87 :(

Compute using the symmetry of P:

$$2\langle \Delta f, g\rangle = 2\sum_x \Delta f(x)\bar{g}(x) = 2\sum_x \sum_y P(x,y)(f(x)-f(y))\bar{g}(x)$$

$$= \sum_{x,y} P(x,y)(f(x)-f(y))\bar{g}(x) + \sum_{y,x} P(y,x)(f(y)-f(x))\bar{g}(y)$$

$$= \sum_{x,y} P(x,y)(f(x)-f(y))(\bar{g}(x)-\bar{g}(y)).$$

We have used that $f, g \in \ell^2$, so that the above sums converge absolutely, and so can be summed together.
Thus, if f is ℓ^2 and harmonic we have that

$$\sum_{x,y} P(x,y)|f(x)-f(y)|^2 = 2\langle \Delta f, f\rangle = 0.$$

Thus, $|f(x)-f(y)|^2 = 0$ for all x, y such that $P(x,y) > 0$. Since P is irreducible (i.e. μ is adapted) this implies that f is constant. :) ✓

Solution to Exercise 1.88 :(

Note that any linear map $z \mapsto \alpha z + \beta$ is still harmonic with respect to this μ.

The dimension is at most 4 since the linear map $f \mapsto (f(-1), f(0), f(1), f(2))$ from the space $\mathsf{HF}(\mathbb{Z}, \mu)$ to $\mathbb{C}^4$ is injective (it has a trivial kernel). Indeed, for any μ-harmonic function f, and any z we have that

$$f(z+2) = 4f(z) - f(z-1) - f(z+1) - f(z-2),$$
$$f(z-2) = 4f(z) - f(z-1) - f(z+1) - f(z+2).$$

So if $f(z-1) = f(z) = f(z+1) = f(z+2) = 0$ for any z then $f \equiv 0$ is identically 0. :) ✓

Solution to Exercise 1.89 :(

It is easy to verify μ-harmonicity.

As for ν, one may check that f, h are ν-harmonic. Also,

$$g * \check{\nu}(x,y) = \tfrac{1}{6}((x+1)^2 - y^2 + (x-1)^2 - y^2 + x^2 - (y+1)^2 + x^2 - (y-1)^2$$
$$+ (x+1)^2 - (y+1)^2 + (x-1)^2 - (y-1)^2)$$
$$= x^2 - y^2 + \tfrac{1}{6}(2-2+1+2x-1-2y+1-2x-1+2y) = x^2 - y^2 = g(x,y),$$
$$k * \check{\nu}(x,y) = \tfrac{1}{6}((x+1)y + (x-1)y + x(y+1) + x(y-1) + (x+1)(y+1) + (x-1)(y-1))$$
$$= xy + \tfrac{1}{6}(y - y + x - x + x + y + 1 - x - y + 1) = xy + \tfrac{1}{3},$$

so g is ν-harmonic, but k is not ν-harmonic. :) ✓

Solution to Exercise 1.90 :(

Let $\varphi: G \to \mathbb{C}$ be a homomorphism. Then, using the symmetry of μ,

$$\sum_y \mu(y)\varphi(xy) = \varphi(x) + \sum_y \mu(y)\tfrac{1}{2}\left(\varphi(y) + \varphi(y^{-1})\right) = \varphi(x).$$

The above sum converges absolutely because μ has finite first moment, and since $|\varphi(xy)| \le |\varphi(x)| + |\varphi(y)| \le \max_{s \in S} |\varphi(s)| \cdot (|x| + |y|)$, where S is the finite symmetric generating set used to determine the metric on G. :) ✓

Solution to Exercise 1.91 :(

For any $x \in G$,

$$\sum_y \nu(y)f(xy) = pf(x) + (1-p)\sum_x \mu(y)f(xy),$$

where the sums on both sides converge absolutely together. :) ✓

Solution to Exercise 1.97 :(

The fact that L is compact is basically the Arzelà–Ascoli theorem. However, let us give a self-contained proof.

The space L with the topology of pointwise convergence is metrizable; for example, one may consider the metric

$$\text{dist}(f, h) = \exp(-R(h, f)), \qquad R(h, f) := \sup\{r \geq 0 : \forall \, |x| \leq r, \, h(x) = f(x)\}.$$

So compactness will follow by showing that any sequence has a converging subsequence.

Let $(f_n)_n$ be a sequence in L. Denote $G = \{x_1, x_2, \dots\}$.

We will inductively construct a sequence of subsets $\mathbb{N} \supset I_1 \supset I_2 \supset I_3 \supset \cdots$, all infinite $|I_j| = \infty$, such that for all $m \geq 1$ the limits $\lim_{I_j \ni k \to \infty} f_k(x_j)$ exist.

Indeed, if $m = 1$ then since $|f_n(x)| \leq |x|$ for all n, the sequence $(f_n(x_1))_n$ is bounded, and thus has a converging subsequence. Let I_1 be the indices of this converging subsequence.

For $m > 1$, given I_{m-1}, we consider $(f_k(x_m))_{k \in I_{m-1}}$. Since this sequence is bounded, it too has a converging subsequence, and we denote by $I_m \subset I_{m-1}$ the indices of this new subsequence.

With this construction, we now write $I_m = (n_k^{(m)})_k$, for each $m \geq 1$. Consider the sequence $h_k = f_{n_k^{(k)}}$.

For any $m \geq 1$, the sequence $(h_k)_{k \geq m}$ is a subsequence of $(f_k)_{k \in I_m}$. Thus, $h(x_m) := \lim_{k \to \infty} h_k(x_m)$ exists.

This shows that $(h_k)_k$ converges pointwise to h, proving that L is compact.

The fact that $x.h(y) = h\left(x^{-1}y\right) - h\left(x^{-1}\right)$ is a left action is easily shown.

Also, if $x.h = h$ for all $x \in G$, then $h(xy) = x^{-1}.h(y) + h(x) = h(y) + h(x)$ for all $x, y \in G$.

If $h : G \to \mathbb{C}$ is a homomorphism, then choose $\alpha = \frac{1}{\max_{s \in S} |h(s)|}$. Then

$$||\nabla_S h||_\infty = \sup_{s \in S} \sup_{x \in G} |h(xs) - h(x)| = \max_{s \in S} |h(s)|,$$

so that $||\nabla_S \alpha h||_\infty = 1$. Hence, $\alpha h \in L$.

Now for the functions b_x. Note that

$$z.b_x(y) = b_x\left(z^{-1}y\right) - b_x\left(z^{-1}\right) = \left|x^{-1}z^{-1}y\right| - \left|x^{-1}z^{-1}\right| = b_{zx}(y).$$

By the triangle inequality, $|b_x(y)| \leq |y|$. So,

$$|b_x(ys) - b_x(y)| = \left|y^{-1}.b_x(s)\right| = |b_{y^{-1}x}(s)| \leq |s|,$$

which implies that $||\nabla_S b_x||_\infty \leq 1$.

Finally, if $b_x = b_y$, then

$$\text{dist}_S(x, y) = b_x(y) + |x| = b_y(y) + |x| = |x| - |y|.$$

Reversing the roles of x, y we have that $\text{dist}_S(x, y) = -\text{dist}_S(x, y)$, implying that $\text{dist}_S(x, y) = 0$, so that $x = y$. :) ✓

Solution to Exercise 1.98 :(

The fact that $|| \cdot ||_{S,k}$ is a semi-norm is easy to verify.

For $f : G \to \mathbb{C}$ and $x \in G$ note that

$$||x.f||_{S,k} = \limsup_{r \to \infty} r^{-k} \sup_{|y| \leq r} \left|f\left(x^{-1}y\right)\right| \leq \limsup_{r \to \infty} (r + |x|)^{-k} \sup_{|z| \leq r+|x|} |f(z)| \cdot \left(\frac{r+|x|}{r}\right)^k = ||f||_{S,k}.$$

Repeating this for x^{-1}, we have that $||x.f||_{S,k} \leq ||f||_{S,k} = \left\|x^{-1}.x.f\right\|_{S,k} \leq ||x.f||_{S,k}$, which implies equality.

It is now immediate that $\text{HF}_k(G, \mu)$ is a G-invariant vector space. :) ✓

Solution to Exercise 1.99 :(

We know that there exists $\kappa > 0$ such that $|x|_T \leq \kappa |x|_S$ for all $x \in G$. Hence,

$$||f||_{S,k} = \limsup_{r \to \infty} r^{-k} \sup_{|x|_S \leq r} |f(x)| \leq \limsup_{r \to \infty} r^{-k} \sup_{|x|_T \leq \kappa r} |f(x)| \leq \kappa^k ||f||_{T,k}. \qquad :) ✓$$

Solution to Exercise 1.102 :(

This is similar to the proof of Proposition 1.5.1.

Let $M \in \text{GL}_n(R)$. Let us recall the *cofactor matrix* $c(M)$ and the *adjugate matrix* $\text{adj}(M)$ given as follows: For every $1 \leq i, j \leq n$ let $M^{i,j}$ be the $(n-1) \times (n-1)$ matrix obtained from M by deleting the ith row and jth column. Define $c(M)$ to be the $n \times n$ matrix given by $c(M)_{i,j} = (-1)^{i+j} \det(M^{i,j})$. It is well known that for any fixed $1 \leq i \leq n$ we have $\det(M) = \sum_{j=1}^{n} M_{i,j} C_{i,j}$. Define $\text{adj}(M) = c(M)^{\tau}$ (the transpose). Thus, $M \text{adj}(M) = \text{adj}(M)M = \det(M) \cdot I$ where I is the $n \times n$ identity matrix.

This implies that if $\det(M)$ is invertible in R, then $M^{-1} = (\det(M))^{-1} \cdot \text{adj}(M)$.   :) ✓

Solution to Exercise 1.103 :(

It is easy to see that $I, -I$ commute with any $A \in \text{GL}_n(\mathbb{Z})$. Thus, $\{I, -I\} \lhd \text{GL}_n(\mathbb{Z})$.

For $A \in \text{GL}_{2n+1}(\mathbb{Z})$ we have that $\det(\det(A) \cdot A) = \det(A)^{2n+1} \cdot \det(A) = 1$. Thus, the map $A \mapsto (\det(A), \det(A) \cdot A)$ is an isomorphism from $\text{GL}_{2n+1}(\mathbb{Z})$ onto $\{-1, 1\} \times \text{SL}_{2n+1}(\mathbb{Z})$.

Also, the map $A \mapsto \{-I, I\}A$ is an isomorphism from $\text{SL}_{2n+1}(\mathbb{Z})$ onto $\text{PGL}_{2n+1}(\mathbb{Z})$.   :) ✓

Solution to Exercise 1.104 :(

Since S is generated by $a = \begin{bmatrix} 1 & 2 \\ 0 & 1 \end{bmatrix}$ and $b = \begin{bmatrix} 1 & 0 \\ 2 & 1 \end{bmatrix}$, it suffices to show that for any matrix $A = \begin{bmatrix} 4k+1 & 2n \\ 2m & 4\ell+1 \end{bmatrix}$, we have that Aa and Ab are both still of this form.

For A as above, compute,

$$Aa = \begin{bmatrix} 4k+1 & 2n \\ 2m & 4\ell+1 \end{bmatrix} \cdot \begin{bmatrix} 1 & 2 \\ 0 & 1 \end{bmatrix} = \begin{bmatrix} 4k+1 & 2(4k+1)+2n \\ 2m & 2 \cdot 2m+4\ell+1 \end{bmatrix} = \begin{bmatrix} 4k+1 & 2(4k+1+n) \\ 2m & 4(m+\ell)+1 \end{bmatrix},$$

which is of the correct form. Similarly,

$$Ab = \begin{bmatrix} 4k+1 & 2n \\ 2m & 4\ell+1 \end{bmatrix} \cdot \begin{bmatrix} 1 & 0 \\ 2 & 1 \end{bmatrix} = \begin{bmatrix} 4k+1+2 \cdot 2n & 2n \\ 2m+2(4\ell+1) & 4\ell+1 \end{bmatrix} = \begin{bmatrix} 4(k+n)+1 & 2n \\ 2(m+4\ell+1) & 4\ell+1 \end{bmatrix},$$

completing the proof.   :) ✓

Solution to Exercise 1.105 :(

Let

$$H = \left\{ A = \begin{bmatrix} 4k+1 & 2n \\ 2m & 4\ell+1 \end{bmatrix} \mid \det(A) = 1, \; k, \ell, n, m \in \mathbb{Z} \right\}.$$

We have already seen that $S \subset H$.

Let $a = \begin{bmatrix} 1 & 2 \\ 0 & 1 \end{bmatrix}$ and $b = \begin{bmatrix} 1 & 0 \\ 2 & 1 \end{bmatrix}$ be the generators of S.

Let $A = \begin{bmatrix} 4k+1 & 2n \\ 2m & 4\ell+1 \end{bmatrix}$. Denote $||A|| = \max\{|4k+1|, |4\ell+1|\}$. Since $A^{-1} = \begin{bmatrix} 4\ell+1 & -2n \\ -2m & 4k+1 \end{bmatrix}$, by possibly replacing A with A^{-1}, we may assume that $|4k+1| \geq |4\ell+1|$, so that $||A|| = |4k+1|$.

We will prove by induction on $||A||$ that if $\det(A) = 1$ then $A \in S$.

The base case is where $||A|| = 1$, which is $k = \ell = 0$. Then $1 = \det(A) = 4(\ell - mn) + 1$ implies that $\ell = nm$, so that either $n = 0$ or $m = 0$. If $n = 0$ then $A = b^m \in S$ and if $m = 0$ then $A = a^n \in S$. This completes the base case.

For $||A|| > 1$ we proceed by induction as follows.

Note that $\det(A) = 1$ implies that $|(4k+1)(4\ell+1)| = |4nm+1|$. If $2\min\{|n|, |m|\} > |4k+1|$ then

$$|4k+1|^2 \geq |(4k+1)(4\ell+1)| \geq 4|nm| - 1 \geq (|4k+1|+1)^2 - 1 > |4k+1|^2,$$

a contradiction! So it must be that

$$2\min\{|n|, |m|\} \leq |4k+1|.$$

Since $2\min\{|n|, |m|\}$ is even, and $|4k+1|$ is odd, equality cannot hold, so we conclude that

$$2\min\{|n|, |m|\} < |4k+1|.$$

We now have two cases.

Case I. $2|n| < |4k+1|$. In this case we see that for some $z \in \{-1, 1\}$ we have $|4(k+zn)+1| < |4k+1|$. Since

$$Ab^z = \begin{bmatrix} 4(k+zn)+1 & 2n \\ 2(m+z(4\ell+1)) & 4\ell+1 \end{bmatrix},$$

if $|4\ell+1| < |4k+1|$, then $||Ab^z|| < ||A||$, and by induction $Ab^z \in S$, implying that $A \in S$ as well.

If $|4\ell + 1| = |4k + 1|$, then we can find $w \in \{-1, 1\}$ such that $|4(\ell + wn) + 1| < |4\ell + 1|$. So the matrix $b^w A b^z$ admits that

$$||b^w A b^z|| = \max\{|4(k + zn) + 1|, |4(\ell + wn) + 1|\} < ||A||.$$

Again by induction $b^w A b^z \in S$ so that $A \in S$ as well.

Case II. $2|m| < |4k + 1|$. Similarly to the previous case, taking $z \in \{-1, 1\}$ such that $|4(k + zm) + 1| < |4k + 1|$, we find that

$$a^z A = \begin{bmatrix} 4(k+zm)+1 & 2(n+z(4\ell+1)) \\ 2m & 4\ell+1 \end{bmatrix}.$$

If $|4\ell + 1| < |4k + 1|$ then $||a^z A|| < ||A||$, so that $a^z A \in S$ by induction, implying that $A \in S$.

If $|4\ell + 1| = |4k + 1|$, then taking $w \in \{-1, 1\}$ such that $|4(\ell + wm) + 1| < |4\ell + 1|$, we obtain that $||a^z A a^w|| = \max\{|4(k + zm) + 1|, |4(\ell + wm) + 1|\} < ||A||$. As before, by induction $a^z A a^w \in S$ so that $A \in S$ as well. :) ✓

Solution to Exercise 1.106 :(

Let $\varphi \colon \mathrm{SL}_2(\mathbb{Z}) \to \mathrm{SL}_2(\mathbb{Z}/4\mathbb{Z})$ be the map given by taking the matrix entries modulo 4. This is easily seen to be a surjective homomorphism.

Let $K = \mathrm{Ker}\varphi \lhd \mathrm{SL}_2(\mathbb{Z})$. By the above exercises, $K \lhd S$. So $[\mathrm{SL}_2(\mathbb{Z}) : S] = [\mathrm{SL}_2(\mathbb{Z})/K : S/K] = [\mathrm{SL}_2(\mathbb{Z}/4\mathbb{Z}) : \varphi(S)] < \infty.$:) ✓

2

Martingales

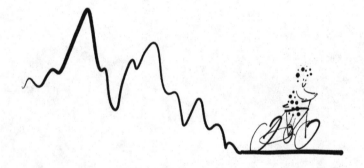

2.1 Conditional Expectation

Martingales are a central object in probability. To define them properly we need to develop the notion of *conditional expectation*.

Proposition 2.1.1 (Existence of conditional expectation) *Let $(\Omega, \mathcal{F}, \mathbb{P})$ be a probability space. Let $\mathcal{G} \subset \mathcal{F}$ be a sub-σ-algebra. Let $X \colon \Omega \to \mathbb{R}$ be an integrable random variable (i.e. $\mathbb{E}|X| < \infty$). Then, there exists an a.s. unique $\mathcal{G}$-measurable and integrable random variable Y such that for all $A \in \mathcal{G}$ we have $\mathbb{E}[X\mathbf{1}_A] = \mathbb{E}[Y\mathbf{1}_A]$.*

Definition 2.1.2 Let $(\Omega, \mathcal{F}, \mathbb{P})$ be a probability space. Let $\mathcal{G} \subset \mathcal{F}$ be a sub-σ-algebra. Let $X \colon \Omega \to \mathbb{R}$ be an integrable random variable (i.e. $\mathbb{E}|X| < \infty$). Denote $\mathbb{E}[X \mid \mathcal{G}]$ to be the (a.s. unique) random variable such that for all $A \in \mathcal{G}$, we have $\mathbb{E}[X\mathbf{1}_A] = \mathbb{E}[\mathbb{E}[X \mid \mathcal{G}]\mathbf{1}_A]$.

For an event $A \in \mathcal{F}$, we denote $\mathbb{P}[A \mid \mathcal{G}] := \mathbb{E}[\mathbf{1}_A \mid \mathcal{G}]$.

If $\mathbb{P}[A] > 0$, we also define

$$\mathbb{E}[X \mid A, \mathcal{G}] := \frac{\mathbb{E}[X\mathbf{1}_A \mid \mathcal{G}]}{\mathbb{P}[A]} \quad \text{and} \quad \mathbb{P}[B \mid A, \mathcal{G}] := \frac{\mathbb{P}[B \cap A \mid \mathcal{G}]}{\mathbb{P}[A]}.$$

It is important to note that conditional expectation produces a random variable and not a number. One may think of $\mathbb{E}[X \mid \mathcal{G}]$ as the "best guess" for X given the information $\mathcal{G}$.

Uniqueness is a simple exercise:

Exercise 2.1 Let $(\Omega, \mathcal{F}, \mathbb{P})$ be a probability space. Let $\mathcal{G} \subset \mathcal{F}$ be a sub-σ-algebra. Let X be an integrable random variable.

Let $Y, Z \colon \Omega \to \mathbb{R}$ be $\mathcal{G}$-measurable random variables, and assume that for any $A \in \mathcal{G}$ the expectations $\mathbb{E}[Y\mathbf{1}_A] = \mathbb{E}[Z\mathbf{1}_A] = \mathbb{E}[X\mathbf{1}_A]$ exist and are equal.

Show that Y, Z are integrable, and that $Y = Z$ a.s.   ▷ solution ◁

We now prove the existence of conditional expectation.

Proof of Proposition 2.1.1 The existence of conditional expectation utilizes a powerful theorem from measure theory: the *Radon–Nykodim theorem*. It states that if μ, ν are σ-finite measures on a measurable space (M, Σ), and if $\nu \ll \mu$ (i.e. for any $A \in \Sigma$, if $\mu(A) = 0$ then $\nu(A) = 0$), then there exists a measurable function $\frac{d\nu}{d\mu}$ such that for any ν-integrable function f, we have that $f\frac{d\nu}{d\mu}$ is μ-integrable and $\int f \, d\nu = \int f \frac{d\nu}{d\mu} \, d\mu$. (See Theorem A.4.6 in Durrett, 2019.)

This is a deep theorem, but from it the existence of conditional expectation is straightforward.

We start with the case where $X \geq 0$. Let $\mu = \mathbb{P}$ on $(\Omega, \mathcal{F})$ and define $\nu(A) = \mathbb{E}[X\mathbf{1}_A]$ for all $A \in \mathcal{G}$. One may easily check that ν is a measure on $(\Omega, \mathcal{G})$ and that $\nu \ll \mathbb{P}|_{\mathcal{G}}$. Thus, there exists a $\mathcal{G}$-measurable function $\frac{d\nu}{d\mu}$ such that for any $A \in \mathcal{G}$, we have $\mathbb{E}[X\mathbf{1}_A] = \nu(A) = \int \mathbf{1}_A d\nu = \int \mathbf{1}_A \frac{d\nu}{d\mu} d\mu$. A $\mathcal{G}$-measurable function is just a random variable measurable with respect to $\mathcal{G}$. So we may take $Y = \frac{d\nu}{d\mu}$, and we have $\mathbb{E}[X\mathbf{1}_A] = \mathbb{E}[Y\mathbf{1}_A]$ for all $A \in \mathcal{G}$.

For a general X (not necessarily nonnegative) we may write $X = X^+ - X^-$ for $X^\pm$ nonnegative. One may check that $Y := \mathbb{E}[X^+ \mid \mathcal{G}] - \mathbb{E}[X^- \mid \mathcal{G}]$ has the required properties.   □

The uniqueness property described in Exercise 2.1 is a good tool for computing the conditional expectation in many cases; usually one "guesses" the correct random variable and verifies it by showing that it admits the properties guaranteeing it is equal to the conditional expectation a.s.

Let us summarize some of the most basic properties of conditional expectation with the following exercises.

Exercise 2.2 Let $(\Omega, \mathcal{F}, \mathbb{P})$ be a probability space, X an integrable random variable, and $\mathcal{G} \subset \mathcal{F}$ a sub-σ-algebra.
Show that if X is $\mathcal{G}$-measurable, then $\mathbb{E}[X \mid \mathcal{G}] = X$ a.s.
Show that if X is independent of $\mathcal{G}$, then $\mathbb{E}[X \mid \mathcal{G}] = \mathbb{E}[X]$ a.s.
Show that if $\mathbb{P}[X = c] = 1$, then $\mathbb{E}[X \mid \mathcal{G}] = c$ a.s.   ▷ solution ◁

Exercise 2.3 Let $(\Omega, \mathcal{F}, \mathbb{P})$ be a probability space, X an integrable random variable, and $\mathcal{G} \subset \mathcal{F}$ a sub-σ-algebra.
Show that $\mathbb{E}[\mathbb{E}[X \mid \mathcal{G}]] = \mathbb{E}[X]$.   ▷ solution ◁

Recall that for $A \in \mathcal{F}$, we defined $\mathbb{P}[A \mid \mathcal{G}] = \mathbb{E}[\mathbf{1}_A \mid \mathcal{G}]$.

Exercise 2.4 Prove Bayes' formula for conditional probabilities:
Show that for any $B \in \mathcal{G}$ and $A \in \mathcal{F}$ with $\mathbb{P}[A] > 0$, we have

$$\mathbb{P}[B \mid A] = \frac{\mathbb{E}[\mathbf{1}_B \mathbb{P}[A \mid \mathcal{G}]]}{\mathbb{P}[A]}.$$ ▷ solution ◁

Exercise 2.5 Show that conditional expectation is linear; that is,

$$\mathbb{E}[aX + Y \mid \mathcal{G}] = a\,\mathbb{E}[X \mid \mathcal{G}] + \mathbb{E}[Y \mid \mathcal{G}] \qquad \text{a.s.}$$

Show that if $X \leq Y$ a.s., then $\mathbb{E}[X \mid \mathcal{G}] \leq \mathbb{E}[Y \mid \mathcal{G}]$ a.s.

Show that if $X_n \nearrow X$ a.s., $X_n \geq 0$ for all n a.s., and X is integrable, then $\mathbb{E}[X_n \mid \mathcal{G}] \nearrow \mathbb{E}[X \mid \mathcal{G}]$.

▷ solution ◁

Exercise 2.6 Let $(\Omega, \mathcal{F}, \mathbb{P})$ be a probability space, X an integrable random variable, and $\mathcal{G} \subset \mathcal{F}$ a sub-σ-algebra.

Show that if Y is $\mathcal{G}$-measurable and $\mathbb{E}\,|XY| < \infty$, then $\mathbb{E}[XY \mid \mathcal{G}] = Y\,\mathbb{E}[X \mid \mathcal{G}]$ a.s.

▷ solution ◁

Exercise 2.7 Let $(\Omega, \mathcal{F}, \mathbb{P})$ be a probability space and X an integrable random variable. Suppose that $(A_n)_n$ is a sequence of pairwise disjoint events such that $\sum_n \mathbb{P}[A_n] = 1$ (i.e. $(A_n)_n$ form an *almost-partition* of Ω). Let $\mathcal{G} = \sigma((A_n)_n)$. Show that for all n,

$$\mathbb{E}[X \mid \mathcal{G}]\mathbf{1}_{A_n} = \frac{\mathbb{E}[X\mathbf{1}_{A_n}]}{\mathbb{P}[A_n]}\mathbf{1}_{A_n} \qquad \text{a.s.}$$

Use this to conclude that

$$\mathbb{E}[X \mid \mathcal{G}] = \sum_n \frac{\mathbb{E}[X\mathbf{1}_{A_n}]}{\mathbb{P}[A_n]} \cdot \mathbf{1}_{A_n} \qquad \text{a.s.}$$

▷ solution ◁

Definition 2.1.3 Let X be an integrable (real-valued) random variable, and Y another random variable, not necessarily real-valued. Define $\mathbb{E}[X \mid Y] := \mathbb{E}[X \mid \sigma(Y)]$.

Exercise 2.8 Show that if X is an integrable random variable, and Y is a random variable taking on countably many values, then

$$\mathbb{E}[X \mid Y] = \sum_{y \in R_Y} \frac{\mathbb{E}\left[X\mathbf{1}_{\{Y=y\}}\right]}{\mathbb{P}[Y = y]}\mathbf{1}_{\{Y=y\}},$$

where $R_Y = \{y \colon \mathbb{P}[Y = y] > 0\}$.

Exercise 2.9 Prove Chebychev's inequality for conditional expectation: Show that if $X \in L^2(\Omega, \mathcal{F}, \mathbb{P})$, then a.s.

$$\mathbb{P}[|X| \geq a \mid \mathcal{G}] \leq a^{-2} \cdot \mathbb{E}\left[X^2 \mid \mathcal{G}\right].$$

Exercise 2.10 Prove Cauchy–Schwarz for conditional expectation: Show that if $X, Y \in L^2(\Omega, \mathcal{F}, \mathbb{P})$, then XY is integrable and a.s.

$$(\mathbb{E}[XY \mid \mathcal{G}])^2 \leq \mathbb{E}\left[X^2 \mid \mathcal{G}\right] \cdot \mathbb{E}\left[Y^2 \mid \mathcal{G}\right].$$

Proposition 2.1.4 (Jensen's inequality) *If φ is a convex function such that $X, \varphi(X)$ are integrable, then a.s.*

$$\mathbb{E}[\varphi(X) \mid \mathcal{G}] \geq \varphi(\mathbb{E}[X \mid \mathcal{G}]).$$

Proof As in the usual proof of Jensen's inequality, we know that $\varphi(x) = \sup_{(a,b) \in S}(ax+b)$ where $S = \left\{(a,b) \in \mathbb{Q}^2 : \forall\, y,\ ay + b \leq \varphi(y)\right\}$. If $(a, b) \in S$, then monotonicity of conditional expectation gives

$$\mathbb{E}[\varphi(X) \mid \mathcal{G}] \geq a\,\mathbb{E}[X \mid \mathcal{G}] + b \qquad \text{a.s.}$$

Taking the supremum over $(a, b) \in S$, since S is countable, we have that

$$\mathbb{E}[\varphi(X) \mid \mathcal{G}] \geq \varphi(\mathbb{E}[X \mid \mathcal{G}]) \qquad \text{a.s.} \qquad \square$$

Proposition 2.1.5 (Tower property) *Let $(\Omega, \mathcal{F}, \mathbb{P})$ be a probability space, X an integrable random variable, and $\mathcal{H} \subset \mathcal{G} \subset \mathcal{F}$ sub-σ-algebras.*
Then, $\mathbb{E}[\mathbb{E}[X \mid \mathcal{G}] \mid \mathcal{H}] = \mathbb{E}[\mathbb{E}[X \mid \mathcal{H}] \mid \mathcal{G}] = \mathbb{E}[X \mid \mathcal{H}]$ a.s.

Proof Note that $\mathbb{E}[X \mid \mathcal{H}]$ is $\mathcal{H}$-measurable and thus $\mathcal{G}$-measurable. So $\mathbb{E}[\mathbb{E}[X \mid \mathcal{H}] \mid \mathcal{G}] = \mathbb{E}[X \mid \mathcal{H}]$ a.s.

For the other assertion, since $\mathbb{E}[X \mid \mathcal{H}]$ is $\mathcal{H}$-measurable, we only need to show the second property. That is, for any $A \in \mathcal{H}$, since $A \in \mathcal{G}$ as well,

$$\mathbb{E}[\mathbb{E}[X \mid \mathcal{H}]\mathbf{1}_A] = \mathbb{E}[\mathbb{E}[X\mathbf{1}_A \mid \mathcal{H}]] = \mathbb{E}[X\mathbf{1}_A]$$
$$= \mathbb{E}[\mathbb{E}[X\mathbf{1}_A \mid \mathcal{G}]] = \mathbb{E}[\mathbb{E}[X \mid \mathcal{G}]\mathbf{1}_A]. \qquad \square$$

2.2 Martingales: Definition and Examples

Definition 2.2.1 Let $(\Omega, \mathcal{F}, \mathbb{P})$ be a probability space. A **filtration** is a sequence $(\mathcal{F}_t)_t$ of nested sub-σ-algebras $\mathcal{F}_t \subset \mathcal{F}_{t+1} \subset \mathcal{F}$.

Example 2.2.2 The basic example of a filtration is one induced by a sequence of random variables. If $(X_t)_t$ is a sequence of random variables in a probability space $(\Omega, \mathcal{F}, \mathbb{P})$, then $\mathcal{F}_t = \sigma(X_0, \ldots, X_t)$ is easily seen to be a filtration. △▽△

Definition 2.2.3 Let $(\Omega, \mathcal{F}, \mathbb{P})$ be a probability space. Let $(\mathcal{F}_t)_t$ be a filtration. Let $(M_t)_t$ be a sequence of complex-valued random variables.

The sequence $(M_t)_t$ is said to be a **martingale with respect to the filtration** $\mathcal{F}_t$ if the following conditions hold: For all t,

- M_t is measurable with respect to $\mathcal{F}_t$,
- $\mathbb{E}|M_t| < \infty$, and
- $\mathbb{E}[M_{t+1} \mid \mathcal{F}_t] = M_t$ a.s.

Exercise 2.11 Let μ be a probability measure on $\mathbb{Z}$ such that for $(U_t)_{t \geq 1}$ i.i.d.-μ, we have $\mathbb{E}[U_t] = 0$ and $\mathbb{E}|U_t| < \infty$. Let $M_0 = 0$ and $M_t := \sum_{k=1}^{t} U_k$. Show that $(M_t)_t$ is a martingale with respect to the filtration $\mathcal{F}_t = \sigma(U_1, \ldots, U_t)$.

What about the filtration $\mathcal{F}_t' = \sigma(M_0, \ldots, M_t)$? ▷ solution ◁

Exercise 2.12 Let $(M_t)_t$ be a martingale with respect to a filtration $(\mathcal{F}_t)_t$. Show that $(M_t)_t$ is also a martingale with respect to the *canonical filtration* $\mathcal{F}_t' = \sigma(M_0, \ldots, M_t)$. ▷ solution ◁

In light of Exercise 2.12, we do not really need to specify the filtration when speaking about a martingale $(M_t)_t$, since we can always refer to the *canonical filtration* $\sigma(M_0, \ldots, M_t)$. Thus, whenever we speak of a martingale without specifying the filtration, we are referring to the canonical filtration.

Exercise 2.13 Let $(X_t)_t$ be the simple random walk on $\mathbb{Z}^d$. That is, $X_t = \sum_{j=1}^{t} U_j$, where U_j are i.i.d. uniform on the standard basis of $\mathbb{Z}^d$ and the inverses.

Show that $M_t = \langle X_t, v \rangle$ is a martingale, where $v \in \mathbb{R}^d$.

Show that $M_t = ||X_t||^2 - t$ is a martingale. ▷ solution ◁

Definition 2.2.4 Let $(\Omega, \mathcal{F}, \mathbb{P})$ be a probability space and $(\mathcal{F}_t)_t$ a filtration. A **stopping time** with respect to the filtration $(\mathcal{F}_t)_t$ is a random variable T with values in $\mathbb{N} \cup \{\infty\}$ such that $\{T \leq t\} \in \mathcal{F}_t$ for every t.

A stopping time with respect to a process $(X_t)_t$ is defined as being a stopping time with respect to the canonical filtration of the process $\sigma(X_0, \ldots, X_t)$.

Usually we will not specify the filtration, since it will be obvious from the context.

Example 2.2.5 Some examples: If $(X_t)_t$ is a process and we take the canonical filtration,

- $T = \inf \{t \colon X_t \in A\}$ is a stopping time;

- $E = \sup\{t: X_t \in A\}$ is typically *not* a stopping time.

△ ▽ △

Exercise 2.14 Show that if $(M_t)_t$ is a martingale and T is a stopping time, then $(M_{T \wedge t})_t$ is also a martingale.   ▷ solution ◁

Exercise 2.15 Show that if T, T' are both stopping times with respect to a filtration $(\mathcal{F}_t)_t$, then so is $T \wedge T'$.   ▷ solution ◁

The relation of probability and harmonic functions is via martingales as the following exercise shows.

Exercise 2.16 Let G be a finitely generated group. Let μ be an adapted probability measure on G. Let $(X_t)_t$ be the μ-random walk.

Show that $f: G \to \mathbb{C}$ is μ-harmonic if and only if $(f(X_t))_t$ is a martingale (with respect to the canonical filtration $\mathcal{F}_t = \sigma(X_0, \ldots, X_t)$).   ▷ solution ◁

2.3 Optional Stopping Theorem

It follows from the definition that $\mathbb{E}[M_t] = \mathbb{E}[M_0]$ for a martingale $(M_t)_t$. We would like to conclude that this also holds for *random* times. However, this is not true in general.

Example 2.3.1 Let $M_t = \sum_{j=1}^t X_j$, where $(X_j)_j$ are all i.i.d. with distribution $\mathbb{P}[X_j = 1] = \mathbb{P}[X_j = -1] = \frac{1}{2}$ (i.e. $(M_t)_t$ is the simple random walk on $\mathbb{Z}$). Let $T = \inf\{t: M_t = 1\}$.

We have seen that $(M_t)_t$ is a martingale and T is a stopping time.

In Section 2.4, we will prove that $T < \infty$ a.s. (i.e. the simple random walk on $\mathbb{Z}$ is recurrent). So M_T is well defined, and actually, by definition $M_T = 1$ a.s. However, $M_0 = 0$ a.s., so $\mathbb{E}[M_T] = 1 \neq 0 = \mathbb{E}[M_0]$.   △ ▽ △

In contrast to the general case, *uniform integrability* is a condition under which $\mathbb{E}[M_T] = \mathbb{E}[M_0]$ for stopping times.

Definition 2.3.2 (Uniform integrability) Let $(X_\alpha)_{\alpha \in I}$ be a collection of random variables. We say that the collection $(X_\alpha)_\alpha$ is **uniformly integrable** if

$$\lim_{K \to \infty} \sup_\alpha \mathbb{E}\left[|X_\alpha| \mathbf{1}_{\{|X_\alpha| > K\}}\right] = 0.$$

Exercise 2.17 Show that if X is integrable, then the collection $X_\alpha := X$ is uniformly integrable.

▷ solution ◁

Exercise 2.18 Show that if $(X_\alpha)_\alpha$ is uniformly integrable, then $\sup_\alpha \mathbb{E}|X_\alpha| < \infty$.

▷ solution ◁

Exercise 2.19 Show that if for some $\varepsilon > 0$ we have $\sup_\alpha \mathbb{E}|X_\alpha|^{1+\varepsilon} < \infty$, then $(X_\alpha)_\alpha$ is uniformly integrable.

▷ solution ◁

Exercise 2.20 Show that if $(\mathcal{F}_\alpha)_\alpha$ is a collection of σ-algebras and X is an integrable random variable, then $(\mathbb{E}[X \mid \mathcal{F}_\alpha])_\alpha$ is uniformly integrable. ▷ solution ◁

Exercise 2.21 Show that if $(X_n)_n$ is uniformly integrable and $X_n \to X$ a.s., then X is integrable.

▷ solution ◁

The following is *not* the strongest form of optional stopping theorems that are possible to prove, but it is sufficient for our purposes.

Theorem 2.3.3 (Optional stopping theorem) *Let $(M_t)_t$ be a martingale and T a stopping time, both with respect to a filtration $(\mathcal{F}_t)_t$.*

We have that $\mathbb{E}|M_T| < \infty$ and $\mathbb{E}[M_T] = \mathbb{E}[M_0]$ if one of the following holds:

- *The stopping time T is a.s. bounded; that is, there exists $t \geq 0$ such that $T \leq t$ a.s.*
- *$T < \infty$ a.s. and $(M_t)_t$ is uniformly integrable and $\mathbb{E}|M_T| < \infty$.*

We will actually see (in Exercise 2.25) that in the last condition, the requirement $\mathbb{E}|M_T| < \infty$ is redundant.

Proof For the first case, if $T \leq t$ a.s., then

$$\mathbb{E}[M_T] = \mathbb{E}\sum_{j=0}^{t-1} \mathbf{1}_{\{T>j\}} \cdot (M_{j+1} - M_j) + \mathbb{E}[M_0].$$

Since $\{T > j\} = \{T \leq j\}^c \in \mathcal{F}_j$, we get that

$$\mathbb{E}\left[(M_{j+1} - M_j)\mathbf{1}_{\{T>j\}}\right] = \mathbb{E}\,\mathbb{E}[(M_{j+1} - M_j)\mathbf{1}_{\{T>j\}} \mid \mathcal{F}_j]$$
$$= \mathbb{E}\left[\mathbf{1}_{\{T>j\}}\,\mathbb{E}[M_{j+1} - M_j \mid \mathcal{F}_j]\right] = 0.$$

So $\mathbb{E}[M_T] = \mathbb{E}[M_0]$ in this case.

For the second case, note that since $\mathbb{E}|M_T| < \infty$, then $\mathbb{E}\left[|M_T|\mathbf{1}_{\{|M_T|>K\}}\right] \to 0$ as $K \to \infty$.

Now, $(M_t)_t$ is uniformly integrable so $\sup_t \mathbb{E}\left[|M_t|\mathbf{1}_{\{|M_t|>K\}}\right] \to 0$ as $K \to \infty$. Thus,

$$\mathbb{E}\left[|M_{T\wedge t}|\mathbf{1}_{\{|M_{T\wedge t}|>K\}}\right] \le \mathbb{E}\left[|M_T|\mathbf{1}_{\{|M_T|>K\}}\mathbf{1}_{\{T\le t\}}\right] + \mathbb{E}\left[|M_t|\mathbf{1}_{\{|M_t|>K\}}\mathbf{1}_{\{T>t\}}\right]$$
$$\le \mathbb{E}\left[|M_T|\mathbf{1}_{\{|M_T|>K\}}\right] + \mathbb{E}\left[|M_t|\mathbf{1}_{\{|M_t|>K\}}\right],$$

so $\sup_t \mathbb{E}\left[|M_{T\wedge t}|\mathbf{1}_{\{|M_{T\wedge t}|>K\}}\right] \to 0$ as $K \to \infty$.

Let

$$\varphi_K(x) = \begin{cases} K & \text{if } x > K, \\ x & \text{if } |x| \le K, \\ -K & \text{if } x < -K. \end{cases}$$

Note that $|\varphi_K(x) - x| \le |x|\mathbf{1}_{\{|x|>K\}}$.

Since $M_{T\wedge t} \to M_T$ a.s. as $t \to \infty$ (because we assumed that $T < \infty$ a.s.), we also have that $\varphi_K(M_{T\wedge t}) \to \varphi_K(M_T)$ a.s. as $t \to \infty$. Since $\varphi_K(M_{T\wedge t}), \varphi_K(M_T)$ are uniformly bounded by K, we can apply dominated convergence to obtain that

$$\lim_{t\to\infty} \mathbb{E}\,|\varphi_K(M_{T\wedge t}) - \varphi_K(M_T)| = 0.$$

Thus,

$$\mathbb{E}\,|M_T - M_{T\wedge t}|$$
$$\le \mathbb{E}\,|\varphi_K(M_T) - M_T| + \mathbb{E}\,|\varphi_K(M_{T\wedge t}) - M_{T\wedge t}| + \mathbb{E}\,|\varphi_K(M_{T\wedge t}) - \varphi_K(M_T)|$$
$$\le \mathbb{E}\left[|M_T|\mathbf{1}_{\{|M_T|>K\}}\right] + \sup_t \mathbb{E}\left[|M_{T\wedge t}|\mathbf{1}_{\{|M_{T\wedge t}|>K\}}\right] + \mathbb{E}\,|\varphi_K(M_{T\wedge t}) - \varphi_K(M_T)|.$$

Taking $t \to \infty$ and then $K \to \infty$, we get that $\mathbb{E}\,|M_T - M_{T\wedge t}| \to 0$.

Since $T \wedge t$ is an a.s. bounded stopping time, $\mathbb{E}[M_{T\wedge t}] = \mathbb{E}[M_0]$. In conclusion,

$$|\mathbb{E}[M_T - M_0]| = |\mathbb{E}[M_T - M_{T\wedge t}]| \to 0. \qquad \square$$

Exercise 2.22 Show that if a stopping time T is a.s. finite and a martingale $(M_t)_t$ is a.s. uniformly bounded (i.e. there exists m such that $|M_t| \le m$ a.s. for all t), then $\mathbb{E}[M_T] = \mathbb{E}[M_0]$.   ▷ solution ◁

Exercise 2.23 Assume that $(X_n)_n, X$ are random variables such that $X_n \to X$ a.s. Show that if $(X_n)_n$ is uniformly integrable, then $X_n \to X$ also in L^1. ▷ solution ◁

Exercise 2.24 Show that for a martingale $(M_t)_t$ and for any a.s. finite stopping time T, we have $\mathbb{E}\,|M_{T\wedge t}| \le \mathbb{E}\,|M_t|$.   ▷ solution ◁

Exercise 2.25 A specific case of the *martingale convergence theorem* states that if $(M_t)_t$ is a martingale with $\sup_t \mathbb{E}\,|M_t| < \infty$, then there exists a random variable M_∞ such that $M_t \to M_\infty$ a.s., and $\mathbb{E}\,|M_\infty| < \infty$. (We will prove the martingale convergence theorem in Theorem 2.6.3.)

Use this to show that if $(M_t)_t$ is a uniformly integrable martingale and T is an a.s. finite stopping time, then $\mathbb{E}\,|M_T| < \infty$ (so this last condition is redundant in the optional stopping theorem).

▷ solution ◁

2.4 Applications of Optional Stopping

Let us give some applications of the optional stopping theorem (OST) to the study of random walks on $\mathbb{Z}$.

We consider $\mathbb{Z} = \langle -1, 1 \rangle$. This is the usual Cayley graph on $\mathbb{Z}$, with neighbors given by adjacent integers. We take the measure $\mu = \frac{1}{2}(\delta_1 + \delta_{-1})$. That is, uniform on $\{-1, 1\}$.

Thus, the μ-random walk $(X_t)_t$ can be represented as $X_t = \sum_{j=1}^t U_j$ where $(U_j)_j$ are i.i.d. and $\mathbb{P}[U_j = 1] = \mathbb{P}[U_j = -1] = \frac{1}{2}$.

First, it is simple to see that $(X_t)_t$ is a martingale (we have already seen this above). Now, let $T_z := \inf\{t: X_t = z\}$. Note that T_z is a stopping time. Also, for $a < 0 < b$ we have the stopping time $T_{a,b} := T_a \wedge T_b$, which is the first exit time of (a, b).

Now, the martingale $\left(M_t = X_{T_{a,b} \wedge t}\right)_t$ is a.s. uniformly bounded (by $|a| \vee |b|$). As an exercise to the reader it is left to show that $T_{a,b} < \infty$ $\mathbb{P}_0$-a.s. Thus,

$$
\begin{aligned}
0 = \mathbb{E}[M_{T_{a,b}}] &= \mathbb{P}[T_a < T_b] \cdot a + \mathbb{P}[T_b < T_a] \cdot b \\
&= \mathbb{P}[T_a < T_b](a - b) + b.
\end{aligned}
$$

We deduce that

$$
\mathbb{P}_0[T_a < T_b] = \frac{b}{b - a}.
$$

Now, note that $(x + X_t)_t$ has the distribution of a random walk started at x. Thus, for all $n > x > 0$,

$$
\mathbb{P}_x[T_0 < T_n] = \mathbb{P}_0[T_{-x} < T_{n-x}] = \frac{n - x}{n} = 1 - \frac{x}{n}.
$$

This is the probability that a gambler starting with x dollars will go bankrupt before reaching n dollars in wealth, and is known as the *gambler's ruin estimate*.

One of the extraordinary facts (albeit classical) about the random walk on $\mathbb{Z}$ is now obtained by taking $n \to \infty$:

$$
\mathbb{P}_x[T_0 = \infty] = \lim_{n \to \infty} \mathbb{P}_x[T_0 > T_n] = 0.
$$

That is, no matter how much money you have entering the casino, you always eventually reach 0 (and this is in the case of a fair game!)

In other words, the random walk on $\mathbb{Z}$ is *recurrent*: it reaches 0 a.s. But how long does it take to reach 0?

Note that since the random walk takes steps of size 1, we have that for $n > x > 0$, under $\mathbb{P}_x$, the event $T_0 > T_n$ implies that $T_0 \geq 2n - x$. Thus,

$$\mathbb{E}_x[T_0] = \sum_{n=0}^{\infty} \mathbb{P}_x[T_0 > n] \geq \sum_{n>x} \mathbb{P}_x[T_0 \geq 2n - x]$$

$$\geq \sum_{n>x} \mathbb{P}_x[T_0 > T_n] = \sum_{n>x} \frac{x}{n} = \infty.$$

So $\mathbb{E}_x[T_0] = \infty$. Indeed the walker reaches 0 a.s., but the time it takes is infinite in expectation. That is, the random walk on $\mathbb{Z}$ is *null-recurrent*. We will expand on the notions of recurrence and null-recurrence in Chapter 3 (and specifically in Section 3.8).

Exercise 2.26 Show that for $a < 0 < b$, we have that $\mathbb{P}_0[T_{a,b} < \infty] = 1$.

In fact, strengthen this to show that for all $a < 0 < b$ there exists a constant $c = c(a, b) > 0$ such that for all t, and any $a < x < b$,

$$\mathbb{P}_x[T_{a,b} > t] \leq e^{-ct}.$$

Conclude that $\mathbb{E}_x[T_{a,b}] < \infty$ for all $a < x < b$.   ▷ solution ◁

Let us now consider a different martingale.

$$\mathbb{E}\left[X_{t+1}^2 \mid X_t\right] = \tfrac{1}{2}(X_t + 1)^2 + \tfrac{1}{2}(X_t - 1)^2 = X_t^2 + 1.$$

So $\left(M_t := X_t^2 - t\right)_t$ is a martingale.

If we apply the OST (Theorem 2.3.3) we get

$$-1 = \mathbb{E}_{-1}[M_{T_0}] = \mathbb{E}_{-1}\left[X_{T_0}^2\right] - \mathbb{E}_{-1}[T_0] = -\mathbb{E}_{-1}[T_0].$$

So $\mathbb{E}_{-1}[T_0] = 1$, a contradiction!

The reason is that we applied the OST in the case where we could not, since $(M_t)_t$ is not necessarily bounded, and this in fact shows that $(M_t)_t$ is not uniformly integrable.

One may note that for $-n < x < n$, under $\mathbb{P}_x$, the martingale $\left(M_{t \wedge T_{-n,n}}\right)_t$ admits

$$|M_{t \wedge T_{-n,n}}| \leq \left|X_{T_{-n,n}}^2 - T_{-n,n}\right| + |X_t^2 - t|\mathbf{1}_{\{T_{-n,n} > t\}} \leq 2n^2 + T_{-n,n} + t\mathbf{1}_{\{T_{-n,n} > t\}}.$$

Thus $\left(\text{using } (a+b)^2 \le 2a^2 + 2b^2\right)$,

$$\mathbb{E}_x\left[\left|M_{t\wedge T_{-n,n}}\right|^2\right] \le 2\,\mathbb{E}_x\left[\left|2n^2 + T_{-n,n}\right|^2\right] + 2t^2\,\mathbb{P}_x[T_{-n,n} > t]$$

$$\le 2\,\mathbb{E}_x\left[\left|2n^2 + T_{-n,n}\right|^2\right] + 2t^2 \cdot e^{-ct},$$

for some $c = c(n) > 0$. This implies that $\sup_t \mathbb{E}_x\left[|M_{t\wedge T_{-n,n}}|^2\right] < \infty$, so $\left(M_{t\wedge T_{-n,n}}\right)_t$ is a uniformly integrable martingale.

Given this, we may apply the OST to get that for any $-n < x < n$,

$$x^2 = \mathbb{E}_x[M_{T_{-n,n}}] = \mathbb{E}_x\left[X^2_{T_{-n,n}}\right] - \mathbb{E}_x[T_{-n,n}] = n^2 - \mathbb{E}_x[T_{-n,n}],$$

so $\mathbb{E}_x[T_{-n,n}] = n^2 - x^2$. Specifically, $\mathbb{E}[T_{-n,n}] = n^2$. This property is sometimes referred to as the random walk on $\mathbb{Z}$ being *diffusive*.

Similarly to the above, one may easily see that the martingale $(M_{t\wedge T_{0,n}})_t$ is $\mathbb{P}_x$-a.s. bounded for any $0 < x < n$. So

$$x^2 + \mathbb{E}_x[T_{0,n}] = \mathbb{E}_x\left[|X_{T_{0,n}}|^2\right] = \mathbb{P}_x[T_0 > T_n] \cdot n^2 = xn.$$

Thus, $\mathbb{E}_x[T_{0,n}] = (n - x)x$.

For general $a < x < b$, note that under $\mathbb{P}_x$, the walk $(X_t)_t$ has the same distribution as $(a + X_t)_t$ under $\mathbb{P}_{x-a}$. Thus, for $a < x < b$,

$$\mathbb{E}_x[T_{a,b}] = \mathbb{E}_{x-a}[T_{0,b-a}] = (b - x)(x - a).$$

2.5 *LP* Maximal Inequality

The goal of this section is to prove the following theorem, which shows how to control the maximum of a martingale up to a certain time, using only the last value.

Theorem 2.5.1 (*LP* maximal inequality) *Let $(M_t)_t$ be a martingale. Then, for any $1 < p < \infty$ and any t,*

$$\mathbb{E}[\max_{k \le t} |M_k|^p] \le (\tfrac{p}{p-1})^p \cdot \mathbb{E}\,|M_t|^p.$$

Proof Let $\mathcal{F}_t = \sigma(M_0, \ldots, M_t)$ be the natural filtration. Let $N_t = \max_{k \le t} |M_k|$.

As a first step we show that for any $r > 0$,

$$r \cdot \mathbb{P}[N_t \ge r] \le \mathbb{E}\left[|M_t|\mathbf{1}_{\{N_t \ge r\}}\right].$$

(This is also known as *Doob's inequality*.) Indeed, fix $r > 0$ and $t > 0$. Let

$$T = \inf\{s \ge 0 : |M_s| \ge r\} \wedge t,$$

which is a stopping time. Since $\{T = s\} \in \mathcal{F}_s$, we have that for all $s \leq t$,

$$\mathbb{E}\left[|M_s|\mathbf{1}_{\{T=s\}}\right] = \mathbb{E}\left[|\,\mathbb{E}[M_t \mid \mathcal{F}_s]|\mathbf{1}_{\{T=s\}}\right] \leq \mathbb{E}\left[|M_t|\mathbf{1}_{\{T=s\}}\right].$$

Summing over $s \leq t$, using that $\mathbb{P}[T \leq t] = 1$, we have

$$\mathbb{E}\,|M_T| \leq \mathbb{E}\,|M_t|.$$

Finally, note that $|M_T|\mathbf{1}_{\{N_t < r\}} = |M_t|\mathbf{1}_{\{N_t < r\}}$ by the definition of N_t and T, so that

$$\mathbb{E}\left[|M_t|\mathbf{1}_{\{N_t \geq r\}}\right] = \mathbb{E}\,|M_t| - \mathbb{E}\left[|M_t|\mathbf{1}_{\{N_t < r\}}\right] \geq \mathbb{E}\left[|M_T|\mathbf{1}_{\{N_t \geq r\}}\right] \geq r \cdot \mathbb{P}[N_t \geq r],$$

because $|M_T|\mathbf{1}_{\{N_t \geq r\}} \geq r\mathbf{1}_{\{N_t \geq r\}}$. This proves Doob's inequality above.

Fix some $R > 0$ and denote $K_t = N_t \wedge R$. Note that $\{K_t \geq r\} = \{N_t \geq r\}$ for $r \leq R$, and $\{K_t \geq r\} = \emptyset$ for $r > R$. Now, for $p > 1$ we integrate Doob's inequality:

$$\mathbb{E}\,|K_t|^p = \int_0^\infty pr^{p-1}\,\mathbb{P}[K_t \geq r]dr \leq \int_0^R pr^{p-2}\,\mathbb{E}\left[|M_t|\mathbf{1}_{\{N_t \geq r\}}\right]dr$$

$$= \mathbb{E}\left[|M_t|\int_0^R pr^{p-2}\mathbf{1}_{\{r \leq N_t\}}dr\right] = \frac{p}{p-1}\,\mathbb{E}\left[|M_t| \cdot |K_t|^{p-1}\right]$$

$$\leq \frac{p}{p-1} \cdot \left(\mathbb{E}\,|M_t|^p\right)^{1/p} \cdot \left(\mathbb{E}\,|K_t|^p\right)^{(p-1)/p},$$

where the last inequality is Hölder's inequality. Recalling that $K_t = N_t \wedge R$, taking $R \to \infty$, and using monotone convergence, we have

$$\left(\mathbb{E}\,|N_t|^p\right)^{1/p} \leq \frac{p}{p-1} \cdot \left(\mathbb{E}\,|M_t|^p\right)^{1/p},$$

which is the required assertion. $\qquad\square$

Exercise 2.27 Let $(X_t)_t$ be a lazy random walk on $\mathbb{Z}$; that is the μ-random walk for $\mu(1) = \mu(-1) = \frac{1}{2}(1 - p)$ and $\mu(0) = p$ for some $p \in [0, 1)$.

Let $M_t = \max_{k \leq t}|X_k|$. Show that $\mathbb{E}\,|M_t|^2 \leq 4t$. ▷ solution ◁

Exercise 2.28 Let $(X_t)_t$ be a lazy random walk on $\mathbb{Z}$; that is the μ-random walk for $\mu(1) = \mu(-1) = \frac{1}{2}(1 - p)$ and $\mu(0) = p$ for some $p \in [0, 1)$. Let $M_t = \max_{k \leq t}|X_k|$. Prove that there exists $C, c > 0$ such that for all $t > 0$ and all $m > 0$,

$$c\exp\left(-C\frac{t}{m^2}\right) \leq \mathbb{P}[M_t \leq m] \leq C\exp\left(-c(1 - p)\frac{t}{m^2}\right). \qquad \text{▷ solution ◁}$$

2.6 Martingale Convergence

One amazing property of martingales is that they converge under appropriate conditions.

Definition 2.6.1 A **sub-martingale** is a process $(M_t)_t$ such that $\mathbb{E}|M_t| < \infty$ and $\mathbb{E}[M_{t+1} \mid M_0, \ldots, M_t] \geq M_t$ for all t.

A **super-martingale** is a process $(M_t)_t$ such that $\mathbb{E}|M_t| < \infty$ and $\mathbb{E}[M_{t+1} \mid M_0, \ldots, M_t] \leq M_t$ for all t.

A process $(H_t)_t$ is called **predictable** (with respect to $(M_t)_t$) if H_t is measurable with respect to $\sigma(M_0, \ldots, M_{t-1})$ for all t.

Of course any martingale is a sub-martingale and a super-martingale.

Exercise 2.29 Show that if $(M_t)_t$ is a sub-martingale then also $X_t := (M_t - a)\mathbf{1}_{\{M_t>a\}}$ is a sub-martingale. ▷ solution ◁

Exercise 2.30 Show that if $(M_t)_t$ is a sub-martingale (respectively, super-martingale) and $(H_t)_t$ is a bounded nonnegative predictable process, then the process

$$(H \cdot M)_t := \sum_{s=1}^{t} H_s (M_s - M_{s-1})$$

is a sub-martingale (respectively, super-martingale).

Show that when $(M_t)_t$ is a martingale and $(H_t)_t$ is bounded and predictable but not necessarily nonnegative, then $(H \cdot M)_t$ is a martingale. ▷ solution ◁

Exercise 2.31 Show that if $(M_t)_t$ is a sub-martingale and T is a stopping time then $(M_{T \wedge t})_t$ is a sub-martingale. ▷ solution ◁

Lemma 2.6.2 (Upcrossing Lemma) *Let $(M_t)_t$ be a sub-martingale. Fix $a < b \in \mathbb{R}$ and let U_t be the number of upcrossings of the interval (a, b) up to time t; more precisely, define: $N_0 = -1$ and inductively*

$$N_{2k-1} = \inf\{t > N_{2k-2} : M_t \leq a\} \quad and \quad N_{2k} = \inf\{t > N_{2k-1} : M_t \geq b\}.$$

Set $U_t = \sup\{k : N_{2k} \leq t\}$.
    Then

$$(b - a) \cdot \mathbb{E}[U_t] \leq \mathbb{E}\left[(M_t - a)\mathbf{1}_{\{M_t>a\}}\right] - \mathbb{E}\left[(M_0 - a)\mathbf{1}_{\{M_0>a\}}\right].$$

Proof Define $X_t = a + (M_t - a)\mathbf{1}_{\{M_t > a\}}$. Set $H_t = \mathbf{1}_{\{\exists\, k:\, N_{2k-1} < t \le N_{2k}\}}$. Note that H_t is $\sigma(X_0, \ldots, X_{t-1})$-measurable, since $H_t = 1$ if and only if $N_{2k-1} \le t-1$ and $N_{2k} > t-1$.

Now, one verifies that

$$\sum_{s=1}^{t} H_s \cdot (X_s - X_{s-1}) \ge \sum_{k=1}^{U_t} \sum_{s=N_{2k-1}+1}^{N_{2k}} (X_s - X_{s-1}) = \sum_{k=1}^{U_t} (M_{N_{2k}} - a) \ge (b-a)U_t.$$

Note that $(X_t)_t$ is a sub-martingale by Exercise 2.29. By Exercise 2.30, since $H_s \in [0,1]$, $A_t := \sum_{s=1}^{t} H_s \cdot (X_s - X_{s-1})$ and $B_t := \sum_{s=1}^{t} (1 - H_s) \cdot (X_s - X_{s-1})$ are also sub-martingales. Specifically, $\mathbb{E}[B_t] \ge \mathbb{E}[B_0] = 0$. We have that

$$(b-a)\,\mathbb{E}[U_t] \le \mathbb{E}[A_t] \le \mathbb{E}[A_t + B_t] = \mathbb{E}[X_t - X_0].$$

This is the required form. $\qquad\qquad\qquad\qquad\qquad\qquad\qquad\qquad\qquad\qquad\qquad\square$

Theorem 2.6.3 (Martingale convergence theorem) *Let $(M_t)_t$ be a sub-martingale such that $\sup_t \mathbb{E}\left[M_t \mathbf{1}_{\{M_t > 0\}}\right] < \infty$. Then there exists a random variable M_∞ such that $M_t \to M_\infty$ a.s. and $\mathbb{E}|M_\infty| < \infty$.*

Proof Since $(M_t - a)\mathbf{1}_{\{M_t > a\}} \le M_t \mathbf{1}_{\{M_t > 0\}} + |a|$, we have by the upcrossing lemma that $(b-a)\,\mathbb{E}[U_t] \le \mathbb{E}\left[M_t \mathbf{1}_{\{M_t > 0\}}\right] + |a|$, where U_t is the number of upcrossings of the interval (a, b). Let $U = U_{(a,b)} = \lim_{t \to \infty} U_t$ be the total number of upcrossings of (a, b). By Fatou's lemma,

$$\mathbb{E}[U] \le \liminf_{t \to \infty} \mathbb{E}[U_t] \le \frac{|a| + \sup_t \mathbb{E}\left[M_t \mathbf{1}_{\{M_t > 0\}}\right]}{b - a} < \infty.$$

Specifically, $U < \infty$ a.s. Since this holds for all $a < b \in \mathbb{R}$, taking a union bound over all $a < b \in \mathbb{Q}$, we have that

$$\mathbb{P}[\exists\, a < b \in \mathbb{Q} : \liminf_{t \to \infty} M_t \le a < b \le \limsup_{t \to \infty} M_t] \le \sum_{a < b \in \mathbb{Q}} \mathbb{P}\left[U_{(a,b)} = \infty\right] = 0.$$

But then, a.s. we have that $\limsup M_t \le \liminf M_t$, which implies that an a.s. limit $M_t \to M_\infty$ exists.

By Fatou's lemma again,

$$\mathbb{E}\left[M_\infty \mathbf{1}_{\{M_\infty > 0\}}\right] \le \liminf_{t \to \infty} \mathbb{E}\left[M_t \mathbf{1}_{\{M_t > 0\}}\right] \le \sup_t \mathbb{E}\left[M_t \mathbf{1}_{\{M_t > 0\}}\right] < \infty.$$

Another application of Fatou's lemma gives

$$\begin{aligned}
\mathbb{E}\left[-M_\infty \mathbf{1}_{\{M_\infty < 0\}}\right] &\le \liminf_{t \to \infty} \left(\mathbb{E}\left[M_t \mathbf{1}_{\{M_t > 0\}}\right] - \mathbb{E}[M_t]\right) \\
&\le \liminf_{t \to \infty} \left(\mathbb{E}\left[M_t \mathbf{1}_{\{M_t > 0\}}\right] - \mathbb{E}[M_0]\right) \\
&\le \sup_t \mathbb{E}\left[M_t \mathbf{1}_{\{M_t > 0\}}\right] - \mathbb{E}[M_0] < \infty.
\end{aligned}$$

Thus

$$\mathbb{E}|M_\infty| = \mathbb{E}\left[M_\infty \mathbf{1}_{\{M_\infty>0\}}\right] - \mathbb{E}\left[M_\infty \mathbf{1}_{\{M_\infty<0\}}\right] < \infty. \qquad \square$$

Exercise 2.32 Show that if $(M_t)_t$ is a super-martingale and $M_t \geq 0$ for all t a.s., then $M_t \to M_\infty$ a.s., for some integrable random variable M_∞. ▷ solution ◁

Exercise 2.33 Show that if $(M_t)_t$ is a uniformly integrable martingale then $M_t \to M_\infty$ a.s. and in L^1 for some integrable M_∞. ▷ solution ◁

Exercise 2.34 Let $(\mathcal{F}_t)_t$ be a filtration. Let X be an integrable random variable. Show that $\mathbb{E}[X \mid \mathcal{F}_t] \to \mathbb{E}[X \mid \mathcal{F}_\infty]$ a.s. and in L^1, where $\mathcal{F}_\infty = \sigma(\bigcup_t \mathcal{F}_t)$. ▷ solution ◁

Exercise 2.35 (Backward martingale convergence theorem) Let $(\sigma_n)_n$ be a non-increasing sequence of σ-algebras. Let X be an integrable random variable. Show that $\mathbb{E}[X \mid \sigma_n] \to \mathbb{E}[X \mid \sigma_\infty]$ a.s. and in L^1, where $\sigma_\infty = \bigcap_n \sigma_n$. (Hint: use the upcrossing lemma.) ▷ solution ◁

2.7 Bounded Harmonic Functions

We will now use the martingale convergence theorem to study the space of the bounded harmonic functions.

Theorem 2.7.1 *Let G be a finitely generated group, and let μ be an adapted probability measure on G. Then, $\dim \mathrm{BHF}(G, \mu) \in \{1, \infty\}$. That is, there are either infinitely many linearly independent bounded harmonic functions or the only bounded harmonic functions are the constants.*

Proof Let h be a bounded harmonic function. Let $(X_t)_t$ be the μ-random walk on G. Then, $(h(X_t))_t$ is a bounded martingale. Thus, $h(X_t) \to L$ a.s. for some integrable random variable L. Hence, $h(X_{t+k}) - h(X_t) \to 0$ a.s. for any k.

Fix $x \in G$. Let $k > 0$ be such that $\mathbb{P}[X_k = x] = \alpha > 0$. Exercise 2.36 proves that $\mathbb{P}[X_{t+k} = X_t x \mid X_t] = \alpha$ a.s. for all t (this is known as the *Markov property*, which will be discussed in Section 3.1, Exercise 3.1). Thus, for any $\varepsilon > 0$,

$$\mathbb{P}[|h(X_t x) - h(X_t)| > \varepsilon] \leq \alpha^{-1} \mathbb{P}[|h(X_{t+k}) - h(X_t)| > \varepsilon] \to 0.$$

Now assume that $\dim \mathrm{BHF}(G, \mu) < \infty$. Then there exists a ball $B = B(1, r)$ such that for all $f, f' \in \mathrm{BHF}(G, \mu)$, if $f|_B = f'|_B$ then $f = f'$. Define a norm on $\mathrm{BHF}(G, \mu)$ by $||f||_B = \max_{x \in B} |f(x)|$. Since all norms on finite-dimensional spaces are equivalent, there exists a constant $K > 0$ such that $||f||_B \leq ||f||_\infty \leq K \cdot ||f||_B$ for all $f \in \mathrm{BHF}(G, \mu)$.

Now, since $||y.h||_\infty = ||h||_\infty$, for any t we have a.s.

$$\inf_{c \in \mathbb{C}} ||h - c||_\infty = \inf_{c \in \mathbb{C}} \left\| X_t^{-1}.h - c \right\|_\infty \leq K \cdot \inf_{c \in \mathbb{C}} \left\| X_t^{-1}.h - c \right\|_B$$

$$\leq K \cdot \inf_{c \in \mathbb{C}} \max_{x \in B} |h(X_t x) - c| \leq K \cdot \max_{x \in B} |h(X_t x) - h(X_t)|.$$

Since this last term converges to 0 in probability, $\inf_{c \in \mathbb{C}} ||h - c||_\infty = 0$ must hold. Thus, h is constant. □

Exercise 2.36 Let $(X_t)_t$ be the μ-random walk for an adapted probability measure μ on a group G. Show that

$$\mathbb{P}[X_{t+k} = X_t x \mid X_t] = \mathbb{P}[X_k = x] \quad \text{a.s.}$$

for all t, k.

▷ solution ◁

Exercise 2.37 Check that $||f||_B$ in the proof of Theorem 2.7.1 is indeed a norm, for the specific B chosen. (In general, it is only a semi-norm.)

The following is a major open problem in the theory of bounded harmonic functions. It basically states that the property of having only constant bounded harmonic functions should not change if we restrict to "nice" random walk measures. We will return to this conjecture in Chapter 6.

Conjecture 2.7.2 Let G be a finitely generated group. Then for any two $\mu, \nu \in \mathrm{SA}(G, 2)$, we have $\dim \mathrm{BHF}(G, \mu) = \dim \mathrm{BHF}(G, \nu)$.

Exercise 2.38 Let G be a finitely generated group and $\mu \in \mathrm{SA}(G, 1)$. Recall the space of Lipschitz harmonic functions $\mathrm{LHF}(G, \mu)$.

Show that, if there exists a nonconstant positive $h \in \mathrm{LHF}(G, \mu)$, then $\dim \mathrm{LHF}(G, \mu) = \infty$. (Hint: consider LHF modulo the constant functions.)

▷ solution ◁

2.8 Solutions to Exercises

Solution to Exercise 2.1 :(

Let $A = \{Y > 0\}$ and $B = \{Y \leq 0\}$. Note that $A, B \in \mathcal{G}$. Thus, $\mathbb{E}[|Y|1_A] = \mathbb{E}[Y1_A] = \mathbb{E}[X1_A] \leq \mathbb{E}[|X|1_A]$, and similarly $\mathbb{E}[|Y|1_B] = -\mathbb{E}[Y1_B] = -\mathbb{E}[X1_B] \leq \mathbb{E}[|X|1_B]$. Thus, $\mathbb{E}|Y| \leq \mathbb{E}|X| < \infty$. So Y is integrable. Similarly for Z.

Now, let $A = \left\{Y - Z > \frac{1}{n}\right\}$. Then since $A \in \mathcal{G}$, we have

$$0 = \mathbb{E}[(Y - Z)1_A] \geq \mathbb{P}[A] \cdot \frac{1}{n},$$

implying that $\mathbb{P}\left[Y > Z + \frac{1}{n}\right] = 0$. A union bound over n implies that $\mathbb{P}[Y > Z] = 0$. Reversing the roles of Y, Z, we have that $\mathbb{P}[Z > Y] = 0$ as well, culminating in $\mathbb{P}[Y \neq Z] = 0$. :) ✓

Solution to Exercise 2.2 :(

When X is $\mathcal{G}$-measurable, since for any $A \in \mathcal{G}$ we have $\mathbb{E}[X1_A] = \mathbb{E}[X1_A]$ trivially, we have that $\mathbb{E}[X \mid \mathcal{G}] = X$ a.s.

If X is independent of $\mathcal{G}$ then for any $A \in \mathcal{G}$ we have $\mathbb{E}[X1_A] = \mathbb{E}[X] \cdot \mathbb{E}[1_A] = \mathbb{E}[\mathbb{E}[X] \cdot 1_A]$. Since a constant random variable is measurable with respect to any σ-algebra (and specifically $\mathbb{E}[X]$ is $\mathcal{G}$-measurable), we have the second assertion.

The third assertion is a direct consequence, since a constant is always independent of $\mathcal{G}$. :) ✓

Solution to Exercise 2.3 :(

Since $\Omega \in \mathcal{G}$ we have

$$\mathbb{E}[X] = \mathbb{E}[X1_\Omega] = \mathbb{E}[\mathbb{E}[X \mid \mathcal{G}]1_\Omega] = \mathbb{E}[\mathbb{E}[X \mid \mathcal{G}]].$$

:) ✓

Solution to Exercise 2.4 :(

Since $B \in \mathcal{G}$,

$$\mathbb{E}[1_B \mathbb{P}[A \mid \mathcal{G}]] = \mathbb{E}[1_A 1_B] = \mathbb{P}[B \mid A] \cdot \mathbb{P}[A].$$

:) ✓

Solution to Exercise 2.5 :(

Linearity: let $Z = a\,\mathbb{E}[X \mid \mathcal{G}] + \mathbb{E}[Y \mid \mathcal{G}]$. So Z is $\mathcal{G}$-measurable. Also, for any $A \in \mathcal{G}$,

$$\mathbb{E}[(aX + Y)1_A] = a\,\mathbb{E}[X1_A] + \mathbb{E}[Y1_A] = a\,\mathbb{E}[\mathbb{E}[X \mid \mathcal{G}]1_A] + \mathbb{E}[\mathbb{E}[Y \mid \mathcal{G}]1_A] = \mathbb{E}[Z1_A].$$

Monotonicity: By linearity it suffices to show that if $X \geq 0$ a.s. then $\mathbb{E}[X \mid \mathcal{G}] \geq 0$ a.s. Indeed, for $X \geq 0$ a.s., let $Y = \mathbb{E}[X \mid \mathcal{G}]$. Then we may consider $\{Y < 0\} \in \mathcal{G}$, and we have that $0 \leq \mathbb{E}\left[X1_{\{Y<0\}}\right] = \mathbb{E}\left[Y1_{\{Y<0\}}\right] \leq 0$, so $Y1_{\{Y<0\}} = 0$ a.s. So if we set $Z = Y1_{\{Y\geq 0\}}$ we have that $Y = Z \geq 0$ a.s.

Monotone convergence: Write $Y_n = \mathbb{E}[X_n \mid \mathcal{G}]$, $Y = \mathbb{E}[X \mid \mathcal{G}]$. By monotonicity above, $Y_n \leq Y_{n+1} \leq Y$ a.s. Let $Z = \lim_n Y_n$, which exists a.s. because the sequence is monotone. Z is $\mathcal{G}$-measurable as a limit of $\mathcal{G}$-measurable random variables. Also, for any $A \in \mathcal{G}$ we have $X_n 1_A \nearrow X1_A$ and $Y_n 1_A \nearrow Z1_A$ a.s. So by monotone convergence, $\mathbb{E}[Z1_A] \searrow \mathbb{E}[Y_n 1_A] \nearrow \mathbb{E}[X_n 1_A] \nearrow \mathbb{E}[X1_A]$. Hence $Z = \mathbb{E}[X \mid \mathcal{G}]$ a.s. :) ✓

Solution to Exercise 2.6 :(

Note that $Y\,\mathbb{E}[X \mid \mathcal{G}]$ is a $\mathcal{G}$-measurable random variable, as a product of two such random variables. So we need to show that for any $A \in \mathcal{G}$ we have

$$\mathbb{E}[XY1_A \mid \mathcal{G}] = \mathbb{E}[\mathbb{E}[X \mid \mathcal{G}]Y1_A] \quad \text{a.s.}$$

If $Y = 1_B$ then this is immediate. If Y is a simple random variable this follows by linearity. For a nonnegative Y we may approximate by simple random variables and use the monotone convergence theorem. For general Y, we may write $Y = Y^+ - Y^-$ where $Y^\pm$ are nonnegative, and use linearity. :) ✓

Solution to Exercise 2.7 :(

Start with the assumption that $X \geq 0$.

Let $Y = \sum_n \frac{\mathbb{E}[X1_{A_n}]}{\mathbb{P}[A_n]} \cdot 1_{A_n}$. It is immediate to verify that Y is $\mathcal{G}$-measurable.

Also, $\mathbb{E}[Y] = \mathbb{E}[X]$, so $Q(B) := \mathbb{E}[X]^{-1} \cdot \mathbb{E}[Y1_B]$ defines a probability measure on $(\Omega, \mathcal{F})$. Similarly, $P(B) := \mathbb{E}[X]^{-1} \cdot \mathbb{E}[X1_B]$ defines a probability measure on $(\Omega, \mathcal{F})$. The system $\{\emptyset, A_n : n \in \mathbb{N}\}$ is a π-system. For any n we have

$$\mathbb{E}[X] \cdot P(A_n) = \mathbb{E}[X1_{A_n}] = \mathbb{E}[Y1_{A_n}] = \mathbb{E}[X] \cdot Q(A_n).$$

Since P, Q are equal on a π-system generating $\mathcal{G}$, they must be equal on all of $\mathcal{G}$ by Dynkin's lemma. That is, for any $B \in \mathcal{G}$ we have

$$\mathbb{E}[X\mathbf{1}_{A_n}\mathbf{1}_B] = \mathbb{E}[X]P(A_n \cap B) = \mathbb{E}[X]Q(A_n \cap B) = \mathbb{E}[Y\mathbf{1}_{A_n}\mathbf{1}_B],$$

which implies, since $A_n \in \mathcal{G}$, that a.s.

$$\mathbb{E}[X \mid \mathcal{G}] \cdot \mathbf{1}_{A_n} = \mathbb{E}[X\mathbf{1}_{A_n} \mid \mathcal{G}] = Y\mathbf{1}_{A_n} = \frac{\mathbb{E}[X\mathbf{1}_{A_n}]}{\mathbb{P}[A_n]} \cdot \mathbf{1}_{A_n}.$$

For general integrable X, decompose $X = X^+ - X^-$. :) ✓

Solution to Exercise 2.11 :(
Since $|M_t| \leq \sum_{k=1}^{t} |U_k|$, we have that M_t is integrable and $\mathcal{F}_t$-measurable.
Note that $M_{t+1} = M_t + U_{t+1}$ and note that U_{t+1} is independent of $\mathcal{F}_t$ and of $\mathcal{F}_t'$. Thus,

$$\mathbb{E}[M_{t+1} \mid \mathcal{F}_t] = M_t + \mathbb{E}[U_{t+1}] = M_t,$$
$$\mathbb{E}[M_{t+1} \mid \mathcal{F}_t'] = M_t + \mathbb{E}[U_{t+1}] = M_t. \qquad \text{:) ✓}$$

Solution to Exercise 2.12 :(
Since $M_0, \ldots, M_t$ are all measurable with respect to $\mathcal{F}_t$, we have that $\mathcal{F}_t' \subset \mathcal{F}_t$ for all t. Thus, by the tower property,

$$\mathbb{E}[M_{t+1} \mid \mathcal{F}_t'] = \mathbb{E}[\mathbb{E}[M_{t+1} \mid \mathcal{F}_t] \mid \mathcal{F}_t'] = \mathbb{E}[M_t \mid \mathcal{F}_t'] = M_t$$

a.s. :) ✓

Solution to Exercise 2.13 :(
Note that in both cases M_t is measurable with respect to $\sigma(X_t) \subset \sigma(X_0, \ldots, X_t)$.
In the first case, denoting by $e_1, \ldots, e_d$ the standard basis of $\mathbb{Z}^d$, then

$$\mathbb{E}[M_{t+1} \mid X_0, \ldots, X_t] = \frac{1}{2d} \sum_{j=1}^{d} \langle X_t + e_j, v \rangle + \langle X_t - e_j, v \rangle = \frac{1}{2d} \sum_{j=1}^{d} 2\langle X_t, v \rangle = M_t.$$

In the second case,

$$\mathbb{E}\left[||X_{t+1}||^2 \mid X_0, \ldots, X_t\right] = \frac{1}{2d} \sum_{j=1}^{d} ||X_t + e_j||^2 + ||X_t - e_j||^2 = \frac{1}{2d} \sum_{j=1}^{d} 2\left(||X_t||^2 + 1\right) = ||X_t||^2 + 1.$$

Thus,

$$\mathbb{E}[M_{t+1} \mid X_0, \ldots, X_t] = ||X_t||^2 + 1 - (t+1) = M_t. \qquad \text{:) ✓}$$

Solution to Exercise 2.14 :(
Because $\{T \leq t\} \in \sigma(M_0, \ldots, M_t)$ and $\{T > t\} = \{T \leq t\}^c \in \sigma(M_0, \ldots, M_t)$, we get that $M_{T \wedge t} = M_t \mathbf{1}_{\{T > t\}} + \sum_{j=0}^{t} M_j \mathbf{1}_{\{T=j\}}$ is measurable with respect to $\sigma(M_0, \ldots, M_t)$. Also,

$$\mathbb{E}|M_{T \wedge t}| = \mathbb{E} \sum_{j=0}^{t-1} \mathbf{1}_{\{T>j\}} \cdot (|M_{j+1}| - |M_j|) + \mathbb{E}|M_0| < \infty.$$

It now suffices to show that

$$\mathbb{E}[M_{T \wedge (t+1)} \mid M_0, \ldots, M_t] = M_{T \wedge t}.$$

Indeed, since $\{T = t\} = \{T \leq t\} \setminus \{T \leq t-1\} \in \sigma(M_0, \ldots, M_t)$, we have that

$$\mathbb{E}[M_{T \wedge (t+1)} \mid M_0, \ldots, M_t] = \mathbb{E}\left[M_{t+1}\mathbf{1}_{\{T>t\}} \mid M_0, \ldots, M_t\right] + \sum_{j=0}^{t} \mathbb{E}\left[M_j\mathbf{1}_{\{T=j\}} \mid M_0, \ldots, M_t\right]$$

$$= \mathbb{E}[M_{t+1} \mid M_0, \ldots, M_t] \cdot \mathbf{1}_{\{T>t\}} + \sum_{j=0}^{t} M_j\mathbf{1}_{\{T=j\}} = M_{T \wedge t}. \qquad \text{:) ✓}$$

Solution to Exercise 2.15 :(
For any t we have that $\{T \wedge T' \le t\} = \{T \le t\} \cup \{T' \le t\} \in \mathcal{F}_t$. :) ✓

Solution to Exercise 2.16 :(
If f is μ-harmonic then for any $x \in G$,

$$\mathbb{E}\left[|f(X_t)|\mathbb{1}_{\{X_{t-1}=x\}}\right] = \sum_y \mu(y)|f(xy)| \cdot \mathbb{P}[X_{t-1} = x] < \infty,$$

so that $f(X_t)$ is integrable for every t. Also,

$$\mathbb{E}[f(X_{t+1}) \mid \mathcal{F}_t] = \mathbb{E}\left[X_t^{-1}.f(U_{t+1}) \mid \mathcal{F}_t\right] = \sum_y \mu(y)X_t^{-1}.f(y) = \sum_y \mu(y)f(X_ty) = f(X_t),$$

which shows that $(f(X_t))_t$ is a martingale.

Now assume that f is such that $(f(X_t))_t$ is a martingale. Since μ is adapted, for any $x \in G$, there exists $t > 0$ such that $\mathbb{P}[X_t = x] > 0$. So,

$$f(x) = \mathbb{E}[f(X_{t+1}) \mid X_t = x] = \sum_y \mu(y)f(xy),$$

which implies that f is harmonic at x. As before, the above sum converges absolutely because $f(X_{t+1})$ is integrable. :) ✓

Solution to Exercise 2.17 :(
Set $Y_K = |X|\mathbb{1}_{\{|X|>K\}}$. So $Y_K \to 0$ a.s. Since $0 \le Y_K \le |X|$ for all K, by dominated convergence we have that

$$\lim_{K\to\infty} \mathbb{E}\left[|X_\alpha|\mathbb{1}_{\{|X_\alpha|>K\}}\right] = \lim_{K\to\infty} \mathbb{E}[Y_K] = 0.$$:) ✓

Solution to Exercise 2.18 :(
Uniform integrability implies that there exists K such that for all α we have $\mathbb{E}\left[|X_\alpha|\mathbb{1}_{\{|X_\alpha|>K\}}\right] < 1$. Since $\mathbb{E}\left[|X_\alpha|\mathbb{1}_{\{|X_\alpha|\le K\}}\right] \le K$, we arrive at

$$\sup_\alpha \mathbb{E}|X_\alpha| \le \sup_\alpha \left(\mathbb{E}\left[|X_\alpha|\mathbb{1}_{\{|X_\alpha|>K\}}\right] + \mathbb{E}\left[|X_\alpha|\mathbb{1}_{\{|X_\alpha|\le K\}}\right]\right) \le 1 + K.$$:) ✓

Solution to Exercise 2.19 :(
Choose $p = 1 + \varepsilon$ and $q = \frac{1+\varepsilon}{\varepsilon}$. Hölder's inequality gives for any α,

$$\begin{aligned}
\mathbb{E}\left[|X_\alpha|\mathbb{1}_{\{|X_\alpha|>K\}}\right] &\le (\mathbb{E}[|X_\alpha|^p])^{1/p} \cdot (\mathbb{P}[|X_\alpha| > K])^{1/q} \\
&\le (\mathbb{E}[|X_\alpha|^p])^{1/p} \cdot (\mathbb{E}[|X_\alpha|^{1+\varepsilon}])^{1/q} \cdot K^{-(1+\varepsilon)/q} \\
&= \mathbb{E}[|X_\alpha|^{1+\varepsilon}] \cdot K^{-\varepsilon}.
\end{aligned}$$

Thus,

$$\lim_{K\to\infty} \sup_\alpha \mathbb{E}\left[|X_\alpha|\mathbb{1}_{\{|X_\alpha|>K\}}\right] \le \sup_\alpha \mathbb{E}\left[|X_\alpha|^{1+\varepsilon}\right] \cdot \lim_{K\to\infty} K^{-\varepsilon} = 0.$$:) ✓

Solution to Exercise 2.20 :(
Let $\varepsilon > 0$. There exists $\delta > 0$ such that if $A \in \mathcal{F}$ and $\mathbb{P}[A] < \delta$ then $\mathbb{E}[|X|\mathbb{1}_A] < \varepsilon$. (Otherwise we could find $(A_n)_n \subset \mathcal{F}$ such that $\mathbb{P}[A_n] < n^{-1}$ and $\mathbb{E}[|X|\mathbb{1}_{A_n}] \ge \varepsilon$. But since X is integrable, $\mathbb{E}[|X|\mathbb{1}_{A_n}] \to 0$ by the dominated convergence theorem, a contradiction.)

Take $K > \delta^{-1}\mathbb{E}|X|$. Then, since $\{\mathbb{E}[|X| \mid \mathcal{F}_\alpha] > K\} \in \mathcal{F}_\alpha$,

$$\begin{aligned}
\mathbb{E}\left[|\mathbb{E}[X \mid \mathcal{F}_\alpha]|\mathbb{1}_{\{|\mathbb{E}[X|\mathcal{F}_\alpha]|>K\}}\right] &\le \mathbb{E}\left[\mathbb{E}[|X| \mid \mathcal{F}_\alpha]\mathbb{1}_{\{\mathbb{E}[|X||\mathcal{F}_\alpha]>K\}}\right] \\
&= \mathbb{E}\left[\mathbb{E}[|X|\mathbb{1}_{\{\mathbb{E}[|X||\mathcal{F}_\alpha]>K\}} \mid \mathcal{F}_\alpha]\right] \\
&= \mathbb{E}\left[|X|\mathbb{1}_{\{\mathbb{E}[|X||\mathcal{F}_\alpha]>K\}}\right].
\end{aligned}$$

Using $A = \{\mathbb{E}[|X| \mid \mathcal{F}_\alpha] > K\}$, we have that

$$\mathbb{P}[A] \le \mathbb{E}[\mathbb{E}[|X| \mid \mathcal{F}_\alpha]] \cdot K^{-1} = \mathbb{E}|X| \cdot K^{-1} < \delta,$$

so $\mathbb{E}[|X|\mathbf{1}_A] < \varepsilon$. This was uniform over α, so we conclude that for all $\varepsilon > 0$ there exists $\delta > 0$ such that if $K > \delta^{-1}\,\mathbb{E}\,|X|$ then

$$\sup_\alpha \mathbb{E}\left[\,|\,\mathbb{E}[X \mid \mathcal{F}_\alpha]|\mathbf{1}_{\{|\mathbb{E}[X|\mathcal{F}_\alpha]|>K\}}\right] < \varepsilon.$$

This is exactly uniform integrability.   :) ✓

Solution to Exercise 2.21   :(
By Fatou's lemma,

$$\mathbb{E}\,|X| = \mathbb{E}[\lim_n |X_n|] \leq \liminf_{n\to\infty} \mathbb{E}\,|X_n| \leq \sup_n \mathbb{E}\,|X_n| < \infty.$$:) ✓

Solution to Exercise 2.22   :(
Note that if $|M_t| \leq m$ a.s., then obviously $(M_t)_t$ is uniformly integrable.

Also, since $T < \infty$ a.s., we have that $|M_{T\wedge t}| \to |M_T|$ a.s. as $t \to \infty$. Thus, $|M_T| \leq m$ a.s., implying that $\mathbb{E}[|M_T|] < \infty$.

By the optional stopping theorem we have that $\mathbb{E}[M_T] = \mathbb{E}[M_0]$.   :) ✓

Solution to Exercise 2.23   :(
This is similar to the proof of Theorem 2.3.3.

Define

$$\varphi_K(x) = \begin{cases} K & \text{if } x > K, \\ x & \text{if } |x| \leq K, \\ -K & \text{if } x < -K. \end{cases}$$

Note that $|\varphi_K(x) - x| \leq |x|\mathbf{1}_{\{|x|>K\}}$. Since $\varphi_K(X_n) \to \varphi_K(X)$ a.s., and $|\varphi_K(X_n)| \leq K$ for all n, by dominated convergence we have that $\varphi_K(X_n) \to \varphi_K(X)$ in L^1. Thus,

$$\mathbb{E}\,|X_n - X| \leq \mathbb{E}\,|\varphi_K(X_n) - \varphi_K(X)| + \mathbb{E}[|X_n|\mathbf{1}_{\{|X_n|>K\}}] + \mathbb{E}[|X|\mathbf{1}_{\{|X|>K\}}],$$

implying that

$$\limsup_{n\to\infty} \mathbb{E}\,|X_n - X| \leq \sup_t \mathbb{E}\left[|X_t|\mathbf{1}_{\{|X_t|>K\}}\right] + \mathbb{E}\left[|X|\mathbf{1}_{\{|X|>K\}}\right].$$

Since $\mathbb{E}\,|X| < \infty$ (as the a.s. limit of a uniformly integrable sequence), this goes to 0 as $K \to \infty$.   :) ✓

Solution to Exercise 2.24   :(
Set $X_t := |M_t| - |M_{T\wedge t}|$. Using Jensen's inequality we have that a.s.

$$\mathbb{E}[|M_{t+1}| \mid M_t, \ldots, M_0] \geq |\mathbb{E}[M_{t+1} \mid M_t, \ldots, M_0]| = |M_t|.$$

(That is, $(|M_t|)_t$ is a *sub-martingale*.) Note that

$$|M_{T\wedge(t+1)}| - |M_{T\wedge t}| = (|M_{t+1}| - |M_t|)\mathbf{1}_{\{T>t\}},$$

so

$$X_{t+1} - X_t = (|M_{t+1}| - |M_t|)\mathbf{1}_{\{T\leq t\}}.$$

Thus,

$$\mathbb{E}[X_{t+1} - X_t \mid M_0, \ldots, M_t] = \mathbf{1}_{\{T\leq t\}} \cdot \mathbb{E}[|M_{t+1}| - |M_t| \mid M_0, \ldots, M_t] \geq 0.$$

(That is, $(X_t)_t$ is also a *sub-martingale*.) Taking expectations we get that $\mathbb{E}[X_t] \geq \mathbb{E}[X_{t-1}] \geq \cdots \geq \mathbb{E}[X_0] = \mathbb{E}[M_0 - M_{T\wedge 0}] = 0$. Hence, $\mathbb{E}\,|M_t| \geq \mathbb{E}\,|M_{T\wedge t}|$, as required.   :) ✓

Solution to Exercise 2.25   :(
Since $\sup_t \mathbb{E}\,|M_{T\wedge t}| \leq \sup_t \mathbb{E}\,|M_t| < \infty$, we have that $M_{T\wedge t} \to M_\infty$ a.s., for some integrable M_∞. But $M_{T\wedge t} \to M_T$ a.s. as well, which implies that $M_T = M_\infty$ a.s., so M_T is integrable.   :) ✓

Solution to Exercise 2.26 :(

Let $K = b - a$. Compute, for any $a < x < b$,

$$P[X_{t+K} \notin (a, b) \mid X_t = x, T_{a,b} > t] \geq P[\forall\, 0 \leq j < K,\ U_{t+j+1} = 1 \mid X_t = x, T_{a,b} > t] = 2^{-K},$$

using the fact that $(U_{t+j+1})_{j=0}^{\infty}$ are all independent of $\mathcal{F}_t$, and that $\{T_{a,b} > t\} = \{T_{a,b} \leq t\}^c \in \mathcal{F}_t$.

Since $T_{a,b} > t + K$ implies that $T_{a,b} > t$ and that $X_{t+K} \in (a, b)$, we may bound

$$P[T_{a,b} > t + K] = \sum_{x=a+1}^{b-1} P[T_{a,b} > t + K \mid X_t = x, T_{a,b} > t] \cdot P[X_t = x, T_{a,b} > t]$$

$$\leq \left(1 - 2^{-K}\right) \cdot \sum_{x=a+1}^{b-1} P[X_t = x, T_{a,b} > t] = \left(1 - 2^{-K}\right) \cdot P[T_{a,b} > t].$$

Inductively we obtain that

$$P[T_{a,b} > Kn] \leq \left(1 - 2^{-K}\right)^n. \qquad \qquad :)\checkmark$$

Solution to Exercise 2.27 :(

This is just the L^p maximal inequality, with $p = 2$, together with the fact that $\mathbb{E}\,|X_t|^2 = (1 - p)t \leq t$, which stems from the OST applied to the martingale $\left(|X_t|^2 - (1 - p)t\right)_t$. $\qquad :)\checkmark$

Solution to Exercise 2.28 :(

We start with the upper bound. Consider $\left(|X_t|^2 - (1 - p)t\right)_t$. This is easily seen to be a martingale. Started at $|x| \leq m$ and up to the stopping time

$$T = T_{m+1} \wedge T_{-m-1} = \inf\{t \geq 0 :\ |X_t| = m + 1\},$$

this is a bounded martingale, so by the OST (Theorem 2.3.3),

$$|x|^2 = \mathbb{E}_x\left[|X_T|^2 - (1 - p)T\right] = (m + 1)^2 - (1 - p)\,\mathbb{E}_x[T].$$

Hence, by Markov's inequality, uniformly over $|x| \leq m$, we have $P_x\left[T > \frac{2}{1-p}(m + 1)^2\right] \leq \frac{1}{2}$.

Let $U_t = X_t - X_{t-1}$ for all $t \geq 1$, so that $(U_t)_{t \geq 1}$ are i.i.d. $-\mu$. Since $(U_{s+k})_{k \geq 1}$ are independent of $\mathcal{F}_s$, for any $|x| \leq m$ and any $t > s$ we have that

$$P_x[T > t \mid \mathcal{F}_s] = \mathbb{1}_{\{T>s\}} \cdot P_x\left[\forall\, 1 \leq k \leq t - s,\ \left|X_s + \sum_{j=1}^{k} U_{s+j}\right| \leq m\right]$$

$$= \mathbb{1}_{\{T>s\}} \cdot P_{X_s}[T > t - s] \leq \mathbb{1}_{\{T>s\}} \cdot \sup_{|y| \leq m} P_y[T > t - s].$$

Thus, for $|x| \leq m$ and any $t > \frac{2}{1-p}(m + 1)^2$,

$$P_x[M_t \leq m] = P_x[T > t] = P_x\left[T > t,\ T > \frac{2}{1-p}(m + 1)^2\right]$$

$$\leq P_x\left[T > \frac{2}{1-p}(m + 1)^2\right] \cdot \sup_{|y| \leq m} P_y\left[T > t - \frac{2}{1-p}(m + 1)^2\right] \leq \cdots \leq 2^{-\left\lfloor \frac{(1-p)t}{2(m+1)^2} \right\rfloor},$$

which gives the desired upper bound.

Now for the lower bound. For $0 \leq x \leq m$ we have

$$P_x[X_t < -m] \leq P_x[|X_t - X_0| \geq m + 1] \leq P[|X_t| \geq m + 1] \leq \frac{\mathbb{E}\,|X_t|^2}{(m + 1)^2} \leq \frac{t}{(m + 1)^2}.$$

Similarly, for $-m \leq x \leq 0$,

$$P_x[X_t > m] \leq \frac{t}{(m + 1)^2}.$$

Under $\mathbb{P}_0$, both X_t and $-X_t$ have the same distribution. Shifting by x, for some $|x| \leq m$, we get that $\mathbb{P}_x[X_t \leq x] = \mathbb{P}[X_t \leq 0] \geq \frac{1}{2}$, and similarly $\mathbb{P}_x[X_t \geq x] \geq \frac{1}{2}$. Recall that by the L^p maximal inequality we know that

$$\mathbb{E}_x |M_t|^2 \leq 4 \mathbb{E}_x |X_t|^2 = 4 \left(x^2 + (1-p)t \right) \leq 4 \left(x^2 + t \right),$$

since $\left(|X_t|^2 - (1-p)t \right)_t$ is a martingale.

Putting all this together we have for any $0 \leq x \leq m$,

$$\mathbb{P}_x[M_t \leq km, \ |X_t| \leq m] \geq \mathbb{P}_x[X_t \leq x] - \mathbb{P}_x[X_t < -m] - \mathbb{P}_x[M_t > km]$$

$$\geq \frac{1}{2} - \frac{t}{(m+1)^2} - \frac{\mathbb{E}_x \left[|M_t|^2 \right]}{k^2 m^2} \geq \frac{1}{2} - \frac{t}{(m+1)^2} - \frac{4 \left(x^2 + t \right)}{k^2 m^2},$$

and similarly for $-m \leq x \leq 0$ we have

$$\mathbb{P}_x[M_t \leq km, \ |X_t| \leq m] \geq \frac{1}{2} - \frac{t}{(m+1)^2} - \frac{4 \left(x^2 + t \right)}{k^2 m^2}.$$

Choosing $k = 8$ and $\ell = \left\lfloor \frac{1}{8} m^2 \right\rfloor$, we arrive at the conclusion that for all $|x| \leq m$ we have

$$\mathbb{P}_x[M_\ell \leq 8m, \ |X_\ell| \leq m] > \frac{1}{4}.$$

Similarly to the proof of the upper bound, for any $|x| \leq m$ and any $t > s$ we have

$$\mathbb{P}_x[M_t \leq 8m, \ |X_s| \leq m \mid \mathcal{F}_s] = \mathbf{1}_{\{M_s \leq 8m, \ |X_s| \leq m\}} \cdot \mathbb{P}_x \left[\forall \, 1 \leq k \leq t-s, \ \left| X_s + \sum_{j=1}^{k} U_{s+j} \right| \leq 8m \right]$$

$$= \mathbf{1}_{\{M_s \leq 8m, \ |X_s| \leq m\}} \cdot \mathbb{P}_{X_s}[|M_{t-s}| \leq 8m]$$

$$\geq \mathbf{1}_{\{M_s \leq 8m, \ |X_s| \leq m\}} \cdot \inf_{|y| \leq m} \mathbb{P}_y[|M_{t-s}| \leq 8m].$$

We obtain that for any $|x| \leq m$ and $t > \ell$,

$$\mathbb{P}_x[M_t \leq 8m] \geq \mathbb{P}_x[M_t \leq 8m, \ |X_\ell| \leq m]$$

$$\geq \mathbb{P}_x[M_\ell \leq 8m, \ |X_\ell| \leq m] \cdot \inf_{|y| \leq m} \mathbb{P}_y[M_{t-\ell} \leq 8m] \geq \cdots \geq 4^{-\lfloor t/\ell \rfloor},$$

which completes the proof of the lower bound.   :) ✓

Solution to Exercise 2.29   :(
The function $\varphi(x) = x \mathbf{1}_{\{x>0\}}$ is convex and nondecreasing, so by Jensen's inequality $\mathbb{E}[\varphi(M_{t+1} - a) \mid M_0, \ldots, M_t] \geq \varphi(\mathbb{E}[M_{t+1} - a \mid M_0, \ldots, M_t]) \geq \varphi(M_t - a)$.   :) ✓

Solution to Exercise 2.30   :(
We write a solution only for the sub-martingale case, since all are very similar.

$$\mathbb{E}[(H \cdot M)_{t+1} - (H \cdot M)_t \mid M_0, \ldots, M_t] = H_{t+1} \, \mathbb{E}[(M_{t+1} - M_t) \mid M_0, \ldots, M_t] \geq 0.$$

We have used the fact that H_{t+1} is $\sigma(M_0, \ldots, M_t)$-measurable.   :) ✓

Solution to Exercise 2.31   :(
Let $H_t = \mathbf{1}_{\{T \geq t\}}$, which is a bounded predictable process. Then, $M_{T \wedge t} = M_0 + \sum_{s=1}^{t} H_s (M_s - M_{s-1})$, which is a sub-martingale by Exercise 2.30.   :) ✓

Solution to Exercise 2.32   :(
The process $X_t := -M_t$ is a sub-martingale and $\sup_t \mathbb{E}\left[X_t \mathbf{1}_{\{X_t > 0\}} \right] \leq 0 < \infty$. So X_t converges a.s., which implies the a.s. convergence of M_t.   :) ✓

Solution to Exercise 2.33 :(
Since $(M_t)_t$ is uniformly integrable,

$$\sup_t \mathbb{E}\left[M_t \mathbf{1}_{\{M_t>0\}}\right] \le \sup_t \mathbb{E}\,|M_t| < \infty,$$

so by martingale convergence $M_t \to M_\infty$ a.s., for some integrable M_∞. By uniform integrability again, $M_t \to M_\infty$ in L^1 as well. :)✓

Solution to Exercise 2.34 :(
Let $M_t = \mathbb{E}[X \mid \mathcal{F}_t]$. Since $(M_t)_t$ is a uniformly integrable martingale, it converges a.s. and in L^1 to some integrable M_∞. Now, for any event A, we have that $M_t \mathbf{1}_A \to M_\infty \mathbf{1}_A$ a.s. and in L^1 as well. Thus, if $A \in \mathcal{F}_n$ for some n,

$$\mathbb{E}[M_\infty \mathbf{1}_A] = \lim_{t\to\infty} \mathbb{E}[M_t \mathbf{1}_A] = \mathbb{E}[X\mathbf{1}_A].$$

Consider the probability measures:

$$\mu(A) := \frac{\mathbb{E}[((M_\infty)^+ + X^-)\mathbf{1}_A]}{\mathbb{E}[(M_\infty)^+ + X^-]} \quad \text{and} \quad \nu(A) := \frac{\mathbb{E}[((M_\infty)^- + X^+)\mathbf{1}_A]}{\mathbb{E}[(M_\infty)^- + X^+]}.$$

Since M_∞, X are integrable, these are indeed probability measures, and μ, ν agree on the π-system $\bigcup_t \mathcal{F}_t$. Thus, by Dynkin's lemma (also known as the $\pi - \lambda$ theorem) μ, ν must agree on all of $\mathcal{F}_\infty$. Hence, $M_\infty = \mathbb{E}[X \mid \mathcal{F}_\infty]$ a.s. :)✓

Solution to Exercise 2.35 :(
Set $X_n = \mathbb{E}[X \mid \sigma_n]$.
  Fix n and consider $M_t := X_{n-t}$ for $t \le n$ and $M_t = X_0$ for $t \ge n$. Then $(M_t)_t$ is a martingale. If U_n is the number of upcrossings of the interval (a, b) by $M_0, \ldots, M_n$, then $(b-a)\,\mathbb{E}[U_n] \le \mathbb{E}\left[(M_n - a)\mathbf{1}_{\{M_n>a\}}\right] = \mathbb{E}\left[(X_0 - a)\mathbf{1}_{\{X_0>a\}}\right]$.
  Now, let U_∞ be the number of upcrossings of the interval (a, b) by $(X_n)_n$. Then $U_n \nearrow U_\infty$, so by monotone convergence, $\mathbb{E}[U_\infty] < \infty$. Exactly as in the proof of the martingale convergence theorem, this holding for all $a < b \in \mathbb{Q}$ implies that $X_n \to X_\infty$ a.s. for some integrable X_∞. Since $(X_n)_n$ is uniformly integrable, we get that $X_n \to X_\infty$ in L^1 as well.
  Set $Y = \limsup_{n\to\infty} X_n$. So $Y = X_\infty$ a.s. Note that for any n, all $(X_t)_{t\ge n}$ are σ_n-measurable, so we have that $Y = \limsup_{n \le t\to\infty} X_t$ is also σ_n-measurable. This implies that Y is measurable with respect to $\sigma_\infty = \bigcap_n \sigma_n$.
  For any $A \in \sigma_\infty$, we have that $X_n \mathbf{1}_A \to X_\infty \mathbf{1}_A$ in L^1, so

$$\mathbb{E}[Y\mathbf{1}_A] = \mathbb{E}[X_\infty \mathbf{1}_A] = \lim_{n\to\infty} \mathbb{E}[\mathbb{E}[X \mid \sigma_n]\mathbf{1}_A] = \lim_{n\to\infty} \mathbb{E}[X\mathbf{1}_A].$$

Thus, $\mathbb{E}[X \mid \sigma_\infty] = Y = X_\infty$ a.s. :)✓

Solution to Exercise 2.36 :(
Write $X_t = U_1 \cdots U_t$ for $(U_t)_{t\ge 1}$ i.i.d.-μ elements.
  Set $Y = X_t^{-1} X_{t+k}$, and note that Y is independent of $\mathcal{F}_t$, and specifically that Y is independent of X_t. Thus,

$$\mathbb{P}[X_{t+k} = X_t x \mid X_t] = \mathbb{P}[Y = x \mid X_t] = \mathbb{P}[Y = x] \quad \text{a.s.}$$

Since $(U_t)_{t\ge 1}$ all have the same distribution, we find that $Y = U_{t+1} \cdots U_{t+k}$ has the same distribution as $X_k = U_1 \cdots U_k$. :)✓

Solution to Exercise 2.38 :(
Let $V = \mathrm{LHF}(G, \mu)/\mathbb{C}$ (modulo the constant functions). Fix some finite symmetric generating set S of G. Assume that $\dim \mathrm{LHF}(G, \mu) < \infty$, so $\dim V < \infty$.
  Recall the Lipschitz semi-norm $||\nabla_S f||_\infty := \sup_{x\in G, s\in S} |f(xs) - f(x)|$. We have that $||\nabla_S f||_\infty = 0$ if and only if f is constant. Also, $||\nabla_S(f + c)||_\infty = ||\nabla_S f||_\infty$ for any constant c. Thus, $||\nabla_S \cdot ||_\infty$ induces a norm on V.
  Another semi-norm on G is given by $||f||_B := \max_{x\in B} |f(x) - f(1)|$, where B is some finite subset. Note that $||f + c||_B = ||f||_B$ for any constant c. Because $\dim \mathrm{LHF}(G, \mu) < \infty$, if $B = B(1, r)$ for r large enough then $|| \cdot ||_B$ is a semi-norm on $\mathrm{LHF}(G, \mu)$ such that $||f||_B = 0$ if and only if f is constant. Thus, $|| \cdot ||_B$ induces a norm on V as well.

Since V is finite dimensional, all norms on it are equivalent. Thus, there exists a constant $K > 0$ such that $||\nabla_S v||_\infty \le K \cdot ||v||_B$ for any $v \in V$. Since these semi-norms are invariant to adding constants, this implies that $||\nabla_S f||_\infty \le K \cdot ||f||_B$ for all $f \in \mathsf{LHF}(G, \mu)$.

Now, let $h \in \mathsf{LHF}(G, \mu)$ be a positive harmonic function. Then, $(h(X_t))_t$ is a positive martingale, implying that it converges a.s. Thus, for any fixed k, we have $h(X_{t+k}) - h(X_t) \to 0$ a.s.

Fix $x \in G$ and let k be such that $\mathbb{P}[X_k = x] = \alpha > 0$. By Exercise 2.36, $\mathbb{P}[X_{t+k} = X_t x \mid X_t] = \alpha$, independently of t. We have that a.s. convergence implies convergence in probability, so for any $\varepsilon > 0$,

$$\mathbb{P}[|h(X_t x) - h(X_t)| > \varepsilon] \le \alpha^{-1} \mathbb{P}[|h(X_{t+k}) - h(X_t)| > \varepsilon] \to 0.$$

So $h(X_t x) - h(X_t) \to 0$ in probability, for any $x \in G$. Since B is a finite ball this implies that $\max_{x \in B} |h(X_t x) - h(X_t)| \to 0$ in probability.

Now we also use the fact that $||\nabla_S (x.h)||_\infty = ||\nabla_S h||_\infty$. Thus, for all t, we have a.s. that

$$||\nabla_S h||_\infty = \left\| \nabla_S \left(X_t^{-1}.h \right) \right\|_\infty \le K \cdot \left\| X_t^{-1}.h \right\|_B = K \cdot \max_{x \in B} |h(X_t x) - h(X_t)|.$$

Since this converges to 0 in probability, we have $||\nabla_S h||_\infty = 0$ and h is constant.   :) ✓

3

Markov Chains

As we have already started to see, random walks play a fundamental role in the study of harmonic functions. In this chapter (and in Chapter 4) we will review fundamentals of random walks on groups and, more generally, *Markov chains*.

3.1 Markov Chains

Recall Section 1.2.

Let G be a countable set. Let $(P(x, y))_{x,y \in G}$ be a matrix (albeit, perhaps, infinite dimensional). We say P is **stochastic** if $P(x, y) \geq 0$ and $\sum_z P(x, z) = 1$, for all $x, y \in G$.

Let ν be a probability measure on G. Let $\mathcal{F}$ be the cylinder σ-algebra on $G^{\mathbb{N}}$. Let $X_t(\omega) = \omega_t$ for $\omega \in G^{\mathbb{N}}$. For $t \geq 0$, let $P_{\nu,t} : \mathcal{F}_t \to [0, 1]$ be the probability measure defined by

$$P_{\nu,t}[C(\{0, 1, \ldots, t\}, \omega)] = \nu(\omega_0) \cdot \prod_{j=1}^{t} P(\omega_{j-1}, \omega_j).$$

One easily verifies that these measures are consistent, and by Kolmogorov's extension theorem there exists a unique probability measure $\mathbb{P}_\nu$ on $(G^{\mathbb{N}}, \mathcal{F})$ such that $\mathbb{P}_\nu[A] = P_t[A]$ for any $A \in \sigma(X_0, \ldots, X_t)$ and any $t \geq 0$.

Definition 3.1.1 A sequence $(X_t)_t$ with distribution $\mathbb{P}_\nu$ as above is called a **Markov chain** with **transition matrix** P and **starting distribution** ν. Sometimes we say that $(X_t)_t$ is **Markov-(ν, P)**.

When $\nu = \delta_x$, we use the notation $\mathbb{P}_x = \mathbb{P}_{\delta_x}$ and write $\mathbb{E}_\nu$, $\mathbb{E}_x$ for the expectation with respect to $\mathbb{P}_\nu$, $\mathbb{P}_x$, respectively.

The set G in which the random process $(X_t)_t$ takes its values is called the **state space**.

Remark 3.1.2 To specify a Markov chain, one can specify either the probability space, or the transition matrix and starting distribution. Somewhat carelessly, we will not distinguish these, and use both as the object designated by the name "Markov chain."

Recall that if P is a $G \times G$ matrix, and $f : G \to \mathbb{R}$, we define $Pf : G \to \mathbb{R}$ by $Pf(x) = \sum_y P(x, y) f(y)$ whenever the sum converges. Similarly, $fP(x) = \sum_y f(y) P(y, x)$ whenever the sum converges.

The following property is called the *Markov property*.

Exercise 3.1 (Markov property) Let P be a stochastic matrix and $(X_t)_t$ be the corresponding Markov chain.

Show that

$$\mathbb{P}[X_{t+1} = y \mid \mathcal{F}_t] = \mathbb{P}[X_{t+1} = y \mid X_t] = P(X_t, y) \quad \text{a.s.}$$

Conclude that for any event $A \in \sigma(X_0, \ldots, X_t)$, any $t, n \geq 0$, and any $x, y \in G$,

$$\mathbb{P}[X_{t+n} = y, \, X_t = x, \, A] = P^n(x, y) \cdot \mathbb{P}[X_t = x, \, A]. \qquad \triangleright \text{solution} \triangleleft$$

Note that this exercise tells us that conditioned on $X_t = x$, the process $(X_{t+n})_n$ is distributed as a Markov chain with transition matrix P started at x. Moreover, conditioned on $X_t = x$ the process $(X_{t+n})_n$ is conditionally independent of $X_0, \ldots, X_t$.

Exercise 3.2 Let $(X_t)_t$ be a Markov chain on G with transition matrix P.

Show that for any bounded nonnegative function $f \colon G \to [0, \infty)$ and any probability measure μ on G, we have that

$$\mathbb{E}_\mu[f(X_t)] = \mu P^t f = \sum_{x,y} \mu(x) P^t(x, y) f(y).$$

Conclude that if $f \colon G \to \mathbb{C}$ is a function such that $\mathbb{E}_\mu |f(X_t)| < \infty$, then $\mathbb{E}_\mu[f(X_t)] = \mu P^t f$.

Exercise 3.3 Let G be a countable set. Let $(X_t)_t$ be a sequence of G-valued random variables. Suppose that for any $t \geq 0$ and all $x, y \in G$, it holds that

$$\mathbb{P}[\forall 0 \leq j \leq t, X_j = x_j, X_{t+1} = y] = \mathbb{P}[\forall 0 \leq j \leq t, X_j = x_j] \cdot P(x, y)$$

for some matrix P.

Show that $(X_t)_t$ is a Markov chain with transition matrix P.

Example 3.1.3 Consider a bored mathematician. She has a (possibly biased) coin and two chairs in her office, say chair a and chair b. Every minute, out of boredom, she tosses the coin. If it comes out heads, she moves to the other chair. Otherwise, she does nothing.

This can be modeled by a Markov chain on the state space $\{a, b\}$. At each time, with some probability $1 - p$ the mathematician does not move, and with probability p she jumps to the other state. The corresponding transition matrix would be $P = \begin{bmatrix} 1 - p & p \\ p & 1 - p \end{bmatrix}$.

What is the probability $\mathbb{P}_a[X_n = b] = ?$ For this we need to calculate P^n.

A complicated way would be to analyze the eigenvalues of P, perhaps using the Jordan form...

But a small trick gives us an easier calculation: Let $\mu_n = P^n(a, \cdot)$. Check that $\mu_{n+1} = \mu_n P$. Consider the vector $\pi = (1/2, 1/2)^\tau$. Then $\pi P = \pi$. Now, consider $a_n = (\mu_n - \pi)(a)$. Since μ_n is a probability measure, we get that $\mu_n(b) = 1 - \mu_n(a)$, so

$$a_n = ((\mu_{n-1} - \pi)P)(a) = (1 - p)\mu_{n-1}(a) + p\mu_{n-1}(b) - 1/2$$
$$= (1 - 2p)(\mu_{n-1} - \pi)(a) + p - \pi(a) + (1 - 2p)\pi(a) = (1 - 2p)a_{n-1}.$$

So $a_n = (1 - 2p)^n a_0 = (1 - 2p)^n \cdot \frac{1}{2}$ and $P^n(a, a) = \mu_n(a) = \frac{1+(1-2p)^n}{2}$. This also implies that $P^n(a, b) = 1 - P^n(a, a) = \frac{1-(1-2p)^n}{2}$.

We see that

$$P^n \to \frac{1}{2}\begin{bmatrix} 1 & 1 \\ 1 & 1 \end{bmatrix} = \begin{bmatrix} \pi & \pi \end{bmatrix}.$$

This was just a practice example, to illustrate some of the properties that we will prove in greater generality.   △▽△

Example 3.1.4 Another basic example is a *random walk on a graph*. Let $G = (V, E)$ be a graph. We may define a stochastic matrix P by $P(x, y) = \frac{1}{\deg(x)}\mathbf{1}_{\{x \sim y\}}$ (recall that $x \sim y$ means that $\{x, y\}$ is an edge in the graph). The corresponding Markov chain is the process that at each step chooses uniformly among the neighbors of the current position and moves to that new position.

Sometimes, one may be interested in the situation where the random walk has some probability of staying in place. This is known as the *lazy random walk*. It requires another parameter to be specified, namely the probability $\alpha \in (0, 1)$ of staying in the same place. In this case, the transition matrix would be $Q(x, y) = \alpha\mathbf{1}_{\{x=y\}} + (1-\alpha)\frac{1}{\deg(x)}\mathbf{1}_{\{x \sim y\}}$. So in matrix form $Q = \alpha I + (1-\alpha)P$. The parameter α is sometimes called the *holding probability*.   △▽△

3.2 Irreducibility

When we speak about graphs, we have the notion of connectivity. We are now interested to generalize this notion to Markov chains. We want to say that a state x is connected to a state y if there is a way to get from x to y; note that for general Markov chains this does not necessarily imply that one can get from y to x.

Definition 3.2.1 Let P be the transition matrix of a Markov chain on G; P is called **irreducible** if for every pair of states $x, y \in G$ there exists $t > 0$ such that $P^t(x, y) > 0$.

A Markov chain is called irreducible if its transition matrix is irreducible.

This means that for any two states x, y, there is a large enough time t such that with positive probability the chain can go from state x to state y in t steps.

Example 3.2.2 Consider the cycle $\mathbb{Z}/n\mathbb{Z}$, for n even. Consider the Markov chain moving $+1$ or -1 (mod n) with equal probability at each step. That is, $P(x, y) = \frac{1}{2}$ for all $|x - y| = 1$ (mod n), and $P(x, y) = 0$ otherwise.

This is an irreducible chain since for any x, y, we have for $t = |x-y|$ (mod n), if $\gamma = (x = x_0, x_1, \ldots, x_t = y)$ is a path of length t from x to y, then

$$P^t(x, y) \geq \mathbb{P}_x[(X_0, \ldots, X_t) = \gamma] = 2^{-t} > 0.$$

Note that at each step, the Markov chain moves from the current position $+1$ or -1 (mod n). Thus, since n is even, at even times the chain must be at even vertices, and at odd times the chain must be at odd vertices.

Thus, it is not true that there exists $t > 0$ such that for all x, y, $P^t(x, y) > 0$.

The main reason for this is that the chain has a *period*: at even times it is on some set, and at odd times on a different set. Similarly, the chain cannot be back at its starting point at odd times, only at even times. △▽△

Definition 3.2.3 Let P be a transition matrix of a Markov chain on G.

- A state x is called **periodic** if $\gcd \{t \geq 1 : P^t(x, x) > 0\} > 1$, and this gcd is called the period of x.
- If $\gcd \{t \geq 1 : P^t(x, x) > 0\} = 1$ then x is called **aperiodic**.
- P is called **aperiodic** if all $x \in G$ are aperiodic. Otherwise P is called periodic.

Example 3.2.4 Note that in the even-length cycle example,

$$\gcd \{t \geq 1 : P^t(x, x) > 0\} = \gcd \{2, 4, 6, \ldots\} = 2. \qquad △▽△$$

Remark 3.2.5 If P is periodic, then there is an easy way to "fix" P to become aperiodic: namely, let $Q = \alpha I + (1 - \alpha)P$ be some "lazy version" of P. Here I is the identity matrix, and $\alpha \in (0, 1)$. Then $Q(x, x) \geq \alpha$ for all x, and thus Q is aperiodic.

Such a Q is called a *lazy version* of P because it basically is the same Markov chain, only that at some times it stays put (with probability α).

Proposition 3.2.6 *Let P be a transition matrix of a Markov chain on state space G.*

- *x is aperiodic if and only if there exists $t(x)$ such that for all $t > t(x)$, $P^t(x, x) > 0$.*
- *If P is irreducible, then P is aperiodic if and only if there exists an aperiodic state x.*
- *Consequently, if P is irreducible and aperiodic, and if G is finite, then there exists t_0 such that for all $t > t_0$ all x, y admit $P^t(x, y) > 0$.*

Proof We start with the first assertion. Assume that x is aperiodic. Let $R = \{t \geq 1 : P^t(x, x) > 0\}$. Since $P^{t+s}(x, x) \geq P^t(x, x)P^s(x, x)$ we get that $t, s \in R$ implies $t + s \in R$; that is, R is closed under addition. A number theoretic result tells us that since $\gcd R = 1$ it must be that R^c is finite.

The other direction is simpler. If R^c is finite, then R contains two consecutive integers $n, n + 1 \in R$, so $\gcd R = \gcd(n, n + 1) = 1$.

For the second assertion, if P is irreducible and x is aperiodic, then let $t(x)$ be such that for all $t > t(x)$, $P^t(x, x) > 0$. For any z, y let $t(z, y)$ be such that $P^{t(z,y)}(z, y) > 0$ (which exists by irreducibility). Then, for any $t > t(y, x) + t(x) + t(x, y)$ we get that

$$P^t(y, y) \geq P^{t(y,x)}(y, x)P^{t-t(y,x)-t(x,y)}(x, x)P^{t(x,y)}(x, y) > 0.$$

So for all large enough t, $P^t(y, y) > 0$, which implies that y is aperiodic. This holds for all y, so P is aperiodic.

The other direction is trivial from the definition.

For the third assertion, for any z, y let $t(z, y)$ be such that $P^{t(z,y)}(z, y) > 0$. Let $T = \max_{z,y} \{t(z, y)\}$. Let x be an aperiodic state and let $t(x)$ be such that for all $t > t(x)$, $P^t(x, x) > 0$. We get that for any $t > 2T + t(x)$ we have that $t - t(z, x) - t(x, z) \geq t - 2T > t(x)$, so

$$P^t(z, y) \geq P^{t(z,x)}(z, x)P^{t-t(z,x)-t(x,z)}(x, x)P^{t(x,z)}(x, z) > 0. \qquad \square$$

Exercise 3.4 Let G be a finite connected graph, and let Q be the lazy random walk on G with holding probability α; that is, $Q = \alpha I + (1 - \alpha)P$ where $P(x, y) = \frac{1}{\deg(x)}$ if $x \sim y$ and $P(x, y) = 0$ if $x \not\sim y$.

Show that Q is aperiodic. Show that for

$$\text{diam}(G) = \max \{\text{dist}(x, y) : x, y \in G\},$$

we have that all $t > \text{diam}(G)$, and all $x, y \in G$ admit $Q^t(x, y) > 0$.

3.3 Random Walks on Groups

The main type of Markov chain we will be studying is the *random walk on G* where G is a finitely generated group.

Let μ be a probability measure on a finitely generated group G. Recall that we defined the μ-random walk as the process $(X_t)_t$ defined by $X_t = X_0 U_1 \cdots U_t$, where X_0 has some distribution on G, and $(U_t)_{t \geq 1}$ are i.i.d.-μ.

Exercise 3.5 Let G be a finitely generated group and let μ be a probability measure on G. Let $P(x, y) = \mu\left(x^{-1}y\right)$.

* Show that the μ-random walk $(X_t)_t$ is a Markov chain with transition matrix P.
* Show that P is irreducible if and only if μ is adapted.
* Show that if $\mu(1) > 0$ then P is aperiodic.

▷ solution ◁

Example 3.3.1 Let G be a finitely generated group, let S be a finite symmetric generating set. Let $\mu(x) = \frac{1}{|S|}\mathbf{1}_{\{x \in S\}}$. This is called the *simple random walk* on the Cayley graph of G with respect to S.

△▽△

Example 3.3.2 Let $\mathbb{Z} = \langle -1, 1 \rangle$, and consider the simple random walk. So $\mu(1) = \mu(-1) = \frac{1}{2}$.

Note that for any sequence of integers, $z_0 = 0, z_1, \ldots, z_n$ such that $|z_{k+1} - z_k| = 1$, we have that

$$\mathbb{P}_0[X_1 = z_1, \ldots, X_n = z_n] = \prod_{k=1}^{n} P(z_k, z_{k+1}) = \prod_{k=1}^{n} \mu(z_{k+1} - z_k) = 2^{-n}.$$

It may be noted that if we set $U_k = X_k - X_{k-1}$, then the sequence $(U_k)_{k \geq 1}$ are i.i.d. with distribution μ. So $X_t = \sum_{k=1}^{t} U_k$ is the sum of i.i.d. random variables.

△▽△

3.4 Stopping Times

If $(X_t)_t$ is a Markov chain on G with transition matrix P, we can define the following. Let $A \subset G$. Define

$$T_A = \inf\{t \geq 0 : X_t \in A\} \qquad \text{and} \qquad T_A^+ = \inf\{t \geq 1 : X_t \in A\}.$$

These are the *hitting time* of A and *return time* to A, respectively. (We use the convention that $\inf \emptyset = \infty$.) If $A = \{x\}$ we write $T_x = T_{\{x\}}$ and similarly $T_x^+ = T_{\{x\}}^+$.

The hitting and return times have the property that their value can be determined by the history of the chain; that is, the event $\{T_A \leq t\}$ is determined by

$(X_0, X_1, \ldots, X_t)$. Recall that this is precisely the definition of a *stopping time* (with respect to the canonical filtration $\mathcal{F}_t = \sigma(X_0, \ldots, X_t)$).

Exercise 3.6 Show that $\{T_A \le t\}$ and $\{T_A^+ \le t\}$ are events in $\mathcal{F}_t = \sigma(X_0, \ldots, X_t)$, for any $t \ge 0$.

That is, show that T_A, T_A^+ are stopping times.

Example 3.4.1 Consider the simple random walk on $\mathbb{Z}^3$. Let

$$T = \sup\{t : X_t = 0\}.$$

This is the *last* time the walk is at 0. One can show that T is a.s. finite, and we will prove this in Exercise 4.39. However, T is not a stopping time, since, for example,

$$\{T \le 0\} = \{\forall\, t > 0, \ X_t \neq 0\} = \bigcap_{t=1}^{\infty} \{X_t \neq 0\} \notin \sigma(X_0). \qquad \triangle \,\triangledown\, \triangle$$

Example 3.4.2 Let $(X_t)_t$ be a Markov chain and let $T = \inf\{t \ge T_A : X_t \in A'\}$, where $A, A' \subset S$. Then T is a stopping time, since

$$\{T \le t\} = \bigcup_{k=0}^{t} \bigcup_{m=0}^{k} \{X_m \in A, X_k \in A'\}. \qquad \triangle \,\triangledown\, \triangle$$

Exercise 3.7 Let T, T' be stopping times. Show that the following holds

- Any constant $t \in \mathbb{N}$ is a stopping time.
- $T \wedge T'$ and $T \vee T'$ are stopping times.
- $T + T'$ is a stopping time. $\triangleright$ solution $\triangleleft$

3.4.1 Conditioning on a Stopping Time

We have already seen important uses for stopping times in the optional stopping theorem (Theorem 2.3.3). Another important property we want is the *strong Markov property*.

For a fixed time t, the Markov property tells us that the process $(X_{t+n})_n$ is a Markov chain with starting distribution X_t, independent of $\sigma(X_0, \ldots, X_t)$. We want to do the same thing for stopping times.

Let T be a stopping time. The information captured by $X_0, \ldots, X_T$, is the σ-algebra $\sigma(X_0, \ldots, X_T)$. This is defined as the collection of all events A such that for all t, $A \cap \{T \le t\} \in \sigma(X_0, \ldots, X_t)$. That is,

$$\sigma(X_0, \ldots, X_T) = \{A : A \cap \{T \le t\} \in \sigma(X_0, \ldots, X_t) \text{ for all } t\}.$$

One can check that this is indeed a σ-algebra.

Exercise 3.8 Show that $\sigma(X_0, \ldots, X_T)$ is a σ-algebra.

Important examples are:

- For any t, $\{T \leq t\} \in \sigma(X_0, \ldots, X_T)$.
- Thus, T is measurable with respect to $\sigma(X_0, \ldots, X_T)$.
- $X_T \mathbf{1}_{\{T < \infty\}}$ is measurable with respect to $\sigma(X_0, \ldots, X_T)$. Indeed, for all $t \geq 0$ and any $x \in G$, we have that $\{X_T = x, T \leq t\} \in \sigma(X_0, \ldots, X_t)$.

Theorem 3.4.3 (Strong Markov property) *Let $(X_t)_t$ be a Markov chain on G with transition matrix P, and let T be a stopping time. For all $t \geq 0$, define $Y_t = X_{T+t}$. Then, conditioned on $T < \infty$ and X_T, the sequence $(Y_t)_t$ is independent of $\sigma(X_0, \ldots, X_T)$, and is a Markov chain started at $Y_0 = X_T$, and with transition matrix P.*

That is, for any $A \in \sigma(X_0, \ldots, X_T)$ and $x \in G$, we have

$$\mathbb{P}[X_{T+t} = y, \ X_T = x, \ A, \ T < \infty] = P^t(x, y) \cdot \mathbb{P}[X_T = x, \ A, \ T < \infty]$$

for all $t \geq 0$.

Note that when writing $X_T = x$, we implicitly assume that this implies that $T < \infty$, since X_∞ is not defined.

Proof The (regular) Markov property tells us that for any $m > k$, and any event $A \in \sigma(X_0, \ldots, X_k)$,

$$\mathbb{P}[X_m = y, \ A, \ X_k = x] = P^{m-k}(x, y) \, \mathbb{P}[A, \ X_k = x].$$

Since $T + t$ is also a stopping time, to prove the theorem it suffices to show that for all t, and any $A \in \sigma(X_0, \ldots, X_T)$,

$$\mathbb{P}[X_{T+t+1} = y, \ X_{T+t} = x, \ A, \ T < \infty] = P(x, y) \, \mathbb{P}[X_{T+t} = x, \ A, \ T < \infty].$$

Note that $A \cap \{T = k\} \in \sigma(X_0, \ldots, X_k) \subset \sigma(X_0, \ldots, X_{k+t})$ for all k, so

$$\mathbb{P}[X_{T+t+1} = y, \ A, \ X_{T+t} = x, \ T < \infty]$$

$$= \sum_{k=0}^{\infty} \mathbb{P}[X_{k+t+1} = y, \ X_{k+t} = x, \ A, \ T = k]$$

$$= \sum_{k=0}^{\infty} P(x, y) \, \mathbb{P}[X_{k+t} = x, \ A, \ T = k]$$

$$= P(x, y) \, \mathbb{P}[X_{T+t} = x, A, \ T < \infty]. \qquad \square$$

3.5 Excursion Decomposition

We now use the strong Markov property to prove the following.

Example 3.5.1 Let P be the transition matrix of an irreducible Markov chain $(X_t)_t$ on G. Fix $x \in G$.

Define inductively the following stopping times: $T_x^{(0)} = 0$ and

$$T_x^{(k)} = \inf \left\{ t \geq T_x^{(k-1)} + 1 : X_t = x \right\}.$$

So $T_x^{(k)}$ is the time of the kth return to x.

Let $V_t(x)$ be the number of visits to x up to time t; that is, $V_t(x) = \sum_{k=1}^{t} \mathbf{1}_{\{X_k = x\}}$.

It is immediate that $V_t(x) \geq k$ if and only if $T_x^{(k)} \leq t$.

Now let us look at the excursions to x: The kth excursion is

$$X \left[T_x^{(k-1)}, T_x^{(k)} \right] = \left(X_{T_x^{(k-1)}}, X_{T_x^{(k-1)}+1}, \ldots, X_{T_x^{(k)}} \right).$$

These excursions are paths of the Markov chain ending at x and starting at x (except, possibly, the first excursion, which starts at X_0).

For $k > 0$ define $\tau_x^{(k)} = T_x^{(k)} - T_x^{(k-1)}$ if $T_x^{(k)} < \infty$ and 0 otherwise. When $T_x^{(k)} < \infty$, this is the length of the kth excursion.

The strong Markov property tells us that conditioned on $T_x^{(k-1)} < \infty$, the excursion $X \left[T_x^{(k-1)}, T_x^{(k)} \right]$, is independent of $\sigma \left(X_0, \ldots, X_{T_x^{(k-1)}} \right)$, and has the distribution of the first excursion $X[0, T_x^+]$ conditioned on $X_0 = x$.   ▵▿▵

The above gives rise to the following relation.

Theorem 3.5.2 *Let P be an irreducible Markov chain on G. Let $V_t(x)$ and $T_x^{(k)}$ be defined as in Example 3.5.1. Then,*

$$(\mathbb{P}_x[T_x^+ < \infty])^k = \mathbb{P}_x[V_\infty(x) \geq k] = \mathbb{P}_x \left[T_x^{(k)} < \infty \right].$$

Consequently,

$$1 + \mathbb{E}_x[V_\infty(x)] = \frac{1}{\mathbb{P}_x[T_x^+ = \infty]},$$

where $1/0 = \infty$.

Proof As in Example 3.5.1, $\{V_\infty(x) \geq k\} = \left\{ T_x^{(k)} < \infty \right\}$. We also saw in the example that for any m,

$$\mathbb{P} \left[T_x^{(m)} < \infty \mid T_x^{(m-1)} < \infty \right] = \mathbb{P} \left[\exists t \geq 1, \ X_{T_x^{(m-1)}+t} = x \mid T_x^{(m-1)} < \infty \right]$$
$$= \mathbb{P}_x[T_x^+ < \infty].$$

Since $\{T_x^{(m)} < \infty\} = \{T_x^{(m)} < \infty, T_x^{(m-1)} < \infty\}$, we can inductively conclude that

$$\mathbb{P}_x\left[T_x^{(k)} < \infty\right] = \mathbb{P}_x\left[T_x^{(k)} < \infty \mid T_x^{(k-1)} < \infty\right] \cdot \mathbb{P}\left[T_x^{(k-1)} < \infty\right]$$

$$= \cdots = (\mathbb{P}_x[T_x^+ < \infty])^k.$$

This proves the first assertion.

The second assertion follows from the fact that

$$1 + \mathbb{E}_x[V_\infty(x)] = \sum_{k=0}^\infty \mathbb{P}_x[V_\infty(x) \geq k] = \frac{1}{1 - \mathbb{P}_x[T_x^+ < \infty]},$$

where this holds even if $\mathbb{P}_x[T_x^+ < \infty] = 1$, interpreting $1/0 = \infty$. □

Exercise 3.9 Let $(X_t)_t$ be a Markov chain on G with transition matrix P. Assume that P is irreducible.

Let $Z \subset G$. Recall T_Z, the first hitting time of the set Z.

Let $x \notin Z$ and define $V = \sum_{t=0}^{T_Z} \mathbf{1}_{\{X_t = x\}}$ (so that $V = V_{T_Z}(x) + \mathbf{1}_{\{X_0 = x\}}$).

Show that under $\mathbb{P}_x$, the random variable V has geometric distribution with mean $1/p$, for $p = \mathbb{P}_x[T_Z < T_x^+]$.   ▷ solution ◁

3.6 Recurrence and Transience

Definition 3.6.1 Let P be a transition matrix of a Markov chain $(X_t)_t$ on G.

A state $x \in G$ is called **recurrent** if $\mathbb{P}_x[T_x^+ < \infty] = 1$. Otherwise, if $\mathbb{P}_x[T_x^+ = \infty] > 0$ we say that x is **transient**.

Excursion decomposition provides us with the following important characterization of recurrence in Markov chains.

Theorem 3.6.2 *The following are equivalent for an irreducible Markov chain* $(X_t)_t$ *on* G:

(1) x *is recurrent.*
(2) $\mathbb{P}_x[V_\infty(x) = \infty] = 1$.
(3) *For any state* y, $\mathbb{P}_x\left[T_y^+ < \infty\right] = 1$.
(4) $\mathbb{E}_x[V_\infty(x)] = \infty$.

Proof (1) ⟹ (2): If x is recurrent, then by definition $\mathbb{P}_x[T_x^+ < \infty] = 1$. So for any k, by Theorem 3.5.2 we have $\mathbb{P}_x[V_\infty(x) \geq k] = 1$. Taking k to infinity, we get that $\mathbb{P}_x[V_\infty(x) = \infty] = 1$.

(2) $\Rightarrow$ (3): Let $y \in G$. Let $E_k = X\left[T_x^{(k-1)}, T_x^{(k)}\right]$ be the kth excursion from x. We assumed that $\mathbb{P}_x\left[\forall\, k\; T_x^{(k)} < \infty\right] = \mathbb{P}_x[V_\infty(x) = \infty] = 1$. So under $\mathbb{P}_x$, all $(E_k)_k$ are independent and identically distributed.

Since P is irreducible, there exists $t > 0$ such that $\mathbb{P}_x[X_t = y,\, t < T_x^+] > 0$ (this is the subject of Exercise 3.10). Thus, we have that $p := \mathbb{P}_x[T_y < T_x^+] \geq \mathbb{P}_x[X_t = y,\, t < T_x^+] > 0$. This implies by the strong Markov property that

$$\mathbb{P}_x\left[T_y < T_x^{(k+1)} \mid T_y > T_x^{(k)},\, T_x^{(k)} < \infty\right] \geq p > 0.$$

So, using the fact that $\mathbb{P}_x\left[\forall\, k\; T_x^{(k)} < \infty\right] = 1$,

$$\mathbb{P}_x\left[T_y \geq T_x^{(k)}\right] = \mathbb{P}_x\left[T_y \geq T_x^{(k)} \mid T_y > T_x^{(k-1)},\, T_x^{(k-1)} < \infty\right] \cdot \mathbb{P}_x\left[T_y > T_x^{(k-1)}\right]$$
$$\leq (1-p) \cdot \mathbb{P}_x\left[T_y \geq T_x^{(k-1)}\right] \leq \cdots \leq (1-p)^k.$$

Thus,

$$\mathbb{P}_x[T_y^+ = \infty] \leq \mathbb{P}_x\left[\forall\, k,\, T_y \geq T_x^{(k-1)}\right] = \lim_{k \to \infty}(1-p)^k = 0.$$

(3) $\Rightarrow$ (1): If for any y we have $\mathbb{P}_x[T_y^+ < \infty] = 1$, then taking $y = x$ shows that x is recurrent.

This shows that (1),(2),(3) are equivalent.

It is obvious that (2) implies (4). Since $\mathbb{P}_x[T_x^+ = \infty] = \frac{1}{\mathbb{E}_x[V_\infty(x)]+1}$, we get that (4) implies (1). $\qquad\square$

Exercise 3.10 Show that if P is irreducible, for any $x \neq y$ there exists $t > 0$ such that $\mathbb{P}_x[X_t = y,\, t < T_x^+] > 0$. ▷ solution ◁

Exercise 3.11 Let P be an irreducible transition matrix of a recurrent Markov chain on state space G.

Show that for any starting distribution v and any non-empty subset $A \subset G$ it holds that $\mathbb{P}_v[T_A^+ < \infty] = 1$. ▷ solution ◁

Example 3.6.3 A gambler plays a fair game. Each round she wins a dollar with probability $1/2$, and loses a dollar with probability $1/2$; all rounds are independent. What is the probability that she never goes bankrupt, if she starts with N dollars?

We have already seen that this is just a simple random walk $(X_t)_t$ on $\mathbb{Z}$. Now, since $X_t = \sum_{k=1}^t U_k$ where $(U_k)_k$ are i.i.d., $\mathbb{P}[U_k = -1] = \mathbb{P}[U_k = 1] = \frac{1}{2}$.

Define

$$R_t = \#\{1 \leq k \leq t : U_k = 1\} = \sum_{k=1}^{t} 1_{\{U_k=1\}},$$

$$L_t = \#\{1 \leq k \leq t : U_k = -1\} = \sum_{k=1}^{t} 1_{\{U_k=-1\}} = t - R_t.$$

So $X_t - X_0 = R_t - L_t = 2R_t - t$. Also, R_t, L_t both have binomial-$(t, \frac{1}{2})$ distribution. Now, it is immediate that

$$\mathbb{P}_0[X_t = 0] = \mathbb{P}_0\left[R_t = \frac{t}{2}\right] = \begin{cases} 0 & t \text{ is odd,} \\ 2^{-t}\binom{t}{t/2} & t \text{ is even.} \end{cases}$$

Stirling's approximation (or other classical computations) tells us that

$$\lim_{t \to \infty} \sqrt{t} 2^{-t}\binom{t}{t/2} = c > 0$$

for some constant $c > 0$. So $\mathbb{P}_0[X_t = 0] \geq ct^{-1/2}$ for some $c' > 0$ and all even $t > 0$. This implies that

$$\mathbb{E}_0[V_\infty(0)] \geq \mathbb{E}_0[V_t(0)] \geq \sum_{k=1}^{\lfloor t/2 \rfloor} \mathbb{P}_0[X_{2k} = 0] \geq c''\sqrt{t}.$$

Thus, taking $t \to \infty$ we get that $\mathbb{E}_0[V_\infty(0)] = \infty$, and so 0 is recurrent.

Note that 0 here was not special, since all vertices look the same. This symmetry implies that $\mathbb{P}_x[T_x^+ < \infty] = 1$ for all $x \in \mathbb{Z}$. Thus, for any N, $\mathbb{P}_N[T_0^+ = \infty] = 0$. That is, no matter how much money the gambler starts with, she will always go bankrupt eventually.

Indeed we have already seen the recurrence of the simple random walk on $\mathbb{Z}$, as an application of the optional stopping theorem (Theorem 2.3.3). There we calculated for $0 < x < n$ that $\mathbb{P}_x[T_0 > T_n] = \frac{x}{n}$. Taking $n \to \infty$ proves that $\mathbb{P}_x[T_0 = \infty] = 0$. ▵▿▵

Corollary 3.6.4 Let P be an irreducible transition matrix on a state space G. Then, for any $x, y \in G$, x is transient if and only if y is transient.

Proof By irreducibility, for any pair of states z, w we can find $t(z, w) > 0$ such that $P^{t(z,w)}(z, w) > 0$.

Fix $x, y \in G$ and suppose that x is transient. For any $t > 0$,

$$P^{t+t(x,y)+t(y,x)}(x, x) \geq P^{t(x,y)}(x, y)P^t(y, y)P^{t(y,x)}(y, x).$$

Thus,

$$\mathbb{E}_y[V_\infty(y)] = \sum_{t=1}^{\infty} P^t(y, y)$$

$$\leq \frac{1}{P^{t(x,y)}(x, y)P^{t(y,x)}(y, x)} \cdot \sum_{t=1}^{\infty} P^{t+t(x,y)+t(y,x)}(x, x) < \infty.$$

So y is transient as well.   □

To conclude, we see that for an irreducible Markov chain *recurrence* and *transience* are properties of the chain and not of specific states.

We say an irreducible Markov chain is **recurrent** if at least one state, and hence all states, are recurrent, and **transient** if all states are transient.

Exercise 3.12 Let P be an irreducible transition matrix on a state space G. Let $Q = \sum_{k=0}^{\infty} \frac{1}{ek!} P^k$.

Show that Q is an irreducible stochastic matrix.

Show that if P is transient if and only if Q is transient.   ▷ solution ◁

∗ **Exercise 3.13** Let P be an irreducible transition matrix on a state space G. For every $x \in G$ let $p_x \in [0, 1)$. Define

$$Q(x, y) = p_x \mathbf{1}_{\{x=y\}} + (1 - p_x)P(x, y).$$

Show that Q is an irreducible Markov chain.

Show that Q is recurrent if and only if P is recurrent.   ▷ solution ◁

3.7 Positive Recurrence

We have seen that recurrence and transience are governed by the return time T_x^+ being finite a.s. or not. A natural question in the recurrent case is whether this time has finite expectation? This has to do with the theory of *stationary distributions*.

Theorem 3.7.1 *Let G be a finitely generated group and let μ be an adapted measure on G. The following are equivalent.*

(1) There exists $x \in G$ such that $\mathbb{E}_x[T_x^+] < \infty$.
(2) For all $x \in G$ we have $\mathbb{E}_x[T_x^+] < \infty$.
(3) G is a finite group.

The goal of this section is to prove this theorem.

Note that if $\mathbb{E}_x[T_x^+] < \infty$ then of course $\mathbb{P}_x[T_x^+ < \infty] = 1$. Thus, it makes sense to state the following definition.

If a state x satisfies $\mathbb{E}_x[T_x^+] < \infty$, we say that x is **positive recurrent**. Otherwise, if x is recurrent but $\mathbb{E}_x[T_x^+] = \infty$, we say that x is **null recurrent**.

3.7.1 Stationary Distributions

Suppose that P is a transition matrix of a Markov chain $(X_t)_t$ on state space G such that for some starting distribution ν, we have that $\mathbb{P}_\nu[X_n = x] \to \pi(x)$ for all $x \in G$, where π is some limiting distribution. One immediately checks that in this case we must have

$$\pi P(x) = \sum_y \lim_{n\to\infty} \nu P^n(y) P(y, x) = \lim_{n\to\infty} \nu P^{n+1}(x) = \pi(x),$$

or $\pi P = \pi$. (That is, π is a left eigenvector for P with eigenvalue 1.)

Definition 3.7.2 Let P be a transition matrix. If π is a distribution satisfying $\pi P = \pi$ then π is called a **stationary distribution**.

Example 3.7.3 Recall from Example 3.13 the two-state chain $P = \begin{bmatrix} 1-p & p \\ p & 1-p \end{bmatrix}$.

We saw that $P \to \frac{1}{2} \cdot \begin{bmatrix} 1 & 1 \\ 1 & 1 \end{bmatrix}$. Indeed, it is simple to check that $\pi = (1/2, 1/2)$ is a stationary distribution in this case. △▽△

Example 3.7.4 Consider a finite graph G. Let P be the transition matrix of a simple random walk on G. So $P(x, y) = \frac{1}{\deg(x)} \mathbf{1}_{\{x\sim y\}}$. Or $\deg(x)P(x, y) = \mathbf{1}_{\{x\sim y\}}$. Thus,

$$\sum_x \deg(x)P(x, y) = \deg(y).$$

So deg is a left eigenvector for P with eigenvalue 1. Since

$$\sum_x \deg(x) = \sum_x \sum_y \mathbf{1}_{\{x\sim y\}} = 2 \sum_{e \in E(G)} = 2|E(G)|,$$

we normalize $\pi(x) = \frac{\deg(x)}{2|E(G)|}$ to get a stationary distribution for P. △▽△

The above stationary distribution has a special property, known as the *detailed balance equation*.

A distribution π is said to satisfy the **detailed balance equation** with respect to a transition matrix P if for all states x, y,

$$\pi(x)P(x, y) = \pi(y)P(y, x).$$

Exercise 3.14 Show that if π satisfies the detailed balance equation, then π is a stationary distribution.

Give an example of an irreducible and aperiodic transition matrix P with a stationary distribution that does not satisfy the detailed balance equation.

▷ solution ◁

3.7.2 Stationary Distributions and Hitting Times

There is a deep connection between stationary distributions and return times. The main result here is the following theorem.

Theorem 3.7.5 *Let P be an irreducible transition matrix on state space G. Then the following are equivalent:*

- *P has a stationary distribution π.*
- *Every x is positive recurrent.*
- *Some x is positive recurrent.*
- *P has a unique stationary distribution, $\pi(x) = \frac{1}{\mathbb{E}_x[T_x^+]}$.*

The proof of this theorem goes through the following few lemmas.

In the next lemma we will consider a function (vector) $v \colon G \to [0, \infty]$. Although it may take the value ∞, since we are only dealing with nonnegative numbers we can write $vP(x) = \sum_y v(y)P(y, x)$ without confusion (with the convention that $0 \cdot \infty = 0$).

Lemma 3.7.6 *Let P be an irreducible transition matrix on a state space G. Let $v \colon G \to [0, \infty]$ be such that $vP = v$. Then:*

- *If there exists a state x such that $v(x) < \infty$ then $v(y) < \infty$ for all states y.*
- *If v is not the zero vector, then $v(y) > 0$ for all states y.*

Note that this implies that if π is a stationary distribution then all the entries of π are strictly positive and finite.

Proof Assume that $v(x) < \infty$. For any t, using the fact that $v \geq 0$,

$$v(x) = \sum_z v(z)P^t(z, x) \geq v(y)P^t(y, x).$$

Thus, for a suitable choice of t, since P is irreducible, we know that $P^t(y, x) > 0$, and so $v(y) \leq \frac{v(x)}{P^t(y,x)} < \infty$.

For the second assertion, if v is not the zero vector, since it is nonnegative, there exists a state x such that $v(x) > 0$. Thus, for any state y and for t such that $P^t(x, y) > 0$, we get

$$v(y) = \sum_z v(z)P^t(z, y) \geq v(x)P^t(x, y) > 0. \qquad \square$$

Recall that for a Markov chain $(X_t)_t$ we denote by $V_t(x) = \sum_{k=1}^t \mathbf{1}_{\{X_k = x\}}$ the number of visits to x.

Lemma 3.7.7 *Let $(X_t)_t$ be an irreducible Markov chain with transition matrix P. Let μ be some starting distribution. Assume T is a stopping time such that*

$$\mathbb{P}_\mu[X_T = x] = \mu(x), \qquad \text{for all } x.$$

Assume further that $1 \leq T < \infty$ $\mathbb{P}_\mu$-a.s. Let $v(x) = \mathbb{E}_\mu[V_T(x)]$.

Then, $vP = v$. Moreover, if $\mathbb{E}_\mu[T] < \infty$ then P has a stationary distribution $\pi(x) = \frac{v(x)}{\mathbb{E}_\mu[T]}$.

Proof The assumptions on T give that for any j,

$$\mathbb{P}_\mu[X_0 = y] = \mu(y) = \mathbb{P}_\mu[X_T = y] = \sum_{j=1}^\infty \mathbb{P}_\mu[X_j = y, T = j],$$

$$\sum_{j=0}^\infty \mathbb{P}_\mu[X_j = y, T > j] = \mathbb{P}_\mu[X_0 = y] + \sum_{j=1}^\infty \mathbb{P}_\mu[X_j = y, T > j]$$

$$= \sum_{j=1}^\infty (\mathbb{P}_\mu[X_j = y, T = j] + \mathbb{P}_\mu[X_j = y, T > j])$$

$$= \sum_{j=1}^\infty \mathbb{P}_\mu[X_j = y, T \geq j] = v(y).$$

Thus we have that

$$v(x) = \sum_{j=1}^\infty \mathbb{P}_\mu[X_j = x, T \geq j] = \sum_{j=0}^\infty \mathbb{P}_\mu[X_{j+1} = x, T > j]$$

$$= \sum_{j=0}^\infty \sum_y \mathbb{P}_\mu[X_{j+1} = x, X_j = y, T > j]$$

$$= \sum_y \sum_{j=0}^\infty \mathbb{P}_\mu[X_j = y, T > j]P(y, x) = (vP)(x).$$

That is, $vP = v$.

Since

$$\sum_x v(x) = \mathbb{E}_\mu \left[\sum_x V_T(x) \right] = \mathbb{E}_\mu[T]$$

if $\mathbb{E}_\mu[T] < \infty$, then $\pi(x) = \frac{v(x)}{\mathbb{E}_\mu[T]}$ defines a stationary distribution.   □

Example 3.7.8 Consider an irreducible Markov chain $(X_t)_t$ with transition matrix P, and let $v(y) = \mathbb{E}_x \left[V_{T_x^+}(y) \right]$. If x is recurrent, then $\mathbb{P}_x$-a.s. we have $1 \le T_x^+ < \infty$, and $\mathbb{P}_x \left[X_{T_x^+} = y \right] = \mathbf{1}_{\{y=x\}} = \mathbb{P}_x[X_0 = y]$. So we conclude that $vP = v$. Since $\mathbb{P}_x$-a.s. $V_{T_x^+}(x) = 1$, we have that $0 < v(x) = 1 < \infty$, so $0 < v(y) < \infty$ for all y.

Note that although it may be that $\mathbb{E}_x \left[T_x^+ \right] = \infty$, that is x is null-recurrent, we still have that for any y, $\mathbb{E}_x \left[V_{T_x^+}(y) \right] < \infty$; that is, the expected number of visits to y until returning to x is finite.

If x is positive recurrent, then $\pi(y) = \frac{\mathbb{E}_x \left[V_{T_x^+}(y) \right]}{\mathbb{E}_x[T_x^+]}$ is a stationary distribution for P.   △▽△

This vector plays a special role, as in the next lemma.

Lemma 3.7.9 *Let P be an irreducible transition matrix. Let $u(y) = \mathbb{E}_x \left[V_{T_x^+}(y) \right]$. Let $v \ge 0$ be a nonnegative vector such that $vP = v$, and $v(x) = 1$. Then, $v \ge u$. Moreover, if x is recurrent, then $v = u$.*

Proof   Let $(X_t)_t$ denote the Markov chain.

If $y = x$ then $v(x) = 1 \ge u(x)$, so we can assume that $y \ne x$.

We will prove by induction that for all t, for any $y \ne x$,

$$\sum_{k=1}^t \mathbb{P}_x \left[X_k = y, T_x^+ \ge k \right] \le v(y). \tag{3.1}$$

Indeed, for $t = 1$ this is just

$$\mathbb{P}_x \left[X_1 = y, T_x^+ \ge 1 \right] = P(x, y) \le \sum_z v(z) P(z, y) = v(y),$$

since $v \ge 0$, $v(x) = 1$, and $y \ne x$.

For general $t > 0$, we rely on the fact that by the Markov property, for any $y \ne x$,

$$\mathbb{P}_x \left[X_{k+1} = y, T_x^+ \ge k + 1 \right] = \sum_{z \ne x} \mathbb{P}_x \left[X_{k+1} = y, X_k = z, T_x^+ \ge k \right]$$

$$= \sum_{z \ne x} \mathbb{P}_x \left[X_k = z, T_x^+ \ge k \right] P(z, y).$$

So by induction,

$$\sum_{k=1}^{t+1} \mathbb{P}_x [X_k = y, T_x^+ \geq k] = P(x, y) + \sum_{k=1}^{t} \mathbb{P}_x [X_{k+1} = y, T_x^+ \geq k + 1]$$

$$= P(x, y) + \sum_{z \neq x} P(z, y) \sum_{k=1}^{t} \mathbb{P}_x [X_k = z, T_x^+ \geq k]$$

$$\leq P(x, y) + \sum_{z \neq x} P(z, y) v(z)$$

$$= \sum_{z} v(z) P(z, y) = v(y).$$

This completes a proof of (3.1) by induction.

Now, one notes that the left-hand side of (3.1) is just the expected number of visits to y started at x, up to time $T_x^+ \wedge t$. Taking $t \to \infty$, using monotone convergence,

$$v(y) \geq \sum_{k=1}^{t} \mathbb{P}_x [X_k = y, T_x^+ \geq k] = \mathbb{E}_x[V_{T_x^+ \wedge t}(y)] \nearrow u(y).$$

This proves that $v \geq u$.

Since x is recurrent, we have $uP = u$, and $u(x) = 1 = v(x)$. We have seen that $v - u \geq 0$, and of course $(v - u)P = v - u$. Until now we have not actually used irreducibility; we will use this to show that $v - u = 0$. Indeed, let y be any state. If $v(y) > u(y)$ then $v - u$ is a nonzero nonnegative left eigenvector for P, so must be positive everywhere. This contradicts $v(x) - u(x) = 0$. So it must be that $v - u \equiv 0$. □

We are now in good shape to prove Theorem 3.7.5.

Proof of Theorem 3.7.5 Assume that π is a stationary distribution for P. Fix any state x. Recall that $\pi(x) > 0$. Define the vector $v(z) = \frac{\pi(z)}{\pi(x)}$. We have that $v \geq 0$, $vP = v$, and $v(x) = 1$. Hence, $v(z) \geq \mathbb{E}_x \left[V_{T_x^+}(z) \right]$ for all z. That is,

$$\mathbb{E}_x [T_x^+] = \sum_{y} \mathbb{E}_x \left[V_{T_x^+}(y) \right] \leq \sum_{y} v(y) = \sum_{y} \frac{\pi(y)}{\pi(x)} = \frac{1}{\pi(x)} < \infty.$$

So x is positive recurrent. This holds for a generic x.

The second bullet of course implies the third.

Now assume some state x is positive recurrent. Let $v(y) = \mathbb{E}_x \left[V_{T_x^+}(y) \right]$. Since x is recurrent, we know that $vP = v$ and $\sum_y v(y) = \mathbb{E}_x[T_x^+] < \infty$. So $\pi = \frac{v}{\mathbb{E}_x[T_x^+]}$ is a stationary distribution for P.

Since P has a stationary distribution, by the first implication all states are positive recurrent. Thus, for any state z, if $v = \frac{\pi}{\pi(z)}$ then $vP = v$ and $v(z) = 1$. So z being recurrent we get that $v(y) = \mathbb{E}_z[V_{T_z^+}(y)]$ for all y. Specifically,

$$\mathbb{E}_z[T_z^+] = \sum_y v(y) = \frac{1}{\pi(z)},$$

which holds for all states z.

For the final implication, if P has a specific stationary distribution, then of course it has a stationary distribution.   □

Corollary 3.7.10 (Stationary distributions are unique) If an irreducible Markov chain P has two stationary distributions π and π', then $\pi = \pi'$.

Exercise 3.15 Let P be an irreducible transition matrix. Show that for positive recurrent states x, y,

$$\mathbb{E}_x\left[V_{T_x^+}(y)\right]\mathbb{E}_y\left[V_{T_y^+}(x)\right] = 1.$$ ▷ solution ◁

Exercise 3.16 Let P be an irreducible transition matrix on a state space G. Let $A \subset G$, and consider the return time T_A^+.

Show that if P is positive recurrent then $\mathbb{E}_v\left[T_A^+\right] < \infty$, for any non-empty subset $A \subset G$ and any starting distribution v.   ▷ solution ◁

Example 3.7.11 Consider the following Markov chain on state space $\mathbb{N}$: the transition probabilities are given by $P(x, x+1) = 1 - P(x, x-1) = p$ for $x > 0$ and $P(0, 1) = 1$.

Assume that $p < \frac{1}{2}$. Define

$$\pi(x) = \begin{cases} c & x = 0, \\ c\left(\frac{p}{1-p}\right)^x \frac{1}{p} & x > 0, \end{cases}$$

for appropriate constant $c > 0$, chosen so that $\sum_x \pi(x) = 1$.

Compute, for $x > 1$,

$$\pi P(x) = \pi(x-1)P(x-1, x) + \pi(x+1)P(x+1, x)$$

$$= \frac{c}{p}\left(\frac{p}{1-p}\right)^{x-1} \cdot \left(p + (1-p)(\frac{p}{1-p})^2\right)$$

$$= \frac{c}{p}\left(\frac{p}{1-p}\right)^{x-1} \cdot \frac{p(1-p)+p^2}{1-p} = \pi(x).$$

Also, $\pi(0) = c = \pi(1)P(1, 0)$, and

$$\pi(1) = \frac{c}{1-p} = c + \frac{cp}{(1-p)^2}(1-p) = \pi(0)P(0, 1) + \pi(2)P(2, 1).$$

So π is a stationary distribution.

Thus, this Markov chain is positive recurrent. In fact, we immediately have the formula $\mathbb{E}_x\left[T_x^+\right] = \frac{1}{\pi(x)}$ (which decreases exponentially in x). Specifically, $\mathbb{E}_1[T_0] = \mathbb{E}_0\left[T_0^+\right] = \frac{1}{c}$.

It is not difficult to compute $c = \frac{1-2p}{2(1-p)}$. △▽△

Theorem 3.7.5 shows that the property of positive recurrence is a property of the whole (irreducible) Markov chain, not just a specific state. So we conclude that we may call an irreducible Markov chain **positive recurrent** if there exists a positive recurrent state, and we call the chain **null recurrent** if there exists a null recurrent state.

We are now ready to prove Theorem 3.7.1.

Proof of Theorem 3.7.1 By Theorem 3.7.5 we have that (1) $\Longleftrightarrow$ (2).

Now, note that if $(X_t)_t$ is a μ-random walk on G with $X_0 = x$, then $(Y_t = yX_t)_t$ is a μ-random walk on G with $Y_0 = yx$. Thus, $\mathbb{E}_x\left[T_x^+\right] = \mathbb{E}_{yx}\left[T_{yx}^+\right]$ for any $y \in G$, which implies that the quantity $\mathbb{E}_x\left[T_x^+\right] = \mathbb{E}\left[T_1^+\right]$ is a constant independent of x.

Hence, the vector $\pi(x) = \frac{1}{\mathbb{E}_x[T_x^+]}$ is a strictly positive left eigenvector of P_μ if and only if $\mathbb{E}_x\left[T_x^+\right] = \mathbb{E}\left[T_1^+\right] < \infty$. Also, since π is a nonzero constant vector, it is a distribution (i.e. $\sum_x \pi(x) = 1$) if and only if G is finite. So by Theorem 3.7.5, $(X_t)_t$ is positive recurrent if and only if G is finite, which is (2) $\Longleftrightarrow$ (3). □

Exercise 3.17 Show that if $(X_t)_t$ is a μ-random walk on a group G with $X_0 = x$, then $(Y_t = yX_t)_t$ is a μ-random walk on G with $Y_0 = yx$.

3.7.3 Summary

Let us sum up what we know so far about irreducible chains. If P is an irreducible transition matrix, then:

- $\mathbb{E}_x[V_\infty(x)] + 1 = \frac{1}{\mathbb{P}_x[T_x^+=\infty]}$.
- For all states x, y, we have that x is transient if and only if y is transient.
- If P is recurrent, the vector $v(z) = \mathbb{E}_x\left[V_{T_x^+}(z)\right]$ is a positive left eigenvector for P, and any nonnegative left eigenvector for P is proportional to v.
- P has a stationary distribution if and only if P is positive recurrent.
- If P is positive recurrent then there exists a unique stationary distribution π satisfying $\pi(x)\,\mathbb{E}_x\left[T_x^+\right] = 1$.

Example 3.7.12 Consider the random walk on $\mathbb{Z}$, given by taking $\mu(1) = p = 1 - \mu(-1)$. So that $P(x, x+1) = p$ and $P(x, x-1) = 1 - p$ for all x.

Suppose $vP = v$. Then, $v(x) = v(x-1)p + v(x+1)(1-p)$, or $v(x+1) = \frac{1}{1-p}(v(x) - pv(x-1))$. Solving such recursions is simple: Set $u_x =$

$\left[v(x+1) \quad v(x)\right]^T$. So $u_{x+1} = \frac{1}{1-p} A u_x$ where

$$A = \begin{bmatrix} 1 & -p \\ 1-p & 0 \end{bmatrix}.$$

Since the characteristic polynomial of A is $\lambda^2 - \lambda + p(1-p) = (\lambda - p)(\lambda - (1-p))$, we get that the eigenvalues of A are p and $1 - p$. One can easily check that A is diagonalizable, and so

$$v(x) = u_x(2) = (1-p)^{-x}(A^x u_0)(2) = (1-p)^{-x} \cdot [0\,1] M D M^{-1} u_0 = a\left(\frac{p}{1-p}\right)^x + b,$$

where D is diagonal with $p, 1 - p$ on the diagonal, and a, b are constants that depend on the matrix M and on u_0 (but are independent of x).

Thus, $\sum_{x \in \mathbb{Z}} v(x)$ will only converge for $a = 0, b = 0$, which gives $v = 0$. That is, there is no stationary distribution, and P is not positive recurrent.

In Exercise 4.27 we will in fact see that P is transient for $p \neq 1/2$, and for $p = 1/2$ we have already seen that P is null-recurrent, since this is the simple random walk on the Cayley graph of $\mathbb{Z}$ with respect to the generating set $\{-1, 1\}$. △▽△

Example 3.7.13 This is a solution to an exercise from Aldous and Fill (2002).

A chess knight moves on a chess board; each step it chooses uniformly among the possible moves. Suppose the knight starts at the corner. What is the expected time it takes the knight to return to its starting point?

At first, this looks difficult...

However, let G be the graph whose vertices are the squares of the chess board, $V(G) = \{1, 2, \ldots, 8\}^2$. Let $x = (1, 1)$ be the starting point of the knight. For edges, we will connect two vertices if the knight can jump from one to the other in a legal move.

Thus, for example, a vertex in the "center" of the board has 8 adjacent vertices. A corner, on the other hand has 2 adjacent vertices. In fact, we can determine the degree of all vertices.

legal moves:

	*		*				
*				*			
		0					
*				*			
	*		*				

$\Rightarrow$

2	3	4	4	4	4	3	2
3	4	6	6	6	6	4	3
4	6	8	8	8	8	6	4
4	6	8	8	8	8	6	4
4	6	8	8	8	8	6	4
4	6	8	8	8	8	6	4
3	4	6	6	6	6	4	3
2	3	4	4	4	4	3	2

Summing all the degrees, one sees that $2|E(G)| = 4 \cdot (4 \cdot 8 + 4 \cdot 6 + 5 \cdot 4 + 2 \cdot 3 + 2) = 4 \cdot 84 = 336$. Thus, the stationary distribution is $\pi(i, j) = \deg(i, j)/336$. Specifically, $\pi(x) = 2/336$ and so $\mathbb{E}_x[T_x^+] = 168$. ▵▽▵

3.7.4 Convergence to Stationarity

We have already seen that if $\mathbb{P}_v[X_t = x] \to \pi(x)$ for all $x \in G$, then π must be a stationary distribution. The following states that the converse is also true for aperiodic irreducible chains.

Theorem 3.7.14 *Let P be an aperiodic irreducible transition matrix of a Markov chain $(X_t)_t$ on a state space G.*

If P is positive recurrent, then for any starting distribution v the limit

$$\pi(x) = \lim_{t \to \infty} \mathbb{P}_v[X_t = x]$$

exists, and π is the stationary distribution of P.

Proof   Since P is positive recurrent, there exists a stationary distribution π.

Let $(X_t)_t$ be the Markov chain with transition matrix P, started with distribution v. Let $(Y_t)_t$ be an independent chain with transition matrix P on the same state space, started with distribution π.

Consider the process $Z_t = (X_t, Y_t) \in G \times G$. This is easily verified to be a Markov chain with transition matrix $Q((x, y), (x', y')) = P(x, x')P(y, y')$.

We use the fact that P is irreducible and aperiodic to prove that Q is irreducible. For any x, y, x', y' there exists t_0 such that for all $t > t_0$ we have

$$Q^t((x, y), (x', y')) = \mathbb{P}_{(x,y)}[(X_t, Y_t) = (x', y')]$$
$$= \mathbb{P}_x[X_t = x'] \cdot \mathbb{P}_y[Y_t = y'] = P^t(x, x')P^t(y, y') > 0.$$

Moreover, one checks readily that $\lambda(x, y) := \pi(x)\pi(y)$ is a stationary distribution for Q. Thus, by Theorem 3.7.5 we know that Q is positive recurrent. Specifically, if we fix some $x \in G$ then by Exercise 3.16 we have that $\mathbb{E}_{(v,\pi)}[T] < \infty$, where $T = \inf\{t \geq 0 : X_t = Y_t = x\}$, and (v, π) is the starting distribution of (X_0, Y_0). Specifically, $\mathbb{P}_{(v,\pi)}[T < \infty] = 1$.

Define

$$A_t = \begin{cases} X_t & \text{if } t \leq T, \\ Y_t & \text{if } t > T, \end{cases} \quad \text{and} \quad B_t = \begin{cases} Y_t & \text{if } t \leq T, \\ X_t & \text{if } t > T. \end{cases}$$

An important observation is that under $\mathbb{P}_{(x,x)}$, both chains $(X_t)_t$ and $(Y_t)_t$ have the exact same distribution and are independent. Thus, under $\mathbb{P}_{(x,x)}$, the chains $(X_t, Y_t)_t$ and $(Y_t, X_t)_t$ have the same distribution. The strong Markov property at time T implies that under $\mathbb{P}_{(v,\pi)}$, the chain $(A_{T+t}, B_{T+t})_t$ is independent of $(A_k, B_k)_{k \leq T}$, and has the same distribution as $(Y_t, X_t)_t$ under $\mathbb{P}_{(x,x)}$. Also, $(A_k, B_k)_{k \leq T} = (X_k, Y_k)_{k \leq T}$ by definition. We conclude that under $\mathbb{P}_{(v,\pi)}$, the following chains have the same distribution:

$$(A_t, B_t)_t = ((X_0, Y_0), \ldots, (X_T, Y_T), (Y_{T+1}, X_{T+1}), \ldots)$$

$$\text{and} \quad ((X_0, Y_0), \ldots, (X_T, Y_T), (X_{T+1}, Y_{T+1}), \ldots) = (X_t, Y_t)_t.$$

Specifically, $(A_t)_t$ and $(X_t)_t$ have the same distribution under $\mathbb{P}_{(v,\pi)}$. So $(A_t)_t$ is a Markov chain with transition matrix P.

Because π is a stationary distribution, $\mathbb{P}_{(v,\pi)}[Y_t = y] = \pi P^t(y) = \pi(y)$. We have that

$$\mathbb{P}_v[X_t = y] = \mathbb{P}_{(v,\pi)}[A_t = y]$$
$$= \mathbb{P}_{(v,\pi)}[X_t = y, \, T \geq t] + \mathbb{P}_{(v,\pi)}[Y_t = y, \, T < t]$$
$$= \mathbb{P}_{(v,\pi)}[X_t = y, \, T \geq t] - \mathbb{P}_{(v,\pi)}[Y_t = y, \, T \geq t] + \pi(y),$$

so

$$|\mathbb{P}_v[X_t = y] - \pi(y)| = |\mathbb{P}_{(v,\pi)}[X_t = y, \, T \geq t] - \mathbb{P}_{(v,\pi)}[Y_t = y, \, T \geq t]|$$
$$\leq \mathbb{P}_{(v,\pi)}[T \geq t] \to 0,$$

as $t \to \infty$, because $\mathbb{P}_{(v,\pi)}[T < \infty] = 1$. Since $P^t(x, y) = \mathbb{P}_x[X_t = y]$ this is a special case of the above with $v = \delta_x$. $\qquad \square$

Remark 3.7.15 Note that if P is transient, then $\sum_{t=0}^{\infty} P^t(x, y) < \infty$, so $P^t(x, y) \to 0$ for all x, y. When we study null recurrent chains, we will see in Theorem 3.8.1 (and Exercise 3.19) that $P^t(x, y) \to 0$ in the null recurrent case.

So we see that for irreducible and aperiodic chains, $(P^t(x, y))_t$ always converges. In the positive recurrent case it converges to the stationary distribution, whereas in the other cases it converges to 0.

3.8 Null Recurrence

We have seen above that an irreducible aperiodic Markov chain converges to a stationary distribution if and only if it is positive recurrent. What about null recurrent chains?

Exercise 3.18 Let P be an irreducible transition matrix. Show that the following are equivalent:

(1) $P^t(x, x) \to 0$ for all x.
(2) $P^t(x, x) \to 0$ for some x.

(3) $P^t(x, y) \to 0$ for some x, y.

(4) $P^t(x, y) \to 0$ for all x, y. ▷ solution ◁

Theorem 3.8.1 *Let P be an irreducible null recurrent Markov chain. Then,*
$P^t(x, y) \to 0$ *for all x, y.*

Proof We will prove this under the additional assumption that P is also aperiodic. Exercise 3.19 shows that this extends to the general irreducible null recurrent case.

Fix x. Set $L = \limsup_{t \to \infty} P^t(x, x)$. By Exercise 3.18, it suffices to show that $L = 0$.

Let $v(y) = \mathbb{E}_x \left[V_{T_x^+}(y) \right]$. We have already seen in Example 3.7.8 that $v > 0$ and $vP = v$.

If $\sum_y v(y) < \infty$ then P would be positive recurrent, so $\sum_y v(y) = \infty$ by assumption. Fix $\varepsilon > 0$ and choose a finite subset A of states such that $v(A) = \sum_{a \in A} v(a) > \varepsilon^{-1}$. Define a probability measure $v_A(y) = 1_{\{y \in A\}} \frac{v(y)}{v(A)}$. Then,

$$v_A P^t(y) \le \frac{1}{v(A)} vP^t(y) = \frac{v(y)}{v(A)} < \varepsilon v(y).$$

Since $v(x) = 1$, we have $v_A P^t(x) < \varepsilon$ for any t.

Now, since P is irreducible and aperiodic, and since A is finite, we know that there exists n such that for all $t \ge n$ we have $P^t(x, a) > 0$ for all $a \in A$. Let $0 < \delta < \varepsilon$ be small enough so that $P^n(x, a) - \delta v_A(a) > 0$ for all $a \in A$. Since $v_A(y) = 0$ for $y \notin A$, we may define a probability measure

$$\eta(y) = \frac{P^n(x, y) - \delta v_A(y)}{1 - \delta}.$$

By Exercise 3.11, $\mathbb{P}_\eta \left[T_x^+ = \infty \right] = 0$. So we may choose k large enough such that $\mathbb{P}_\eta \left[T_x^+ > k \right] < \delta^2$. Then for $t > k$,

$$\mathbb{P}_\eta[X_t = x] = \sum_{j=1}^{t} \mathbb{P}_\eta \left[T_x^+ = j \right] \cdot P^{t-j}(x, x)$$

$$\le \sum_{j=1}^{k} \mathbb{P}_\eta \left[T_x^+ = j \right] \cdot P^{t-j}(x, x) + \mathbb{P}_\eta \left[T_x^+ > k \right],$$

so

$$\limsup_{t \to \infty} \eta P^t(x) = \limsup_{t \to \infty} \mathbb{P}_\eta[X_t = x] < L + \delta^2.$$

However,

$$P^{t+n}(x,x) = \sum_y P^n(x,y)P^t(y,x) = (1-\delta)\eta P^t(x) + \delta v_A P^t(x)$$

$$< (1-\delta)\eta P^t(x) + \delta\varepsilon,$$

so taking a lim sup, we get that $L \le (1-\delta)\left(L + \delta^2\right) + \delta\varepsilon$, implying that

$$L \le (1-\delta)\delta + \varepsilon < 2\varepsilon.$$

Since this holds for arbitrary $\varepsilon > 0$, we get that $\limsup_{t\to\infty} P^t(x,x) = 0$.   □

Exercise 3.19 Show that if P is irreducible and null recurrent, then $P^t(x,y) \to 0$ for any x,y.

▷ solution ◁

3.9 Finite Index Subgroups

In this section we will apply the optional stopping theorem and the basic theory of Markov chains to investigate the behavior of harmonic functions when moving to a finite index subgroup.

Let G be a finitely generated group with a symmetric adapted probability measure μ. Let $H \le G$ be a finite index subgroup.

Given the μ-random walk on G, we may consider the random walk on the coset space $(HX_t)_t$. Indeed, it is a Markov process: if $Hx = Hy$ then $x^{-1}H = y^{-1}H$ so

$$\mathbb{P}[HX_{t+1} = Hz \mid HX_t = Hx] = \mathbb{P}\left[U_{t+1} \in x^{-1}Hz\right] = \mathbb{P}\left[U_{t+1} \in y^{-1}Hz\right]$$

$$= \mathbb{P}[HX_{t+1} = Hz \mid HX_t = Hy].$$

So $(HX_t)_t$ is a Markov chain on G/H, which is a finite set. The transition matrix is given by

$$P(Hx, Hy) = \mathbb{P}[HX_1 = Hy \mid HX_0 = Hx] = \mathbb{P}\left[X_1 \in x^{-1}Hy\right].$$

Let us show that the uniform distribution is stationary for this Markov chain. For $n = [G : H]$, let $x_1, \ldots, x_n \in G$ be such that $G = \biguplus_{j=1}^n x_j^{-1}H$. Note that $Hx_j = Hx_i$ if and only if $x_j^{-1}H = x_i^{-1}H$, so that $G/H = \{Hx_1, \ldots, Hx_n\}$. Because $G = Gy$, we know that $G = \biguplus_{j=1}^n x_j^{-1}Hy$ for any $y \in G$. Hence,

$$\sum_{j=1}^n P(Hx_j, Hy) = \sum_{j=1}^n \mathbb{P}\left[X_1 \in x_j^{-1}Hy\right] = 1.$$

This Markov chain is irreducible because μ is adapted; indeed, $P^t(Hx, Hy) \geq \mathbb{P}_x[X_t = y] > 0$ for large enough t.

We have proved above, in the theory of finite state space Markov chains, that the stationary distribution is unique for an irreducible finite chain, and is exactly equal to the inverse of the expected return time; that is, we have shown the following proposition.

Proposition 3.9.1 *Let G be a finitely generated group and let μ be an adapted probability measure on G. Let $H \leq G$ be a finite index subgroup $[G : H] < \infty$. Recall $T_H^+ := \inf\{t \geq 1 : X_t \in H\}$, the return time to H, where $(X_t)_t$ is the μ-random walk on G. Then,*

$$\mathbb{E}_1\left[T_H^+\right] = [G : H].$$

Proof If we consider the random walk on G/H by projecting $(HX_t)_t$, we have that $T_H^+ = \inf\{t \geq 1 : HX_t = H\}$, which is the first return time of the projected walk to the origin. So $\mathbb{E}_1\left[T_H^+\right]$ is the reciprocal of the stationary distribution at $H = H1$, which by the above is exactly the number of states; that is, $\mathbb{E}_1\left[T_H^+\right] = [G : H]$. □

Exercise 3.20 Let G be a finitely generated group and let μ be an adapted probability measure on G. Let $H \leq G$ be a finite index subgroup $[G : H] < \infty$. Show that there exist constants $C, c > 0$ such that for any $x \in G$ and all t,

$$\mathbb{P}_x\left[T_H^+ > t\right] \leq Ce^{-ct}.$$

▷ solution ◁

Definition 3.9.2 Let G be a finitely generated group and μ an adapted probability measure on G. Let $H \leq G$ be a finite index subgroup $[G : H] < \infty$. Define a probability measure on H, called the **hitting measure** with respect to μ, by

$$\mu_H(x) := \mathbb{P}_1[X_{T_H^+} = x],$$

where T_H^+ is the return time to H.

This measure arises naturally in the study of harmonic functions. We will see that natural spaces of μ-harmonic functions on G are isomomorphic to the corresponding spaces of μ_H-harmonic functions on H, via the restriction function. Exercise 3.21 provides an example.

Exercise 3.21 Show that if $H \leq G$ is a finite index subgroup, $[G : H] < \infty$, and $f \in \mathrm{BHF}(G, \mu)$, then $f|_H \in \mathrm{BHF}(H, \mu_H)$.

▷ solution ◁

When studying the spaces $\mathrm{BHF}(G, \mu)$, we would often like to consider properties that are "geometric" in nature. That is, they should be invariant to changing the geometry slightly, for example passing to a finite index subgroup.

A naive choice may be to only consider measures with finite support. But this fails to be preserved when passing to a finite index subgroup.

Exercise 3.22 Consider the Cayley graph of $\mathbb{Z}^2$ with respect to the standard generating set $\{\pm(0, 1), \pm(1, 0)\}$. Let μ be the uniform measure on this generating set. Let $H = \{(2x, 2y) : x, y \in \mathbb{Z}\}$.

Show that H is a finite index subgroup in $\mathbb{Z}^2$.

Show that μ_H is not finitely supported. ▷ solution ◁

Contrary to the above, the next theorem shows that the category of symmetric, adapted, and exponential tail measures is invariant under moving to hitting measures on finite index subgroups, and is thus a natural choice for the geometric study of groups via their harmonic functions.

Exercise 3.23 Let G be a finitely generated group and μ an adapted probability measure on G. Let $H \leq G$ be a finite index subgroup $[G : H] < \infty$.

Show that the hitting measure μ_H is also adapted (to H).

Show that if μ is symmetric then μ_H is also symmetric. ▷ solution ◁

Theorem 3.9.3 *Let G be a finitely generated group and let μ be an adapted probability measure on G with an exponential tail. Let $H \leq G$ be a subgroup of finite index $[G : H] < \infty$. Let μ_H be the hitting measure on H.*

Then μ_H is also adapted and has an exponential tail.

Proof We have that μ_H is adapted to H because μ is adapted to G by Exercise 3.23. We only need to show that μ_H has an exponential tail.

By Exercise 3.20, we may choose $\delta > 0$ so that for some constant $K > 0$ we have $\mathbb{P}\left[T_H^+ \geq t\right] \leq K e^{-2\delta t}$ for any $t \geq 1$.

Let $(X_t)_t$ denote the μ-random walk, and let $U_t = X_{t-1}^{-1} X_t$ be the independent "jumps" of the walk. Since μ has an exponential tail we have that $\mathbb{E}\left[e^{\varepsilon |U_1|}\right] < \infty$ for some $\varepsilon > 0$. By dominated convergence, $\mathbb{E}\left[e^{\varepsilon |U_1|}\right] \to 1$ as $\varepsilon \to 0$. So we may choose $\varepsilon > 0$ small enough so that $\mathbb{E}\left[e^{2\varepsilon |U_1|}\right] \leq e^{\delta}$, for $\delta > 0$ as above.

Using Cauchy–Schwarz appropriately, we can calculate:

$$\mathbb{E}\left[e^{\varepsilon |X_{T_H^+}|}\right] \leq \sum_{t=1}^{\infty} \mathbb{E}\left[\mathbf{1}_{\{T_H^+ = t\}} \prod_{j=1}^{t} e^{\varepsilon |U_j|}\right]$$

$$\leq \sum_{t=1}^{\infty} \sqrt{\mathbb{P}\left[T_H^+ = t\right]} \cdot \left(\mathbb{E}\left[e^{2\varepsilon |U_1|}\right]\right)^{t/2} \leq \sum_{t=1}^{\infty} \sqrt{K} e^{-\delta t} \cdot e^{\delta t/2} < \infty.$$

So μ_H has an exponential tail. ☐

3.9.1 Random Walks with k Moments

We have seen that if $\mu \in SA(G, \infty)$, then $\mu_H \in SA(H, \infty)$ when $[G : H] < \infty$. In this section we will prove the slightly more complicated result stating that if $\mu \in SA(G, k)$ then $\mu_H \in SA(H, k)$ when $[G : H] < \infty$. That is, the property of the kth moment is preserved when passing to a finite index subgroup.

We require two lemmas first.

Lemma 3.9.4 *Let G be a finitely generated group and μ an adapted measure on G. Assume that μ has finite kth moment. Let $(X_t)_t$ be a μ-random walk on G. Then, there exists a constant $C = C(k, \mu) > 0$ such that for every $t \geq 0$, and for any $x \in G$,*

$$\mathbb{E}_x\left[|X_t|^k\right] \leq C \cdot (t + |x|)^k.$$

Proof By the triangle inequality we have $|X_t| \leq |X_0| + \sum_{j=1}^{t} |U_j|$, where $U_j = X_{j-1}^{-1} X_j$ is the jump at time j. So

$$\mathbb{E}\left[|X_t|^k\right] \leq \mathbb{E}\left[\left(\sum_{j=1}^{t} |U_j|\right)^k\right] = \sum_{\vec{j} \in \{1, \ldots, t\}^k} \mathbb{E}\left[\prod_{i=1}^{k} |U_{j_i}|\right]. \tag{3.2}$$

Note that by Jensen's inequality,

$$\max_{1 \leq n \leq k} \left(\mathbb{E}[|U_1|^n]\right)^{k/n} \leq \mathbb{E}\left[|U_1|^k\right] =: C_k.$$

It follows that

$$\mathbb{E}\left[\prod_{i=1}^{k} |U_{j_i}|\right] \leq C_k \text{ for all } \vec{j} \in \{1, \ldots, t\}^k. \tag{3.3}$$

It now follows from (3.2) and (3.3) that

$$\mathbb{E}_x |X_t|^k = \mathbb{E}|xX_t|^k \leq \mathbb{E}\left[(|x| + |X_t|)^k\right]$$

$$= \sum_{j=0}^{k} \binom{k}{j} |x|^j \cdot \mathbb{E}|X_t|^{k-j} \leq \sum_{j=0}^{k} \binom{k}{j} |x|^j \cdot C_{k-j} t^{k-j}.$$

Taking $C = \max_{j \leq k} C_j$ we get that

$$\mathbb{E}_x |X_t|^k \leq C \cdot (t + |x|)^k. \qquad \square$$

Lemma 3.9.5 *Let G be a finitely generated group and μ an adapted measure on G. Assume that μ has finite kth moment. Let $H \leq G$ be a subgroup of finite index $[G : H] < \infty$. Let $(X_t)_t$ denote the μ-random walk.*

Then, there exist i.i.d. nonnegative, integer-valued random variables $(R_t)_t$, also independent of T_H^+, such that a.s.

$$\left|X_{T_H^+}\right| \le \max_{1 \le t \le T_H^+} |X_t| \le |X_0| + \sum_{t=1}^{T_H^+} R_t,$$

and such that $\mathbb{E}\left[R_1^k\right] < \infty$.

Proof   Let $n = [G : H]$ and choose a set $g_1, \ldots, g_n$ of representatives for the cosets of H, so that $G = \biguplus_{j=1}^n Hg_j$. Let $1 \le m \le n$ be such that $Hg_m = H$. For all $t \ge 1$ and $1 \le i, j \le n$, let $S(t, i, j)$ be independent elements of G, with distribution given by

$$\mathbb{P}[S(t, i, j) = x] = \begin{cases} \frac{\mu(x)}{\mu(g_i^{-1}Hg_j)} \cdot \mathbf{1}_{\{x \in g_i^{-1}Hg_j\}} & \mu\left(g_i^{-1}Hg_j\right) > 0, \\ \mathbf{1}_{\{x=1\}} & \mu\left(g_i^{-1}Hg_j\right) = 0. \end{cases}$$

For each $t \ge 1$ set $R_t = \max\{|S(t, i, j)| : 1 \le i, j \le n\}$. It is Exercise 3.24 to verify that $\mathbb{E}\left[R_t^k\right] = \mathbb{E}\left[R_1^k\right] < \infty$ (because μ has finite kth moment).

Let $(I_t)_t$ be a Markov chain on $\{1, \ldots, n\}$ with transition matrix

$$\mathbb{P}[I_{t+1} = j \mid I_t = i] = P(i, j) = \mu\left(g_i^{-1}Hg_j\right)$$

started at $I_0 = 1$. Choose $(I_t)_t$ so that it is independent of $(S(t, i, j))_{t,i,j}$. Recalling that $Hg_m = H$, define

$$T = \inf\{t \ge 1 : I_t = m\}.$$

Since R_t is measurable with respect to $\sigma(S(t, i, j) : 1 \le i, j \le n)$, we get that $(R_t)_t$ is independent of $(I_t)_t$. Also, since $(S(t, i, j))_{t,i,j}$ are independent, and since T is measurable with respect to $\sigma(I_t : t \ge 0)$, we conclude that $(R_t)_t, T$ are all independent, as required.

Now, consider the sequence $Y_0 = X_0$ and $Y_t = Y_{t-1}S(t, I_{t-1}, I_t)$ for $t \ge 1$. Exercise 3.26 shows that $(Y_t)_t$ is a Markov chain on G with transition matrix

$$\mathbb{P}[Y_{t+1} = y \mid Y_t = x] = \mu\left(x^{-1}y\right),$$

so that $(Y_t)_t$ has the distribution of a μ-random walk on G. Exercise 3.25 shows that $I_t = i$ if and only if $Y_t \in Hg_i$. Thus,

$$T = \inf\{t \ge 1 : I_t = m\} = \inf\{t \ge 1 : Y_t \in H\},$$

which implies that T and T_H^+ have the same distribution.

Finally, note that

$$\max_{1 \leq t \leq T} |Y_t| \leq |Y_0| + \sum_{t=1}^{T} |S(t, I_t, I_{t-1})| \leq |Y_0| + \sum_{t=1}^{T} R_t. \qquad \square$$

Exercise 3.24 Show that $\mathbb{E}[R_1^k] < \infty$ in the proof of Lemma 3.9.5. ▷ solution ◁

Exercise 3.25 Show that in the proof of Lemma 3.9.5 $I_t = i$ if and only if $Y_t \in Hg_i$.

▷ solution ◁

Exercise 3.26 Show that $(Y_t)_t$ from the proof of Lemma 3.9.5 is a Markov chain on G with transition matrix $\mathbb{P}[Y_{t+1} = y \mid Y_t = x] = \mu\left(x^{-1}y\right)$. ▷ solution ◁

Using Lemma 3.9.5, we can now deduce that the kth moment property is preserved when passing to the hitting measure of a finite index subgroup.

Corollary 3.9.6 Let $H \leq G$ be a subgroup of finite index $[G : H] < \infty$ in a finitely generated group G.

If μ is an adapted probability measure on G with finite kth moment, then the hitting measure μ_H is also an adapted probability measure on H with finite kth moment.

Proof Let $(X_t)_t$ denote the μ-random walk. Let $T = T_H^+$. So X_T has law μ_H. We need to prove that $\mathbb{E}\left[|X_T|^k\right] < \infty$.

By Lemma 3.9.5, $\mathbb{P}$-a.s., $|X_T| \leq \sum_{t=1}^{T} R_t$ for some i.i.d. $(R_t)_t$, independent of T, all nonnegative integer-valued random variables. Also, $\mathbb{E}\left[R_1^k\right] < \infty$, and by Exercise 3.20, $\mathbb{E}\left[T^k\right] < \infty$.

Using that T is independent of $(R_t)_t$, we have that

$$\mathbb{E}\left[|X_T|^k\right] \leq \mathbb{E}\left[\left|\sum_{t=1}^{T} R_t\right|^k\right] = \sum_{s=1}^{\infty} \mathbb{P}[T = s] \cdot \mathbb{E}\left[\left|\sum_{t=1}^{s} R_t\right|^k\right]$$

$$\leq \sum_{s=1}^{T} \mathbb{P}[T = s] \cdot \sum_{\vec{j} \in \{1,\dots,s\}^k} \mathbb{E}\left[\prod_{i=1}^{k} R_{j_i}\right]$$

$$\leq \sum_{s=1}^{T} \mathbb{P}[T = s] \cdot \sum_{\vec{j} \in \{1,\dots,s\}^k} \mathbb{E}\left[R_1^k\right] = \mathbb{E}\left[T^k\right] \cdot \mathbb{E}\left[R_1^k\right] < \infty.$$

We have used Jensen's inequality, so that $\mathbb{E}\left[R_1^m\right] \leq \mathbb{E}\left[R_1^k\right]^{m/k}$ for all $m \leq k$. $\qquad \square$

Exercise 3.27 Let G be a finitely generated group, μ an adapted probability measure on G with finite kth moment, and $H \leq G$ a subgroup of finite index $[G : H] < \infty$.

Show that there exists $C = C(k, \mu, H) > 0$ such that for any $x \in G$, we have
$$\mathbb{E}_x \left[|X_{T_H^+}|^k \right] \leq C \cdot (1 + |x|)^k. \qquad \qquad \text{▷ solution ◁}$$

3.9.2 HF_n Restricted to Finite Index Subgroups

Recall that $\mathsf{HF}_k(G, \mu)$ is the space of μ-harmonic functions, with growth bounded by a polynomial of degree at most k. We now investigate the change in HF_k as we move from a group to a finite index subgroup.

Theorem 3.9.7 shows that μ-harmonic functions on G correspond bijectively to μ_H-harmonic functions on H, even in the unbounded case. It is a generalization of Exercise 3.21.

Theorem 3.9.7 *Let G be a finitely generated group. Let $k \geq 0$ and let μ be an adapted probability measure on G with finite kth moment. Let $H \leq G$ be a subgroup of finite index $[G : H] < \infty$.*

Then, the restriction of any $f \in \mathsf{HF}_k(G, \mu)$ to H is in $\mathsf{HF}_k(H, \mu_H)$. Conversely, any $\tilde{f} \in \mathsf{HF}_k(H, \mu_H)$ is the restriction of a unique $f \in \mathsf{HF}_k(G, \mu)$.

Thus, the restriction map is a linear bijection (isomorphism) of the vector spaces $\mathsf{HF}_k(G, \mu) \cong \mathsf{HF}_k(H, \mu_H)$.

Proof **Step I: Extension** Let $\tilde{f} \in \mathsf{HF}_k(H, \mu_H)$. Define $f : G \to \mathbb{C}$ by
$$f(x) := \mathbb{E}_x[\tilde{f}(X_T)],$$
where $T = T_H$ is the hitting time of H and $(X_t)_t$ is a μ-random walk on G. Note that this coincides with $\tilde{f}$ for $x \in H$, since $T = 0$ a.s. when starting at $X_0 = x \in H$. Also, for $x \notin H$ we have that $f(x) = \mathbb{E}_x \left[\tilde{f}\left(X_{T_H^+}\right) \right]$.

We now wish to show that f is well defined (i.e. the expectation $\mathbb{E}_x |\tilde{f}(X_T)|$ is finite), and that $f \in \mathsf{HF}_k(G, \mu)$.

Note that since $[G : H] < \infty$ and G is finitely generated, then so is H (Exercise 1.61). Also, with respect to any choices of finite symmetric generating sets $G = \langle S \rangle$ and $H = \langle \tilde{S} \rangle$, the corresponding metrics are bi-Lipschitz. Namely, there exist $C > 1$ so that for any $x \in H$ we have that $C^{-1}|x|_S \leq |x|_{\tilde{S}} \leq C \cdot |x|_S$ (see Exercise 1.75). Thus, since $X_T \in H$ and $\tilde{f} \in \mathsf{HF}_k(H, \mu_H)$, we have that $|\tilde{f}(X_T)| \leq C \cdot \left(1 + |X_T|^k\right)$ for some constant $C > 0$. Also, by Exercise 3.27 we have that $\mathbb{E}_x |X_T|^k \leq C(1 + |x|)^k$ for some constant $C > 0$ (because μ has finite kth moment). This proves that $\mathbb{E}_x |\tilde{f}(X_T)| < \infty$, so that f is well defined, and also that $|f(x)| \leq C(1 + |x|)^k$, for some (perhaps adjusted) constant $C > 0$ and all $x \in G$. Exercise 3.28 shows that f is indeed μ-harmonic. So we conclude that $f \in \mathsf{HF}_k(G, \mu)$.

To recap: for every $\tilde{f} \in \mathsf{HF}_k(H, \mu_H)$ the extension $f(x) := \mathbb{E}_x[\tilde{f}(X_T)]$ is a well-defined function $f \in \mathsf{HF}_k(G, \mu)$, with $f|_H \equiv \tilde{f}$.

Step II: Restriction To show that this is indeed a unique extension, it suffices to show that if $f|_H \equiv 0$ and $f \in \mathsf{HF}_k(G, \mu)$ then $f \equiv 0$ on all of G.

Indeed, if $f \in \mathsf{HF}_k(G, \mu)$ then $(f(X_{T \wedge t}))_t$ is a martingale. This martingale is uniformly integrable. Indeed, there is a constant $C > 0$ such that $|f(y)| \leq C|y|^k$ for all $1 \neq y \in G$. If we denote $M = \max_{1 \leq t \leq T} |X_t|$, then we know from Lemma 3.9.5 that we can find i.i.d. nonnegative integer-valued random variables $(R_t)_t$, with finite kth moment, also independent of T, such that $M \leq |X_0| + \sum_{t=1}^{T} R_t$. Thus,

$$\mathbb{E}_x\left[|f(X_{T \wedge t})| \mathbf{1}_{\{|f(X_{T \wedge t})| > K\}}\right] \leq C \mathbb{E}_x\left[M^k \mathbf{1}_{\{CM^k > K\}}\right].$$

By Jensen's inequality, as in the solution to Exercise 3.27, one may show that

$$\mathbb{E}_x\left[M^k\right] \leq \sum_{s=1}^{\infty} \mathbb{P}[T = s] \sum_{m=0}^{k} \binom{k}{m} \mathbb{E}\left[\left|\sum_{t=1}^{s} R_t\right|^m\right] \cdot |x|^{k-m}$$

$$\leq \sum_{s=1}^{\infty} \mathbb{P}[T = s] \sum_{m=0}^{k} \binom{k}{m} s^m \mathbb{E}\left[R_1^m\right] |x|^{k-m} = \mathbb{E}\left[(TR_1 + |x|)^k\right]$$

$$\leq |x|^k \cdot \mathbb{P}[R_1 = 0] + \mathbb{E}\left[T^k\right] \cdot \mathbb{E}\left[R_1^k\right] \cdot (1 + |x|)^k < \infty.$$

Since $\mathbf{1}_{\{CM^k > K\}} \to 0$ as $K \to \infty$, by dominated convergence we obtain that

$$\lim_{K \to \infty} \sup_t \mathbb{E}_x\left[|f(X_{T \wedge t})| \mathbf{1}_{\{|f(X_{T \wedge t})| > K\}}\right] \leq C \cdot \lim_{K \to \infty} \mathbb{E}_x\left[M^k \mathbf{1}_{\{CM^k > K\}}\right] = 0.$$

This proves that $(f(X_{T \wedge t}))_t$ is a uniformly integrable martingale.

Thus, we may apply the optional stopping theorem (Theorem 2.3.3) to obtain $\mathbb{E}_x[f(X_T)] = f(x)$. If $f(x) = 0$ for all $x \in H$, then $f(X_T) = 0$ a.s., and so $f \equiv 0$ on all of G.

This shows that the linear map $f \mapsto f|_H$ is injective on $\mathsf{HF}_k(G, \mu)$. □

Exercise 3.28 Let G be a finitely generated group. Let $k \geq 0$ and let μ be an adapted probability measure on G with finite kth moment. Let $H \leq G$ be a subgroup of finite index $[G : H] < \infty$.

Show that if $\tilde{f} \in \mathsf{HF}_k(H, \mu_H)$ then $f(x) = \mathbb{E}_x[\tilde{f}(X_{T_H})]$ defines a μ-harmonic function.

▷ solution ◁

Exercise 3.29 Let G be a finitely generated group and let μ be a probability measure on G. Let $H \leq G$ be a subgroup such that the hitting measure μ_H is well defined; that is, $T_H^+ < \infty$ a.s.

Show that (G, μ) is recurrent if and only if (H, μ_H) is recurrent.   ▷ solution ◁

For examples of finite index subgroups recall Exercises 1.68, 1.72, and 1.73.

Exercise 3.30 Consider the infinite dihedral group

$$D_\infty = \left\langle x, y \mid y^2, \, yxyx \right\rangle.$$

Let $\mu = \frac{1}{3}(\delta_x + \delta_{x^{-1}} + \delta_y)$ $\left(\text{i.e. uniform on the standard generators} x, x^{-1}, y\right)$.

Show that $H = \langle x \rangle$ is a subgroup of index $[D_\infty : H] = 2$ that is isomorphic to $H \cong \mathbb{Z}$.

What is μ_H in this case?

Show that $\dim \mathrm{HF}_1(D_\infty, \mu) \geq 2$.   ▷ solution ◁

3.10 Solutions to Exercises

Solution to Exercise 3.1 :(

Let $\mathcal{F}_t = \sigma(X_0, \ldots, X_t)$. Let $R_t = \{x \in G : \mathbb{P}[X_t = x] > 0\}$. Using Exercise 2.8, we have that a.s.

$$
\begin{aligned}
\mathbb{P}[X_{t+1} = y \mid \mathcal{F}_t] &= \sum_{\substack{x_j \in R_j \\ 0 \leq j \leq t}} \mathbb{P}[X_{t+1} = y \mid \forall\, 0 \leq j \leq t, \, X_j = x_j] \mathbb{1}_{\{\forall\, 0 \leq j \leq t, \, X_j = x_j\}} \\
&= \sum_{\substack{x_j \in R_j \\ 0 \leq j \leq t}} \frac{\nu(x_0) \cdot \prod_{j=1}^t P(x_{j-1}, x_j) \cdot P(x_t, y)}{\nu(x_0) \cdot \prod_{j=1}^t P(x_{j-1}, x_j)} \prod_{j=0}^t \mathbb{1}_{\{X_j = x_j\}} \\
&= \sum_{x_t \in R_t} P(x_t, y) \mathbb{1}_{\{X_t = x_t\}} = \mathbb{P}[X_{t+1} = y \mid X_t] = P(X_t, y).
\end{aligned}
$$

This implies that for any bounded function $\varphi : G \to [0, \infty)$, we have a.s.

$$\mathbb{E}[\varphi(X_{t+1}) \mid \mathcal{F}_t] = \sum_y \mathbb{P}[X_{t+1} = y \mid \mathcal{F}_t]\varphi(y) = P\varphi(X_t).$$

Inductively, we can now show that

$$
\begin{aligned}
\mathbb{E}[\varphi(X_{t+n}) \mid \mathcal{F}_t] &= \mathbb{E}\left[\mathbb{E}[\varphi(X_{t+n}) \mid \mathcal{F}_{t+n-1}] \mid \mathcal{F}_t\right] \\
&= \mathbb{E}[P\varphi(X_{t+n-1}) \mid \mathcal{F}_t] = \cdots = P^n \varphi(X_t).
\end{aligned}
$$

When $\varphi = \delta_y$ we get $P^n \varphi = P^n(\cdot, y)$.

Now, if $A \in \mathcal{F}_t$ then

$$
\begin{aligned}
\mathbb{P}[X_{t+n} = y, \, A, \, X_t = x] &= \mathbb{E}\left[\mathbb{P}[X_{t+n} = y \mid \mathcal{F}_t]\mathbb{1}_A \mathbb{1}_{\{X_t = x\}}\right] = \mathbb{E}\left[P^n(X_t, y)\mathbb{1}_A \mathbb{1}_{\{X_t = x\}}\right] \\
&= P^n(x, y) \cdot \mathbb{P}[A, X_t = x].
\end{aligned}
$$

:)✓

Solution to Exercise 3.5 :(

Since $(U_t)_t$ are i.i.d.-μ and independent of the starting distribution, we have that for $x_0, \ldots, x_t \in G$,

$$\mathbb{P}_{x_0}[\forall\, 0 \leq j \leq t,\, X_j = x_j] = \mathbb{P}_{x_0}\left[X_0 = x_0,\ \forall\, 1 \leq j \leq t,\, U_j = (x_{j-1})^{-1}x_j\right]$$

$$= \prod_{j=1}^{t} \mu\left((x_{j-1})^{-1}x_j\right) = \prod_{j=1}^{t} P(x_{j-1}, x_j).$$

This shows that $(X_t)_t$ is a Markov chain with transition matrix P.

Assume that P is irreducible. Let $x \in G$. There exists $t > 0$ such that $\mathbb{P}[X_t = x] = P^t(1, x) > 0$. Under $\mathbb{P}$ we have that $X_t = U_1 \cdots U_t$ for i.i.d.-μ elements $(U_k)_k$. Thus, there must exist $u_1, \ldots, u_t \in \mathrm{supp}(\mu)$ such that $u_1 \cdots u_t = x$. This holds for any $x \in G$, so $\mathrm{supp}(\mu)$ generates G.

Now, assume that μ is adapted. Let $x, y \in G$. Write $x^{-1}y = u_1 \cdots u_t$ for some $u_1, \ldots, u_t \in \mathrm{supp}(\mu)$. Thus,

$$P^t(x, y) = \mathbb{P}_x[X_t = y] \geq \mathbb{P}_x[X_0 = x, X_1 = xu_1, \ldots, X_t = xu_1 \cdots u_t] = \prod_{j=1}^{t} \mu(u_j) > 0.$$

Finally, if $\mu(1) > 0$, then $P(x, x) \geq \mu(1) > 0$ for any $x \in G$, which immediately implies that x is aperiodic.

:) ✓

Solution to Exercise 3.7 :(

Since $\{t \leq k\} \in \{\emptyset, \Omega\}$ (the trivial σ-algebra), we get that $\{t \leq k\} \in \sigma(X_0, \ldots, X_k)$ for any k. So constants are stopping times.

For the minimum:

$$\{T \wedge T' \leq t\} = \{T \leq t\} \bigcup \{T' \leq t\} \in \sigma(X_0, \ldots, X_t).$$

The maximum is similar:

$$\{T \vee T' \leq t\} = \{T \leq t\} \bigcap \{T' \leq t\} \in \sigma(X_0, \ldots, X_t).$$

For the addition,

$$\{T + T' \leq t\} = \bigcup_{k=0}^{t} \{T = k, T' \leq t - k\}.$$

Since $\{T = k\} = \{T \leq k\} \backslash \{T \leq k-1\} \in \sigma(X_0, \ldots, X_k)$, we get that $T + T'$ is a stopping time.   :) ✓

Solution to Exercise 3.9 :(

Let $T_x^{(0)} = 0$ and $T_x^{(k+1)} = \inf\left\{t > T_x^{(k)} : X_t = x\right\}$ be the successive times the chain is at x.

Let $A_k = \left\{T_x^{(k)} < T_Z\right\}$. The strong Markov property tells us that for any $k \geq 1$,

$$\mathbb{P}_x[V \geq k + 1] = \mathbb{P}_x[A_k] = \mathbb{P}_x[A_k \mid A_{k-1}] \cdot \mathbb{P}_x[A_{k-1}]$$

$$= \cdots = (\mathbb{P}_x[A_1])^k,$$

and $\mathbb{P}_x[V \geq 1] = 1 = (\mathbb{P}_x[A_1])^0$ trivially.

This is exactly a geometric distribution; that is, one obtains that for any $k \geq 1$,

$$\mathbb{P}_x[V = k] = \mathbb{P}_x[V \geq k] - \mathbb{P}_x[V \geq k + 1] = (\mathbb{P}_x[A_1])^{k-1}(1 - \mathbb{P}_x[A_1]).$$

Note that this gives the correct mean since $1 - \mathbb{P}[A_1] = \mathbb{P}_x\left[T_Z \leq T_x^+\right] = \mathbb{P}_x\left[T_Z < T_x^+\right]$ (as $x \notin Z$ so $T_Z \neq T_x^+$).

:) ✓

Solution to Exercise 3.10 :(

There exists n such that $P^n(x, y) > 0$ (because P is irreducible). Thus, there is a sequence $x = x_0, x_1, \ldots, x_n = y$ such that $P(x_j, x_{j+1}) > 0$ for all $0 \le j < n$. Let $m = \max\{0 \le j < n : x_j = x\}$, and let $t = n - m$ and $y_j := x_{m+j}$ for $0 \le j \le t$. Then we have the sequence $x = y_0, \ldots, y_t = y$ so that $y_j \ne x$ for all $0 < j \le t$, and we know that $P(y_j, y_{j+1}) > 0$ for all $0 \le j < t$. Thus,

$$\mathbb{P}_x[X_t = y, \ t < T_x^+] \ge \mathbb{P}_x[\forall\, 0 \le j \le t, \ X_j = y_j] = P(y_0, y_1) \cdots P(y_{t-1}, y_t) > 0. \qquad :)\checkmark$$

Solution to Exercise 3.11 :(

Note that if $y \in A$ then $T_A^+ \le T_y^+$, so that it suffices to prove that $\mathbb{P}_\nu[T_y^+ < \infty] = 1$ for any $y \in G$ and any ν.
  Also,

$$\mathbb{P}_\nu[T_y^+ < \infty] = \sum_x \nu(x)\, \mathbb{P}_x\left[T_y^+ < \infty\right] = \sum_x \nu(x) = 1$$

when the chain is recurrent.   :)$\checkmark$

Solution to Exercise 3.12 :(

Q is easily seen to be a stochastic matrix since

$$\sum_y Q(x, y) = \sum_{k=0}^{\infty} \frac{1}{ek!} \sum_y P^k(x, y) = 1.$$

Also, since $Q(x, y) \ge e^{-1} P(x, y)$ for all $x, y \in G$, we obtain inductively that $Q^t(x, y) \ge e^{-t} P^t(x, y)$ for all $x, y \in G$ and all t, implying that Q is irreducible (because P is).
  Let $(\Lambda_t)_{t=1}^{\infty}$ be i.i.d. Poisson of mean 1 and let $S_t = \sum_{j=1}^{t} \Lambda_j$ (with $S_0 = 0$). If $(X_t)_t$ is the Markov chain with transition matrix P, independent of $(\Lambda_t)_t$, then one verifies that $(Y_t = X_{S_t})_t$ is a Markov chain with transition matrix Q. Indeed,

$$\mathbb{P}[Y_{t+1} = y, \ Y_t = x] = \mathbb{P}\left[X_{S_t + \Lambda_{t+1}} = y, \ X_{S_t} = x\right]$$

$$= \sum_{s,k=0}^{\infty} \mathbb{P}[S_t = s] \cdot \mathbb{P}[\Lambda_{t+1} = k] \cdot \mathbb{P}[X_{s+k} = y, \ X_s = x]$$

$$= \sum_{s=0}^{\infty} \mathbb{P}[S_t = s] \cdot \sum_{k=0}^{\infty} \frac{1}{ek!} P^k(x, y) = Q(x, y).$$

For every t define $L_t = \{k : S_k = t\}$. Note that if $k, k + n \in L_t$ then $S_{k+n} = S_k$, so $\Lambda_{k+1} = \cdots = \Lambda_{k+n} = 0$. Also,

$$\{L_t \ne \emptyset, \ \min L_t = k\} \in \sigma(\Lambda_1, \ldots, \Lambda_k).$$

Thus, for any $\ell \ge 1$,

$$\mathbb{P}[|L_t| \ge \ell] = \sum_{k=0}^{\infty} \mathbb{P}[|L_t| \ge \ell, \ \min L_t = k]$$

$$\le \sum_{k=0}^{\infty} \mathbb{P}[L_t \ne \emptyset, \ \min L_t = k, \ \forall\, 1 \le j \le \ell - 1, \ \Lambda_{k+j} = 0]$$

$$= \sum_{k=0}^{\infty} \mathbb{P}[L_t \ne \emptyset, \ \min L_t = k] \cdot e^{1-\ell} \le e^{1-\ell}.$$

This implies that

$$\mathbb{E}[|L_t|] = \sum_{\ell=0}^{\infty} \mathbb{P}[|L_t| > \ell] \le \sum_{\ell=0}^{\infty} e^{-\ell} < 2.$$

Now, we may bound the number of visits of $(Y_t)_t$ to a specific site x by

$$\sum_{k=0}^{\infty} 1_{\{Y_k=x\}} \le \sum_{t=0}^{\infty} 1_{\{X_t=x\}} \cdot |L_t|.$$

Hence

$$\sum_{t=0}^{\infty} \mathbb{P}[Y_t = x] \le \sum_{t=0}^{\infty} \mathbb{P}[X_t = x] \cdot \mathbb{E}\,|L_t| \le 2 \cdot \sum_{t=0}^{\infty} \mathbb{P}[X_t = x].$$

Also, note that since S_t is Poisson of mean t, and since $\int_0^{\infty} e^{-t} t^n \, dt = n!$, we have

$$\sum_{t=0}^{\infty} \mathbb{P}[S_t = n] = \sum_{t=0}^{\infty} e^{-t} \frac{t^n}{n!} \ge c,$$

for some constant $c > 0$ independent of n. Summing over n we have

$$2 \sum_{t=0}^{\infty} \mathbb{P}[X_t = x] \ge \sum_{t=0}^{\infty} \mathbb{P}[Y_t = x] = \sum_{t=0}^{\infty} \sum_{n=0}^{\infty} \mathbb{P}[S_t = n] \cdot \mathbb{P}[X_n = x] \ge c \cdot \sum_{n=0}^{\infty} \mathbb{P}[X_n = x],$$

where we have used the independence of X_t, L_t.
Thus Q is transient if and only if P is transient. :)✓

Solution to Exercise 3.13 :(
Let $(U_t)_{t=1}^{\infty}$ be i.i.d. random variable, each uniform on $[0, 1]$. Define a Markov chain (Y_t, N_t) on $G \times \mathbb{N}$ as follows. Given $Y_t = x$, $N_t = n$, let

$$(Y_{t+1}, N_{t+1}) = \begin{cases} (x, n) & \text{if } U_{t+1} \le p_x, \\ (y, n+1) & \text{if } U_t > p_x, \end{cases} \text{ with probability } P(x, y).$$

One checks easily that $(Y_t)_t$ is a Markov chain with transition matrix Q.
Also, note that $N_t \le N_{t+1} \le N_t + 1$ for all t, and

$$\mathbb{P}[N_{t+1} = N_t + 1 \mid Y_t = x] = 1 - p_x > 0.$$

So we may define

$$T_n = \inf\{t \ge 0 : N_t = n\},$$

with $T_0 = 0$ and $T_n < \infty$ a.s. Set $X_n = Y_{T_n}$. One then can also easily check that $(X_n)_n$ is a Markov chain with transition matrix P.
Moreover, this coupling gives that if $X_n = x$ for infinitely many n then $Y_t = x$ for infinitely many t.
Conversely, note that for any $T_n \le t < T_{n+1}$, we have that $Y_{T_n} = Y_t$. So if $Y_t = x$ for infinitely many t and $X_n = x$ for only finitely many n, it must be that $T_{n+1} - T_n = \infty$ for some n, which happens with probability 0.
We conclude that $(X_n)_n$ is recurrent if and only if $(Y_t)_t$ is recurrent. :)✓

Solution to Exercise 3.14 :(
If $\pi(x)P(x, y) = \pi(y)P(y, x)$ for all $x, y \in G$, then

$$\pi P(y) = \sum_x \pi(x)P(x, y) = \sum_x \pi(y)P(y, x) = \pi(y).$$

For the example, consider

$$P = \begin{bmatrix} \frac{1}{4} & \frac{1}{4} & \frac{1}{2} \\ \frac{1}{2} & \frac{1}{4} & \frac{1}{4} \\ \frac{1}{4} & \frac{1}{4} & \frac{1}{4} \end{bmatrix},$$

which is easily seen to have the uniform distribution on three states, $\pi(x) = \frac{1}{3}$, as a stationary distribution. However, if $\pi(x)P(x, y) = \pi(y)P(y, x)$, in this case $P(x, y) = P(y, x)$, which is easily seen to be false. :)✓

Solution to Exercise 3.15 :(

Let $v(z) = \mathbb{E}_x\left[V_{T_x^+}(z)\right]$ and $u(z) = \mathbb{E}_y\left[V_{T_y^+}(z)\right]$. Since x, y are positive recurrent we know that P has a unique stationary distribution given by $\pi(z) = \frac{1}{\mathbb{E}_z[T_z^+]}$. Also, since x, y are recurrent we know that $\pi(x)v$ and $\pi(y)u$ are stationary distributions, as in Example 3.7.8. But then

$$\pi(x)v(y) = \pi(y)u(y) = \pi(y) \qquad \text{and} \qquad \pi(x) = \pi(x)v(x) = \pi(y)u(x),$$

so that

$$v(y)u(x) = \frac{\pi(y)}{\pi(x)} \cdot \frac{\pi(x)}{\pi(y)} = 1. \hspace{4cm} :)\checkmark$$

Solution to Exercise 3.16 :(

If $x \in A$, then $T_A^+ \le T_x^+$. Also, $\mathbb{E}_v\left[T_x^+\right] = \sum_y v(y)\,\mathbb{E}_y\left[T_x^+\right]$. So it suffices to show that $\mathbb{E}_y\left[T_x^+\right] < \infty$ for all $x, y \in G$.

If $y = x$ this is immediate from Theorem 3.7.5.

For $x \ne y \in G$, by Exercise 3.10 there exists $t > 0$ such that $\mathbb{P}_x\left[X_t = y, T_x^+ > t\right] > 0$. The Markov property and positive recurrence give that

$$\infty > \mathbb{E}_x\left[T_x^+\right] \ge \mathbb{P}_x\left[X_t = y, T_x^+ > t\right] \cdot \mathbb{E}_y\left[T_x^+ + t\right],$$

so $\mathbb{E}_y\left[T_x^+\right] < \infty$. $\hspace{5cm} :)\checkmark$

Solution to Exercise 3.18 :(

(1) $\Rightarrow$ (2), (4) $\Rightarrow$ (1) are trivial.

(2) $\Rightarrow$ (3) follows from the fact that given x, y there exists k such that $P^k(y, x) > 0$, so as $t \to \infty$,

$$P^t(x, y) \le \frac{P^{t+k}(x, x)}{P^k(y, x)} \to 0.$$

(3) $\Rightarrow$ (4): Let z, w be such that $P^t(z, w) \to 0$. Then, for any x, y there exist k, n such that $P^k(z, x) > 0$ and $P^n(y, w) > 0$. Thus,

$$P^t(x, y) \le \frac{P^{t+k+n}(z, w)}{P^k(z, x)P^n(y, w)} \to 0. \hspace{3cm} :)\checkmark$$

Solution to Exercise 3.19 :(

By Exercise 3.18 it suffices to show that $P^t(x, x) \to 0$ for some x.

Fix x. Let $n = \gcd\{t > 0 : P^t(x, x) > 0\}$. Let $Q = P^n$. Note that $\{t > 0 : P^t(x, x) > 0\} \subset \{nt : t > 0\}$, so $\limsup_{t\to\infty} P^t(x, x) \le \limsup_{t\to\infty} Q^t(x, x)$.

Set $Y = \{y : \exists\, t \ge 0,\ Q^t(x, y) > 0\}$. Then Q restricted to Y is an irreducible Markov chain.

Moreover, by definition of gcd, we know that

$$\gcd\{t > 0 : Q^t(x, x) > 0\} = \gcd\{t > 0 : P^{nt}(x, x) > 0\}.$$

Moreover, the definition of gcd implies that

$$\gcd\{t > 0 : Q^t(x, x) > 0\} = \gcd\{t > 0 : P^{nt}(x, x) > 0\} = 1.$$

So Q restricted to Y is aperiodic and irreducible, and thus $Q^t(x, x) \to 0$ by Theorem 3.8.1. $\hspace{1cm} :)\checkmark$

Solution to Exercise 3.20 :(

Consider the projected random walk $(HX_t)_t$. Let $Hx \in G/H$. Because this chain is irreducible, there exists $k > 0$ such that $P^{k-1}(Hx, H) > 0$. Thus,

$$\mathbb{P}_x\left[T_H^+ < k\right] \ge \mathbb{P}_x[X_{k-1} \in H] \ge \mathbb{P}_{Hx}[HX_{k-1} = H] = P^{k-1}(Hx, H).$$

Since there are only finitely many different cosets, we may find k and $\alpha > 0$ such that for any $x \in G$, we have $\mathbb{P}_x\left[T_H^+ < k\right] \ge \alpha$.

Now, for any $t > k$, using the Markov property,

$$\mathbb{P}_x\left[T_H^+ > t\right] \le \mathbb{P}_x\left[T_H^+ > t - k\right] \cdot \sup_y \mathbb{P}_y\left[T_H^+ \ge k\right] \le \mathbb{P}_x\left[T_H^+ > t - k\right] \cdot (1 - \alpha)$$

$$\le \cdots \le (1 - \alpha)^{\lfloor t/k \rfloor}.$$

:)✓

Solution to Exercise 3.21 :(

Let $T = T_H^+$. We have seen that $T_H^+ < \infty$ a.s. If $f \in \mathrm{BHF}(G, \mu)$ then $(f(X_t))_t$ is a bounded martingale. Thus, by the optional stopping theorem (Theorem 2.3.3), $f(x) = \mathbb{E}_x[f(X_T)]$ for all $x \in G$.

Note that for any $x, y \in H$, since $x^{-1}H = H$ and $x^{-1}y \in H$, we have that

$$\mathbb{P}_x[X_T = y] = \sum_{t=1}^{\infty} \mathbb{P}_x[X_0 = x, \ X_t = y, \ \forall 1 \le j < t, \ X_j \notin H]$$

$$= \sum_{t=1}^{\infty} \mathbb{P}\left[X_0 = 1, \ X_t = x^{-1}y, \ \forall 1 \le j < t, \ X_j \notin H\right]$$

$$= \mathbb{P}\left[X_T = x^{-1}y\right].$$

Thus, for $x \in H$,

$$f(x) = \mathbb{E}_x[f(X_T)] = \sum_y \mathbb{P}_x[X_T = y]f(y) = \sum_y \mathbb{P}\left[X_T = x^{-1}y\right]f(y)$$

$$= \sum_y \mathbb{P}[X_T = y]f(xy) = \sum_y \mu_H(y)f(xy).$$

So $f|_H \in \mathrm{BHF}(H, \mu_H)$.

:)✓

Solution to Exercise 3.22 :(

Consider the homomorphism $\varphi \colon \mathbb{Z}^2 \to \{0, 1\}^2$ given by $\varphi(x, y) = (x \pmod 2, y \pmod 2)$. We have $H = \mathrm{Ker}\varphi$, so $H \lhd \mathbb{Z}^2$ with $[\mathbb{Z}^2 : H] = 4$.

We have that μ_H is not finitely supported since for any $(2x, 2y) \in H$, we can find a path in the Cayley graph of $\mathbb{Z}^2$ from $(0, 0)$ to $(2x, 2y)$ that avoids H until reaching $(2x, 2y)$. For example, if $x > 0, y > 0$, then we take the path

$$\gamma = ((0, 0), (1, 0), (1, 1), (1, 2), \ldots, (1, 2y - 1), (2, 2y - 1), \ldots, (2x, 2y - 1), (2x, 2y)).$$

Similarly in the other cases. Thus, $\mu_H(2x, 2y) > 0$ for all $(2x, 2y) \in H$.

:)✓

Solution to Exercise 3.23 :(

Let $(X_t)_t$ denote the μ-random walk, and let $U_t = X_{t-1}^{-1}X_t$ be the independent "jumps" of the walk.

For any $x \in H$ we can write $x = s_1 \cdots s_n$ for $s_1, \ldots, s_n \in \mathrm{supp}(\mu)$ because μ is adapted.

Define $x_0 = 1$ and $x_j = s_1 \cdots x_j$. Let $J = \{0 \le j \le n : x_j \in H\}$. Write $J = \{0 = j_0 < j_1 < \cdots < j_k = n\}$. For $1 \le \ell \le k$ set

$$u_\ell = \left(x_{j_{\ell-1}}\right)^{-1} x_{j_\ell} = s_{j_{\ell-1}+1} \cdots s_{j_\ell}.$$

So $u_1 = x_{j_1}$ and $x = u_1 \cdots u_k$. For any $1 \le \ell \le k$ it is simple to see inductively that $u_\ell \in H$. Moreover, since $x_{j_{\ell-1}+1}, \ldots, x_{j_\ell-1} \notin H$, we have that

$$\mathbb{P}\left[X_{T_H^+} = u_\ell\right] \ge \mathbb{P}\left[U_1 = s_{j_{\ell-1}+1}, \ldots, U_{j_\ell-j_{\ell-1}} = s_{j_\ell}\right] = \prod_{i=1}^{j_\ell-j_{\ell-1}} \mu(s_{j_i}) > 0.$$

Thus, $u_\ell \in \mathrm{supp}(\mu_H)$ for all ℓ. As $x = u_1 \cdots u_k$, this shows that H is generated by $\mathrm{supp}(\mu_H)$.

Now, if μ is symmetric, then for any n, $(X_t)_{t=1}^n$, and $(X_t^{-1})_{t=1}^n$ have the same distribution. Thus, for any $x \in H$,

$$\mathbb{P}\left[X_{T_H^+} = x\right] = \sum_{t=1}^{\infty} \mathbb{P}[X_t = x, \ \forall 1 \leq j \leq t-1, \ X_j \notin H]$$

$$= \sum_{t=1}^{\infty} \mathbb{P}\left[X_t = x^{-1}, \ \forall 1 \leq j \leq t-1, \ X_j \notin H\right] = \mathbb{P}\left[X_{T_H^+} = x^{-1}\right]. \qquad :)\checkmark$$

Solution to Exercise 3.24 :(
Since μ has finite kth moment, it follows that as long as $\mu\left(g_i^{-1}Hg_j\right) > 0$, we have

$$\mathbb{E}\left[|S(1,i,j)|^k\right] = \mu\left(g_i^{-1}Hg_j\right)^{-1} \cdot \mathbb{E}\left[|X_1|^k \mathbf{1}_{\{X_1 \in g_i^{-1}Hg_j\}}\right] \leq \mu\left(g_i^{-1}Hg_j\right)^{-1} \cdot \mathbb{E}\left[|X_1|^k\right] < \infty.$$

Since $R_1^k \leq \sum_{i,j \leq n} |S(1,i,j)|^k$, we have $\mathbb{E}\left[R_1^k\right] \leq n^2 \cdot \mathbb{E}\left[|X_1|^k\right] < \infty.$ $\qquad :)\checkmark$

Solution to Exercise 3.25 :(
This is by induction on t. For $t = 0$ this is by definition, since $I_0 = 1$ and $Hg_1 = Hg$ and $Y_0 = X_0 = g$.

For $t > 0$, note that $S(t,i,j) \in g_i^{-1}Hg_j$ if $\mu\left(g_i^{-1}Hg_j\right) > 0$, and $S(t,i,j) = 1$ otherwise. Also, by the induction hypothesis, $Y_{t-1} \in Hg_{I_{t-1}}$. By definition of $(I_t)_t$, it must be that $\mu\left(g_{I_{t-1}}^{-1}Hg_{I_t}\right) > 0$, hence, $S(t,I_{t-1},I_t) \in g_{I_{t-1}}^{-1}Hg_{I_t} = Y_{t-1}^{-1}Hg_{I_t}$. This implies that $Y_t = Y_{t-1}S(t,I_{t-1},I_t) \in Hg_{I_t}$, completing the induction. $\qquad :)\checkmark$

Solution to Exercise 3.26 :(
Note that $I_t = j$ for $1 \leq j \leq n$ such that $Y_t \in Hg_j$. Let $x, y \in G$ and let $1 \leq i, j \leq n$ be such that $x \in Hg_i$ and $y \in Hg_j$. Recall that $S(t,i,j)$ is independent of $(I_t)_t$. For any t, since I_t, I_{t-1} are measurable with respect to Y_t, Y_{t-1},

$$\mathbb{P}[Y_{t+1} = y \mid Y_t = x, \ Y_{t-1}, \ldots, Y_0]$$

$$= \mathbb{P}[S(t+1, I_t, I_{t+1}) = x^{-1}y \mid I_t, I_{t-1}, Y_t = x, Y_{t-1}, \ldots, Y_0]$$

$$= \mathbb{P}[I_{t+1} = j \mid I_t, I_{t-1}, Y_t = x, Y_{t-1}, \ldots, Y_0] \cdot \mathbb{P}[S(t+1,i,j) = x^{-1}y]$$

$$= \mu\left(g_i^{-1}Hg_j\right) \frac{\mu\left(x^{-1}y\right)}{\mu\left(g_i^{-1}Hg_j\right)} = \mu\left(x^{-1}y\right). \qquad :)\checkmark$$

Solution to Exercise 3.27 :(
Fix $x \in G$. As in the proof of Corollary 3.9.6 we know that

$$\mathbb{E}_x\left[|X_T|^k\right] \leq \mathbb{E}\left[\left|\,|x| + \sum_{t=1}^{T} R_t\right|^k\right],$$

where $(R_t)_t$, $T = T_H^+$ are independent, positive, and integer-valued, and $\mathbb{E}\left[R_t^k\right] = \mathbb{E}\left[R_1^k\right] < \infty$.

Similarly to the proof of Corollary 3.9.6, this last expectation can be bounded as follows. First, for any m, using Jensen's inequality, we have the bound

$$\mathbb{E}\left[\left|\sum_{t=1}^{s} R_t\right|^m\right] \leq \sum_{\vec{j} \in \{1, \ldots, s\}^m} \mathbb{E}\left[\prod_{i=1}^{m} R_{j_i}\right] \leq \sum_{\vec{j} \in \{1, \ldots, s\}^m} \mathbb{E}[R_1^m] = s^m \, \mathbb{E}[R_1^m].$$

Then,

$$\mathbb{E}_x\left[|X_T|^k\right] \le \sum_{s=1}^{\infty}\mathbb{P}[T=s]\sum_{m=0}^{k}\binom{k}{m}\mathbb{E}\left[\left|\sum_{t=1}^{s}R_t\right|^m\right]\cdot|x|^{k-m}$$

$$\le \sum_{s=1}^{\infty}\mathbb{P}[T=s]\sum_{m=0}^{k}\binom{k}{m}s^m\,\mathbb{E}\left[R_1^m\right]|x|^{k-m} = \mathbb{E}\left[(TR_1+|x|)^k\right]$$

$$\le |x|^k\,\mathbb{P}[R_1=0] + \mathbb{E}\left[T^k\right]\cdot\mathbb{E}\left[R_1^k\right]\cdot(1+|x|)^k. \qquad :)\checkmark$$

Solution to Exercise 3.28 :(

First, note that if $x \notin H$ then $\mathbb{P}_x$-a.s. we have $T_H = T_H^+$, so that $f(x) = \mathbb{E}_x\left[\tilde{f}\left(X_{T_H^+}\right)\right]$ for $x \notin H$.

On the other hand, when $x \in H$, we have, using the fact that $\tilde{f}$ is μ_H harmonic, $f(x) = \tilde{f}(x) = \mathbb{E}_x\left[\tilde{f}\left(X_{T_H^+}\right)\right]$.

Finally, using $T_H^+ \ge 1$ and the Markov property, we have

$$f(x) = \mathbb{E}_x\left[\tilde{f}\left(X_{T_H^+}\right)\right] = \sum_y \mu(y)\,\mathbb{E}_x\left[\tilde{f}\left(X_{T_H^+}\right) \mid X_1 = xy\right]$$

$$= \sum_y \mu(y)\,\mathbb{E}_{xy}\left[\tilde{f}\left(X_{T_H}\right)\right] = \sum_y \mu(y)f(xy). \qquad :)\checkmark$$

Solution to Exercise 3.29 :(

Let $(X_t)_t$ denote the μ-random walk. Let $\tau_0 = 0$ and inductively,

$$\tau_{n+1} = \inf\{t > \tau_n : X_t \in H\}.$$

So $(Y_n := X_{\tau_n})_n$ is a μ_H-random walk on H.

If (H, μ_H) is recurrent, then $Y_n = 1$ infinitely many times a.s. So $X_t = 1$ infinitely many times as well, since $(Y_n)_n$ is a subsequence of $(X_t)_t$.

Conversely, if $X_t = 1$ infinitely many times a.s., then since $1 \in H$, we have that

$$\{t : X_t = 1\} \subset \{\tau_n : n \ge 0\},$$

implying that $Y_n = 1$ infinitely many times as well. $\qquad :)\checkmark$

Solution to Exercise 3.30 :(

It is easy to see that any element in D_∞ can be uniquely written as $y^\varepsilon x^z$ for $\varepsilon \in \{0, 1\}$ and $z \in \mathbb{Z}$. This shows that $H \cong \mathbb{Z}$ and $[D_\infty : H] = 2$.

Let $(X_t)_t$ be the μ-random walk. Starting from the origin, note that with probability $\frac{2}{3}$ we have $X_1 \in \{x, x^{-1}\}$. With the remaining probability $\frac{1}{3}$ we have $X_1 = y$.

Now, given that $X_t = yx^z$ for some $z \in \mathbb{Z}$, we have that with probability $\frac{1}{3}$ we get $X_{t+1} = X_t y = x^{-z} \in H$. The other probability of $\frac{2}{3}$ is split over two events, $\{X_{t+1} = X_t x\}$ and $\{X_{t+1} = X_t x^{-1}\}$, each with probability $\frac{1}{3}$, and in both these events we have $X_{t+1} \notin H$. This shows that for any $t > 0$,

$$\mathbb{P}\left[T_H^+ > t+1 \mid T_H^+ > t\right] = \tfrac{2}{3}.$$

Thus, inductively,

$$\mathbb{P}\left[T_H^+ > t\right] = \mathbb{P}\left[T_H^+ > t \mid T_H^+ > t-1\right]\cdot\mathbb{P}\left[T_H^+ > t-1\right] = \cdots = \left(\tfrac{2}{3}\right)^{t-1}\mathbb{P}\left[T_H^+ > 1\right] = \left(\tfrac{2}{3}\right)^{t-1}\tfrac{1}{3}.$$

Also, $\mathbb{P}\left[T_H^+ = 1\right] = \tfrac{2}{3}$.

One may also note that conditioned on $X_1 = y$, we have that $X\left[1, T_H^+ - 1\right]$ is just the process that moves by x, x^{-1} with probability $\frac{1}{2}$ each. (In the last step, one must have the final jump with y into H.)

Thus, we may describe the distribution of μ_H as follows. Let $B_t \sim \text{Bin}(t, \frac{1}{2})$. Note that $Z_t = 2B_t - t$ has the distribution of a simple random walk on $\mathbb{Z}$ of t steps. Then,

$$\mu_H(x^z) = \sum_{t=1}^{\infty} \mathbb{P}\left[T_H^+ = t\right] \cdot \mathbb{P}\left[X_t = x^z \mid T_H^+ = t\right] = \frac{1}{3}\mathbf{1}_{\{z=1\}} + \frac{1}{3}\mathbf{1}_{\{z=-1\}} + \sum_{t=2}^{\infty} \left(\frac{1}{3}\right)^2 \left(\frac{2}{3}\right)^{t-2} \cdot \mathbb{P}[Z_{t-2} = -z].$$

Finally, since the restriction map from D_∞ to H is a linear isomorphism of $\text{HF}_1(D_\infty, \mu)$ and $\text{HF}_1(H, \mu_H)$, it suffices to show that the latter has dimension at least 2. This has already been shown in a previous exercise, but for completeness, recall that the maps $x^z \mapsto \alpha z + \beta$ are μ_H-harmonic (actually for any symmetric measure).

:) ✓

4
Networks and Discrete Analysis

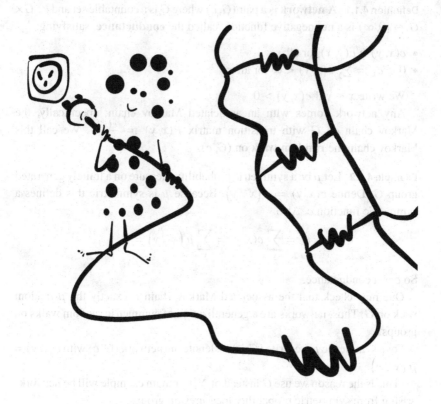

The theory of random walks is deeply connected to the theory of electrical networks. In this chapter, we will go through the basics of the theory. The main application at the end of the chapter is that transience or recurrence is, in essence, a group property. More precisely, these properties remain stable when changing between different symmetric, adapted measures with finite second moment.

4.1 Networks

Definition 4.1.1 A **network** is a pair (G, c) where G is a countable set and $c : G \times G \to [0, \infty)$ is a nonnegative function, called the **conductance**, satisfying:

- $c(x, y) = c(y, x)$ for all x, y.
- $0 < c_x := \sum_y c(x, y) < \infty$ for all x.

We write $x \sim y$ if $c(x, y) > 0$.

Any network comes with an associated Markov chain; specifically, the Markov chain on G with transition matrix $P(x, y) := \frac{c(x,y)}{c_x}$. We call this Markov chain the **random walk** on (G, c).

Example 4.1.2 Let μ be a symmetric probability measure on a finitely generated group G. Define $c(x, y) = \mu\left(x^{-1}y\right)$. Because μ is symmetric this defines a symmetric function c. Also,

$$c_x = \sum_y c(x, y) = \sum_y \mu\left(x^{-1}y\right) = 1.$$

So c is a conductance.

One may check that the associated Markov chain is exactly the μ-random walk on G. Thus, networks are a generalization of symmetric random walks on groups.

For a symmetric μ we write (G, μ) to denote the network (G, c) with $c(x, y) = \mu\left(x^{-1}y\right)$.

(This is the reason we use G instead of N; our main example will be networks arising from symmetric probability measures on groups.)   △▽△

Throughout, given some network (G, c), we will use $(X_t)_t$ to denote the random walk on (G, c). As in the case of random walks on groups, $\mathbb{P}_x, \mathbb{E}_x$ will denote probability measure and expectation $\left(\text{on } G^{\mathbb{N}}\right)$ conditioned on $X_0 = x$. Similarly, $\mathbb{P}_\nu, \mathbb{E}_\nu$ are conditioned on X_0 having law ν. P will denote the transition matrix of the Markov chain; that is, $P(x, y) = \frac{c(x,y)}{c_x}$.

Definition 4.1.3 A network (G, c) is called **connected** if the associated transition matrix $P(x, y) = \frac{c(x,y)}{c_x}$ is irreducible.

Exercise 4.1 Let (G, c) be a connected network. Let $Z \subset G$ be a subset such that $G \setminus Z$ is finite. Recall $T_Z = \inf\{t \geq 0 : X_t \in Z\}$ where $(X_t)_t$ is the random walk on (G, c).

Show that T_Z has an exponential tail; that is, there exists $K, \delta > 0$ such that for all t and all $x \in G$,

$$\mathbb{P}_x[T_Z > t] \leq Ke^{-\delta t}.$$

▷ solution ◁

From now on, unless otherwise specified, by *network* we will always refer to a connected network.

4.2 Gradient and Divergence

Let (G, c) be a network. Let $E = E(G, c) := \{(x, y) \in G \times G : c(x, y) > 0\}$. We write $x \sim y$ to denote the fact that $(x, y) \in E(G, c)$ (this implicitly depends on the specific conductance c).

We consider two spaces of functions. For $f \colon G \to \mathbb{C}$ and $F \colon E \to \mathbb{C}$, define

$$\|f\|_c^2 := \sum_x c_x |f(x)|^2, \qquad \|F\|_c^2 := \sum_{(x,y) \in E} \frac{1}{c(x,y)} |F(x,y)|^2,$$

and let

$$\ell^2(G, c) = \{f \colon G \to \mathbb{C} : \|f\|_c < \infty\},$$
$$\ell^2(E(G, c)) = \{F \colon E(G, c) \to \mathbb{C} : \|F\|_c < \infty\}.$$

Definition 4.2.1 Define

$$f, f' \in \ell^2(G, c), \qquad \langle f, f' \rangle_c := \sum_x c_x f(x) \overline{f'(x)},$$

and

$$F, F' \in \ell^2(E(G, c)), \qquad \langle F, F' \rangle_c := \sum_{(x,y) \in E(G,c)} \frac{1}{c(x,y)} F(x, y) \overline{F'(x, y)}.$$

The next exercise shows the classical fact that $\ell^2(G, c)$ and $\ell^2(E(G, c))$ are Hilbert spaces with inner product given by the above forms, respectively.

Exercise 4.2 Let N be a countable set. Let $m: N \rightarrow (0, \infty)$ be a positive function.

Show that for $f, g: N \rightarrow \mathbb{C}$, the form $\langle f, g \rangle := \sum_{x \in N} m(x) f(x) \overline{g(x)}$ defines an inner product structure when restricted to

$$f, g \in \ell^2(N, m) = \{f: G \rightarrow \mathbb{C} : \langle f, f \rangle < \infty\}.$$

Show that, in fact, $\ell^2(N, m)$ with this inner product is a Hilbert space. ▷ solution ◁

Remark 4.2.2 Note that if μ is a symmetric probability measure and $c(x, y) = \mu\left(x^{-1}y\right)$, then $c_x = 1$ for all x. So actually $\ell^2(G, \mu) = \ell^2(G) = \{f: G \rightarrow \mathbb{C} : \sum_x |f(x)|^2 < \infty\}$.

Definition 4.2.3 Let (G, c) be a network. For functions $f: G \rightarrow \mathbb{C}$ and $F: E(G, c) \rightarrow \mathbb{C}$ define two operators:

$$\nabla f(x, y) := c(x, y)(f(x) - f(y)), \qquad \mathrm{div} F(x) := \sum_{y \sim x} \frac{1}{c_x}(F(x, y) - F(y, x)).$$

Exercise 4.3 Show that $\nabla: \ell^2(G, c) \rightarrow \ell^2(E(G, c))$ and $\mathrm{div}: \ell^2(E(G, c)) \rightarrow \ell^2(G, c)$, and that these are bounded linear operators. ▷ solution ◁

The following is an extremely useful identity, sometimes called *integration by parts*, or *Green's identity*.

Theorem 4.2.4 Let (G, c) be a network. For any $f \in \ell^2(G, c)$ and $F \in \ell^2(E(G, c))$, we have

$$\langle f, \mathrm{div} F \rangle_c = \langle \nabla f, F \rangle_c.$$

That is, the operators ∇, div are adjoints of each other.

Proof Compute:

$$\langle f, \mathrm{div} F \rangle_c = \sum_x c_x f(x) \sum_{y \sim x} \frac{1}{c_x} (\overline{F(x, y)} - \overline{F(y, x)})$$

$$= \sum_{(x,y) \in E(G,c)} f(x) \overline{F(x, y)} - \sum_{(x,y) \in E(G,c)} f(x) \overline{F(y, x)}$$

$$= \sum_{(x,y) \in E(G,c)} \frac{1}{c(x,y)} c(x, y)(f(x) - f(y)) \overline{F(x, y)} = \langle \nabla f, F \rangle_c.$$

$\square$

4.3 Laplacian

Recall that a network (G, c) comes with an associated Markov chain with transition matrix $P(x, y) = \frac{c(x,y)}{c_x}$. The Laplace operator, or **Laplacian**, is defined to be $\Delta = I - P$ where I is the identity operator.

Exercise 4.4 Show that Δ is a self-adjoint bounded operator on $\ell^2(G, c)$, and that $\Delta = \frac{1}{2}\mathrm{div}\nabla$.
▷ solution ◁

Recall that a function is called **harmonic** at x if $\Delta f(x) = 0$, and **harmonic** if $\Delta f \equiv 0$.

Define a bilinear form on $\ell^2(G, c)$ by

$$\mathcal{E}(f, g) := \langle \nabla f, \nabla g \rangle_c = \sum_{x,y} c(x, y)(f(x) - f(y))\overline{(g(x) - g(y))}.$$

Exercise 4.5 Prove that $\mathcal{E}$ is a nonnegative definite bilinear form. That is, show that for $f, g \in \ell^2(G, c)$ and $\alpha \in \mathbb{C}$,

- $\mathcal{E}(\alpha f + g, h) = \alpha \mathcal{E}(f, h) + \mathcal{E}(g, h)$,
- $\mathcal{E}(f, g) = \overline{\mathcal{E}(g, f)}$,
- $\mathcal{E}(f, f) \geq 0$.

Show that $\mathcal{E}(f, f) = 0$ if and only if f is constant.

Integration by parts shows that $\mathcal{E}(f, g) = 2\langle \Delta f, g \rangle_c$ for $f, g \in \ell^2(G, c)$. The quantity

$$\mathcal{E}(f, f) = 2\langle \Delta f, f \rangle_c = \sum_{x,y} c(x, y)|f(x) - f(y)|^2$$

is usually called the **Dirichlet energy** of f.

Exercise 4.6 Let (G, c) be a connected network. Show that if $f \in \ell^2(G, c)$ and f is harmonic, then f is constant.
▷ solution ◁

4.4 Path Integrals

Let (G, c) be a network. A **finite path** in (G, c) is a finite sequence $\gamma = (\gamma_0, \gamma_1, \ldots, \gamma_n)$ of elements of G satisfying $c(\gamma_j, \gamma_{j+1}) > 0$ for all $0 \leq j < n$. If $\gamma_0 = x, \gamma_n = y$, we write $\gamma : x \to y$. The **length** of a path $\gamma = (\gamma_0, \gamma_1, \ldots, \gamma_n)$ is $|\gamma| = n$. If $\gamma : x \to x$ (i.e. $\gamma_0 = \gamma_{|\gamma|}$), we say that γ is a **cycle**.

Accordingly, an **infinite path** in (G, c) is an infinite sequence $(\gamma_0, \gamma_1, \dots)$ such that $c(\gamma_j, \gamma_{j+1}) > 0$ for all $j \geq 0$. The length of an infinite path is $|\gamma| = \infty$.

A **path** in (G, c) is either an infinite path or a finite path.

If $\gamma = (\gamma_0, \gamma_1, \dots, \gamma_n)$ is a finite path, the **reversal** of γ is defined to be the path $\check{\gamma} := (\gamma_n, \gamma_{n-1}, \dots, \gamma_1, \gamma_0)$. We also define the reversal a function $F \colon E(G, c) \to \mathbb{C}$ to be $\check{F}(x, y) = F(y, x)$.

Exercise 4.7 Show that $\operatorname{div}\check{F} = -\operatorname{div}F$.

Show that $\|\check{F}\|_c = \|F\|_c$.

If $F = \check{F}$ we say that F is **symmetric**. If $F = -\check{F}$ we say that F is **anti-symmetric**.

Exercise 4.8 Show that (G, c) is connected if and only if for all $x, y \in G$ there exists a finite path $\gamma \colon x \to y$.   ▷ solution ◁

For a finite path γ in a network (G, c) and a function $F \in \ell^2(E(G, c))$, define

$$\oint_\gamma F := \sum_{j=0}^{|\gamma|-1} F(\gamma_j, \gamma_{j+1}) \cdot \frac{1}{c(\gamma_j, \gamma_{j+1})}.$$

Exercise 4.9 Let $F \in \ell^2(E(G, c))$ and let γ be a finite path in (G, c). Show that:

- $\oint_{\check{\gamma}} F = \oint_\gamma \check{F}$.
- If $F = \nabla f$ then $\oint_\gamma F = f(\gamma_0) - f(\gamma_{|\gamma|})$.   ▷ solution ◁

Proposition 4.4.1 *Suppose that (G, c) is a connected network. If $\nabla f = \nabla g$ for some $f, g \in \ell^2(G, c)$ then there exists a constant $\eta \in \mathbb{C}$ such that $f \equiv g + \eta$.*

Proof Let $x, y \in G$ and let $\gamma \colon x \to y$ (which is possible since (G, c) is connected). Then,

$$f(x) - f(y) = \oint_\gamma \nabla f = \oint_\gamma \nabla g = g(x) - g(y).$$

Thus, $f(x) - g(x) = f(y) - g(y) =: \eta$ for all x, y.   □

A function $F \in \ell^2(E(G, c))$ is said to respect **Kirchhoff's cycle law** if for any cycle $\gamma \colon x \to x$ we have $\oint_\gamma F = 0$ (a cycle is just a path starting and ending at the same point). For example, if $F = \nabla f$ then by the above it respects Kirchhoff's cycle law. As we shall see, this is the only example.

Proposition 4.4.2 *Let (G, c) be a connected network. Then $F \in \ell^2(E(G, c))$ respects Kirchhoff's cycle law if and only if there exists $f \in \ell^2(G, c)$ such that $F = \nabla f$.*

Proof First note that $F = -\check{F}$ since for any x, y with $c(x, y) > 0$, we may consider the cycle (x, y, x) to get

$$0 = \oint_{(x,y,x)} F = c(x, y)(F(x, y) + F(y, x)).$$

If $\alpha: x \to y$ and $\beta: x \to y$ are two paths, then we may concatenate them to get a path $\gamma = \alpha\check{\beta} = (\alpha_0, \ldots, \alpha_{|\alpha|}, \beta_{|\beta|-1}, \ldots, \beta_0)$. Note that $\gamma: x \to x$, so

$$\oint_\alpha F - \oint_\beta F = \oint_\alpha F + \oint_{\check{\beta}} F = \oint_\gamma F = 0.$$

Thus, for any $\alpha: x \to y$ we have that the quantity $\oint_\alpha F$ does not depend on α, but only on x, y.

Now, fix some $o \in G$ and for any x define $f(x) := \oint_\alpha F$ for some path $\alpha: x \to o$ (so $f(o) = 0$). This is well defined by the above. It is immediate that if $\gamma: y \to o$ and $\gamma_1 = x$ then denoting $\gamma' = (\gamma_1, \ldots, \gamma_{|\gamma|})$ we have $\gamma': x \to o$ and

$$f(x) - f(y) = \oint_{\gamma'} F - \oint_\gamma F = -\tfrac{1}{c(x,y)}F(y, x) = \tfrac{1}{c(x,y)}F(x, y). \qquad \square$$

4.5 Voltage and Current

Definition 4.5.1 Let (G, c) be a network. Let A, Z be disjoint subsets of G. A **flow** on (G, c) from A to Z is a function $F: E(G, c) \to \mathbb{R}$ satisfying $F = -\check{F}$ (i.e. F is anti-symmetric) and $\mathrm{div}F(x) = 0$ for all $x \notin A \cup Z$.

A function having 0 divergence is sometimes said to satisfy *Kirchhoff's node law*.

If $A = \{a\}$ we say that F is a flow from a to Z. If $Z = \emptyset$ we say that F is a flow from A to infinity. If $A = \{a\}$ and $\mathrm{div}F(a) = 1$ we say that F is a **unit flow** from a to Z (or, respectively, from a to infinity in the case that $Z = \emptyset$).

Example 4.5.2 Let (G, c) be a network and let A, Z be disjoint subsets of G. Let $v: G \to \mathbb{R}$ be a function. Assume that $\Delta v(x) = 0$ for all $x \notin A \cup Z$ (that is, v is harmonic off $A \cup Z$). Let $I = \nabla v$. Then, I is anti-symmetric and for any $x \notin A \cup Z$ we have $\mathrm{div}I(x) = 2\Delta v(x) = 0$. So I is a flow from A to Z. △▽△

This example is a central one.

Definition 4.5.3 Let (G, c) be a network. Let A, Z be disjoint subsets of G.

A **voltage** from A to Z is a function $v : G \to \mathbb{R}$ that is harmonic outside $A \cup Z$; that is, $\Delta v(x) = 0$ for any $x \notin A \cup Z$.

If, in addition, $v(a) = 1$ for all $a \in A$ and $v(z) = 0$ for all $z \in Z$, we say that v is a **unit voltage** from A to Z.

If v is a voltage, the function $I = \nabla v$ is called the **current** (from A to Z) induced by v.

Exercise 4.10 Show that any current I is a flow that satisfies Kirchhoff's cycle law.

Show that if F is a flow from A to Z satisfying Kirchhoff's cycle law, then F is a current.

▷ solution ◁

Theorem 4.5.4 (Maximum principle) *Let A, Z be disjoint subsets of a connected network (G, c) with $G \backslash (A \cup Z)$ finite.*

Let v be a voltage from A to Z. If $\sup_x v(x) = v(z) < \infty$ then there exists $y \in A \cup Z$ such that $v(y) = \sup_x v(x)$. That is, the maximal value of a voltage (if it exists) is always attained on the "boundary" $A \cup Z$.

Proof Let v be any voltage from A to Z. Let $T = T_{A \cup Z} = \inf\{t : X_t \in A \cup Z\}$, where $(X_t)_t$ is the random walk on (G, c).

Now, if $\sup_x v(x) < \infty$, then $(v(X_{T \wedge t}))_t$ is a martingale, bounded from above. By the Martingale convergence theorem (Theorem 2.6.3), $v(X_{T \wedge t}) \xrightarrow{\text{a.s.}} M_\infty$ as $t \to \infty$, for some integrable random variable M_∞. However, $v(X_{T \wedge t}) \xrightarrow{\text{a.s.}} v(X_T)$ as well. So $v(X_T)$ is integrable.

Using that $G \backslash (A \cup Z)$ is finite, define $b = \max\{|v(x)| : x \notin A \cup Z\}$. Then the event $|v(X_{T \wedge t})| > b$ implies the event $\{T \le t\}$. Thus, for $K > b$ we have

$$\mathbb{E}_x \left[|v(X_{T \wedge t})| \mathbf{1}_{\{|v(X_{T \wedge t})| > K\}} \right] \le \mathbb{E}_x \left[|v(X_T)| \mathbf{1}_{\{|v(X_T)| > K\}} \right] \le \mathbb{E}_x |v(X_T)| < \infty.$$

Taking a supremum over t and then a limit $K \to \infty$, we see that $(v(X_{T \wedge t}))_t$ is uniformly integrable.

Since $G \backslash (A \cup Z)$ is finite and (G, c) is connected, we know that $T < \infty$ a.s. (In fact, we have seen that T has an exponential tail.) The optional stopping theorem (Theorem 2.3.3) now gives that $v(x) = \mathbb{E}_x[v(X_T)]$ for any $x \in G$. Hence, if $v(z) = \sup_x v(x)$ then it must be that there is some y with $\mathbb{P}_z[X_T = y] > 0$ such that $v(y) \ge v(z)$.   □

Exercise 4.11 Let (G, c) be a connected network. Let v be a voltage from A to Z. (We do not necessarily assume that $G \backslash (A \cup Z)$ is finite.) Assume that $T = T_{A \cup Z} < \infty$, $\mathbb{P}_x$-a.s., for all x. Assume further that $v(X_T)$ is integrable under $\mathbb{P}_x$ for all x. Also, assume that $\sup_{x \notin A \cup Z} |v(x)| < \infty$.

Show that $v(x) = \mathbb{E}_x[v(X_T)]$ for all x.

Conclude that v satisfies a maximum principle.

Show that v also satisfies a minimum principle; that is, if $\inf_x v(x) = v(z) > -\infty$, then there exists $y \in A \cup Z$ such that $v(y) = \inf_x v(x)$. ▷ solution ◁

Exercise 4.12 Let A, Z be disjoint subsets of a connected network (G, c) with $G \backslash (A \cup Z)$ finite.

Show that the only voltage from A to Z that is 0 on $A \cup Z$ is the identically 0 function.

Show that if two voltages from A to Z have the same values on $A \cup Z$, then they are equal.

Conclude that there is a unique unit voltage from A to Z. ▷ solution ◁

If (G, c) is a network, we can consider the function $r(x, y) = \frac{1}{c(x,y)}$; r is called the **resistance**. Note that if v is a voltage and I the induced current then $v(x) - v(y) = I(x, y)r(x, y)$, which is reminiscent of classical physics.

4.6 Effective Conductance

Definition 4.6.1 Let (G, c) be a network. Let $a \notin Z$, and assume $G \backslash Z$ is finite. Let v be the unit voltage from a to Z. The **effective conductance** between a and Z is defined to be

$$C_{\text{eff}}(a, Z) = \frac{c_a}{2} \text{div} I(a) = \sum_x I(a, x) = \sum_x c(a, x)(v(a) - v(x)) = c_a \Delta v(a).$$

The **effective resistance** is defined to be $\mathcal{R}_{\text{eff}}(a, Z) = \frac{1}{C_{\text{eff}}(a,Z)}$.

Exercise 4.13 Let (G, c) be a network. Let $a \notin Z$, and assume $G \backslash Z$ is finite. Let v be a voltage from a to Z such that $v|_Z \equiv 0$ and $v(a) \neq 0$. Show that $C_{\text{eff}}(a, Z) = \frac{c_a}{v(a)} \Delta v(a)$. ▷ solution ◁

Exercise 4.14 Let (G, c) be a network. Let $a \notin Z$, and assume $G \backslash Z$ is finite. Let v be a unit voltage from a to Z. Show that $2C_{\text{eff}}(a, Z) = \mathcal{E}(v, v)$. ▷ solution ◁

Exercise 4.15 Let (G, c) be a network. Let $a \notin Z$. Assume that $G \backslash Z$ is finite. Show that $C_{\text{eff}}(a, Z) = c_a \, \mathbb{P}_a \left[T_Z < T_a^+ \right]$.

▷ solution ◁

Theorem 4.6.2 *Let (G, c) be a network. Let $(G_n)_n$ be an exhaustion of G; that is, a nondecreasing sequence $G_n \subset G_{n+1}$ of finite subsets of G such that $G = \bigcup_n G_n$. Let $Z_n = G \backslash G_n$. Assume that $a \in \bigcap_n G_n$.*
Then the limit

$$C_{\text{eff}}(a, \infty) := \lim_{n \to \infty} C_{\text{eff}}(a, Z_n)$$

exists and does not depend on the specific choice of exhaustion. In fact,

$$C_{\text{eff}}(a, \infty) = c_a \, \mathbb{P}_a \left[T_a^+ = \infty \right] .$$

Proof Since $G_n \subset G_{n+1}$, the sequence $(\{T_{Z_n} < T_a^+\})_n$ is a decreasing sequence of events. Thus,

$$c_a \cdot \mathbb{P}_a \left[T_a^+ = \infty \right] = \lim_{n \to \infty} c_a \cdot \mathbb{P}_a \left[T_{Z_n} < T_a^+ \right] = \lim_{n \to \infty} C_{\text{eff}}(a, Z_n). \qquad \square$$

$C_{\text{eff}}(a, \infty)$ is called the **effective conductance to infinity**. We define the **effective resistance to infinity** by $\mathcal{R}_{\text{eff}}(a, \infty) = \frac{1}{C_{\text{eff}}(a, \infty)} = \lim_{n \to \infty} \mathcal{R}_{\text{eff}}(a, Z_n)$.
We then immediately obtain the following characterization of recurrence for networks.

Corollary 4.6.3 A connected network (G, c) is recurrent if and only if $\mathcal{R}_{\text{eff}}(a, \infty) = \infty$ for some a.

4.7 Thompson's Principle and Rayleigh Monotonicity

In this section we may want to consider effective conductance of different networks on the same set of points G. Hence, if $(G, c), (G, c')$ are two networks, we use the notation $C_{\text{eff}}(a, Z \mid c), C_{\text{eff}}(a, Z \mid c')$ to specify the conductance function used.

Theorem 4.7.1 (Thompson's principle) *Let (G, c) be a network. Let A, Z be disjoint subsets such that $G \backslash (A \cup Z)$ is finite. Let v be the unit voltage from A to Z. For any function $f : G \to \mathbb{R}$ such that $f|_A \equiv 1$ and $f|_Z \equiv 0$, we have $\mathcal{E}(f, f) \geq \mathcal{E}(v, v)$.*

That is, the function minimizing the Dirichlet energy is the voltage.

Proof Since $f - v|_{A \cup Z} \equiv 0$ and since v is harmonic off $A \cup Z$, we have that $(f - v)\Delta v \equiv 0$. Also, $f - v \in \ell^2(G, c)$. Thus,

$$\tfrac{1}{2}\mathcal{E}(f - v, v) = \langle \Delta(f - v), v \rangle = \langle f - v, \Delta v \rangle = 0.$$

This implies that

$$\mathcal{E}(f, f) = \mathcal{E}(f - v + v, f - v + v)$$
$$= \mathcal{E}(f - v, f - v) + \mathcal{E}(v, v) - 2\mathcal{E}(f - v, v) \geq \mathcal{E}(v, v). \qquad \square$$

Theorem 4.7.2 (Thompson's principle, dual form) *Let (G, c) be a network. Let $a \notin Z$ be such that $G \backslash Z$ is finite. Let v be a voltage from a to Z such that $v|_Z \equiv 0$ and such that the current $I = \nabla v$ is a unit flow from a to Z. Then, for any unit flow F from a to Z, we have $\|F\|_c \geq \|I\|_c$.*

Proof Since $\operatorname{div}(F - I)|_{G \backslash Z} \equiv 0$ and $v|_Z \equiv 0$, we have that $\operatorname{div}(F - I) \cdot v \equiv 0$. Thus,

$$\langle F - I, I \rangle_c = \langle F - I, \nabla v \rangle_c = \langle \operatorname{div}(F - I), v \rangle_c = 0.$$

So

$$\|F\|_c^2 = \|F - I + I\|_c^2 = \|F - I\|_c^2 + \|I\|_c^2 - 2\langle F - I, I \rangle_c \geq \|I\|_c^2. \qquad \square$$

Theorem 4.7.3 (Rayleigh monotonicity) *Let $(G, c), (G, c')$ be two networks on the same set of points. Let $a \notin Z$ such that $G \backslash Z$ is finite. Suppose that $c(x, y) \leq c'(x, y)$ for all x, y. Then,*

$$C_{\mathrm{eff}}(a, Z \mid c) \leq C_{\mathrm{eff}}(a, Z \mid c').$$

Proof Let v be the unit voltage from a to Z with respect to the network (G, c) and let u be the unit voltage from a to Z with respect to the network (G, c'). By Exercise 4.14 we know that $\mathcal{E}_c(v, v) = 2C_{\mathrm{eff}}(a, Z \mid c)$ and $\mathcal{E}_{c'}(u, u) = 2C_{\mathrm{eff}}(a, Z \mid c')$. Here, $\mathcal{E}_c, \mathcal{E}_{c'}$ are the Dirichlet energies with respect to the conductances c, c', respectively.

We now use Thompson's principle (Theorem 4.7.1), which tells us that v minimizes $\mathcal{E}_c$ over functions with value 1 on a and 0 on Z. Hence,

$$2C_{\mathrm{eff}}(a, Z \mid c) = \mathcal{E}_c(v, v) \leq \mathcal{E}_c(u, u) = \sum_{x,y} c(x, y)|u(x) - u(y)|^2$$

$$\leq \sum_{x,y} c'(x, y)|u(x) - u(y)|^2 = \mathcal{E}_{c'}(u, u) = 2C_{\mathrm{eff}}(a, Z \mid c').$$

$\square$

Perhaps the most important application of Rayleigh monotonicity is the following corollary. It states that increasing the conductances of a network (in any way imaginable) cannot change a transient network into a recurrent one.

Corollary 4.7.4 Let $(G, c), (G, c')$ be two connected networks on the same set of points. Suppose that $c(x, y) \leq c'(x, y)$ for all x, y.

If (G, c) is transient then (G, c') is transient.

Exercise 4.16 Prove Corollary 4.7.4. ▷ solution ◁

Exercise 4.17 Let (G, c) be a network. Let $G' \subset G$ and let $c'(x, y) = c(x, y)\mathbf{1}_{\{x, y \in G'\}}$. Assume that both (G, c) and (G', c') are connected.

Show that if (G', c') is transient then (G, c) is transient. ▷ solution ◁

4.8 Green Function

Definition 4.8.1 Let (G, c) be a connected network. Let $Z \subset G$ be a subset, possibly empty. Define

$$g_Z(x, y) = \sum_{t=0}^{\infty} \mathbb{P}_x[X_t = y, \, t < T_Z] = \mathbb{E}_x \sum_{t < T_Z} \mathbf{1}_{\{X_t = y\}}.$$

This is the expected number of visits to y before hitting Z, started at x. It may be that g_Z is infinite (for example, if $Z = \emptyset$ and (G, c) is recurrent). We call g_Z the **Green function**. When $Z = \emptyset$ we write $g(x, y) = g_\emptyset(x, y) = \sum_{t=0}^{\infty} \mathbb{P}_x[X_t = y]$.

Theorem 4.8.2 *Let (G, c) be a connected network. Let $Z \subset G$. For any $x, y \in G$ we have that*

$$g_Z(x, y) = \frac{\mathbb{P}_x\left[T_y < T_Z\right]}{\mathbb{P}_y\left[T_y^+ \geq T_Z\right]},$$

where $T_Z = \infty$ if $Z = \emptyset$.

Proof Fix $x, y \in G$. Define

$$V = \sum_{t=0}^{T_Z-1} \mathbf{1}_{\{X_t = y\}},$$

which is the number of visits to y before hitting Z (if $T_Z = 0$ we define $V = 0$).

Since $\{t < T_Z \wedge T_y, \, X_t = y\} = \emptyset$, the strong Markov property at time $T_y \wedge T_Z$ gives that

$$\mathbb{E}_x[V] = \mathbb{E}_x \sum_{T_y \wedge T_Z \leq t < T_Z} \mathbf{1}_{\{X_t = y\}} = \mathbb{P}_x\left[T_y \wedge T_Z < T_Z\right] \cdot \mathbb{E}_y[V].$$

So $g_Z(x, y) = \mathbb{P}_x \left[T_y < T_Z \right] \cdot g_Z(y, y)$.

Similarly, since $\mathbb{P}_y[X_0 = y] = 1$, then if $y \notin Z$ the strong Markov property at time $T_y^+ \wedge T_Z$ gives

$$\mathbb{E}_y[V] = 1 + \mathbb{E}_y \sum_{T_y^+ \wedge T_Z \leq t < T_Z} \mathbf{1}_{\{X_t = y\}} = 1 + \mathbb{P}_y \left[T_y^+ \wedge T_Z < T_Z \right] \cdot \mathbb{E}_y[V],$$

so that for $y \notin Z$,

$$g_Z(y, y) = \frac{1}{\mathbb{P}_y \left[T_y^+ \geq T_Z \right]}.$$

For $y \in Z$ it is obvious that $g_Z(y, y) = 0$ because $\mathbb{P}_y$-a.s. we have $T_Z = 0$. Also, for $y \in Z$ we have $\mathbb{P}_y \left[T_y^+ \geq T_Z \right] = 1$ and $\mathbb{P}_x \left[T_y < T_Z \right] = 0$ for any $x \in G$. □

It is useful to note the relationship $g_Z(x, y) = \mathbb{P}_x \left[T_y < T_Z \right] \cdot g_Z(y, y)$. Also, compare $g_Z(y, y) = \frac{1}{\mathbb{P}_y[T_y^+ \geq T_Z]}$ to what we obtained from the *excursion decomposition* method; see Theorem 3.5.2.

Corollary 4.8.3 Let (G, c) be a connected network and $a \notin Z \subset G$ for $G \backslash Z$ finite. Then,

$$c_a \mathcal{R}_{\text{eff}}(a, Z) = g_Z(a, a), \qquad c_a \mathcal{R}_{\text{eff}}(a, \infty) = g(a, a).$$

Exercise 4.18 Let (G, c) be a connected network. Show that the following are equivalent.

(1) $g(x, x) < \infty$ for some $x \in G$.
(2) $g(x, y) < \infty$ for some $x, y \in G$.
(3) $g(x, y) < \infty$ for all $x, y \in G$.
(4) The random walk on (G, c) is transient. ▷ solution ◁

Compare this to Theorem 3.6.2.

Exercise 4.19 Let (G, c) be a connected network. Show that if $Z \neq \emptyset$ then $g_Z(x, y) < \infty$ for any $x, y \in G$. ▷ solution ◁

Proposition 4.8.4 *Let (G, c) be a connected network and let $Z \subset G$. Assume that $g_Z(x, y) < \infty$ for all x, y.*

For any $x \notin Z$ the function $h(z) = g_Z(z, x)$ admits $\Delta h(z) = \mathbf{1}_{\{z = x\}}$ for all $z \notin Z$.

Proof Fix $x \notin Z$. Let $h(z) = g_Z(z, x)$. Note that for $t > 0$ and $z \notin Z$,

$$\mathbb{P}_z[X_t = x, \, t < T_Z] = \sum_y P(z, y) \, \mathbb{P}_y[X_{t-1} = x, \, t - 1 < T_Z].$$

So if $z \notin Z$,

$$h(z) = \mathbf{1}_{\{x=z\}} + \sum_{t=1}^{\infty} \mathbb{P}_z[X_t = x, \, t < T_Z]$$

$$= \mathbf{1}_{\{x=z\}} + \sum_{t=1}^{\infty} \sum_y P(z, y) \, \mathbb{P}_y[X_{t-1} = x, \, t - 1 < T_Z]$$

$$= \mathbf{1}_{\{x=z\}} + \sum_y P(z, y) \sum_{t=0}^{\infty} \mathbb{P}_y[X_t = x, \, t < T_Z] = \mathbf{1}_{\{x=z\}} + Ph(z).$$

So $\Delta h(z) = \mathbf{1}_{\{x=z\}}$. $\square$

Proposition 4.8.5 *Let (G, c) be a connected network and let $Z \subset G$.*
For all $x, y \in G$ we have $c_x g_Z(x, y) = c_y g_Z(y, x)$.

Proof If $g_Z(x, y) = \infty$ for some x, y then there is nothing to prove.
So assume that $g_Z(x, y) < \infty$.
Note that $c_x P(x, y) = c_y P(y, x)$ for all x, y. Thus, for any sequence $x = x_0, x_1, \ldots, x_n = y$,

$$c_x \, \mathbb{P}_x[X_1 = x_1, \ldots, X_n = x_n] = c_y \, \mathbb{P}_y[X_1 = x_{n-1}, \ldots, X_n = x_0].$$

That is, for any finite path $\gamma : x \rightarrow y$,

$$c_x \, \mathbb{P}_x[(X_0, \ldots, X_n) = \gamma] = c_y \, \mathbb{P}_y[(X_0, \ldots, X_n) = \check{\gamma}].$$

Now, compute for any $t > 0$, if $x \notin Z$ and $y \notin Z$, then

$$c_x \, \mathbb{P}_x[X_t = y, \, t < T_Z] = \sum_{\substack{\gamma : x \rightarrow y \\ |\gamma| = t, \, \gamma \cap Z = \emptyset}} c_x \, \mathbb{P}_x[(X_0, \ldots, X_t) = \gamma]$$

$$= \sum_{\substack{\gamma : x \rightarrow y \\ |\gamma| = t, \, \gamma \cap Z = \emptyset}} c_y \, \mathbb{P}_y[(X_0, \ldots, X_t) = \check{\gamma}]$$

$$= \sum_{\substack{\gamma : y \rightarrow x \\ |\gamma| = t, \, \gamma \cap Z = \emptyset}} c_y \, \mathbb{P}_y[(X_0, \ldots, X_t) = \gamma]$$

$$= c_y \, \mathbb{P}_y[X_t = x, \, t < T_Z].$$

(This technique is known as *path reversal*.) Summing over t, we have that $c_x g_Z(x, y) = c_y g_Z(y, x)$ for all $x, y \notin Z$.
Also, if $x \in Z$ then $g_Z(x, y) = 0 = g_Z(y, x)$. $\square$

Example 4.8.6 Let (G, c) be a connected network. Let $Z \subset G$ be such that $G \setminus Z$ is finite. Let $a \notin Z$. Let $v(x) = \frac{1}{2}g_Z(x, a)$. So v is harmonic off $Z \cup \{a\}$; that is, v is a voltage from a to Z. Also, $v|_Z \equiv 0$. Note that the induced current $I = \nabla v$ satisfies $\mathrm{div} I(a) = 2\Delta v(a) = 1$, so I is a unit flow from a to Z. Recall that by Thompson's principle, I minimizes the norm $\| \cdot \|_c$ over all unit flows from a to Z.

△▽△

We now provide a probabilistic interpretation of current and effective resistance.

Proposition 4.8.7 *Let (G, c) be a connected network. Let $a \notin Z \subset G$ be such that $G \setminus Z$ is finite.*

For $x, y \in G$, denote by $V_{x,y}$ the number of times the random walk traverses from x to y up to hitting Z; that is,

$$V_{x,y} = \sum_{t=1}^{T_Z} \mathbf{1}_{\{X_{t-1}=x, \, X_t=y\}}.$$

(Note that if $c(x, y) = 0$ then $V_{x,y} = 0$.) Then,

$$\mathbb{E}_a[V_{x,y} - V_{y,x}] = I(x, y) \cdot \mathcal{R}_{\mathrm{eff}}(a, Z),$$

where v is the unit voltage from a to Z, and $I = \nabla v$ is the current.

Proof Let $v(x) = \frac{g_Z(x,a)}{g_Z(a,a)}$. Then v is a unit voltage from a to Z. We have already seen that

$$g_Z(a, a) = \frac{1}{\mathbb{P}_a[T_Z \leq T_a^+]} = c_a \mathcal{R}_{\mathrm{eff}}(a, Z).$$

Also, the strong Markov property at time $T_Z \wedge t$ gives for any $(x, y) \in E(G, c)$,

$$\mathbb{E}_a[V_{x,y}] = \sum_{t=1}^{\infty} \mathbb{P}_a[X_{t-1} = x, \, X_t = y, \, t \leq T_Z]$$

$$= \sum_{t=0}^{\infty} \mathbb{P}_a[X_t = x, \, X_{t+1} = y, \, t < T_Z]$$

$$= \sum_{t=0}^{\infty} \mathbb{P}_a[X_t = x, \, t < T_Z] \cdot \mathbb{P}_x[X_1 = y]$$

$$= g_Z(a, x) P(x, y) = \frac{1}{c_a} c(x, y) g_Z(x, a) = \mathcal{R}_{\mathrm{eff}}(a, Z) c(x, y) v(x).$$

□

4.9 Finite Energy Flows

Recall that a flow from a to infinity in a network (G, c) is a function $F: E(G, c) \to \mathbb{R}$ such that $F = -\check{F}$ (F is anti-symmetric) and $\text{div}F(a) \neq 0$ and $\text{div}F(x) = 0$ for all $x \neq a$. If, in addition, $\text{div}F(a) = 1$ then F is called a unit flow from a to infinity.

The **energy** of a flow F is just $\|F\|_c^2 = \langle F, F \rangle_c = \sum_{x \sim y} r(x, y)|F(x, y)|^2$; here, $r(x, y) = \frac{1}{c(x,y)}$ is the resistance.

Theorem 4.9.1 *Let (G, c) be a connected network. Then (G, c) is transient if and only if there exists a finite energy flow. That is, there exists $a \in G$ and a flow F from a to infinity such that $F \in \ell^2(E(G, c))$.*

Proof Suppose that F is a finite energy flow from a to infinity. By multiplying F by a constant, we may assume without loss of generality that $\text{div}F(a) = 1$.

Let $(G_n)_n$ be an exhaustion of G, that is, a nondecreasing sequence $G_n \subset G_{n+1}$ of finite subsets with $G = \bigcup_n G_n$, such that $a \in G_1$. Let $Z_n = G \setminus G_n$. Let $v_n(x) = \frac{1}{2}\mathfrak{g}_{Z_n}(x, a)$. Note that $I_n = \nabla v_n$ is a unit flow from a to Z_n. Since F is a unit flow from a to Z_n, by Thompson's principle we have $\|I_n\|_c \leq \|F\|_c < \infty$ for all n.

Moreover,

$$\|I_n\|_c^2 = \mathcal{E}(v_n, v_n) = 2\langle \Delta v_n, v_n \rangle_c = 2 \sum_x c_x \Delta v_n(x) v_n(x)$$

$$= c_a v_n(a) = c_a \frac{1}{2}\mathfrak{g}_{Z_n}(a, a) = \frac{c_a}{2\mathbb{P}_a[T_{Z_n} \leq T_a^+]} = \frac{c_a^2}{2}\mathcal{R}_{\text{eff}}(a, Z_n).$$

Thus,

$$\mathcal{R}_{\text{eff}}(a, \infty) = \lim_{n \to \infty} \mathcal{R}_{\text{eff}}(a, Z_n) \leq \frac{2}{c_a^2}\|F\|_c^2 < \infty.$$

So (G, c) is transient.

For the other direction, assume that (G, c) is transient. Let $v(x) = \mathbb{P}_x[T_a < \infty]$, and let $I = \nabla v$. Note that $\text{div}I = 2\Delta v$ so $\text{div}I(x) = 0$ if $x \neq a$ and since $v(a) = 1$,

$$\text{div}I(a) = 2\Delta v(a) = 2 \sum_x P(a, x)(1 - v(x))$$

$$= 2 \sum_x P(a, x) \mathbb{P}_x[T_a = \infty] = \mathbb{P}_a[T_a^+ = \infty] > 0,$$

by transience. So I is a flow from a to infinity.

To estimate the energy of I, let $(G_n)_n$ be an exhaustion of G as above, with $a \in G_1$ and $Z_n = G \setminus G_n$. Let $v_n(x) = \mathbb{P}_x[T_a < T_{Z_n}]$. So $v_n(x) \nearrow v(x)$ for all

x, and $v_n(a) = v(a) = 1$. Thus, for any x, y we have $\nabla v_n(x, y) \to I(x, y)$. By Fatou's lemma we now have

$$\|I\|_c^2 = \sum_{x,y} r(x, y)|I(x, y)|^2 = \sum_{x,y} r(x, y) \lim_{n \to \infty} |\nabla v_n(x, y)|^2$$

$$\leq \liminf_{n \to \infty} \sum_{x,y} r(x, y)|\nabla v_n(x, y)|^2 = \liminf_{n \to \infty} \|\nabla v_n\|_c^2.$$

Since

$$g_{Z_n}(x, a) = v_n(x) \cdot g_{Z_n}(a, a),$$

we get that

$$\|\nabla v_n\|_c^2 = 2 \langle \Delta v_n, v_n \rangle_c = 2 \sum_x c_x \Delta v_n(x) v_n(x) = \frac{2 c_a}{g_{Z_n}(a, a)}.$$

Because $g_{Z_n}(a, a) \to g(a, a)$ we conclude that $\|I\|_c^2 \leq \frac{2 c_a}{g(a, a)} < \infty$. □

4.10 Paths and Summable Intersection Tails

Let (G, c) be a network. Recall that an infinite path is a sequence $\gamma = (\gamma_0, \gamma_1, \ldots)$ such that $c(\gamma_j, \gamma_{j+1}) > 0$ for all j. A path γ is called **simple** if $\gamma_j \neq \gamma_i$ for all $i \neq j$. Let Γ_a be the collection of all infinite simple paths γ with $\gamma_0 = a$.

For two infinite simple paths α, β, let

$$E(\alpha, \beta) = \{(x, y) : \exists\, i, j,\ \alpha_i = \beta_j = x,\ \alpha_{i+1} = \beta_{j+1} = y\}.$$

That is, $E(\alpha, \beta)$ is the set of all (x, y) such that $c(x, y) > 0$ and both α and β traverse (x, y). Recall the resistance $r(x, y) = \frac{1}{c(x,y)}$, and let

$$R(\alpha, \beta) = \sum_{(x,y) \in E(\alpha, \beta)} r(x, y).$$

A probability measure λ on Γ_a is called **SIT** (summable intersection tails) if the following holds: For α, β, two independent paths each of law λ, we have

$$\mathbb{E}_{\lambda \otimes \lambda}[R(\alpha, \beta)] < \infty.$$

That is, the total resistance of edges traversed by two independent paths has finite expectation. (Here $\lambda \otimes \lambda$ is the product measure on $\Gamma_a \times \Gamma_a$ of two independent paths.)

A network (G, c) is called SIT if there exist $a \in G$ and a probability measure λ on Γ_a such that λ is SIT.

Theorem 4.10.1 *A connected network (G, c) is transient if and only if (G, c) is SIT.*

The proof of the theorem is split into two parts. The simpler part is given by Proposition 4.10.2.

Proposition 4.10.2 *If (G, c) is SIT then (G, c) is transient.*

Proof Assume that (G, c) is SIT. Let a be such that some probability measure λ on Γ_a is SIT. Let α be a random path drawn from λ. Define

$$F(x, y) = \mathbb{P}_\lambda[\exists j, \, \alpha_j = x, \, \alpha_{j+1} = y] - \mathbb{P}_\lambda[\exists j, \, \alpha_j = y, \, \alpha_{j+1} = x].$$

Note that $F = -\check{F}$.

Since α is simple, we have that $\alpha_0 = a$ and $\alpha_j \neq a$ for any $j > 0$. Thus, $F(a, x) = \mathbb{P}_\lambda[\alpha_1 = x]$, and

$$\operatorname{div}F(a) = \frac{2}{c_a} \sum_x F(a, x) = \frac{2}{c_a}.$$

Since α is a simple path,

$$\sum_{y \sim x} \mathbb{P}_\lambda[\exists j, \, \alpha_j = x, \, \alpha_{j+1} = y] = \mathbb{P}_\lambda[\exists j, \, \alpha_j = x],$$

and

$$\sum_{y \sim x} \mathbb{P}_\lambda[\exists j, \, \alpha_j = y, \, \alpha_{j+1} = x] = \mathbb{P}_\lambda[\exists j, \, \alpha_{j+1} = x].$$

So for any $x \neq a$ we get that

$$\operatorname{div}F(x) = \frac{2}{c_x} \sum_{y \sim x} (\mathbb{P}_\lambda[\exists j, \, \alpha_j = x, \, \alpha_{j+1} = y] - \mathbb{P}_\lambda[\exists j, \, \alpha_j = y, \, \alpha_{j+1} = x]) = 0.$$

Hence, F is a flow from a to infinity.

Finally, since

$$|F(x, y)|^2 \leq (\mathbb{P}_\lambda[\exists j, \, \alpha_j = x, \, \alpha_{j+1} = y])^2 + (\mathbb{P}_\lambda[\exists j, \, \alpha_j = y, \, \alpha_{j+1} = x])^2,$$

summing over all edges we obtain

$$\|F\|_c^2 \leq 2 \sum_{x \sim y} r(x, y)(\mathbb{P}_\lambda[\exists j, \, \alpha_j = x, \, \alpha_{j+1} = y])^2$$

$$= 2 \sum_{x \sim y} r(x, y)\, \mathbb{P}_{\lambda \otimes \lambda}[(x, y) \in E(\alpha, \beta)] = 2\, \mathbb{E}_{\lambda \otimes \lambda}[R(\alpha, \beta)] < \infty.$$

So F has finite energy and hence (G, c) is transient. $\qquad\square$

We now move to the proof of the second part of Theorem 4.10.1, namely that transience implies SIT.

For a flow F from a to infinity, we say that a cycle $\gamma: x \to x$ is a **positive flow cycle** if $F(\gamma_j, \gamma_{j+1}) > 0$ for all $0 \le j < |\gamma|$. Note that if F admits some positive flow cycle $\gamma: x \to x$, then

$$\oint_\gamma F = \sum_{j=0}^{|\gamma|-1} F(\gamma_j, \gamma_{j+1}) \frac{1}{c(\gamma_j, \gamma_{j+1})} > 0.$$

So F does not respect Kirchhoff's cycle law.

Lemma 4.10.3 *If (G, c) is a connected transient network, then there exists a unit flow F from some $a \in G$ to infinity such that no cycle is a positive flow cycle. Moreover, $F(x, a) \le 0$ for all $x \in G$ (i.e. there is no flow into a).*

Proof Define $v(x) = \frac{1}{2}g(x, a)$ and $F = \nabla v$. So $F = -\check{F}$, and

$$\mathrm{div} F(x) = 2\Delta v(x) = \mathbf{1}_{\{x=a\}}.$$

So F is a unit flow. Also,

$$g(x, a) = \mathbb{P}_x[T_a < \infty] \cdot g(a, a),$$

so $F(x, a) \le 0$ for all x.

Finally, since $F = \nabla v$ we know that F respects Kirchhoff's cycle law, so F cannot admit a positive flow cycle. $\qquad\square$

Let F be a unit flow in (G, c) from a to infinity. Suppose that F does not admit any positive flow cycle. We define a new Markov chain on G, which is different from the random walk with transition matrix P. Define

$$f(x) = \sum_{y:F(x,y)>0} F(x, y)$$

and

$$Q(x, y) = \begin{cases} \frac{1}{f(x)}F(x, y) & F(x, y) > 0, \\ 0 & F(x, y) \le 0. \end{cases}$$

In order to differentiate from the random walk, denoted by $(X_t)_t$, we will use $(Y_t)_t$ to denote the Markov chain with transition matrix Q, and we will use $\mathbb{Q}_x$ to denote the corresponding probability measures. Note that T_x will still denote hitting times for this chain.

Claim 4.10.4 Since the flow F does not admit positive flow cycles, we get that the chain $(Y_t)_t$ is a simple path in the network (G, c), $\mathbb{Q}_x$-a.s. for any x.

Proof First, note that $(Y_t)_t$ is indeed a path in (G, c), because $Q(x, y) > 0$ only if $c(x, y) > 0$.

If $(Y_t)_t$ is not a simple path with some positive probability, then there must exist a point x and a time $t > 0$ such that $Q^t(x, x) = \mathbb{Q}_x[Y_t = x] > 0$. Thus, there is a cycle $\gamma : x \to x$ such that

$$0 < \mathbb{Q}_x[(Y_0, \ldots, Y_t) = \gamma] = \prod_{j=0}^{|\gamma|-1} Q(\gamma_j, \gamma_{j+1})$$

$$= \prod_{j=0}^{|\gamma|-1} \frac{1}{f(\gamma_j)} F(\gamma_j, \gamma_{j+1}) \mathbf{1}_{\{F(\gamma_j, \gamma_{j+1}) > 0\}}.$$

This implies that $F(\gamma_j, \gamma_{j+1}) > 0$ for all $0 \leq j < |\gamma|$, meaning that γ is a positive flow cycle, a contradiction! □

Claim 4.10.5 For all $x \in G$ we have

$$\sum_{j=0}^{\infty} Q^j(a, x) = \mathbb{Q}_a[T_x < \infty] \leq \tfrac{c_a}{2} f(x).$$

Proof Since $F = -\check{F}$,

$$f(x) = \sum_{y : F(x, y) > 0} F(x, y) = \frac{2}{c_x} \mathrm{div} F(x) - \sum_{y : F(x, y) < 0} F(x, y)$$

$$= \frac{2}{c_a} \mathbf{1}_{\{x = a\}} + \sum_{y : F(y, x) > 0} F(y, x).$$

Thus,

$$(fQ)(x) = \sum_y f(y) Q(y, x) = \sum_{y : F(y, x) > 0} F(y, x) = f(x) - \frac{2}{c_a} \delta_a(x).$$

Iterating this inductively,

$$(fQ^t)(x) = \left(fQ^{t-1} \right)(x) - \frac{2}{c_a} \left(\delta_a Q^{t-1} \right)(x)$$

$$= \cdots = f(x) - \frac{2}{c_a} \sum_{j=0}^{t-2} Q^j(a, x) - \frac{2}{c_a} Q^{t-1}(a, x)$$

$$= f(x) - \frac{2}{c_a} \sum_{j=0}^{t-1} Q^j(a, x).$$

Since fQ^t has only nonnegative entries, taking $t \to \infty$ we arrive at

$$\sum_{j=0}^{\infty} Q^j(a, x) \le \tfrac{c_a}{2} f(x).$$

Now, let A be the event that $(Y_t)_t$ is a simple path, and recall that $Q_a[A] = 1$. Note that for any x, the events $(\{Y_j = x\} \cap A)_{j \ge 0}$ are mutually disjoint. Thus,

$$\tfrac{c_a}{2} f(x) \ge \sum_{j=0}^{\infty} Q^j(a, x) = \sum_{j=0}^{\infty} Q_a[Y_j = x, A]$$

$$= Q_a[\exists\, j \ge 0,\ Y_j = x] = Q_a[T_x < \infty]. \qquad \square$$

Claim 4.10.6 For all $x \sim y$ we have

$$Q_a[\exists\, j \ge 0,\ Y_j = x,\ Y_{j+1} = y] \le \tfrac{c_a}{2} F(x, y)\mathbf{1}_{\{F(x,y)>0\}}.$$

Proof As before, let A be the event that $(Y_t)_t$ is a simple path, and recall that $Q_a[A] = 1$. Since the events $(\{Y_j = x,\ Y_{j+1} = y\} \cap A)_{j \ge 0}$ are disjoint, we use the previous claim to compute:

$$Q_a[\exists\, j \ge 0,\ Y_j = x,\ Y_{j+1} = y] = \sum_{j=0}^{\infty} Q^j(a, x) Q(x, y)$$

$$= \sum_{j=0}^{\infty} Q^j(a, x) \tfrac{1}{f(x)} F(x, y)\mathbf{1}_{\{F(x,y)>0\}} \le \tfrac{c_a}{2} F(x, y)\mathbf{1}_{\{F(x,y)>0\}}. \qquad \square$$

Claim 4.10.7 If $\|F\|_c < \infty$ and F does not admit a positive flow cycle, then the measure Q_a is SIT.

Proof Since F does not admit a positive flow cycle, Q_a is indeed supported on simple paths. Since F has finite energy, if $Y = (Y_t)_t$, $Z = (Z_t)_t$ are two independent chains under Q_a, we have that

$$Q_a \otimes Q_a[\exists\, j, i \ge 0,\ Y_j = x,\ Y_{j+1} = y,\ Z_i = x,\ Z_{i+1} = y] \le |\tfrac{c_a}{2} F(x, y)|^2.$$

So

$$\mathbb{E}_{Q_a \otimes Q_a}[R(Y, Z)] \le \tfrac{c_a^2}{4} \cdot \sum_{x \sim y} r(x, y)|F(x, y)|^2 = \tfrac{c_a^2}{4} \cdot \|F\|_c^2 < \infty. \qquad \square$$

This completes the construction of a SIT measure from the finite energy flow F above, which was obtained under the assumption that the network (G, c) is transient.

Combined with Proposition 4.10.2, this proves Theorem 4.10.1.

Exercise 4.20 Let (G, c) be a network. Let Γ_a^* be the set of all infinite paths γ (not necessarily simple) that start at $\gamma_0 = a$ and that visit any point only finitely many times. That is, for all $x \in G$,

$$\sum_j \mathbf{1}_{\{\gamma_j = x\}} < \infty.$$

We call Γ_a^* the set of **transient paths**.

Show that for any $\gamma \in \Gamma_a^*$ there exists an infinite simple path α such that for all $j \geq 0$, there exists $k \geq 0$ with $\gamma_k = \alpha_j$, $\gamma_{k+1} = \alpha_{j+1}$. That is, any edge that α traverses, γ also traverses.   ▷ solution ◁

Exercise 4.21 Let (G, c) be a network. Let Γ_a^* be the set of transient paths from Exercise 4.20.

Suppose that λ^* is a probability measure on Γ_a^* such that $\mathbb{E}_{\lambda^* \otimes \lambda^*} R(\alpha, \beta) < \infty$.

Show that there exists a SIT measure λ on Γ_a.   ▷ solution ◁

Exercise 4.22 Let (G, c) be a connected network. Let $x, y \in G$. Show that if there is a SIT measure on Γ_x, then there exists a SIT measure on Γ_y.   ▷ solution ◁

Exercise 4.23 Let $\varepsilon > 0$. Let (G, c) be a network, and define conductances $c'(x, y) = c(x, y) + \varepsilon$ for all $(x, y) \in E(G, c)$ and $c'(x, y) = 0$ for $(x, y) \notin E(G, c)$. (Note that $c'(x, y) > 0$ if and only if $c(x, y) > 0$. That is, $E(G, c) = E(G, c')$.)

Show that if (G, c) is transient then (G, c') is transient.

Show that if $\inf_{(x,y) \in E(G,c)} c(x, y) > 0$ and (G, c') is transient, then (G, c) is transient.   ▷ solution ◁

Exercise 4.24 Let $\varepsilon > 0$. Let (G, c) be a network, and define conductances $c'(x, y) = c(x, y) + \varepsilon$ for all $(x, y) \in E(G, c)$ and $c'(x, y) = 0$ for $(x, y) \notin E(G, c)$.

Give an example of a network (G, c) and $\varepsilon > 0$ such that (G, c) is recurrent and (G, c') is transient.   ▷ solution ◁

Exercise 4.25 This is called the *Nash-Williams criterion*.

Let (G, c) be a network, and fix some $a \in G$. Assume that $(K_n)_n$ is a sequence of subsets of edges $K_n \subset E(G, c)$ that are pairwise disjoint, $K_n \cap K_m = \emptyset$ for all $n \neq m$.

Assume that for all n the set K_n is a *cutset* for a; that is, any infinite simple path starting from a must traverse some edge in K_n. Precisely, for any $\gamma \in \Gamma_a$ and any n, there exists j such that $(\gamma_j, \gamma_{j+1}) \in K_n$.

Prove that for any probability measure λ on Γ_a, and α, β independent paths of law λ, we have that

$$\mathbb{E}_{\lambda \otimes \lambda}[R(\alpha, \beta)] \geq \sum_n \frac{1}{c(K_n)},$$

where $c(K_n) = \sum_{(x,y) \in K_n} c(x, y)$.

▷ solution ◁

Exercise 4.26 Show that the simple random walk on $\mathbb{Z}^2$ is recurrent. That is, the μ-random walk on $\mathbb{Z}^2$ for μ uniform on $\{\pm(1,0), \pm(0,1)\}$.

▷ solution ◁

Exercise 4.27 Consider the group $\mathbb{Z}$ with the nonsymmetric measure $\mu(1) = 1 - \mu(-1) = p$, for some $p \in (0, 1)$.

Show that the μ-random walk can be realized as the random walk on a network $(\mathbb{Z}, c)$ where conductances are given by $c(x, x + 1) = (\frac{p}{1-p})^x$.

Show that if $p \neq \frac{1}{2}$ then $(\mathbb{Z}, \mu)$ is transient.

▷ solution ◁

4.11 Capacity

Recall that for a network (G, c) we consider the spaces $\ell^2(G, c)$ and $\ell^2(E(G, c))$. Also, we denote by $\ell^0(G, c) = \{f : G \to \mathbb{C} : |\text{supp}(f)| < \infty\}$ the subspace of finitely supported functions. This space does not actually depend on c, so sometimes we will just write $\ell^0(G)$.

Exercise 4.28 Show that $\ell^0(G, c)$ is dense in $\ell^2(G, c)$ under the norm $\| \cdot \|_c$.

▷ solution ◁

Recall the *Dirichlet energy* of a function $f \in \ell^2(G, c)$, defined by $\mathcal{E}(f, f) = \langle \nabla f, \nabla f \rangle_c = 2 \langle \Delta f, f \rangle_c$.

Definition 4.11.1 Let (G, c) be a network. Let $A \subset G$ be a finite subset. Define the **capacity** of A to be

$$\text{cap}(A) = \inf \left\{ \tfrac{1}{2} \mathcal{E}(f, f) : f|_A \equiv 1, \ f \in \ell^0(G, c) \right\}.$$

Also, denote $\text{cap}(x) = \text{cap}(\{x\})$.

Theorem 4.11.2 *Let (G, c) be a network. For any $a \in G$ we have that* $\text{cap}(a) = \frac{c_a}{g(a,a)} = c_a \mathbb{P}_a[T_a^+ = \infty]$.

Consequently, the following are equivalent:

(1) (G, c) is transient.

(2) There exists $x \in G$ such that $\mathrm{cap}(x) > 0$.

(3) For all $x \in G$ we have $\mathrm{cap}(x) > 0$.

Proof Let $Z \subset G$ be such that $G \backslash Z$ is finite. Fix $a \notin Z$ and let $v(x) = \frac{g_Z(x,a)}{g_Z(a,a)}$.
Then $v(a) = 1$, $\mathrm{supp}(v) \subset G \backslash Z$, so it is easy to calculate that

$$\mathcal{E}(v, v) = 2 \langle \Delta v, v \rangle_c = 2 c_a \Delta v(a) \cdot v(a) = \frac{2 c_a}{g_Z(a,a)}.$$

Thus, $g_Z(a, a) \leq \frac{c_a}{\mathrm{cap}(a)}$. Taking a monotone sequence $(Z_n)_n$, $Z_n \supset Z_{n+1}$,
$\bigcap_n Z_n = \emptyset$, we arrive at $g(a, a) \leq \frac{2 c_a}{\mathrm{cap}(a)}$.

Conversely, v above is a unit voltage from a to Z, so by Thompson's principle
(Theorem 4.7.1), for any $f : G \to \mathbb{C}$ such that $f(a) = 1$ and $\mathrm{supp}(f) \subset G \backslash Z$,
we have $\mathcal{E}(f, f) \geq \mathcal{E}(v, v) = \frac{2 c_a}{g_Z(a,a)}$. Again, taking a monotone sequence
$(Z_n)_n$, $Z_n \supset Z_{n+1}$, $\bigcap_n Z_n = \emptyset$, we arrive at

$$\mathrm{cap}(a) = \inf \left\{ \tfrac{1}{2} \mathcal{E}(f, f) : f|_A \equiv 1, \ f \in \ell^0(G, c) \right\} \geq \inf_n \frac{c_a}{g_{Z_n}(a,a)} \geq \frac{c_a}{g(a,a)}. \quad \square$$

Exercise 4.29 Show that

$$\mathrm{cap}(A) = \inf \left\{ \tfrac{1}{2} \mathcal{E}(v_Z, v_Z) : v_Z(x) = \mathbb{P}_x[T_A < T_Z], \ A \subset Z^c, \ |Z^c| < \infty \right\}.$$

▷ solution ◁

Definition 4.11.3 Let (G, c) be a transient network, and let $A \subset G$ be a finite
subset. Define the **equilibrium measure** of A by

$$e_A(x) = \mathbf{1}_{\{x \in A\}} \mathbb{P}_x \left[T_A^+ = \infty \right].$$

Exercise 4.30 Show that in a transient network (G, c), for any finite subset A,

$$\sum_y g(x, y) e_A(y) = \mathbb{P}_x[T_A < \infty].$$

(Hint: consider $L = \sup\{t \geq -1 : X_t \in A\}$.)

▷ solution ◁

Exercise 4.31 Show that for a finite subset A in a network (G, c),

$$\mathrm{cap}(A) = \sum_x c_x e_A(x).$$

▷ solution ◁

Exercise 4.32 Show that for finite subsets A, B in a network (G, c), the capacity
satisfies $\mathrm{cap}(A \cup B) \leq \mathrm{cap}(A) + \mathrm{cap}(B)$.

▷ solution ◁

Exercise 4.33 Show that if $A \subset B$ are finite subsets in a network (G, c) then $\mathrm{cap}(A) \leq \mathrm{cap}(B)$. ▷ solution ◁

Exercise 4.34 Let A be a finite subset in a transient network (G, c). Define

$$\Sigma_{\geq}(A) := \left\{ 0 \leq \varphi \in \ell^0(G, c) \mid \forall\, x \in A : \sum_y g(x, y)\varphi(y) \geq 1 \right\},$$

$$\Sigma_{\leq}(A) := \left\{ 0 \leq \varphi \leq \mathbf{1}_A \mid \forall\, x \in A : \sum_y g(x, y)\varphi(y) \leq 1 \right\}.$$

Show that

$$\mathrm{cap}(A) = \min_{\varphi \in \Sigma_{\geq}(A)} \sum_x c_x \varphi(x) = \max_{\varphi \in \Sigma_{\leq}(A)} \sum_x c_x \varphi(x).$$ ▷ solution ◁

Exercise 4.35 Let (G, c) be a transient network. Let A be a finite subset. Define $g(x, A) = \sum_{a \in A} g(x, a)$.
Prove that

$$\inf_{a \in A} g(a, A) \leq \frac{\sum_{a \in A} c_a}{\mathrm{cap}(A)} \leq \sup_{a \in A} g(a, A).$$ ▷ solution ◁

4.12 Transience and Recurrence of Groups

In this section we apply the relation of Dirichlet energy, capacity, and transience to prove that transience/recurrence is essentially a group property.

Theorem 4.12.1 *Let G be a finitely generated group. Let $\mu, \nu \in \mathrm{SA}(G, 2)$ be two symmetric, adapted probability measures on G with finite second moment.*

Then, the μ-random walk is transient if and only if the ν-random walk is transient.

In light of this theorem we may state the following definition.

Definition 4.12.2 A finitely generated group G is called **recurrent** if some (and hence, any) symmetric, adapted random walk on G with finite second moment is a recurrent Markov chain.
Otherwise, we say that G is **transient**.

We have already seen that $\mathbb{Z}, \mathbb{Z}^2$ are recurrent (see Section 2.4, Example 3.6.3, and Exercise 4.26). Thus, by Exercise 3.29, any group that is virtually $\mathbb{Z}$ or $\mathbb{Z}^2$ is recurrent. In Theorem 9.7.1, we will classify all (finitely generated) recurrent groups, and we will see that finite groups, virtually $\mathbb{Z}$ and virtually $\mathbb{Z}^2$ are the only examples.

Exercise 4.27 shows that there exist nonsymmetric random walks on $\mathbb{Z}$ that are transient. In Exercise 4.44 we will give an example of $\mu \in \mathsf{SA}\left(\mathbb{Z}^2, 2 - \varepsilon\right)$ such that $\left(\mathbb{Z}^2, \mu\right)$ is transient. These examples show that the conditions in Theorem 4.12.1 cannot be relaxed in general.

Exercise 4.36 Let G be a finitely generated group, and let μ be any symmetric adapted probability measure on G. Assume that the μ-random walk is recurrent.

Show that there exists a finitely supported, symmetric, adapted random walk ν on G such that the ν-random walk is recurrent. Show that ν can be chosen so that $\mathsf{supp}(\nu) \subset \mathsf{supp}(\mu)$. ▷ solution ◁

Our main tool to prove Theorem 4.12.1 is the notion of a *canonical path ensemble*.

Let $(G, c), (G, c')$ be two networks on the same state space G. A **canonical path ensemble** is a collection of finite paths $\Gamma = (\gamma^{xy})_{(x,y) \in E(G,c)}$, one for each edge $(x, y) \in E(G, c)$, such that $\gamma^{xy} \colon x \to y$ is a path in (G, c'). For a canonical path ensemble Γ, define the **maximal load**

$$L_\Gamma(c, c') = \sup_{(z,w) \in E(G,c')} \frac{1}{c'(z,w)} \sum_{x,y} c(x, y) |\gamma^{xy}| \mathbf{1}_{\{(z,w) \in \gamma^{xy}\}}.$$

Furthermore, define $L(c, c') = \inf L_\Gamma(c, c')$ where the infimum is over all possible choices for canonical path ensembles $\Gamma = (\gamma^{xy})_{(x,y) \in E(G,c)}$.

Lemma 4.12.3 *Let $(G, c), (G, c')$ be two networks on the same state space G. Then, for any $f \in \ell^2(G, c) \cap \ell^2(G, c')$ we have $\mathcal{E}_c(f, f) \le L(c, c') \cdot \mathcal{E}_{c'}(f, f)$.*

Proof If $\Gamma = (\gamma^{xy})_{(x,y) \in E(G,c)}$ is a canonical path ensemble, for any $f \in \ell^2(G, c) \cap \ell^2(G, c')$, we have

$$\mathcal{E}_c(f, f) = \sum_{x,y} c(x, y) |f(x) - f(y)|^2$$

$$\le \sum_{x,y} c(x, y) |\gamma^{xy}| \sum_{j=0}^{|\gamma^{xy}|-1} |f(\gamma_j^{xy}) - f(\gamma_{j+1}^{xy})|^2$$

$$= \sum_{z,w} c'(z, w) |f(z) - f(w)|^2 \cdot \sum_{x,y} \frac{c(x,y)|\gamma^{xy}|}{c'(z,w)} \mathbf{1}_{\{(z,w) \in \gamma^{xy}\}}$$

$$\le L_\Gamma(c, c') \cdot \mathcal{E}_{c'}(f, f).$$

Taking the infimum over all possible choices for Γ proves the lemma. □

Proof of Theorem 4.12.1 Assume that (G, μ) is transient. We wish to show that (G, ν) is transient. For this it suffices to bound the ratio between respective Dirichlet energies (by the capacity criterion for transience, Theorem 4.11.2).

If (G, ν) is recurrent, then by Exercise 4.36 we may find a finitely supported, symmetric and adapted probability measure $\tilde{\nu}$ such that $(G, \tilde{\nu})$ is also recurrent. By replacing ν with $\tilde{\nu}$, we may assume without loss of generality that ν has a finite support S. We use this generating set S to determine the metric on G.

For any $x \in G$ write $x = s_1 \cdots s_{|x|}$ for $s_j \in S$ and let $\alpha^x : 1 \to x$ be the path $\alpha^x = (1, s_1, s_1 s_2, \ldots, s_1 \cdots s_{|x|-1}, x)$.

For $x, y \in G$ define $\gamma^{x,y} : x \to y$ by setting $\gamma^{x,y} = x\alpha^{x^{-1}y}$, where $x\alpha = (x\alpha_0, \ldots, x\alpha_{|\alpha|})$. Note that $\gamma^{1,y} = \alpha^y$. The collection $\Gamma = (\gamma^{x,y})_{(x,y) \in E(G,\mu)}$ is our choice of canonical path ensemble.

An important property we will use is that for any edge $(z, w) \in E(G, \nu)$ in the network induced by ν, if $(z, w) \in \gamma^{x,y}$, then also $(\nu^{-1}z, \nu^{-1}w) \in E(G, \nu)$ and $(\nu^{-1}z, \nu^{-1}w) \in \nu^{-1}\gamma^{x,y} = \gamma^{\nu^{-1}x, \nu^{-1}y}$. Using the fact that μ has finite second moment,

$$
L_\Gamma(\mu, \nu) = \sup_{\substack{z \in \text{supp}(\nu) \\ w \in G}} \frac{1}{\nu(z)} \sum_{x,y} \mu\left(x^{-1}y\right) |\gamma^{x,y}| \mathbf{1}_{\{(w, wz) \in \gamma^{x,y}\}}
$$

$$
= \sup_{\substack{z \in \text{supp}(\nu) \\ w \in G}} \frac{1}{\nu(z)} \sum_{y} \mu(y) |\alpha^y| \sum_{x} \mathbf{1}_{\{(1,z) \in w^{-1}x\alpha^y\}}
$$

$$
\leq \sup_{\substack{z \in \text{supp}(\nu) \\ w \in G}} \frac{1}{\nu(z)} \sum_{y} \mu(y) |\alpha^y| \cdot |\alpha^y| \leq \sup_{z \in \text{supp}(\nu)} \frac{1}{\nu(z)} \cdot \mathbb{E}_\mu\left[|X_1|^2\right] < \infty,
$$

where we have used that for any z, w, y,

$$
\sum_{x} \mathbf{1}_{\{(1,z) \in w^{-1}x\alpha^y\}} = \sum_{x} \mathbf{1}_{\{(x, xz) \in \alpha^y\}} \leq |\alpha^y|. \qquad \square
$$

Exercise 4.37 Let μ be the uniform measure on $\{-1, 1\}$. Let $(X_t)_t$ be the μ-random walk on $\mathbb{Z}$.

Show that the transition matrix for $(X_t)_t$ is given by $P(z, z+1) = P(z, z-1) = \frac{1}{2}$.

Show that there exists some constant $C > 0$ such that for all $t \geq 1$,

$$
C^{-1} t^{-1/2} \leq \mathbb{P}[X_{2t} = 0] \leq C t^{-1/2}. \tag{4.1}
$$

Show that $\mathbb{P}[X_{2t+1} = 0] = 0$.

Conclude that $\mathbb{Z}$ is recurrent (we have already proven this fact). ▷ solution ◁

Exercise 4.38 Let μ be the uniform measure on $\{-1, 0, 1\}$. Let $(X_t)_t$ be the μ-random walk on $\mathbb{Z}$.

Show that the transition matrix for $(X_t)_t$ is given by $P(z, z+1) = P(z, z-1) = P(z, z) = \frac{1}{3}$.

Show that there exists some constant $C > 0$ such that for all $t \geq 1$,

$$C^{-1} t^{-1/2} \leq \mathbb{P}[X_t = 0] \leq C t^{-1/2}. \tag{4.2}$$

▷ solution ◁

Exercise 4.39 Consider the set $S = \{-1, 0, 1\}^d$ as a subset of $\mathbb{Z}^d$. That is, S is the set of all d-dimensional vectors taking values in $\{-1, 0, 1\}$. Let μ be the uniform measure on S.

Show that $\mu \in \mathsf{SA}\left(\mathbb{Z}^d, \infty\right)$.

Let $(Z_t)_t$ denote the μ-random walk. Denote $Z_t = \left(X_t^1, \ldots, X_t^d\right)$ for every t, where $X_t^j \in \mathbb{Z}$.

Show that $\left(X_t^1\right)_t, \ldots, \left(X_t^d\right)_t$ are independent random walks on $\mathbb{Z}$, each with transition matrix $P(z, z + 1) = P(z, z - 1) = P(z, z) = \frac{1}{3}$.

Conclude that $\mathbb{Z}^d$ is recurrent if and only if $d \leq 2$.

▷ solution ◁

4.13 Additional Exercises

Let (G, c) be a network. Let $f, g \in \ell^2(G, c)$ be functions such that $f \cdot g \in \ell^2(G, c)$. Define $\Gamma(f, g) \colon G \to \mathbb{C}$ by

$$\Gamma(f, g) := \tfrac{1}{2}\left(f \cdot \Delta \bar{g} + \bar{g} \cdot \Delta f - \Delta(f \cdot \bar{g})\right).$$

Exercise 4.40 Show that $\Gamma(f, g) = \overline{\Gamma(g, f)}$.

Show that $\Gamma(\alpha f + h, g) = \alpha \Gamma(f, g) + \Gamma(h, g)$, where $\alpha \in \mathbb{C}$.

Show that

$$\Gamma(f, g)(x) = \tfrac{1}{2} \sum_y P(x, y)(f(x) - f(y))\overline{(g(x) - g(y))}.$$

Conclude that $\Gamma(f, f) \geq 0$.

Show that

$$\mathcal{E}(f, g) = 2 \langle \Gamma(f, g), 1 \rangle_c,$$

where 1 is the constant function.

▷ solution ◁

Exercise 4.41 Show that

$$\mathcal{E}\left(f, g^2\right) = 4 \langle \Gamma(f, g), g \rangle_c.$$

▷ solution ◁

Exercise 4.42 Show that for any x,

$$|\Gamma(f,g)(x)|^2 \leq \Gamma(f,f)(x) \cdot \Gamma(g,g)(x).$$

Conclude that

$$\|\Gamma(f,g)\|_c^2 \leq \langle \Gamma(f,f), \Gamma(g,g) \rangle_c .$$

▷ solution ◁

In the following exercises we work to provide an example of a transient symmetric random walk on the recurrent group $\mathbb{Z}^2$. In light of Theorem 4.12.1 and Exercise 4.39, this random walk cannot have finite second moment. This example will show that Theorem 4.12.1 is tight in the sense that the number of moments cannot be relaxed.

Exercise 4.43 Fix $\alpha \in (1,2)$. Let $\zeta(\alpha) = \sum_{n=1}^{\infty} n^{-\alpha}$. Define a probability measure $\nu = \nu_\alpha$ on $\mathbb{N}$ by $\nu(n) = \mathbf{1}_{\{n \geq 1\}} \frac{1}{\zeta(\alpha)} n^{-\alpha}$. (This is sometimes called the ζ-distribution.)

Let $(U_k)_k$ be i.i.d.-ν, and $T_n = \sum_{k=1}^{n} U_k$.

Show that for $\varepsilon > 2 - \alpha$ we have that $\mathbb{E}\left[U_1^{1-\varepsilon}\right] < \infty$ but $\mathbb{E}\left[U_1^{1-\varepsilon'}\right] = \infty$ for $\varepsilon' \leq 2 - \alpha$.

Show that there exists a constant $C = C(\alpha) > 0$ such that $\mathbb{E}\left[\frac{1}{T_n}\right] \leq Cn^{\alpha-3}$. (Hint: consider $\delta > 0$ and define $B_n(\delta) = \#\{1 \leq k \leq n : U_k \geq n^\delta\}$. Use Chebychev's inequality for the binomial random variable $B_n(\delta)$.)

▷ solution ◁

Exercise 4.44 Fix $\alpha \in (1,2)$. Let $\zeta(\alpha) = \sum_{n=1}^{\infty} n^{-\alpha}$. Define a probability measure $\nu = \nu_\alpha$ on $\mathbb{N}$ by $\nu(n) = \mathbf{1}_{\{n \geq 1\}} \frac{1}{\zeta(\alpha)} n^{-\alpha}$. (This is sometimes called the ζ-distribution.)

Let $(U_k)_k$ be i.i.d.-ν, and $T_n = \sum_{k=1}^{n} U_k$. Let $(X_t, Y_t)_t$ be two independent simple random walks on $\mathbb{Z}$, independent of $(U_k)_k$ as well.

Consider the process $Z_n := (X_{T_n}, Y_{T_n}) \in \mathbb{Z}^2$.

Show that $(Z_n)_n$ is a μ-random walk, for some $\mu \in \text{SA}\left(\mathbb{Z}^2, 2 - \varepsilon\right)$ whenever $\varepsilon > 4 - 2\alpha$.

Show that $\mathbb{P}[Z_n = \vec{0}] \leq Cn^{-\beta}$ for some constants $C = C(\alpha) > 0$, $\beta = \beta(\alpha) > 1$ and all $n \geq 1$. Conclude that $\left(\mathbb{Z}^2, \mu\right)$ is transient.

▷ solution ◁

Exercise 4.45 Let G be a finitely generated group. Let $(X_t)_t$ be a μ-random walk for an adapted symmetric probability measure μ on G.

Define the *range* of the random walk to be

$$R_t = \{X_0, \ldots, X_t\}.$$

Show that

$$\lim_{t \to \infty} \frac{1}{t} \mathbb{E} |R_t| = \mathbb{P} \left[T_1^+ = \infty \right].$$

▷ solution ◁

Exercise 4.46 Let $\mu \in \mathsf{SA}(\mathbb{Z}, 1)$. Let $(X_t)_t$ be the μ-random walk, and let $R_t = \{X_0, \ldots, X_t\}$.

Show that $\frac{1}{t} |R_t| \to 0$ a.s.

Use Exercise 4.45 to show that $(\mathbb{Z}, \mu)$ is recurrent.

▷ solution ◁

4.14 Remarks

A more comprehensive treatment of network theory can be found in Aldous and Fill (2002) and Lyons and Peres (2016). Much of this chapter is based on the latter.

Benjamini, Pemantle & Peres introduced the notion of *exponential intersection tails*, or EIT, in Benjamini et al. (1998). Their motivation was transience of *percolation clusters*, see their paper for details.

Recall the framework from Section 4.10, expecially the definitions of Γ_a, $E(\alpha, \beta)$, $R(\alpha, \beta)$.

A probability measure λ on Γ_a is called **EIT** (exponential intersection tails) if the following holds. For α, β two independent paths each of law λ, we have

$$\text{there exists } \varepsilon > 0, \qquad \mathbb{E}_{\lambda \otimes \lambda} \left[e^{\varepsilon R(\alpha, \beta)} \right] < \infty.$$

That is, the distribution of the total resistance of edges traversed by two independent paths has an exponential tail. A network is called **EIT** if it admits some EIT measure on Γ_a for some $a \in G$.

In Benjamini et al. (1998) it was shown that if a graph G is EIT then there exists some $p < 1$ such that *Bernoulli-p percolation clusters* on G are transient.

The relation between SIT and transience, namely Theorem 4.10.1, was known but seems to have been unpublished until recently. The following is currently still open however.

Conjecture 4.14.1 Let G be a Cayley graph of a transient group. Then G is EIT.

That is, we conjecture that the seemingly much weaker property of SIT (finite expectation) implies EIT (exponential tails) in Cayley graphs.

All exponential growth Cayley graphs are EIT, as a result of Lyons (1995). One may combine Gromov's theorem (Theorem 9.0.1) and the methods in Chapter 9 with Benjamini et al. (1998) to show that all transient Cayley graphs of polynomial growth are EIT (see also Raoufi and Yadin, 2017).

4.15 Solutions to Exercises

Solution to Exercise 4.1 :(

Let $T = T_Z$. Since (G, c) is connected, for any $a \notin Z$ there exists $t(a)$ such that

$$\mathbb{P}_a[T \le t(a)] \ge \mathbb{P}_a[X_{t(a)} \in Z] > 0.$$

Let $t_* = \max\{t(a) \;:\; a \notin Z\}$ (using that $G\backslash Z$ is finite). Then,

$$\min_{a \notin Z} \mathbb{P}_a[T \le t_*] \ge \min_{a \notin Z} \mathbb{P}_a[X_{t(a)} \in Z] =: \alpha > 0.$$

Thus, for any $t > t_*$ and any $x \in G$, we have, using the Markov property,

$$\mathbb{P}_x[T > t] = \mathbb{P}_x[T > t \mid T > t - t_*] \cdot \mathbb{P}_x[T > t - t_*]$$

$$\le \sup_y \mathbb{P}_y[T > t_*] \cdot \mathbb{P}_x[T > t - t_*] \le (1 - \alpha) \cdot \mathbb{P}_x[T > t - t_*]$$

$$\le \cdots \le (1 - \alpha)^{\lfloor t/t_* \rfloor}. \qquad\qquad :)\checkmark$$

Solution to Exercise 4.2 :(

We only sketch the proof, as it is a very classical fact, basically coming from the fact that $\ell^2(N, m) = L^2\left(N, 2^N, m\right)$, considering m as a (not necessarily finite) measure on N.

The fact that this is an inner product is very easy to prove. To show that $\ell^2(N, m)$ is a Hilbert space, we only need to show that the induced metric is complete.

To this end, let $(f_n)_n$ be a Cauchy sequence in $\ell^2(N, m)$. We wish to find a limit in $\ell^2(N, m)$ for this sequence.

One then proceeds in a few steps:

- Show that for any $x \in N$ we have that $(f_n(x))_n$ is a Cauchy sequence in $\mathbb{C}$, and therefore the pointwise limit $f(x) := \lim f_n(x)$ exists for all x. (This uses the fact that $m(x) > 0$.)
- Show that $M := \sup_n \|f_n\| < \infty$.
- Enumerate $N = \{x_1, x_2, \dots\}$ (N is countable by assumption). Use the above to bound $\sum_{j=1}^r |f(x_j) - f_n(x_j)|^2$ as follows.
  Fix $\varepsilon > 0$. Let n_0 be large enough so that $\|f_n - f_m\|^2 < \varepsilon$ for all $n, m \ge n_0$. For $r > 0$ let $n_r \ge n_0$ be large enough so that $|f(x_j) - f_m(x_j)|^2 < \frac{\varepsilon}{r}$ for all $j \le r$ and $m \ge n_r$. Then for any $n \ge n_0$, for any $r > 0$, and for any $m \ge n_r$,

$$\sum_{j=1}^r |f(x_j) - f_n(x_j)|^2 \le 2 \sum_{j=1}^r |f_m(x_j) - f_n(x_j)|^2 + 2 \sum_{j=1}^r |f(x_j) - f_m(x_j)|^2 < 4\varepsilon.$$

- This culminates in showing that $\|f - f_n\| \to 0$. $\qquad\qquad :)\checkmark$

Solution to Exercise 4.3 :(

Linearity is immediate.

If $f \in \ell^2(G, c)$ then

$$\|\nabla f\|_c^2 = \sum_{x \sim y} \frac{1}{c(x,y)} |c(x,y)(f(x) - f(y))|^2$$

$$= \sum_{x,y} c(x,y)|f(x) - f(y)|^2 \le \sum_x c_x |f(x)|^2 + \sum_y c_y |f(y)|^2 + 2\sum_{x,y} c(x,y)|f(x)| \cdot |f(y)|.$$

The Cauchy–Schwarz inequality tells us that

$$\left(\sum_{x,y} c(x,y)|f(x)| \cdot |f(y)|\right)^2 \le \sum_{x,y} c(x,y)|f(x)|^2 \cdot \sum_{x,y} c(x,y)|f(y)|^2 = \|f\|_c^2 \cdot \|f\|_c^2.$$

All in all, this gives the bound

$$\|\nabla f\|_c^2 \le 4\|f\|_c^2 < \infty.$$

Also, if $F \in \ell^2(E(G, c))$ then similarly,

$$||\text{div} F||_c^2 = \sum_x c_x \left| \sum_{y \sim x} \frac{1}{c_x} (F(x, y) - F(y, x)) \right|^2$$

$$\leq \sum_{x \sim y} \frac{1}{c_x} \left(|F(x, y)|^2 + |F(y, x)|^2 + 2|F(x, y)| \cdot |F(y, x)| \right)$$

$$\leq 4 \sum_{x \sim y} \frac{1}{c_x} |F(x, y)|^2 \leq 4||F||_c^2,$$

where we have used Cauchy–Schwarz again, and the fact that $c_x \geq c(x, y)$. :) ✓

Solution to Exercise 4.4 :(

Since $\nabla f(y, x) = -\nabla f(x, y)$ we have:

$$\tfrac{1}{2} \text{div} \nabla f(x) = \tfrac{1}{2} \frac{1}{c_x} \sum_{y \sim x} (\nabla f(x, y) - \nabla f(y, x)) = \frac{1}{c_x} \sum_{y \sim x} \nabla f(x, y)$$

$$= \sum_{y \sim x} P(x, y)(f(x) - f(y)) = \Delta f(x).$$

Now, Δ is self-adjoint just because ∇, div are adjoints of each other; indeed, for any $f, g \in \ell^2(G, c)$,

$$2 \langle \Delta f, g \rangle_c = \langle \nabla f, \nabla g \rangle_c = 2 \langle f, \Delta g \rangle_c .$$:) ✓

Solution to Exercise 4.6 :(

We have that

$$\sum_{x,y} c(x, y) |f(x) - f(y)|^2 = \mathcal{E}(f, f) = 2 \langle \Delta f, f \rangle_c = 0.$$

Since the sum on the left-hand side is of nonnegative terms, all terms must be 0. Thus, $f(x) = f(y)$ for all $x \sim y$. Since (G, c) is connected, f is constant. If $\sum_x c_x = \infty$ this is only possible in $\ell^2(G, c)$ when f is 0. :) ✓

Solution to Exercise 4.8 :(

The main observation here is that

$$P^n(x, y) = \sum_{(x = \gamma_0, \ldots, \gamma_n = y) \in G^{n+1}} \prod_{j=0}^{n-1} P(\gamma_j, \gamma_{j+1}).$$

Assume that for all $x, y \in G$ there exists a finite path $\gamma : x \to y$. Then, for some specific $x, y \in G$, let $\gamma = (x = \gamma_0, \ldots, \gamma_n = y)$ be such a path. Note that

$$P^n(x, y) \geq \prod_{j=0}^{n-1} P(\gamma_j, \gamma_{j+1}) = \prod_{j=0}^{n-1} \frac{c(\gamma_j, \gamma_{j+1})}{c_{\gamma_j}} > 0.$$

So P is irreducible, meaning that G is connected.

Conversely, assume that P is irreducible. Let $x, y \in G$. So there exists n for which $P^n(x, y) > 0$. This implies that there exists a finite sequence $(x = \gamma_0, \ldots, \gamma_n = y) \in G^{n+1}$ such that

$$\prod_{j=0}^{n-1} P(\gamma_j, \gamma_{j+1}) = \prod_{j=0}^{n-1} \frac{c(\gamma_j, \gamma_{j+1})}{c_{\gamma_j}} > 0.$$

But this implies in turn that $c(\gamma_j, \gamma_{j+1}) > 0$ for all $0 \leq j < n$. So $\gamma : x \to y$ is a path from x to y. :) ✓

Solution to Exercise 4.9 :(

The first bullet is immediate.

For the second bullet, just compute the telescopic sum:

$$\oint_\gamma \nabla f = \sum_{j=0}^{|\gamma|-1} f(\gamma_j) - f(\gamma_{j+1}) = f(\gamma_0) - f(\gamma_{|\gamma|}).$$:) ✓

Solution to Exercise 4.10 :(
If I is a current then it satisfies Kirchhoff's cycle law by Exercise 4.9.

If F is a flow from A to Z satisfying Kirchhoff's cycle law, then Proposition 4.4.2 tells us that $F = \nabla v$ for some v. For any $x \notin A \cup Z$ we have that

$$\Delta v(x) = \tfrac{1}{2}\mathrm{div} F(x) = 0,$$

which shows that v is a voltage from A to Z, implying that F is a current. :) ✓

Solution to Exercise 4.11 :(
As in the proof of Theorem 4.5.4, take $b = \sup_{x\notin A\cup Z} |v(x)| < \infty$, and note that for $K > b$ we have

$$\mathbb{E}_x \left[|v(X_{T\wedge t})| \mathbf{1}_{\{|v(X_{T\wedge t})|>K\}} \right] \le \mathbb{E}_x \left[|v(X_T)| \mathbf{1}_{\{|v(X_T)|>K\}} \right] \le \mathbb{E}_x |v(X_T)| < \infty,$$

so that $(v(X_{T\wedge t}))_t$ is a uniformly integrable martingale. Since $T < \infty$ a.s., by the OST we have $v(x) = \mathbb{E}_x[v(X_T)]$ for all x.

If $\sup_x v(x) = v(z) < \infty$ for some z, then since $v(z) = \mathbb{E}_z[v(X_T)]$, there must exist some y such that $\mathbb{P}_z[X_T = y] > 0$ and $v(y) \ge v(z)$.

The minimum principle for v follows by applying the maximum principle to $-v$. :) ✓

Solution to Exercise 4.12 :(
Let v be a voltage from A to Z such that $v(x) = 0$ for all $x \in A \cup Z$. Since $G\backslash(A \cup Z)$ is finite, we get that $v \in \ell^2(G, c)$. Note that $\Delta v(x) \cdot v(x) = 0$ for all $x \in G$. Thus,

$$\mathcal{E}(v, v) = 2\langle \Delta v, v \rangle = 2 \sum_x c_x \Delta v(x) v(x) = 0.$$

But

$$0 = \mathcal{E}(v, v) = \sum_{x,y} c(x, y) |v(x) - v(y)|^2,$$

implying that v is constant (because (G, c) is connected). So $v \equiv 0$.

Let u, v be two voltages from A to Z. Then $u - v$ is a voltage from A to Z that is identically 0 on $A \cup Z$. So we get that $v = u$. :) ✓

Solution to Exercise 4.13 :(
Just note that $u = \frac{v}{v(a)}$ is a unit voltage from a to Z. :) ✓

Solution to Exercise 4.14 :(
Compute:

$$\mathcal{E}(v, v) = 2\langle \Delta v, v \rangle = 2 \sum_x c_x \Delta v(x) v(x) = 2 c_a \Delta v(a).$$:) ✓

Solution to Exercise 4.15 :(
Consider $v(x) = \mathbb{P}_x[T_a < T_Z]$. It is simple to verify using the Markov property that v is a voltage from a to Z. Moreover, $v(a) = 1$ and $v(z) = 0$ for all $z \in Z$. So v is the unique unit voltage from a to Z.

Since $v(a) - v(x) = 1 - \mathbb{P}_x[T_a < T_Z] = \mathbb{P}_x[T_Z < T_a]$,

$$C_{\mathrm{eff}}(a, Z) = c_a \Delta v(a) = c_a \sum_x P(a, x)(v(a) - v(x))$$

$$= c_a \sum_x P(a, x)\,\mathbb{P}_x[T_Z < T_a] = c_a\,\mathbb{P}_a[T_Z < T_a^+].$$

In the last line we have used the Markov property at time 1. :) ✓

Solution to Exercise 4.16 :(
Let $(Z_n)_n$ be a decreasing sequence $Z_n \supset Z_{n+1}$ of subsets such that $|G \backslash Z_n| < \infty$ for all n, and such that $a \notin Z_0$. Thus,

$$C_{\text{eff}}(a, \infty \mid c) = \lim_{n \to \infty} C_{\text{eff}}(a, Z_n \mid c) \leq \lim_{n \to \infty} C_{\text{eff}}(a, Z_n \mid c') = C_{\text{eff}}(a, \infty \mid c').$$

Since effective conductance to infinity is positive if and only if the network is transient, we have that if (G, c) is transient, then $0 < C_{\text{eff}}(a, \infty \mid c) \leq C_{\text{eff}}(a, \infty \mid c')$, so that (G, c') is also transient.   :) ✓

Solution to Exercise 4.17 :(
Because $c'(x, y) \leq c(x, y)$ for all $x, y \in G$, this is a direct application of Rayleigh monotonicity, Theorem 4.7.3.   :) ✓

Solution to Exercise 4.18 :(
$(1) \Rightarrow (2)$ is trivial, and so is $(3) \Rightarrow (1)$.
   For $(2) \Rightarrow (4)$, recall that $g(x, y) = \mathbb{P}_x[T_y < \infty] \cdot g(y, y)$, for all x, y. Since (G, c) is connected, $\mathbb{P}_x[T_y < \infty] > 0$ for all x, y. Thus, if $g(x, y) < \infty$ for some x, y, then $g(y, y) < \infty$. But $g(y, y) = \frac{1}{\mathbb{P}_y[T_y^+ = \infty]}$,
so it must be that $\mathbb{P}_y[T_y^+ = \infty] > 0$, implying that (G, c) is transient.
   Finally, for $(4) \Rightarrow (3)$, assume that (G, c) is transient, and let $x, y \in G$. Since $g(x, y) \leq g(y, y)$ it suffices to prove that $g(y, y) < \infty$. As $g(y, y) = \frac{1}{\mathbb{P}_y[T_y^+ = \infty]}$ this follows from transience because $\mathbb{P}_y[T_y^+ = \infty] > 0$.

                     :) ✓

Solution to Exercise 4.19 :(
If the random walk is transient then $g_Z(x, y) \leq g(x, y) < \infty$.

   If the random walk is recurrent, then we have seen that $g_Z(x, y) \leq g_Z(y, y) = \frac{1_{\{y \notin Z\}}}{\mathbb{P}_y[T_Z \leq T_y^+]}$. If $y \in Z$ then this is 0 so assume that $y \notin Z$. Since the random walk is recurrent, we have that $\mathbb{P}_y[T_z < \infty] = 1$ for any $z \in Z$. This implies that there is a finite path $\gamma : y \to z$ in the network (G, c). By considering the last time this path is at y, we may assume without loss of generality that $\gamma_k \neq y$ for all $1 \leq k \leq |\gamma|$. Hence

$$\mathbb{P}_y\left[T_y^+ \geq T_Z\right] \geq \mathbb{P}_y\left[T_z < T_y^+\right] \geq \mathbb{P}_y[\forall\, 0 \leq k \leq |\gamma|,\ X_k = \gamma_k] > 0.   \text{:) ✓}$$

Solution to Exercise 4.20 :(
Erasing the loops of γ provides such a path α, as follows:
   Let $A = \{\gamma_j : j \geq 0\}$ be the set of point visited by γ. For every $x \in A$ define

$$t(x) = \min\{j \geq 0 : \gamma_j = x\}, \qquad \ell(x) = \max\{j \geq 0 : \gamma_j = x\},$$

which are well defined because γ is a transient path. Note that $\gamma_{t(x)} = \gamma_{\ell(x)} = x$. By definition, $\ell(\gamma_j) \geq j$.
   Define inductively, $\alpha_0 = a$ and for all $n > 0$ define $\alpha_n = \gamma_{\ell(\alpha_{n-1})+1}$. We have that $(\alpha_n)_{n=0}^\infty$ is a subsequence of $(\gamma_j)_{j=0}^\infty$. Note that

$$\ell(\alpha_n) = \ell\left(\gamma_{\ell(\alpha_{n-1})+1}\right) \geq \ell(\alpha_{n-1}) + 1 > \ell(\alpha_{n-1}).$$

If $\alpha_n = \alpha_k$ for some $n > k$, then we would have $\ell(\alpha_n) = \ell(\alpha_k)$, contradicting the above, so $\alpha_n \neq \alpha_k$ for all $n \neq k$. Also, for any n, note that $\alpha_{n+1} = \gamma_{\ell(\alpha_n)+1}$ and $\gamma_{\ell(\alpha_n)} = \alpha_n$, so that α_n, α_{n+1} is always an edge, and one that is traversed by γ as well.
   We conclude that α is an infinite simple path with the required properties.   :) ✓

Solution to Exercise 4.21 :(
For $\gamma \in \Gamma_a^*$, fix some infinite simple path $\tilde{\gamma} \in \Gamma_a$ such that for any x, y with $c(x, y) > 0$, if $\tilde{\gamma}_j = x, \tilde{\gamma}_{j+1} = y$ then $\gamma_k = x, \gamma_{k+1} = y$ for some k.
   Now, given the measure λ^*, define the measure λ by taking a random α with law λ^*, and letting λ be the law of $\tilde{\alpha}$.
   For two independent such paths α, β, we have that

$$(x, y) \in E(\tilde{\alpha}, \tilde{\beta}) \quad \Rightarrow \quad (x, y) \in E(\alpha, \beta),$$

so that a.s. $R(\tilde{\alpha}, \tilde{\beta}) \leq R(\alpha, \beta)$. Thus, λ is SIT.   :) ✓

Solution to Exercise 4.22 :(

Let λ be a SIT measure on Γ_x. Let α, β be two independent paths in Γ_x with law λ.

Let $\gamma: y \to x$ be some finite path (using that (G, c) is connected). Define $\tilde{\alpha} = \gamma\alpha$ and $\tilde{\beta} = \gamma\beta$. Then $\tilde{\alpha}, \tilde{\beta}$ are independent paths in Γ_y^* with

$$R(\tilde{\alpha}, \tilde{\beta}) \le \sum_{j=1}^{|\gamma|} r(\gamma_{j-1}, \gamma_j) + R(\alpha, \beta).$$

By Exercise 4.21 we get that there exists a SIT measure on Γ_y. :) ✓

Solution to Exercise 4.23 :(

If (G, c) is transient, then that (G, c') is transient just follows from Rayleigh monotonicity (Theorem 4.7.3).

If (G, c') is transient, then there exists a SIT measure λ on paths in (G, c') started at some $a \in G$. Note that since $E(G, c) = E(G, c')$, we have that paths in (G, c') are also paths in (G, c). If α, β are two independent paths sampled from the distribution λ, in the network (G, c) we then have

$$\mathbb{E}_{\lambda \otimes \lambda}[R(\alpha, \beta)] = \sum_{(x,y) \in E(G,c)} \mathbb{P}_{\lambda \otimes \lambda}[(x, y) \in E(\alpha, \beta)] \frac{1}{c(x, y)}$$

$$= \sum_{(x,y) \in E(G,c')} \mathbb{P}_{\lambda \otimes \lambda}[(x, y) \in E(\alpha, \beta)] \frac{1}{c'(x, y)} \cdot \frac{c(x,y) + \varepsilon}{c(x,y)}$$

$$\le \sum_{(x,y) \in E(G,c')} \mathbb{P}_{\lambda \otimes \lambda}[(x, y) \in E(\alpha, \beta)] \frac{1}{c'(x, y)} \cdot (1 + \tfrac{\varepsilon}{\eta}),$$

where $\eta = \inf_{(x,y) \in E(G,c)} c(x, y)$. Thus, when $\eta > 0$, we have that λ is a SIT measure on the network (G, c) as well, implying that (G, c) is transient. :) ✓

Solution to Exercise 4.24 :(

Let $G = \{0, 1\}^*$ be the set of all finite words in the letters 0, 1, including the empty word, denoted by $\star$. For a non-empty word $\omega = \omega_1 \cdots \omega_n$, write $\hat{\omega} = \omega_1 \cdots \omega_{n-1}$ for the unique word that is obtained by removing the last letter of ω. Write $|\omega|$ for the length of the word ω, with $|\star| = 0$.

Define conductances $c(\omega, \omega 1) = c(\omega, \omega 0) = 2^{-|\omega|}$.

Let $(X_t)_t$ be the random walk on the network (G, c). One easily checks that

$$\mathbb{P}[|X_{t+1}| = n + 1 \mid |X_t| = n] = \mathbb{P}[|X_{t+1}| = n - 1 \mid |X_t| = n] = \tfrac{1}{2},$$

$$\mathbb{P}[|X_{t+1}| = 1 \mid |X_t| = 0] = 1.$$

So $(|X_t|)_t$ is a simple random walk on the natural numbers $\mathbb{N}$, and is easily seen to be recurrent.

Now, consider the conductances $c'(x, y) = c(x, y) + 1$ for any $(x, y) \in E(G, c)$. We will show this new network admits a SIT measure. Choose a random path α inductively as follows. Set $\alpha_0 = \star$. Given α_n, let $\alpha_{n+1} = \alpha_n 1$ or $\alpha_{n+1} = \alpha_n 0$ with probability $\tfrac{1}{2}$ each. Let λ be the law of this random path. We claim that λ is a SIT measure. Indeed, consider two independent samples, α, β. Note that if $\alpha_n \ne \beta_n$ then $\alpha_{n+1} \ne \beta_{n+1}$, by construction. So if $N = \min\{n : \alpha_n \ne \beta_n\}$, then since $c'(x, y) \ge 1$ for $(x, y) \in E(G, c')$,

$$R(\alpha, \beta) \le \sum_{k=1}^{N} \frac{1}{c'(\alpha_{k-1}, \alpha_k)} \le N.$$

Also, since α, β are independent,

$$\mathbb{P}[\alpha_{n+1} \ne \beta_{n+1} \mid \alpha_n = \beta_n] = \tfrac{1}{2}.$$

Thus,

$$\mathbb{E}_{\lambda \otimes \lambda}[R(\alpha, \beta)] \le \mathbb{E}_{\lambda \otimes \lambda}[N] = 2 < \infty.$$

So λ is indeed a SIT measure on (G, c'), and (G, c') is transient. :) ✓

Solution to Exercise 4.25 :(

Let $\alpha, \beta \in \Gamma_a$ be two independent random paths with law λ. We denote $\{(x, y) \in \alpha\} = \{\exists j, (\alpha_j, \alpha_{j+1}) = $

$(x, y)\}$. Since α must traverse some edge in K_n, by Cauchy–Schwarz,

$$1 \leq \left(\sum_{(x,y)\in K_n} \mathbb{P}_\lambda[(x,y) \in \alpha] \right)^2 \leq \sum_{(x,y)\in K_n} r(x,y)(\mathbb{P}_\lambda[(x,y) \in \alpha])^2 \cdot \sum_{(x,y)\in K_n} c(x,y).$$

As $(K_n)_n$ are pairwise disjoint,

$$\mathbb{E}_{\lambda\otimes\lambda}[R(\alpha,\beta)] \geq \sum_n \sum_{(x,y)\in K_n} r(x,y)(\mathbb{P}_\lambda[(x,y) \in \alpha])^2$$

$$\geq \sum_n \frac{1}{\sum_{(x,y)\in K_n} c(x,y)} = \sum_n \frac{1}{c(K_n)}. \qquad :) \checkmark$$

Solution to Exercise 4.26 :(
For ever $n \geq 1$, define

$$K_n = \{((\varepsilon n, j), (\varepsilon(n+1), j)), ((j, \varepsilon n), (j, \varepsilon(n+1))) : -n \leq j \leq n, \ \varepsilon \in \{-1, 1\}\}.$$

That is, K_n is the collection of edges emanating from a $n \times n$ box. One easily checks that $(K_n)_n$ are pairwise disjoint cutsets for 0, and that $c(K_n) = 2n + 1$. Since $\sum_n \frac{1}{c(K_n)} = \infty$, we conclude that if λ is any probability measure on Γ_0, then by Exercise 4.25, λ cannot be SIT.

By Exercise 4.22 and Theorem 4.10.1 we see that $\left(\mathbb{Z}^2, \mu\right)$ is recurrent. $\qquad :) \checkmark$

Solution to Exercise 4.27 :(
Consider conductances given by $c(x, x+1) = \left(\frac{p}{1-p}\right)^x$. Note that if P is the transition matrix of the induced random walk then

$$P(x, x+1) = \frac{c(x, x+1)}{c(x, x+1) + c(x-1, x)} = \frac{p^x}{p^x + (1-p)p^{x-1}} = p = \mu(1),$$

and

$$P(x, x-1) = \frac{c(x-1, x)}{c(x, x+1) + c(x-1, x)} = 1 - p = \mu(-1).$$

So the induced random walk on $(\mathbb{Z}, c)$ is exactly the μ-random walk.

Now consider the (degenerate) probability measure λ on simple infinite paths in $(\mathbb{Z}, c)$ started at 0, given by choosing (deterministically) the path $\gamma = (0, 1, 2, \ldots)$ if $p > \frac{1}{2}$ and $\gamma = (0, -1, -2, \ldots)$ if $p < \frac{1}{2}$. Note that for $q = \max\{p, 1-p\}$,

$$R(\gamma, \gamma) = \sum_{n=0}^\infty \left(\frac{1-q}{q}\right)^n < \infty,$$

because $q > 1 - q$. Thus, we have a SIT on $(\mathbb{Z}, c)$ implying that $(\mathbb{Z}, c)$ is transient by Theorem 4.10.1. $\qquad :) \checkmark$

Solution to Exercise 4.28 :(
Let $f \in \ell^2(G, c)$. Let $\varepsilon > 0$. We know that $\sum_x c_x |f(x)|^2 < \infty$, so there exists a finite set $F \subset G$ such that $\sum_{x\notin F} c_x |f(x)|^2 < \varepsilon$. Define $g = f\mathbf{1}_F$. Then it is easy to check that $||f - g||_c \leq \sum_{x\notin F} c_x |f(x)|^2 < \varepsilon$.
$\qquad :) \checkmark$

Solution to Exercise 4.29 :(
This follows from Thompson's principle (Theorem 4.7.1), which states that if $f|_A \equiv 1$ and supported on Z^c, then $\mathcal{E}(v_Z, v_Z) \leq \mathcal{E}(f, f)$ (since v_Z is the unit voltage from A to Z). $\qquad :) \checkmark$

Solution to Exercise 4.30 :(
Define

$$L = \sup\{t \geq -1 : X_t \in A\}.$$

This is *not* a stopping time. Note that $\{L = -1\} = \{T_A = \infty\}$. For any $t \geq 0$, by the Markov property,

$$\mathbb{P}_x[L = t, X_t = y] = \mathbf{1}_{\{y \in A\}} \mathbb{P}_x[X_t = y, \forall s > 0, X_{t+s} \notin A] = \mathbf{1}_{\{y \in A\}} \cdot \mathbb{P}_x[X_t = y] \cdot \mathbb{P}_y\left[T_A^+ = \infty\right]$$

$$= \mathbb{P}_x[X_t = y] \cdot e_A(y).$$

Thus, summing over $t \geq 0$ and y,

$$\mathbb{P}_x[T_A < \infty] = \sum_{t=0}^{\infty} \sum_y \mathbb{P}[L = t, X_t = y]$$

$$= \sum_{t=0}^{\infty} \sum_y \mathbb{P}_x[X_t = y] \cdot e_A(y) = \sum_y g(x, y) e_A(y). \qquad :) \checkmark$$

Solution to Exercise 4.31 :(
Let $v_Z(x) = \mathbb{P}_x[T_A < T_Z]$ for some $A \cap Z = \emptyset$ with $|Z^c| < \infty$. Then,

$$\tfrac{1}{2}\mathcal{E}(v_Z, v_Z) = \langle \Delta v_Z, v_Z \rangle_c = \sum_{a \in A} c_a \Delta v_Z(a).$$

For any $a \in A$ we have

$$\Delta v_Z(a) = \sum_x P(a, x)(1 - v_Z(x)) = \mathbb{P}_a\left[T_A^+ \geq T_Z\right].$$

Taking an infimum over all Z with $A \subset Z^c$ and $|Z^c| < \infty$, we arrive at

$$\mathrm{cap}(A) = \inf \tfrac{1}{2}\mathcal{E}(v_Z, v_Z) = \inf \sum_{a \in A} c_a \mathbb{P}_a\left[T_A^+ \geq T_Z\right] = \sum_{a \in A} c_a \mathbb{P}_a\left[T_A^+ = \infty\right]. \qquad :) \checkmark$$

Solution to Exercise 4.32 :(
We have $\mathbb{P}_x\left[T_B^+ = \infty\right] \leq \mathbb{P}_x\left[T_A^+ = \infty\right]$ for all $A \subset B$.
  Thus,

$$\mathrm{cap}(A \cup B) = \sum_{x \in A \cup B} c_x \mathbb{P}_x\left[T_{A \cup B}^+ = \infty\right]$$

$$\leq \sum_{x \in A} c_x \mathbb{P}_x\left[T_A^+ = \infty\right] + \sum_{x \in B \setminus A} c_x \mathbb{P}_x\left[T_B^+ = \infty\right] \leq \mathrm{cap}(A) + \mathrm{cap}(B). \qquad :) \checkmark$$

Solution to Exercise 4.33 :(
Note that

$$\left\{f \in \ell^0(G, c) : f|_B \equiv 1\right\} \subset \left\{f \in \ell^0(G, c) : f|_A \equiv 1\right\}$$

whenever $A \subset B$. $\qquad :) \checkmark$

Solution to Exercise 4.34 :(
Let $\varphi \in \Sigma_{\geq}(A)$. Then,

$$\mathrm{cap}(A) \leq \sum_{x \in A} c_x e_A(x) \sum_y g(x, y) \varphi(y) = \sum_y c_y \varphi(y) \sum_x g(y, x) e_A(x) \leq \sum_y c_y \varphi(y).$$

The minimum over $\Sigma_{\geq}(A)$ is attained by $\varphi = e_A$.
  Now, for $\varphi \in \Sigma_{\leq}(A)$, we use the fact that φ is supported in A to obtain

$$\mathrm{cap}(A) \geq \sum_{x \in A} c_x e_A(x) \sum_y g(x, y) \varphi(y)$$

$$= \sum_y c_y \varphi(y) \sum_x g(y, x) e_A(x) = \sum_{y \in A} \mathbb{P}_y[T_A < \infty] c_y \varphi(y) = \sum_y c_y \varphi(y).$$

Again, the maximum on $\Sigma_{\leq}(A)$ is obtained by taking $\varphi = e_A$. $\qquad :) \checkmark$

Solution to Exercise 4.35 :(

This follows by taking $\varphi_a = \frac{1_A}{g(a,A)}$ for different $a \in A$. If $a, b \in A$ are such that $g(a, A) = \inf_{x \in A} g(x, A)$ and $g(b, A) = \sup_{x \in A} g(x, A)$, then $\varphi_a \in \Sigma_\geq(A)$ and $\varphi_b \in \Sigma_\leq(A)$. So

$$\frac{\sum_{x \in A} c_x}{g(b, A)} = \sum_x c_x \varphi_b(x) \leq \mathrm{cap}(A) \leq \sum_x c_x \varphi_a(x) = \frac{\sum_{x \in A} c_x}{g(a, A)}. \qquad \text{:) } \checkmark$$

Solution to Exercise 4.36 :(

Let $S \subset \mathrm{supp}(\mu)$ be a finite symmetric generating set such that $1 \notin S$, which can be found since μ is symmetric and adapted. Define

$$\nu(x) = \mu(x) + \tfrac{1}{|S|} - \tfrac{1}{|S|} \sum_{s \in S} \mu(s)$$

for $x \in S$ and $\nu(x) = 0$ for $x \notin S$. Obviously, ν is symmetric, adapted with finite support S.

Now, consider the network on G with conductances $c(x, y) = \mu\left(x^{-1}y\right) 1_{\{x^{-1}y \in S\}}$. This network obviously has conductances bounded by those of μ. By Rayleigh monotonicity (Theorem 4.7.3), since (G, μ) is recurrent, (G, c) is also recurrent.

Moreover, the conductances for the network (G, ν) are given by adding the constant $\varepsilon = \tfrac{1}{|S|} - \tfrac{1}{|S|} \sum_{s \in S} \mu(s)$ to every positive conductance in (G, c). Since

$$\inf_{(x,y) \in E(G,c)} c(x, y) = \min_{s \in S} \mu(s) > 0,$$

we have seen in Exercise 4.23 that if (G, ν) is transient then so is (G, c). Since (G, c) is recurrent, it must be that (G, ν) is recurrent as well. $\qquad \text{:) } \checkmark$

Solution to Exercise 4.37 :(

The transition matrix is just $P(x, y) = \mu(x - y) = 1_{\{x-y \in \{-1,1\}\}} \tfrac{1}{2}$.

To prove (4.1), we write $X_t = X_0 + \sum_{k=1}^t U_k$ where $(U_k)_{k \geq 1}$ are i.i.d.-μ elements. Set $R_t = \sum_{k=1}^t 1_{\{U_k = 1\}}$ and $L_t = \sum_{k=1}^t 1_{\{U_k = -1\}}$. So $R_t + L_t = t$ and $X_t = X_0 + R_t - L_t$. Thus,

$$\mathbb{P}[X_t = 0] = \mathbb{P}[R_t = L_t] = \mathbb{P}[R_t = t/2] = \begin{cases} 0 & \text{if } t = 2n + 1, \\ \binom{2n}{n} 2^{-2n} & \text{if } t = 2n, \end{cases}$$

because $R_t \sim \mathrm{Bin}\left(t, \tfrac{1}{2}\right)$. Stirling's Approximation tells us that for all $n \geq 1$,

$$1 < \frac{n! e^n}{\sqrt{2\pi n} n^n} < 2.$$

Thus, for all $n \geq 1$,

$$\tfrac{1}{4} < \binom{2n}{n} 2^{-2n} \sqrt{\pi n} < 2.$$

This proves (4.1).

We have that $\mathbb{Z}$ is recurrent since $\mu \in \mathrm{SA}(\mathbb{Z}, \infty)$ and

$$\sum_{t=0}^{\infty} \mathbb{P}[X_t = 0] = \infty. \qquad \text{:) } \checkmark$$

Solution to Exercise 4.38 :(

The transition matrix computation is immediate.

We first prove (4.2). Let $(U_t)_{t \geq 1}$ be i.i.d.-μ elements, so that $X_t = X_0 + \sum_{k=1}^t U_k$, and define $M = \{t \geq 1 : U_t \in \{-1, 1\}\}$. Write $M = \{t_1 < t_2 < \cdots\}$, and define $Y_0 = X_0$ and inductively $Y_{n+1} = Y_n + U_{t_{n+1}}$. It is immediate that $Y_n = X_{t_n}$ for all n (where $t_0 = 0$). Also, for $t_n < t < t_{n+1}$ we have that $X_t = Y_n$, because $U_{t_n+1} = \cdots = U_t = 0$.

Now, $(Y_n)_n$ is easily seen to be a simple random walk on $\mathbb{Z}$; that is, the ν-random walk, for ν uniform on $\{-1, 1\}$. By Exercise 4.37, we know that there exists $C > 0$ such that for all $n \geq 0$ and $C^{-1}(n+1)^{-1/2} \leq \mathbb{P}[Y_{2n} = 0 \mid M] \leq C(n+1)^{-1/2}$. Also, $\mathbb{P}[Y_{2n+1} = 0 \mid M] = 0$ for all $n \geq 0$.

Finally, for any $t \geq 0$, set $N(t) = n$ such that $t_n \leq t < t_{n+1}$. It is simple to see that $N(t)$ has Binomial- $\left(t, \frac{2}{3}\right)$ distribution. Since $N(t)$ is a function of the set M, we conclude that

$$P[X_t = 0] = \sum_{n=0}^{t} P[N(t) = n] \cdot P[Y_n = 0] = \sum_{0 \leq n \leq t/2} P[N(t) = 2n] \cdot P[Y_{2n} = 0].$$

Thus,

$$C^{-1} \, \mathbb{E}\left[1_{\{N(t) \text{ is even}\}} (N(t) + 1)^{-1/2}\right] \leq P[X_t = 0] \leq C \, \mathbb{E}\left[(N(t) + 1)^{-1/2}\right].$$

These last quantities are bounded as follows.
For the lower bound, note that for any $t > 1$,

$$P[N(t) \text{ is even}] \geq P[N(t-1) \text{ is even}, U_t = 0] + P[N(t-1) \text{ is odd}, U_t \neq 0]$$
$$= \tfrac{1}{3} P[N(t-1) \text{ is even}] + \tfrac{2}{3} P[N(t-1) \text{ is odd}]$$
$$= \tfrac{2}{3} - \tfrac{1}{3} P[N(t-1) \text{ is even}] \geq \tfrac{1}{3}.$$

Thus,

$$P[X_t = 0] \geq C^{-1} \tfrac{1}{3} (t+1)^{-1/2}.$$

For the upper bound, by Chebychev's inequality,

$$P\left[N(t) < \tfrac{1}{3}t\right] \leq P\left[|N(t) - \tfrac{2}{3}t| > \tfrac{1}{3}t\right] \leq \frac{2}{t},$$

so that

$$P[X_t = 0] \leq C \, \mathbb{E}\left[(N(t)+1)^{-1/2}\right] \leq C \, P\left[N(t) \geq \tfrac{1}{3}t\right]\left(\tfrac{1}{3}t+1\right)^{-1/2} + C \, P\left[N(t) < \tfrac{1}{3}t\right]$$
$$\leq C\sqrt{3}(t+3)^{-1/2} + \frac{2C}{t}. \qquad \qquad \text{:)} \checkmark$$

Solution to Exercise 4.39   :(
We have that S generates $\mathbb{Z}^d$ as it contains the standard basis for $\mathbb{R}^d$. Also, μ is obviously symmetric, and has finite support, so has an exponential tail as well.

To show that $\left(X_t^1\right)_t, \ldots, \left(X_t^d\right)_t$ are independent Markov chains with transition matrix P, one notes that for any $\vec{z}_k = \left(x_k^1, \ldots, x_k^d\right) \in \mathbb{Z}^d$,

$$P[Z_{t+1} = \vec{z}_{t+1} \mid Z_t = \vec{z}_t] = \mu(z_{t+1} - z_t) = 1_{\{z_{t+1} - z_t \in S\}} 3^{-d}$$
$$= \prod_{j=1}^{d} 1_{\left\{x_{t+1}^j - x_t^j \in \{-1,0,1\}\right\}} \tfrac{1}{3} = \prod_{j=1}^{d} P\left(x_t^j, x_{t+1}^j\right).$$

This immediately gives the identity

$$P[Z_t = 0] = \prod_{j=1}^{d} P\left[X_t^j = 0\right] = \left(P\left[X_t^1 = 0\right]\right)^d.$$

Thus, using Exercise 4.38, we have that when $d > 2$,

$$\sum_{t=0}^{\infty} P[Z_t = 0] \leq 1 + C \sum_{t=1}^{\infty} t^{-d/2} < \infty,$$

implying that $\mathbb{Z}^d$ is transient for $d > 2$.
For $d \leq 2$ we find that

$$\sum_{t=0}^{\infty} P[Z_t = 0] \geq C^{-1} \sum_{t=1}^{\infty} t^{-d/2} = \infty,$$

implying that $\mathbb{Z}^d$ is recurrent for $d \leq 2$. \qquad \qquad \text{:)} \checkmark

Solution to Exercise 4.40 :(

The first two assertions are very easy.

The third assertion follows from the following computation. For $f, g \in \ell^2(G, c)$ with $fg \in \ell^2(G, c)$, and for any $x \in G$,

$$\sum_y P(x, y)(f(x) - f(y))\overline{(g(x) - g(y))} = \bar{g}(x) \cdot \Delta f(x) - f(x) \cdot P\bar{g}(x) + P(f\bar{g})(x)$$

$$= (\bar{g} \cdot \Delta f + f \cdot \Delta \bar{g} - \Delta(f\bar{g}))(x) = 2\Gamma(f, g)(x).$$

The final assertion follows from

$$\mathcal{E}(f, g) = \sum_{x,y} c(x, y)(f(x) - f(y))\overline{(g(x) - g(y))}$$

$$= 2\sum_x c_x \frac{1}{2} \sum_y P(x, y)(f(x) - f(y))\overline{(g(x) - g(y))}$$

$$= 2 \langle \Gamma(f, g), 1 \rangle_c .$$

:) ✓

Solution to Exercise 4.41 :(

Compute:

$$4 \langle \Gamma(f, g), g \rangle_c = \sum_x c_x \bar{g}(x) \sum_y P(x, y)(f(x) - f(y))\overline{(g(x) - g(y))}$$

$$+ \sum_y c_y \bar{g}(y) \sum_x P(y, x)(f(y) - f(x))\overline{(g(y) - g(x))}$$

$$= \sum_{x,y} c(x, y)(f(x) - f(y))\overline{(g(x) - g(y))}(g(x) + g(y)) = \mathcal{E}\left(f, g^2\right). \quad :) ✓$$

Solution to Exercise 4.42 :(

This is just a straightforward application of Cauchy–Schwarz.

:) ✓

Solution to Exercise 4.43 :(

We will make repeated use of the following estimate: for any $r \geq 1$,

$$\sum_{n=r}^{\infty} n^{-\alpha} \geq \int_r^{\infty} \xi^{-\alpha} d\xi = \frac{1}{\alpha-1} r^{1-\alpha},$$

$$\sum_{n=r}^{\infty} n^{-\alpha} \leq r^{-\alpha} + \int_r^{\infty} \xi^{-\alpha} d\xi \leq \frac{\alpha}{\alpha-1} r^{1-\alpha}.$$

We have that

$$\mathbb{E}\left[U_1^{1-\varepsilon}\right] = \frac{1}{\zeta(\alpha)} \sum_{n=1}^{\infty} n^{1-\varepsilon-\alpha},$$

which converges for $\varepsilon > 2 - \alpha$ and is infinite for $\varepsilon \leq 2 - \alpha$.

Also, setting $p = \mathbb{P}\left[U_1 \geq n^{\delta}\right]$ we have that $B_n(\delta) \sim \text{Bin}(n, p)$. Note that

$$1 \leq \zeta(\alpha)(\alpha - 1)pn^{\delta(\alpha-1)} \leq \alpha.$$

By Chebychev's inequality with $\lambda = \frac{1}{2}\mathbb{E}[B_n(\delta)]$,

$$\mathbb{P}[|B_n(\delta) - np| > \lambda] \leq \frac{\text{Var}[B_n(\delta)]}{\lambda^2} \leq 4(np)^{-1}.$$

Note that $T_n \geq B_n(\delta) \cdot n^{\delta}$, so

$$\mathbb{P}\left[T_n < \frac{1}{2}n^{1+\delta}p\right] \leq \mathbb{P}[B_n < np - \lambda] \leq 4(np)^{-1}.$$

Since $np \geq c(\alpha)n^{1-\delta(\alpha-1)}$ for some constant $c(\alpha) > 0$, we find that (by perhaps updating the constant $c(\alpha) > 0$)

$$\mathbb{E}\left[\frac{1}{T_n}\right] \leq 4(np)^{-1}n^{-1} + 2(np)^{-1}n^{-\delta} \leq c(\alpha)n^{-1+\delta(\alpha-1)} \cdot \left(n^{-1} + n^{-\delta}\right).$$

Choosing $\delta = 1$ completes the proof.   :)✓

Solution to Exercise 4.44 :(

Let $\mu(x, y) = \mathbb{P}\left[(X_{T_1}, Y_{T_1}) = (x, y)\right]$. It is immediate from the independence of $(U_k)_k$, $(X_t)_t$, $(Y_t)_t$ that $(Z_n)_n$ is a μ-random walk. It is also easy to see that μ is symmetric.

To show that μ has the proper moments, let $\varepsilon > 2 - \alpha$. Consider the process $M_t = |X_t|^2 - t$, which is easily seen to be a martingale. So by Jensen's inequality,

$$\mathbb{E}\left[|X_t|^{2-2\varepsilon}\right] \leq \left(\mathbb{E}\left[|X_t|^2\right]\right)^{1-\varepsilon} = t^{1-\varepsilon}.$$

Similarly $\mathbb{E}\left[|Y_t|^{2-2\varepsilon}\right] \leq t^{1-\varepsilon}$. Finally, since $|(x, y)| \leq |x| + |y|$ for all $(x, y) \in \mathbb{Z}^2$, we get that

$$|Z_1|^{2-2\varepsilon} \leq \left(2|X_t|^2 + 2|Y_t|^2\right)^{1-\varepsilon} \leq 2(|X_t|^{2-2\varepsilon} + |Y_t|^{2-2\varepsilon}).$$

Since T_1 is independent of $(X_t)_t$ and $(Y_t)_t$, we have that

$$\mathbb{E}\left[|Z_1|^{2-2\varepsilon}\right] \leq 2\mathbb{E}\left[T_1^{1-\varepsilon}\right] < \infty,$$

by Exercise 4.43, as long as $\varepsilon > 2 - \alpha$. Thus, $\mu \in \mathsf{SA}\left(\mathbb{Z}^2, 2 - \varepsilon\right)$ for all $\varepsilon > (4 - 2\alpha)$.

Exercise 4.37 tells us that $\mathbb{P}[(X_t, Y_t) = (0, 0)] \leq Ct^{-1}$ for some constant $C > 0$, and all $t \geq 1$. By Exercise 4.43, for some constants $C' = C'(\alpha) > 0$ and $\beta = \beta(\alpha) > 1$,

$$\mathbb{P}[Z_n = \vec{0}] \leq \mathbb{E}\,\mathbb{P}[(X_{T_n}, Y_{T_n}) = (0, 0) \mid T_n] \leq C\,\mathbb{E}\left[\frac{1}{T_n}\right] \leq C'n^{-\beta}.$$

As this is summable over n we have the transience of μ.   :)✓

Solution to Exercise 4.45 :(

Let $\tilde{R}_t = \{X_1, \ldots, X_t\}$ (so $\tilde{R}_0 = \emptyset$).

Let $T_x^+ = \inf\{t > 0 : X_t = x\}$. If $x \neq 1$ then we have that

$$\mathbb{P}_1\left[T_x^+ = t\right] = \mathbb{P}_x\left[T_x^+ > t, \ X_t = 1\right] = \mathbb{P}_1\left[T_1^+ > t, \ X_t = x^{-1}\right].$$

The first equality comes from reversing paths – the path $s_1 \cdots s_t$ and $(s_t)^{-1} \cdots (s_1)^{-1}$ have the same distribution because μ is symmetric. The second comes from the fact that we can translate the starting point by x^{-1}.

Now, the event $X_t \notin \tilde{R}_{t-1}$ is the event that there exists x such that $T_x^+ = t$. Thus,

$$\mathbb{P}_1[X_t \notin \tilde{R}_{t-1}] = \mathbb{P}_1\left[T_1^+ = t\right] + \sum_{x \neq 1} \mathbb{P}_1\left[T_x^+ = t\right]$$

$$= \mathbb{P}_1\left[T_1^+ = t\right] + \sum_{x \neq 1} \mathbb{P}_1\left[T_1^+ > t, \ X_t = x^{-1}\right] = \mathbb{P}_1\left[T_1^+ \geq t\right].$$

Since $\mathbb{P}_1\left[T_1^+ \geq t\right] \to \mathbb{P}\left[T_1^+ = \infty\right]$, we have that the Césaro limit is the same:

$$\frac{1}{t}\,\mathbb{E}\,|\tilde{R}_t| = \frac{1}{t}\sum_{k=1}^{t}\mathbb{P}[X_k \notin \tilde{R}_{k-1}] \to \mathbb{P}\left[T_1^+ = \infty\right].$$

Since $0 \leq |R_t| - |\tilde{R}_t| \leq 1$ we are done.   :)✓

Solution to Exercise 4.46 :(

Let $(U_t)_{t=1}^{\infty}$ be i.i.d.-μ. Then $X_t = \sum_{k=1}^{t} U_k$ is a μ-random walk.

Fix $\varepsilon > 0$. Let

$$\tau_\varepsilon = \inf\{t \geq 0 : \ \forall s \geq t, \ |X_s| \leq \varepsilon s\},$$

with inf $\emptyset = \infty$. The law of large numbers tells us that $\frac{1}{t} X_t \to 0$ a.s., which implies that $\mathbb{P}[\tau_\varepsilon < \infty] = 1$. Note that a.s.

$$|R_t| \leq \tau_\varepsilon + |\mathbb{Z} \cap [-t\varepsilon, t\varepsilon]| \leq \tau_\varepsilon + 2t\varepsilon + 1.$$

Thus,

$$\mathbb{P}\left[\limsup_{t \to \infty} \frac{1}{t} |R_t| > 2\varepsilon\right] \leq \mathbb{P}[\tau_\varepsilon = \infty] = 0.$$

As $\varepsilon > 0$ was arbitrary, we have that $\frac{1}{t} |R_t| \to 0$ a.s. As $|R_t| \leq t + 1$, dominated convergence implies that

$$\mathbb{P}\left[T_1^+ = \infty\right] = \lim_{t \to \infty} \frac{1}{t} \mathbb{E} |R_t| = 0.$$

So the random walk is recurrent.   :) ✓

PART II

Results and Applications

5

Growth, Dimension, and Heat Kernel

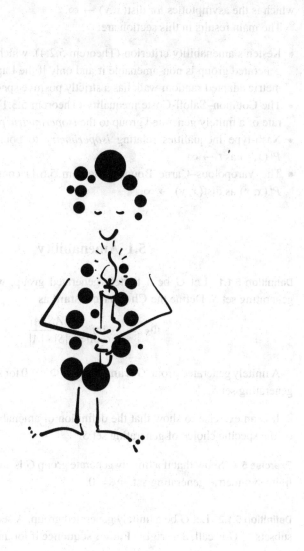

In this chapter we will connect the *volume growth* of the group to the decay of the probabilities $P^t(x, y)$. The quantity $P^t(x, y)$ is sometimes called the *heat kernel*, where the term *off-diagonal* is used for $y \neq x$ and *on-diagonal* is used when $x = y$. One is usually interested in the asymptotics of this sequence as $t \to \infty$ for fixed x, y. We will see that the growth of the group provides bounds for the decay of the heat kernel.

This should be contrasted to the Varopolous–Carne bound (Theorem 5.6.1), which is the asymptotics for $\text{dist}(x, y) \to \infty$.

The main results in this section are:

- Kesten's amenability criterion (Theorem 5.2.4), which states that a finitely generated group is non-amenable if and only if the Laplacian of some symmetric adapted random walk has a strictly positive spectral gap.
- The Coulhon–Saloff-Coste inequality (Theorem 5.3.1), relating the growth rate of a finitely generated group to the *isoperimetric profile*.
- Nash-type inequalities relating *isoperimetry* to bounds on the decay of $P^t(x, y)$ as $t \to \infty$.
- The Varopolous–Carne Bound (Theorem 5.6.1) controlling the decay of $P^t(x, y)$ as $\text{dist}(x, y) \to \infty$.

5.1 Amenability

Definition 5.1.1 Let G be a finitely generated group, with finite symmetric generating set S. Define the **Cheeger constant** as

$$\Phi_S := \inf_{A \subset G: |A| < \infty} \frac{|AS \setminus A|}{|S| \cdot |A|}.$$

A finitely generated group G is **amenable** if $\Phi_S = 0$ for some finite symmetric generating set S.

It is an exercise to show that the definition of amenability does not depend on the specific choice of generating set S.

Exercise 5.1 Show that if a finitely generate group G is amenable then for every finite symmetric generating set, $\Phi_S = 0$.   ▷ solution ◁

Definition 5.1.2 Let G be a finitely generated group. A sequence $(F_n)_n$ of finite subsets of G is called a **(right) Følner sequence** if for any $x \in G$ we have

$$\lim_{n \to \infty} \frac{|F_n x \triangle F_n|}{|F_n|} = 0.$$

Exercise 5.2 Show that there exists a Følner sequence in G if and only if G is amenable.

▷ solution ◁

Exercise 5.3 Show that G is amenable if and only if there exists a sequence $(F_n)_n$ of finite subsets $F_n \subset G$ such that for any $x \in G$ we have

$$\lim_{n \to \infty} \frac{|x F_n \triangle F_n|}{|F_n|} = 0.$$

(Such a sequence $(F_n)_n$ is sometimes called a *left Følner sequence*.) ▷ solution ◁

Exercise 5.4 For a probability measure μ on G and μ-random walk $(X_t)_t$, define the **Cheeger constant**

$$\Phi_\mu := \inf_{A \subset G : |A| < \infty} \frac{1}{|A|} \sum_{a \in A} \mathbb{P}_a[X_1 \notin A].$$

Show that if μ is uniform measure on a finite symmetric generating set S then $\Phi_\mu = \Phi_S$.

Show that for an adapted symmetric measure μ, $\Phi_\mu = 0$ if and only if G is amenable.

▷ solution ◁

Exercise 5.5 Let μ be a symmetric adapted probability measure on G and $(X_t)_t$ the μ-random walk. Show that for any finite set $A \subset G$,

$$\sum_{a \in A} \mathbb{P}_a[X_1 \notin A] = \sum_{x \notin A^{-1}} \mathbb{P}_x[X_1 \in A],$$

where $A^{-1} = \{a^{-1} : a \in A\}$.

▷ solution ◁

Definition 5.1.3 Let (G, c) be a network. For a subset $A \subset G$ define $c(A) = \sum_{a \in A} c_a$. Define the **Cheeger constant** of the network to be

$$\Phi = \Phi_{(G,c)} = \inf_{\substack{A \subset G \\ c(A) < \infty}} \frac{1}{c(A)} \sum_{\substack{a \in A \\ b \notin A}} c(a, b).$$

Exercise 5.6 Show that in a connected network (G, c), the Cheeger constant is

$$\Phi = \Phi_{(G,c)} = \inf_{\substack{A \subset G \\ c(A) < \infty}} \frac{1}{c(A)} \sum_{a \in A} c_a \, \mathbb{P}_a[X_1 \notin A],$$

where $(X_t)_t$ is the induced random walk on the network (G, c).

Exercise 5.7 Show that if μ is a symmetric adapted probability measure on a finitely generated group G, and if $c(x, y) = \mu\left(x^{-1}y\right)$, then the definitions of Cheeger constants coincide $\left(\text{i.e. } \Phi_{(G,c)} = \Phi_\mu\right)$.

5.2 Spectral Radius

Let (G, c) be a connected network. Recall that we have a transition matrix $P(x, y) = \frac{c(x,y)}{c_x}$, which provides a self-adjoint operator on $\ell^2(G, c)$ via $Pf(x) = \sum_y P(x, y) f(y)$ for all $f \in \ell^2(G, c)$. As an operator on $\ell^2(G, c)$, P has an operator norm $||P|| = ||P||_{2 \to 2} = \sup_{||f||_c = 1} ||Pf||_c$.

Exercise 5.8 Show that $||P|| \leq 1$ (i.e. P is a contraction on $\ell^2(G, c)$). ▷ solution ◁

Exercise 5.9 Show that in a connected network (G, c), the limit

$$\rho = \rho(P) = \rho(G, c) := \limsup_{t \to \infty} \left(P^t(x, y)\right)^{1/t}$$

does not depend on x, y.   ▷ solution ◁

Definition 5.2.1 For a connected network (G, c), the quantity

$$\rho = \rho(P) = \rho(G, c) := \limsup_{t \to \infty} \left(P^t(x, y)\right)^{1/t}$$

is called the **spectral radius** of (G, c) (or of P).

When the network is given by a symmetric adapted probability measure μ on a group G, we write $\rho = \rho(\mu)$ for the spectral radius.

Exercise 5.10 Let (G, c) be a connected network, and let $P(x, y) = \frac{c(x,y)}{c_x}$ be the corresponding transition matrix.

Fix $x, y \in G$ and define

$$g(x, y|\zeta) = \sum_{t=0}^{\infty} P^t(x, y)\zeta^t.$$

Show that this series has radius of convergence $\rho(G, c)^{-1}$.

We now connect the operator norm to the spectral radius.

Proposition 5.2.2 *Let (G, c) be a connected network, and let $P(x, y) = \dfrac{c(x,y)}{c_x}$ be the corresponding transition matrix.*
     Then $\|P\| = \rho(P)$.

Proof We have that $c_x P^t(x, y) = \left\langle P^t \delta_y, \delta_x \right\rangle_c \leq \|P\|^t \cdot \|\delta_y\|_c \cdot \|\delta_x\|_c = c_y c_x \|P\|^t$, so that $\rho \leq \|P\|$.

The other direction is slightly more involved. First, let $0 \neq f : G \to \mathbb{R}$ have finite support. Then for every $\varepsilon > 0$, there exists t_0 such that for all $t > t_0$ and any x, y in the support of f we have that $P^t(x, y) \leq (\rho + \varepsilon)^t$. Thus, for $t > t_0$,

$$\|P^t f\|_c^2 = \left\langle P^{2t} f, f \right\rangle_c = \sum_{x,y} c_x P^{2t}(x, y) f(y) f(x)$$

$$\leq (\rho + \varepsilon)^{2t} \cdot \sum_{x,y} c_x f(y) f(x) \mathbf{1}_{\{f(y)f(x)>0\}}.$$

Since f has finite support, taking the tth root and $t \to \infty$ we obtain that

$$\limsup_t \|P^t f\|^{1/t} \leq \rho + \varepsilon \text{ for any } \varepsilon > 0,$$

implying that for any function $f : G \to \mathbb{R}$ of finite support,

$$\limsup_t \|P^t f\|^{1/t} \leq \rho.$$

Using the self-adjointness of P and Cauchy–Schwarz,

$$\left\|P^{t+1} f\right\|_c^2 = \left\langle P^t f, P^{t+2} f \right\rangle_c \leq \|P^t f\|_c \cdot \left\|P^{t+2} f\right\|_c.$$

Setting $a_t := \|P^t f\|_c$ and $b_t = \dfrac{a_t}{a_{t-1}}$, we have that $(b_t)_t$ is a nondecreasing sequence bounded by 1 (as P is a contraction). Such a sequence converges, and its limit is equal to the Cesàro limit; that is,

$$\sup_t b_t = \lim_{t \to \infty} b_t = \lim_{t \to \infty} (a_t)^{1/t} \leq \rho$$

(using that $a_t = b_t \cdots b_1 \cdot a_0$). Thus, $\dfrac{\|Pf\|_c}{\|f\|_c} = b_1 \leq \rho$, for any $f : G \to \mathbb{R}$ with finite support.

It is an exercise to show that $\|Pf\|_c \leq \rho \|f\|_c$ for any $f \in \ell^2(G, c)$ given this holds for finitely supported real-valued functions. $\qquad\square$

Exercise 5.11 Let (G, c) be a connected network, and let $P(x, y) = \frac{c(x,y)}{c_x}$ be the corresponding transition matrix. Assume that $\tilde{\rho} > 0$ is some constant such that for any finitely supported real-valued function $f : G \to \mathbb{R}$ we have $\|Pf\|_c \leq \tilde{\rho}\|f\|_c$. Show that this holds for all $f \in \ell^2(G, c)$, implying that $\|P\| \leq \tilde{\rho}$.

▷ solution ◁

Exercise 5.12 Let (G, c) be a connected network, and let $P(x, y) = \frac{c(x,y)}{c_x}$ be the corresponding transition matrix. Show that

$$\rho(P) = \sup_{0 \neq f \in \ell^0(G,c)} \frac{\langle Pf, f \rangle_c}{\langle f, f \rangle_c} = \sup_{\substack{f : G \to \mathbb{R} \\ 0 \neq f \in \ell^0(G,c)}} \frac{\langle Pf, f \rangle_c}{\langle f, f \rangle_c}.$$

▷ solution ◁

Exercise 5.13 Let (G, c) be a connected network, and let $P(x, y) = \frac{c(x,y)}{c_x}$ be the corresponding transition matrix. Let $h \in \ell^2(G, c)$ such that $Ph = \lambda h$.

Show that $|\lambda| \leq \rho(P)$.

Show that if $\lambda < \rho(P)$ and h is nonnegative, then $h \equiv 0$.

▷ solution ◁

We now move to prove some relationship between the spectral radius ρ and amenability.

Proposition 5.2.3 (ℓ^1-**Sobolev inequality**) *Let (G, c) be a network and let $\Phi = \Phi_{(G,c)}$ be the Cheeger constant. Then, for any $f : G \to \mathbb{R}$ of finite support we have that*

$$\Phi \cdot \sum_x c_x |f(x)| \leq \sum_{x,y} c(x, y)|f(x) - f(y)|.$$

Proof Recall that $c(A) = \sum_{a \in A} c_a$. For $t > 0$ let $S_t = \{x : f(x) > t\}$, which is a finite set, because f has finite support. Note that

$$\int_0^\infty c(S_t)dt = \sum_x c_x f(x) \mathbf{1}_{\{f(x) \geq 0\}}.$$

We have

$$\int_0^\infty \mathbf{1}_{\{f(y) \leq t < f(x)\}}dt = |f(x) - f(y)|\mathbf{1}_{\{f(x) \geq f(y)\}}.$$

Also,

$$\sum_{x,y} c(x,y)|f(x) - f(y)|\mathbf{1}_{\{f(x)\geq f(y)\}} = \frac{1}{2}\sum_{x,y} c(x,y)|f(x) - f(y)|\mathbf{1}_{\{f(x)\geq f(y)\}}$$

$$+ \frac{1}{2}\sum_{x,y} c(y,x)|f(y) - f(x)|\mathbf{1}_{\{f(y)\geq f(x)\}}$$

$$= \frac{1}{2}\sum_{x,y} c(x,y)|f(x) - f(y)|.$$

Thus,

$$\Phi \cdot \sum_x c_x f(x)\mathbf{1}_{\{f(x)\geq 0\}} = \int_0^\infty \Phi \cdot c(S_t) dt$$

$$\leq \int_0^\infty \sum_{\substack{x\in S_t \\ y\notin S_t}} c(x,y) dt = \int_0^\infty \sum_{x,y} c(x,y)\mathbf{1}_{\{f(y)\leq t < f(x)\}} dt$$

$$= \frac{1}{2}\sum_{x,y} c(x,y)|f(x) - f(y)|. \tag{5.1}$$

Finally, applying this to $-f$ as well, we have that

$$\Phi \cdot \sum_x c_x|f(x)| = \Phi \cdot \sum_x c_x f(x)\mathbf{1}_{\{f(x)\geq 0\}} + \Phi \cdot \sum_x c_x(-f(x))\mathbf{1}_{\{-f(x)\geq 0\}}$$

$$\leq \sum_{x,y} c(x,y)|f(xy) - f(x)|. \qquad \square$$

Exercise 5.14 Show that if $0 \leq f \in \ell^0(G, c)$ is a nonnegative function of finite support, then we can improve the ℓ^1-Sobolev inquality by a factor of 2:

$$2\Phi_{(G,c)} \cdot \sum_x c_x f(x) \leq \sum_{x,y} c(x,y)|f(x) - f(y)|.$$

Exercise 5.15 Recall *integration by parts*, Theorem 4.2.4.

Let (G, c) be a connected network, and let $P(x, y) = \frac{c(x,y)}{c_x}$ be the corresponding transition matrix.

Show that for any $f \in \ell^2(G, c)$,

$$2\langle(I - P)f, f\rangle_c = \sum_{x,y} c(x,y)|f(x) - f(y)|^2,$$

$$2\langle(I + P)f, f\rangle_c = \sum_{x,y} c(x,y)|f(x) + f(y)|^2. \qquad \triangleright \text{ solution } \triangleleft$$

Theorem 5.2.4 (Kesten's amenability criterion) *Let (G, c) be a connected network. Let $\rho = \rho(G, c)$ and $\Phi = \Phi_{G,c}$ Then,*

$$\tfrac{1}{2}\Phi^2 \le 1 - \sqrt{1 - \Phi^2} \le 1 - \rho \le \Phi.$$

Remark 5.2.5 Note that this says that a group is amenable if and only if $\rho = 1$.

Proof $\tfrac{1}{2}\Phi^2 \le 1 - \sqrt{1 - \Phi^2}$ is easy, and is just an inequality for all numbers in $[0, 1]$.

To prove $1 - \rho \le \Phi$, note that if $A \subset G$ is a finite set then for $f = \mathbf{1}_A$ we have $\langle f, f \rangle_c = c(A)$, and by Exercise 5.15,

$$\langle \Delta f, f \rangle_c = \tfrac{1}{2} \sum_{x,y} c(x, y) \left(\mathbf{1}_{\{x \in A\}} - \mathbf{1}_{\{y \in A\}} \right)^2 = \sum_{x,y} c(x, y) \mathbf{1}_{\{x \in A\}} \mathbf{1}_{\{y \notin A\}},$$

where $\Delta = I - P$. So

$$\frac{\langle \Delta f, f \rangle_c}{\langle f, f \rangle_c} = \frac{1}{c(A)} \sum_{\substack{a \in A \\ b \notin A}} c(a, b).$$

For any $\varepsilon > 0$ we can find some finite subset A_ε such that $\frac{\langle \Delta \mathbf{1}_{A_\varepsilon}, \mathbf{1}_{A_\varepsilon} \rangle_c}{\langle \mathbf{1}_{A_\varepsilon}, \mathbf{1}_{A_\varepsilon} \rangle_c} \le \Phi + \varepsilon$. Thus,

$$\rho = \sup_{0 \ne f \in \ell^0(G,c)} \frac{\langle Pf, f \rangle_c}{\langle f, f \rangle_c} \ge 1 - \frac{\langle \Delta \mathbf{1}_{A_\varepsilon}, \mathbf{1}_{A_\varepsilon} \rangle_c}{\langle \mathbf{1}_{A_\varepsilon}, \mathbf{1}_{A_\varepsilon} \rangle_c} \ge 1 - \Phi - \varepsilon.$$

Taking $\varepsilon \to 0$ completes the inequality $1 - \rho \le \Phi$.

Next, we need to prove that $\rho^2 \le 1 - \Phi^2$. Let $f : G \to \mathbb{R}$ be a function of finite support. By Exercise 5.15,

$$2\langle f, f \rangle_c + 2\langle Pf, f \rangle_c = \sum_{x,y} c(x, y) |f(xy) + f(x)|^2.$$

Applying the improved Sobolev inequality for nonnegative functions, Exercise 5.14, to $g = |f|^2$,

$$2\Phi \langle f, f \rangle_c = 2\Phi \sum_x c_x g(x) \le \sum_{x,y} c(x, y) |f(x)^2 - f(y)^2|$$

$$= \sum_{x,y} c(x, y) |f(x) - f(y)| \cdot |f(x) + f(y)|.$$

So by Cauchy–Schwarz and Exercise 5.15,

$$4\Phi^2 \langle f, f \rangle_c^2 \le \sum_{x,y} c(x, y)(f(x) - f(y))^2 \cdot \sum_{x,y} c(x, y)(f(x) + f(y))^2$$

$$= 4\langle \Delta f, f \rangle_c \cdot (\langle f, f \rangle_c + \langle Pf, f \rangle_c).$$

Since $f \in \ell^0(G, c)$, we have $\langle Pf, f \rangle_c \le \rho \cdot \langle f, f \rangle_c$. We conclude that for any $f \in \ell^0(G, c)$,

$$\Phi^2 \le \frac{\langle \Delta f, f \rangle_c}{\langle f, f \rangle_c} \cdot (1 + \rho) = \left(1 - \frac{\langle Pf, f \rangle_c}{\langle f, f \rangle_c}\right)(1 + \rho).$$

By Exercise 5.12, ρ is the supremum of $\frac{\langle Pf, f \rangle_c}{\langle f, f \rangle_c}$ over $f \colon G \to \mathbb{R}$ of finite support, we get that $\Phi^2 \le (1 - \rho)(1 + \rho) = 1 - \rho^2$. $\square$

5.3 Isoperimetric Dimension

Let G be a finitely generated group and S a finite symmetric generating set. Recall the *Cheeger constant*

$$\Phi_S = \inf_{A \subset G : |A| < \infty} \frac{|AS \backslash A|}{|S| \cdot |A|}.$$

We have seen that G is amenable if and only if $\Phi_S = 0$; Φ_S measures the size of $AS \backslash A$ (the "boundary" of A) relative to the volume of A.

For amenable groups we require a finer way of measuring the relationship between the boundary and volume of sets. The Coulhon–Saloff-Coste inequality quantifies the fact that if the growth of a group is not small, then the ratio between boundary and volume cannot be too small.

Theorem 5.3.1 (Coulhon–Saloff-Coste inequality) *Let G be a finitely generated group, generated by a finite symmetric generating set S. Let $B_r = B(1, r)$ be the radius r ball in the metric induced by S. Define the* inverse growth rate *by*

$$r(n) = \inf\{r : |B_r| \ge n\}.$$

Then, for any finite subset $A \subset G$ we have

$$|\partial A| \ge \frac{|A|}{2r(2|A|)}.$$

Here, $\partial A = \{a \in A : \exists s \in S, \ as \notin A\}$ is the inner vertex boundary.

Proof Fix a finite subset $A \subset G$.

Note that $x \in A, xy \notin A$ is equivalent to $x \in A \backslash Ay^{-1}$. Writing $y = s_1 \cdots s_{|y|}$ for $s_j \in S$ and $y_0 = 1$ and $y_j = s_1 \cdots s_j$, we have that $x \in A \backslash Ay^{-1}$ if and only if there exists $1 \le j \le |y|$ such that $xy_{j-1} \in A \backslash As_j^{-1} \subset \partial A$. Thus, $|A \backslash Ay^{-1}| \le |y| \cdot |\partial A|$.

Let $r = r(2|A|)$. So $|B_r| \ge 2|A|$. Now, let U be a random element in B_r, chosen uniformly. Then, for any $x \in A$ we have

$$\mathbb{P}[xU \in A] = \frac{|xB_r \cap A|}{|B_r|} \le \frac{|A|}{|B_r|} \le \frac{1}{2},$$

so

$$\tfrac{1}{2} \le \mathbb{P}[xU \notin A] = \mathbb{P}\left[x \in A \backslash AU^{-1}\right].$$

Summing over all $x \in A$ and using linearity of expectation,

$$\frac{|A|}{2} \le \mathbb{E}\left|A \backslash AU^{-1}\right| \le \mathbb{E}[\mathrm{dist}_S(U, 1)] \cdot |\partial A| \le r \cdot |\partial A|.$$

Definition 5.3.2 Let (G, c) be a network. We say that (G, c) has **isoperimetric dimension** at least $d \in [1, \infty)$ if there exists a constant $\kappa > 0$ such that for any finite subset $A \subset G$, we have

$$\sum_{x \in A, y \notin A} c(x, y) \ge \kappa \cdot \left(\sum_{x \in A} c_x\right)^{(d-1)/d}.$$

We write $\dim^{\mathrm{iso}}(G, c)$ for the supremum over all such $d \in [1, \infty)$.

Exercise 5.16 Show that for a network (G, c) we have

$$\dim^{\mathrm{iso}}(G, c) = \liminf_{c(A) \to \infty} \frac{\log c(A)}{\log \frac{c(A)}{c(A, A^c)}}.$$

Here, $c(A) = \sum_{x \in A} c_x = c(A, G)$ and $c(A, B) = \sum_{x \in A, y \in B} c(x, y)$.   ▷ solution ◁

Exercise 5.17 Show that if S, S' are two finite symmetric generating sets for a finitely generated group G, then $\dim^{\mathrm{iso}}(\Gamma(G, S)) = \dim^{\mathrm{iso}}(\Gamma(G, S'))$ where $\Gamma(G, T)$ denotes the Cayley graph of G with respect to a generating set T.

▷ solution ◁

In light of this exercise, we write $\dim^{\mathrm{iso}}(G)$ for the isoperimetric dimension of the group G considered as a network with conductance 1 on every edge of some Cayley graph.

Exercise 5.18 Show that for any non-amenable finitely generated group G we have that $\dim^{\mathrm{iso}}(G) = \infty$.   ▷ solution ◁

Exercise 5.19 Let Γ be an infinite connected graph and consider Γ as a network with conductances $c(x, y) = \mathbf{1}_{\{x \sim y\}}$. Show that if $\dim^{\mathrm{iso}}(\Gamma) \ge d \ge 1$ then there exists a constant $\alpha > 0$ such that for any $x \in \Gamma$ we have $c(B(x,r)) \geq \alpha \cdot r^d$ for all $r \geq 1$. (Here $B(x,r)$ is the ball of radius r around x.) ▷ solution ◁

Exercise 5.20 Show that $\dim^{\text{iso}}(\mathbb{Z}^d) = d$. ▷ solution ◁

Recall in a network (G, c) we can consider different ℓ^p norms; namely, we define

$$\|f\|_p^p = \|f\|_{c,p}^p := \sum_x c_x |f(x)|^p,$$

where $\| \cdot \|_p$ is a norm when $p \geq 1$. We also define $\|f\|_\infty = \sup_x |f(x)|$. Let $\ell^p(G, c)$ be the space of all $f : G \to \mathbb{C}$ with $\|f\|_p < \infty$.

The following generalizes Proposition 5.2.3.

Theorem 5.3.3 (Sobolev inequality) *Let (G, c) be a network.*
Let $p > 1$ and $d = \frac{p}{p-1}$. Then, the Sobolev inequality

$$\|f\|_p \leq K_p \cdot \sum_{x,y} |\nabla f(x,y)|, \qquad \text{for all } f \in \ell^0(G, c), \tag{5.2}$$

holds for some constant $K_p > 0$ if and only if the d-dimensional isoperimetric inequality

$$c(A, A^c) \geq \kappa_d c(A)^{(d-1)/d}, \qquad \text{for all } A \subset G, |A| < \infty, \tag{5.3}$$

holds for some constant $\kappa_d > 0$.

Proof For a finite subset $A \subset G$, note that

$$\|\mathbf{1}_A\|_p^p = \sum_x c_x |\mathbf{1}_A(x)|^p = c(A),$$

$$\sum_{x,y} |\nabla \mathbf{1}_A(x,y)| = 2 \sum_{x,y} c(x,y) \mathbf{1}_{\{x \in A, \, y \notin A\}} = 2c(A, A^c).$$

So (5.2) implies (5.3) with $\kappa_d = \frac{2}{K_p}$.

For the other direction, let $f \in \ell^0(G, c)$. Since

$$\sum_{x,y} |\nabla f(x,y)| \geq \sum_{x,y} c(x,y) ||f(x)| - |f(y)||,$$

by replacing f with $|f|$ we may assume that $f : G \to \mathbb{R}$ is a nonnegative function $f \geq 0$.

For $t \geq 0$ let $S_t = \{x : f(x) > t\}$. S_t is a finite set since f has finite support.

Using the Fubini–Tonelli Theorem,

$$\sum_{x,y} c(x,y)|f(y) - f(x)| = 2 \sum_{x,y:f(y)<f(x)} c(x,y)(f(x) - f(y))$$

$$= 2 \sum_{x,y:f(y)<f(x)} c(x,y) \int_0^\infty \mathbf{1}_{[f(y),f(x))}(t)dt$$

$$= 2 \int_0^\infty \sum_{x,y} c(x,y)\mathbf{1}_{\{x\in S_t,\ y\notin S_t\}}dt$$

$$\geq 2\kappa_d \int_0^\infty \left(\sum_{x\in S_t} c_x\right)^{1/p} dt = 2\kappa_d \int_0^\infty (c(S_t))^{1/p}dt,$$

using (5.3). On the other hand, since $\varphi(t) := c(S_t)^{1/p}$ is a nonincreasing function,

$$\|f\|_p^p = \sum_x c_x f(x)^p = \sum_x c_x \int_0^\infty \mathbf{1}_{\{f(x)>t\}} pt^{p-1}dt$$

$$= \int_0^\infty \varphi(t)^p pt^{p-1}dt \leq p \int_0^\infty \left(\int_0^t \varphi(x)dx\right)^{p-1} \varphi(t)dt$$

$$\leq p \left(\int_0^\infty \varphi(x)dx\right)^{p-1} \cdot \int_0^\infty \varphi(t)dt = p \left(\int_0^\infty c(S_t)^{1/p}dt\right)^p,$$

again by Fubini–Tonelli.

In conclusion, for all $f \in \ell^0(G,c)$,

$$\|f\|_p \leq p^{1/p} \int_0^\infty c(S_t)^{1/p}dt \leq \frac{p^{1/p}}{2\kappa_d} \cdot \sum_{x,y} |\nabla f(x,y)|,$$

establishing (5.2).   □

5.4 Nash Inequality

We now proceed to consider the isoperimetric dimension of a network with a finer resolution.

Definition 5.4.1 Let $\mathcal{I} \colon [0,\infty) \to [0,\infty)$ be a nondecreasing function. Let $\kappa > 0$ be a positive real number. We say that a network (G,c) satisfies a $(\mathcal{I}, \kappa)$-**isoperimetric inequality** (or sometimes we just say that (G,c) is $(\mathcal{I}, \kappa)$-isoperimetric), if for any finite subset $A \subset G$, it holds that $c(A, A^c) \geq \kappa \cdot \mathcal{I}(c(A))$.

If (G, c) satisfies a $(\mathcal{I}, \kappa)$-isoperimetric inequality with $\mathcal{I}(t) = t^{(d-1)/d}$, we say that (G, c) satisfies a d-**dimensional isoperimetric inequality**.

Note that $\dim^{\mathrm{iso}}(G, c) \geq d$ if and only if for any $n < d$ (G, c) satisfies an n-dimensional isoperimetric inequality. So this is a finer notion than isoperimetric dimension.

Definition 5.4.2 Let $\mathcal{N} : [0, \infty) \to [0, \infty)$ be a nondecreasing function. We say that (G, c) satisfies a $\mathcal{N}$-**Nash inequality** if for any $f \in \ell^0(G, c)$,

$$\|f\|_2^2 \leq \mathcal{N}\left(\frac{\|f\|_1^2}{\|f\|_2^2}\right) \cdot \mathcal{E}(f, f).$$

Recall that the Sobolev inequality (Theorem 5.3.3) gives a relationship between isoperimetric dimension and an inequality comparing $\|f\|_p$ to $\|\nabla f\|_1$. In a similar fashion, the following theorem relates isoperimetric inequalities to Nash inequalities.

Theorem 5.4.3 *Suppose (G, c) satisfies a $(\mathcal{I}, \kappa)$-isoperimetric inequality, with $\mathcal{I}$ such that $t \mapsto \frac{t}{\mathcal{I}(t)}$ is nondecreasing.*

Then (G, c) satisfies a $\mathcal{N}$-Nash inequality, with $\mathcal{N}(t) = \frac{2}{\kappa^2} \cdot \frac{(4t)^2}{\mathcal{I}(4t)^2}$.

Proof Note that since $\|f\|_p = \| |f| \|_p$ and $\mathcal{E}(f, f) \geq \mathcal{E}(|f|, |f|)$, it suffices to prove the inequality for nonnegative f.

So let $f \in \ell^0(G, c)$ such that $f \geq 0$. For $t \geq 0$ let $S_t = \{x : f(x) > t\}$. Note that S_t decreases as t increases, so $\frac{c(S_t)}{\mathcal{I}(c(S_t))}$ decreases as well. Since S_t is finite, we may now compute that

$$f(x) = \int_0^\infty \mathbf{1}_{\{f(x) > t\}} dt = \int_0^\infty \mathbf{1}_{\{x \in S_t\}} dt,$$

$$f(x) - f(y) = \int_0^\infty \left(\mathbf{1}_{\{x \in S_t,\ y \notin S_t\}} - \mathbf{1}_{\{y \in S_t,\ x \notin S_t\}}\right) dt,$$

$$\|f\|_1 = \sum_x c_x f(x) = \int_0^\infty c(S_t) dt \leq \kappa^{-1} \cdot \int_0^\infty \frac{c(S_t)}{\mathcal{I}(c(S_t))} \cdot c(S_t, S_t^c) dt$$

$$\leq \frac{c(S_0)}{\kappa \mathcal{I}(c(S_0))} \cdot \int_0^\infty \sum_{x \in S_t, y \notin S_t} c(x, y) dt$$

$$= \frac{c(S_0)}{\kappa \mathcal{I}(c(S_0))} \cdot \sum_{x,y} \mathbf{1}_{\{f(y) < f(x)\}} c(x, y)(f(x) - f(y))$$

$$= \frac{c(S_0)}{\kappa \mathcal{I}(c(S_0))} \cdot \frac{1}{2} \sum_{x,y} c(x, y)|f(x) - f(y)|.$$

Note that $S_0 = \{x : f(x)^2 > 0\}$. Considering $f^2 \geq 0$ in the above, we have by Cauchy–Schwarz that

$$\left\| f^2 \right\|_1^2 \leq \left(\frac{c(S_0)}{2\kappa I(c(S_0))} \cdot \sum_{x,y} c(x,y)|f(x)^2 - f(y)^2| \right)^2$$

$$\leq \frac{c(S_0)^2}{4\kappa^2 I(c(S_0))^2} \cdot \sum_{x,y} c(x,y)|f(x) - f(y)|^2 \cdot \sum_{x,y} c(x,y)(|f(x)| + |f(y)|)^2$$

$$\leq \frac{c(S_0)^2}{4\kappa^2 I(c(S_0))^2} \cdot \mathcal{E}(f,f) \cdot 4\|f\|_2^2 = \tfrac{1}{2}\mathcal{N}\left(\tfrac{1}{4}c(S_0)\right) \cdot \mathcal{E}(f,f) \cdot \|f^2\|_1.$$

Using that $f \geq 0$, Markov's inequality is just

$$t c(S_t) = t \cdot \sum_x c_x \mathbf{1}_{\{f(x) > t\}} \leq \sum_x c_x f(x) = \|f\|_1.$$

Now, fix $\lambda > 0$ and consider $f_\lambda := (f - \lambda)_+ = \max\{f - \lambda, 0\}$. So $f^2 \leq f_\lambda^2 + 2\lambda f$ (because $f \geq 0$). Also, $\mathrm{supp}(f_\lambda) = S_\lambda$ and

$$\mathcal{E}(f_\lambda, f_\lambda) = \sum_{x,y \in S_\lambda} c(x,y)|f(x) - f(y)|^2 \leq \mathcal{E}(f,f).$$

Thus, for any $\lambda > 0$,

$$\|f\|_2^2 = \sum_x c_x f(x)^2 \leq \left\| f_\lambda^2 \right\|_1 + 2\lambda\|f\|_1$$

$$\leq \tfrac{1}{2}\mathcal{N}\left(\tfrac{1}{4}c(S_\lambda)\right) \cdot \mathcal{E}(f,f) + 2\lambda\|f\|_1 \quad \leq \tfrac{1}{2}\mathcal{N}\left(\tfrac{\|f\|_1}{4\lambda}\right) \cdot \mathcal{E}(f,f) + 2\lambda\|f\|_1.$$

We choose $\lambda = \frac{\|f\|_2^2}{4\|f\|_1}$, so that

$$\|f\|_2^2 \leq \tfrac{1}{2}\mathcal{N}\left(\frac{\|f\|_1^2}{\|f\|_2^2}\right) \cdot \mathcal{E}(f,f) + \tfrac{1}{2}\|f\|_2^2. \qquad \square$$

5.5 Operator Theory for the Heat Kernel

We now continue our investigation of the operator $P(x,y) = \frac{c(x,y)}{c_x}$ as an operator between the spaces $\ell^p(G,c) = \{f : G \to \mathbb{C} : \|f\|_p < \infty\}$ (we will need to be careful with the choices of p, q for which this operator is well defined). Recall that

$$\|f\|_p^p = \sum_x c_x |f(x)|^p \quad \text{and} \quad \|f\|_\infty = \sup_x |f(x)|.$$

Our main focus will be for $p \in \{1, 2, \infty\}$.

Recall that for a linear operator $T \colon \ell^p(G, c) \to \ell^q(G, c)$, we define the $(p \to q)$ operator norm

$$\|T\|_{p \to q} = \sup_{0 \neq f \in \ell^p(G,c)} \frac{\|Tf\|_q}{\|f\|_p}.$$

With these definitions, we see that $\|f\|_c = \|f\|_2$ and $\rho(P) = \|P\| = \|P\|_{2 \to 2}$.

Exercise 5.21 Show that

$$\|P^t\|_{1 \to 2}^2 = \sup_{x,y} \frac{P^{2t}(x, y)}{c_y} = \sup_{x} \frac{P^{2t}(x, x)}{c_x}.$$ ▷ solution ◁

Exercise 5.22 Let Q be an operator with $\|Q\|_{2 \to 2} \leq 1$. Let $f \in \ell^2(G, c)$. Show that the sequence $(\|Q^t f\|_2)_t$ is nonincreasing.   ▷ solution ◁

Exercise 5.23 Let Q be a self-adjoint operator on the Hilbert space $\ell^2(G, c)$. Let $f \in \ell^2(G, c)$. Show that the sequence $\left(\|Q^t f\|_2^2 - \|Q^{t+1} f\|_2^2 \right)_t$ is nonincreasing.   ▷ solution ◁

As one consequence of the Nash inequality we obtain the following theorem.

Theorem 5.5.1 *Consider the random walk $(X_t)_t$ on a connected network (G, c). If (G, c) satisfies a $(\mathcal{I}, \kappa)$-isoperimetric inequality with $\mathcal{I}(t) = t^{(d-1)/d}$, then for any $x, y \in G$, there exists a constant $K = K(d, \kappa) > 0$ such that for any $t \geq 2$,*

$$\mathbb{P}_x[X_t = y] \leq K c_y t^{-d/2}.$$

Proof Let P be the transition matrix $P(x, y) = \frac{c(x,y)}{c_x}$ and let $Q = \frac{1}{2}(I + P)$ be the lazy version of P. Note that for any x, y,

$$Q^2(x, y) = \frac{1}{4} \left(I + 2P + P^2 \right)(x, y) \geq \frac{1}{2} P(x, y).$$

Because Q is also self-adjoint,

$$\mathcal{E}(f, f) = \sum_{x,y} c_x P(x, y) |f(x) - f(y)|^2$$

$$\leq 2 \sum_{x,y} c_x Q^2(x, y) |f(x) - f(y)|^2 = 4 \left\langle \left(I - Q^2 \right) f, f \right\rangle$$

$$= 4 \|f\|_2^2 - 4 \|Qf\|_2^2.$$

By Theorem 5.4.3, for any $f \in \ell^0(G, c)$ we have a Nash-type inequality:

$$\|f\|_2^2 \leq \mathcal{N} \left(\frac{\|f\|_1^2}{\|f\|_2^2} \right) \cdot \left(\|f\|_2^2 - \|Qf\|_2^2 \right), \tag{5.4}$$

where $\mathcal{N}(t) = \frac{8}{\kappa^2} \cdot (4t)^{2/d}$ (and κ is the constant from the d-dimensional isoperimetric inequality).

Fix $f \in \ell^0(G)$ with $\|f\|_1 = 1$. Let $f_t = Q^t f$ and let $\xi(t) = \|f_t\|_2^2$. Recall that Q is a contraction and that $f_0 = f \in \ell^2(G, c)$, so $f_t \in \ell^2(G, c)$.

Fix $\varepsilon > 0$, and let $g_{t,\varepsilon} \in \ell^0(G, c)$ be such that $\|g_{t,\varepsilon} - f_t\|_2 < \varepsilon$ and $\|g_{t,\varepsilon}\|_1 \leq \|f_t\|_1$. For example, this can be achieved by taking $g_{t,\varepsilon} = \mathbf{1}_A f_t$ for some finite but large enough subset $A \subset G$.

Because $c_x Q^t(x, y) = c_y Q^t(y, x)$,

$$\|f_t\|_1 = \sum_x c_x \left| \sum_y Q^t(x, y) f(y) \right| \leq \sum_{x,y} Q^t(y, x) c_y |f(y)| = \|f\|_1 = 1.$$

By (5.4),

$$\|g_{t,\varepsilon}\|_2^{2(d+2)/d} \leq \frac{8 \cdot 4^{2/d}}{\kappa^2} \cdot \left(\|g_{t,\varepsilon}\|_2^2 - \|Q g_{t,\varepsilon}\|_2^2 \right).$$

Taking $\varepsilon \to 0$, we arrive at

$$\xi(t)^{(d+2)/d} \leq M \cdot (\xi(t) - \xi(t+1)) \qquad \text{for} \qquad M = \frac{8 \cdot 4^{2/d}}{\kappa^2}. \tag{5.5}$$

Now, by Exercises 5.22 and 5.23, the sequences $(\xi(t))_t$ and $(\xi(t) - \xi(t+1))_t$ are both nonincreasing. We thus may interpolate ξ to be a smooth function $\xi \colon [0, \infty) \to [0, \infty)$ such that ξ is nonincreasing and convex. We conclude for any $t > 0$ that $\xi(t+1) - \xi(t) \geq \xi'(t)$. Plugging this into (5.5) we have the differential inequality

$$1 \leq -M \cdot \xi'(t) \cdot \xi(t)^{-(d+2)/d} = \frac{dM}{2} \cdot \frac{\partial}{\partial t} \left(\xi(t)^{-2/d} \right),$$

which by integrating implies that

$$t \leq \frac{dM}{2} \cdot \left(\xi(t)^{-2/d} - \xi(0)^{-2/d} \right) \leq \frac{dM}{2} \xi(t)^{-2/d}.$$

Hence, for the constant $K = \frac{2}{dM} > 0$ we get that for any $f \in \ell^0(G)$,

$$\|Q^t f\|_2^2 = \xi(t) \leq (Kt)^{-d/2}.$$

Choosing $f = \frac{1}{c_x} \delta_x$ we arrive at

$$\frac{Q^{2t}(x, x)}{c_x} = (c_x)^{-2} \cdot \left\langle Q^{2t} \delta_x, \delta_x \right\rangle_c = \|Q^t f\|_2^2 \leq (Kt)^{-d/2}.$$

Finally, since $c_x P^{2t}(x, x) = \|P^t \delta_x\|_2^2$, we have, by Exercise 5.22, that $\left(P^{2t}(x, x) \right)_t$ is a nonincreasing sequence. Thus,

$$Q^{2t}(x, x) = 2^{-2t} \sum_{j=0}^{2t} \binom{2t}{j} P^j(x, x) \geq 2^{-2t} \sum_{k=0}^{t} \binom{2t}{2k} P^{2t}(x, x) \geq \tfrac{1}{2} P^{2t}(x, x),$$

where we have used that

$$2^{-2t} \sum_{k=0}^{t} \binom{2t}{2k} = \mathbb{P}\left[\text{Bin}\left(2t, \tfrac{1}{2}\right) \text{ is even}\right] = \tfrac{1}{2},$$

which is a simple exercise to prove (see Exercise 5.29).

Since $\|P\| \leq 1$, we have by Cauchy–Schwarz that

$$c_x P^{2t+1}(x, y) = \left\langle P^t \delta_y, P^{t+1} \delta_x \right\rangle_c \leq \|P^t \delta_y\|_2 \cdot \|P^{t+1} \delta_x\|_2$$

$$\leq \|P^t \delta_y\|_2 \cdot \|P^t \delta_x\|_2 = \sqrt{c_y P^{2t}(y, y) \cdot c_x P^{2t}(x, x)}.$$

Similarly,

$$c_x P^{2t}(x, y) \leq \sqrt{c_y P^{2t}(y, y) \cdot c_x P^{2t}(x, x)}.$$

Hence both $c_x P^{2t}(x, y)$ and $c_x P^{2t+1}(x, y)$ are bounded by $2c_x c_y (Kt)^{-d/2}$. Dividing by c_x, we obtain, for any $t \geq 2$,

$$P^t(x, y) \leq 2^{1+d/2} c_y (K(t-1))^{-d/2},$$

which completes the proof.   □

Since any infinite network satisfies a 1-dimensional isoperimetric inequality (with implicit constant $\kappa = 1$), we obtain the following corollary.

Corollary 5.5.2 There is a universal constant $K > 0$ such that for any infinite network (G, c) the induced random walk $(X_t)_t$ satisfies

$$\mathbb{P}_x[X_t = y] \leq K c_y t^{-1/2}$$

for all $t \geq 2$ and all x, y.

Exercise 5.24 Let G be a finitely generated group. Assume that $|B_r| \geq \alpha r^d$ for all r and some constant $\alpha > 0$, where $B_r = B_S(1, r)$ is the ball of radius r around 1 in some fixed Cayley graph of G.

Show that for any symmetric, adapted μ-random walk $(X_t)_t$ on G we have that

$$\mathbb{P}_x[X_t = y] \leq C t^{-d/2}$$

for some constant $C = C(\alpha, d, \mu) > 0$ and any $t > 0$.   ▷ solution ◁

Let G be a finitely generated group, with a finite symmetric generating set S. Recall that $B_S(x, r)$ denotes the ball of radius r about x in the Cayley graph with respect to S. We say that G has **exponential growth** if there exists $c = c_S > 0$

such that for all $r > 0$ we have $|B_S(1, r)| \geq e^{cr}$. (We will revisit the notion of *growth* in the future, in Section 8.1.)

Exercise 5.25 Show that the definition of exponential growth does not depend on the specific choice of finite symmetric generating set S (although the implicit constant c does).

▷ solution ◁

Exercise 5.26 Show that if G is a group of exponential growth, then for any symmetric, adapted measure μ on G we have that the network (G, μ) satisfies an $(\mathcal{I}, \kappa)$-isoperimetric inequality, where $\mathcal{I}(t) = \frac{t}{\log(t)}$ and $\kappa = \kappa_\mu > 0$ is some constant.

▷ solution ◁

Exercise 5.27 Show that if G is a group of exponential growth, then for any symmetric, adapted measure μ on G we have that the μ-random walk $(X_t)_t$ satisfies

$$\mathbb{P}[X_t = 1] \leq 2 \exp\left(-c_\mu t^{1/3}\right).$$

▷ solution ◁

Example 5.5.3 We now provide an example of a finitely supported random walk on an exponential growth group for which $\mathbb{P}[X_t = 1] \geq c e^{-ct^{1/3}}$ for some $c > 0$ and all $t > 0$, showing that the bound obtained from the Nash inequality cannot be improved for general exponential growth groups.

This example will be expanded upon in Section 6.9.

Consider

$$\Sigma = \bigoplus_{\mathbb{Z}} \{0, 1\} = \{\sigma : \mathbb{Z} \to \{0, 1\} : |\mathrm{supp}(\sigma)| < \infty\},$$

with point-wise addition modulo 2. So $(\sigma + \tau)(z) = \sigma(z) + \tau(z) \pmod 2$.

$\mathbb{Z}$ acts on Σ via $z.\sigma(x) = \sigma(x - z)$.

We may construct a group via this action, which is actually the *semi-direct product*: $L = L(\mathbb{Z}) = \mathbb{Z} \ltimes \Sigma$ (see Exercise 1.68). Recall that this is the group whose elements are $\mathbb{Z} \times \Sigma$ and multiplication is given by

$$(x, \sigma)(y, \tau) = (x + y, \sigma + x.\tau).$$

One can easily verify that this constitutes a group.

Now, let $S = \{(1, 0), (-1, 0), (0, \delta_0), (0, 0)\}$, which is easily seen to be a generating set for $L(\mathbb{Z})$.

It is simple to verify that for a general element $(z, \sigma) \in L$, multiplying on the right by a generator $(\pm 1, 0)$ will change the $\mathbb{Z}$-coordinate by ± 1, and not affect the Σ-coordinate. Multiplying on the right by $(0, \delta_0)$ will change the Σ-

coordinate by flipping the value of σ at position z. Multiplying by the identity element $(0, 0)$ does nothing of course.

Let μ be the uniform measure on S, and let $(X_t, \sigma_t)_t$ denote the μ-random walk on L.

A μ-random walk on L can be thought of as follows: a "lamplighter" is walking on the integers $\mathbb{Z}$, where a "lamp" is placed at each integer. The $\mathbb{Z}$-coordinate is the position of the lamplighter, and $\sigma(x)$ gives the state of the lamp placed at x (either 1 or 0). At each step, the lamplighter either moves left or right, without changing any lamps, or the lamplighter switches the state of the lamp at the current position, or the lamplighter does nothing, each of these four possibilities with equal probability.

Define

$$Q_t = \{x \in \mathbb{Z} : \exists\, 0 \le k \le t - 1,\ X_{k+1} = X_k = x\}.$$

That is, Q_t is the set of all lamps such that the lamplighter has stayed at that lamp for at least one time step up to t. Note that Q_t is measurable with respect to $\sigma(X_0, X_1, \ldots, X_t)$.

It is quite intuitive that every time the lamplighter stays at some lamp, there is equal chance that the lamp state is switched or that nothing happens. This is independent of other times steps. Thus, the states of the lamps in Q_t should have the distribution of independent Bernoulli random variables. This is the content of the next exercise.

Exercise 5.28 Show that for (L, μ) and Q_t as above, for any $\xi : Q_t \to \{0, 1\}$, we have

$$\mathbb{P}[\sigma_t(x) = \xi(x) \text{ for all } x \in Q_t \mid (X_n)_n] = 2^{-|Q_t|} \quad \text{a.s.} \qquad \triangleright \text{solution} \triangleleft$$

In the solution of Exercise 5.28 one may require the following.

Exercise 5.29 Show that for $X \sim \text{Bin}\left(n, \frac{1}{2}\right)$ the probability that X is even is $\frac{1}{2}$.

$$\triangleright \text{solution} \triangleleft$$

The set Q_t is a bit complicated, and it will be useful to note that $Q_t \subset R_{t-1}$ where $R_t = \{X_0, \ldots, X_t\}$.

Now, for any $z \in \mathbb{Z}$ with $|z| \le m + 1$, we have

$$\mathbb{P}[(X_{2t}, \sigma_{2t}) = (z, 0)] \ge \mathbb{P}[X_{2t} = z] \cdot \mathbb{E}\,\mathbb{P}[\sigma_{2t}(x) = 0 \ \forall\, x \in Q_{2t} \mid X_{2t} = z]$$

$$= \mathbb{E}\left[2^{-|Q_{2t}|}\mathbf{1}_{\{X_{2t}=z\}}\right] \ge \mathbb{P}[|Q_{2t}| \le m,\ X_{2t} = z] \cdot 2^{-m}$$

$$\ge \mathbb{P}[|R_{2t-1}| \le m,\ X_{2t} = z] \cdot 2^{-m}.$$

Summing over $|z| \le m + 1$, and noting that in a group the heat kernel is always maximized on the diagonal, we arrive at

$$\mathbb{P}[(X_{2t}, \sigma_{2t}) = (0,0)] \geq \frac{1}{2m+3} \sum_{|z| \leq m+1} \mathbb{P}[|R_{2t-1}| \leq m, X_{2t} = z] \cdot 2^{-m}$$

$$= \mathbb{P}[|R_{2t-1}| \leq m] \cdot \frac{1}{2m+3} \cdot 2^{-m}.$$

One notes that $(X_t)_t$ has the distribution of a lazy random walk on $\mathbb{Z}$, so if we define $M_t = \max_{k \leq t} |X_k|$, we may estimate

$$\mathbb{P}[|R_{2t-1}| \leq 2m + 1] \geq \mathbb{P}[M_{2t} \leq m].$$

We have seen in Exercise 2.28 that for the lazy random walk on $\mathbb{Z}$,

$$\mathbb{P}[M_{2t} \leq m] \geq c \exp\left(-c\frac{t}{m^2}\right),$$

for some constant $c > 0$ and all $t > 0$. So, perhaps by modifying the constant $c > 0$, we may conclude with the bound

$$\mathbb{P}[(X_{2t}, \sigma_{2t}) = (0,0)] \geq \sup_{m \geq 1} c \exp(-c\frac{t}{m^2} - (2m + 1)\log 2 - \log(4m + 5)),$$

and by choosing $m = c't^{1/3}$ for an appropriate $c' > 0$, we arrive at the required bound

$$\mathbb{P}[(X_{2t}, \sigma_{2t}) = (0,0)] \geq c'' \exp\left(-c''t^{1/3}\right). \qquad \triangle \triangledown \triangle$$

5.6 The Varopolous–Carne Bound

We have seen how to bound the heat kernel $P^t(x, y)$ as a function of t (time), using Nash-type inequalities and volume growth. We now bound $P^t(x, y)$ as a function of space, not time; that is, the bounds will depend on the distance between x and y, and fixed t.

We begin with some motivation from some classical results on martingale concentration.

Exercise 5.30 (Hoeffding's inequality) Let X be a (real-valued) random variable, with $\mathbb{E}[X] = 0$ and $|X| \leq 1$ a.s.

Show that for any $\varepsilon > 0$ we have

$$\mathbb{E}\left[e^{\varepsilon X}\right] \leq \exp\left(\frac{\varepsilon^2}{2}\right).$$

(Hint: the function $x \mapsto e^{\varepsilon x}$ is convex.)   ▷ solution ◁

Exercise 5.31 (Azuma's inequality) Let $(M_t)_t$ be a martingale. Assume that $|M_{t+1} - M_t| \leq b$ a.s. for all t (i.e. the martingale has *a.s. bounded differences*).

Show that for any $\lambda > 0$ we have

$$\mathbb{P}[M_t - M_0 \geq \lambda] \leq \exp\left(-\tfrac{\lambda^2}{2b^2 t}\right)$$

for all $t > 0$. (Hint: bound $\mathbb{E}[\exp(\varepsilon M_t)]$.)

▷ solution ◁

Exercise 5.32 Let μ be a symmetric and adapted measure on $\mathbb{Z}$ of finite support. Let $(X_t)_t$ be the μ-random walk. Show that there exists $c = c_\mu > 0$ such that for all $t > 0$ and all $x \in \mathbb{Z}$,

$$\mathbb{P}[X_t = x] \leq \exp\left(-c\tfrac{|x|^2}{t}\right).$$

▷ solution ◁

Compare this to Exercise 2.28.

We see that it is difficult for the random walk on $\mathbb{Z}$ to go further than order $\sqrt{t}$ at time t. (Compare this with the central limit theorem.) However, in larger graphs or groups, a priori it may be that to reach a vertex x in t steps there are many more possible paths, so that the probability to be at x at time t could be larger than $\exp\left(-cr^2/t\right)$, where r is the distance between x and the origin. We will now see that (perhaps surprisingly) this is actually never the case.

Theorem 5.6.1 (Varopolous–Carne bound) *Let G be a finitely generated group, and fix some finite symmetric generating set S of G. Let $\mu \in \mathrm{SA}(G, \infty)$ and let $(X_t)_t$ be the μ-random walk. Then there exists $c = c_{S,\mu} > 0$ such that for all t,*

$$\mathbb{P}_x[X_t = y] \leq \begin{cases} e^{-\frac{1}{2}(d-1)c} & t < (d-1)c + 1, \\ \exp\left(-\tfrac{(d-1)^2 c^2}{2(t-1)}\right) & t \geq (d-1)c + 1, \end{cases}$$

where $d = \mathrm{dist}_S(x, y) = \left|x^{-1}y\right|_S$.
Specifically, there is a constant $c' = c'_{S,\mu}$ such that for all $t > 0$,

$$\mathbb{P}_x[X_t = y] \leq \exp\left(-\tfrac{c'|x^{-1}y|^2}{t}\right).$$

First, we require a variation on Hoeffding's inequality that holds for unbounded random variables.

Lemma 5.6.2 *Let X be a (real-valued) random variable of mean $\mathbb{E}[X] = 0$, and assume that $C, c > 0$ are such that for all $r > 0$ we have $\mathbb{P}[|X| > r] \leq Ce^{-cr}$. Then, for any $0 < \lambda < c$ we have*

$$\mathbb{E}\left[e^{\lambda X}\right] \leq \exp\left(\tfrac{C\lambda^2}{c(c-\lambda)}\right).$$

Proof It is quite simple to show that for any integer $k > 0$,

$$\mathbb{E}\left[|X|^k\right] = \int_0^\infty \mathbb{P}\left[|X|^k > \xi\right] d\xi = \int_0^\infty \mathbb{P}[|X| > \xi] k \xi^{k-1} d\xi$$

$$\leq C \cdot \int_0^\infty e^{-c\xi} k \xi^{k-1} d\xi = \tfrac{Ck}{c} \cdot \mathbb{E}\left[\mathrm{Exp}(c)^{k-1}\right] = C \cdot k! \cdot c^{-k}.$$

Since $\mathbb{E}[X] = 0$,

$$\mathbb{E}\left[e^{\lambda X}\right] = \sum_{k=0}^\infty \frac{\lambda^k \mathbb{E}\left[X^k\right]}{k!} \leq 1 + C \cdot \sum_{k=2}^\infty (\tfrac{\lambda}{c})^k = 1 + C \cdot \tfrac{\lambda^2}{c(c-\lambda)} \leq \exp\left(\tfrac{C\lambda^2}{c(c-\lambda)}\right),$$

as long as $|\lambda| < c$. $\qquad\qquad \square$

Exercise 5.33 Let μ be a symmetric and adapted probability measure on a finitely generated group G. Let $(X_t)_t$ be the μ-random walk.

Let $x, y \in G$. Let $t > 0$ be such that $\mathbb{P}_x[X_t = y] > 0$.

Show that $\mathbb{P}_y[X_t = x] = \mathbb{P}_x[X_t = y] > 0$.

Show that for any z and any $0 < s < t$,

$$\mathbb{P}_x[X_s = z \mid X_t = y] = \mathbb{P}_y[X_{t-s} = z \mid X_t = x].$$

Proof of the Varopolous–Carne bound, Theorem 5.6.1 Fix x, y and let $d = \mathrm{dist}(x, y) = \left|x^{-1}y\right|$. Let $\varphi : G \to \mathbb{R}$ be any function that is 1-Lipschitz $\|\nabla\varphi\|_\infty \leq 1$ and $\varphi(x) = 0, \varphi(y) = d$. For example, one may take $\varphi(z) = \mathrm{dist}(x, z)$. Set $\delta(z) = \mathbb{E}_z[\varphi(X_1) - \varphi(X_0)] = \mathbb{E}_z[\varphi(X_1)] - \varphi(z)$.

Now, let $(X_t)_t$ be a random walk started at $X_0 = x$, and $(Y_t)_t$ be an independent random walk started at $Y_0 = y$. Define the process $M_1 = 0$ and

$$M_t = \varphi(X_t) - \varphi(Y_t) + \varphi(Y_1) - \varphi(X_1) - \sum_{j=1}^{t-1} (\delta(X_j) - \delta(Y_j))$$

$$= \sum_{j=1}^{t-1} (\varphi(X_{j+1}) - \mathbb{E}[\varphi(X_{j+1}) \mid X_j]) - \sum_{j=1}^{t-1} (\varphi(Y_{j+1}) - \mathbb{E}[\varphi(Y_{j+1}) \mid Y_j]).$$

It is simple to check that $(M_t)_t$ is a martingale. In fact,

$$M_{t+1} - M_t = \varphi(X_{t+1}) - \mathbb{E}[\varphi(X_{t+1}) \mid X_t] + \mathbb{E}[\varphi(Y_{t+1}) \mid Y_t] - \varphi(Y_{t+1}).$$

Denote

$$Z_t = \varphi(X_t) - \mathbb{E}[\varphi(X_t) \mid X_{t-1}] \quad \text{and} \quad W_t = \mathbb{E}[\varphi(Y_t) \mid Y_{t-1}] - \varphi(Y_t).$$

Let $\mathcal{F}_t = \sigma(M_1, \ldots, M_t)$.

Because φ is 1-Lipschitz, and because $\mu \in \text{SA}(G, \infty)$, we have that there exists $c > 0$ such that for any t and any $r > 0$,

$$\mathbb{P}[|Z_t| > r \mid \mathcal{F}_{t-1}, W_t] \le \mathbb{P}\left[|(X_{t-1})^{-1}X_t| > r\right] \le e^{-2cr},$$

$$\mathbb{P}[|W_t| > r \mid \mathcal{F}_{t-1}, Z_t] \le \mathbb{P}\left[|(Y_{t-1})^{-1}Y_t| > r\right] \le e^{-2cr}.$$

Also, it is immediate that $\mathbb{E}[Z_t \mid \mathcal{F}_{t-1}, W_t] = \mathbb{E}[W_t \mid \mathcal{F}_{t-1}, Z_t] = 0$. Thus, by the variation on Hoeffding's inequality, namely Lemma 5.6.2, for any $0 < \lambda \le c$ we have

$$\mathbb{E}\left[e^{\lambda Z_t} \mid \mathcal{F}_{t-1}, W_t\right] \le \exp\left(\tfrac{\lambda^2}{2c^2}\right) \quad \text{and} \quad \mathbb{E}\left[e^{\lambda W_t} \mid \mathcal{F}_{t-1}, Z_t\right] \le \exp\left(\tfrac{\lambda^2}{2c^2}\right).$$

Since $M_t = \sum_{j=2}^t (Z_t + W_t)$, we conclude that for $0 < \lambda \le c$,

$$\mathbb{E}\left[e^{\lambda M_t}\right] \le \exp\left(\tfrac{\lambda^2}{c^2} \cdot (t-1)\right).$$

Let $\mathcal{E} = \{X_t = y, \ Y_t = x\}$. Note that because μ is symmetric, $\mathbb{P}[\mathcal{E}] = \mathbb{P}[X_t = y \mid X_0 = x] \cdot \mathbb{P}[Y_t = x \mid Y_0 = y] = (\mathbb{P}_x[X_t = y])^2$. So it suffices to bound this probability.

Since,

$$\mathbb{P}_x[X_s = z \mid X_t = y] = \mathbb{P}_y[Y_{t-s} = z \mid Y_t = x],$$

we find that on the event $\mathcal{E}$ we have

$$\sum_{j=1}^{t-1}(\delta(X_j) - \delta(Y_j)) = 0.$$

Thus, on the event $\mathcal{E}$ we have $M_t \ge 2(d-1)$. Hence,

$$\exp\left(\tfrac{\lambda^2}{c^2}(t-1)\right) \ge \mathbb{E}\left[e^{\lambda M_t}\mathbf{1}_{\mathcal{E}}\right] \ge e^{2\lambda(d-1)} \cdot \mathbb{P}[\mathcal{E}],$$

implying that

$$\mathbb{P}[\mathcal{E}] \le \exp\left(\tfrac{(t-1)}{c^2}\lambda^2 - 2(d-1)\lambda\right).$$

Optimizing over λ, we would like to choose $\lambda = \frac{(d-1)c^2}{t-1}$. We still require that $\lambda \le c$ however.

When $t - 1 < (d-1)c$, we choose $\lambda = c$ resulting in a bound of

$$\mathbb{P}_x[X_t = y] = \sqrt{\mathbb{P}[\mathcal{E}]} \le \exp\left(-(d-1)\tfrac{c}{2}\right).$$

For $t - 1 \ge (d-1)c$, we choose $\lambda = \frac{(d-1)c^2}{t-1} \le c$ to obtain

$$\mathbb{P}_x[X_t = y] = \sqrt{\mathbb{P}[\mathcal{E}]} \le \exp\left(-\tfrac{(d-1)^2 c^2}{2(t-1)}\right). \qquad \square$$

5.7 Additional Exercises

Exercise 5.34 Let G be a finitely generated group, generated by a finite symmetric set S.

Let $\mu \in SA(G, \infty)$, and let $(X_t)_t$ denote the μ-random walk. Recall the *Green Function* (from Section 4.8):

$$g(x, y) = \sum_{t=0}^{\infty} \mathbb{P}[X_t = y].$$

Assume G has exponential growth.

Show that there exists some constant $C = C_{S,\mu} > 0$ such that for all $x, y \in G$,

$$g(x, y) \leq C \exp\left(-C\sqrt{|x^{-1}y|_S}\right).$$

▷ solution ◁

Exercise 5.35 Show that if G is a non-amenable finitely generated group and $\mu \in SA(G, 1)$, then $\liminf_{t\to\infty} \frac{|X_t|}{t} > 0$ a.s.

▷ solution ◁

Note that Fatou's Lemma tells us that in the setting above,

$$\liminf_{t\to\infty} \frac{\mathbb{E}|X_t|}{t} \geq \mathbb{E}\left[\liminf_{t} \frac{|X_t|}{t}\right] > 0.$$

This is not an equivalence; that is, there are amenable groups of positive speed, for example some lamp-lighter groups, see Section 6.9.

Exercise 5.36 Let G be an infinite finitely generated group. Show that for any $\varepsilon > 0$, there exists $\mu \in SA\left(G, \frac{1}{3} - \varepsilon\right)$ such that (G, μ) is transient. (Hint: recall Exercises 4.43 and 4.44.)

▷ solution ◁

This is not the best possible. It can be shown that for any finitely generated group G and any $\varepsilon > 0$, there is $\mu \in SA(G, 1 - \varepsilon)$ such that (G, μ) is transient. However, this requires methods outside the scope of those presented here.

5.8 Remarks

Kesten was one of the founding fathers of random walks on general groups (departing from the more classical Euclidean setting considered by Pólya, 1921). He proved the amenability criterion, Theorem 5.2.4, in his PhD thesis (see also Kesten, 1959).

Bounds of the Varopolous–Carne type, Theorem 5.6.1, were first proved by Varopoulos (1985) and Carne and Varopoulos (1985). Many proofs in the literature are specialized to μ of finite support. The proof presented is an elegant probabilistic argument by Rémi Peyre (2008), which is useful for the generalization to exponential tail measures, even with infinite support.

Theorem 5.3.1 was proved by by Coulhon and Saloff-Coste (1993).

Our treatment of isoperimetric inequalities, Nash inequalities, and heat kernel bounds is based on Woess (2000).

A more careful analysis of heat kernel decay in polynomial growth Cayley graphs can be used to prove that the expected number of visits to a ball of radius r in the graph is at most $O\left(r^2\right)$ (when the graph is transient). It is expected that this phenomenon is universal, at least for transient transitive graphs.

Conjecture 5.8.1 Let G be a transient Cayley graph. Fix some $x \in G$ and let $B_r = B(x, r)$ be the ball of radius r around G. Recall the *Green function*, $g(x, y)$, which is the expected number of visits to y started at x.

Then, there exists a constant $C = C(G) > 0$ such that for all $r \geq 0$,

$$g(x, B_r) = \sum_{y \in B_r} g(x, y) \leq Cr^2.$$

To our knowledge, the state of the art at the time of writing is a bound in Lyons et al. (2017):

$$g(x, B_r) \leq Cr^2 \sqrt{\log |B_r|} \leq C'r^{5/2}$$

(the last inequality from the fact that Cayley graphs grow at most exponentially). This is also related to results we will prove in the following chapters.

In Chapter 9, Theorem 9.4.1 tells us that on a Cayley graph the simple random walk $(X_t)_t$ is always at least *diffusive*; that is $\mathbb{E}\left[|X_t|^2\right] \geq ct$. Thus, for $T_r = \inf\{t \geq 0 : |X_t| > r\}$, we have that

$$ct \leq \mathbb{E}\left[|X_t|^2\right] \leq \mathbb{P}[|X_t| \leq r] \cdot r^2 + \mathbb{P}[|X_t| > r] \cdot t^2,$$

so that

$$\mathbb{P}[|X_t| > r] \geq \frac{ct - r^2}{t^2},$$

which can be made at least some fixed $\varepsilon > 0$ by choosing $t = Cr^2$ for appropriate $C > 0$. Thus, with constant probability at least $\varepsilon > 0$, the walk exits a ball of radius Kr by time $t = r^2$. As the walk is transient, one would intuitively expect the random walk to now escape and never return to the ball of radius r with some positive probability. Thus, the expected total number of visits to the ball should be at most of order r^2. This is not a proof, of course.

5.9 Solutions to Exercises

Solution to Exercise 5.1 :(

Let S, T be two finite symmetric generating sets for a group G. Assume that $\Phi_S = 0$.

Let $A \subset G$ be a finite subset. Assume that $a \in A$ and $at \notin A$ for some $t \in T$. Then, there exist $s_1, \ldots, s_k \in S$ such that $|t|_S = k$ and $t = s_1 \cdots s_k$. Since $a \in A$ and $at \notin A$, it must be that $as_1 \ldots s_{j-1} \in A$ and $as_1 \cdots s_j \notin A$ for some $1 \le j \le k$, where $s_0 = 1$. Thus,

$$\sum_a \mathbf{1}_{\{a \in A\}} \mathbf{1}_{\{at \notin A\}} \le \sum_a \sum_{j=1}^{k} \mathbf{1}_{\{as_1 \cdots s_{j-1} \in A\}} \mathbf{1}_{\{as_1 \cdots s_j \notin A\}}$$

$$\le |t|_S \cdot \sum_a \sum_{s \in S} \mathbf{1}_{\{a \in A\}} \mathbf{1}_{\{as \notin A\}} = |t|_S \cdot |AS \backslash A|.$$

Summing this over T, we have that

$$|AT \backslash T| \le \max_{t \in T} |t|_S \cdot |T| \cdot |AS \backslash A|.$$

Taking an infimum over finite subsets A, we see that

$$0 \le \Phi_T \le \frac{\max_{t \in T} |t|_S \cdot |T| \cdot |S|}{|T|} \cdot \inf_{|A| < \infty} \frac{|AS \backslash A|}{|S| \cdot |A|} \le \frac{\max_{t \in T} |t|_S \cdot |T| \cdot |S|}{|T|} \cdot \Phi_S = 0. \qquad :) \checkmark$$

Solution to Exercise 5.2 :(

Assume that $(F_n)_n$ is a Følner sequence. Fix some finite symmetric generating set S for G.

Let $\varepsilon > 0$. Take n large enough so that for all $s \in S$ we have

$$\frac{|F_n s \triangle F_n|}{|F_n|} < \varepsilon.$$

Then,

$$|F_n S \backslash F_n| \le \sum_{s \in S} |F_n s \triangle F_n| < |S| \cdot |F_n| \cdot \varepsilon,$$

so $\Phi_S < \varepsilon$. Since this holds for all $\varepsilon > 0$, we have that G is amenable.

For the other direction, assume that G is amenable. Let S be a finite symmetric generating set. We know that $\Phi_S = 0$. So there is a sequence of finite subsets $(F_n)_n$ such that

$$\lim_{n \to \infty} \frac{|F_n S \backslash F_n|}{|F_n|} = 0.$$

Note that for any $x, y \in G$ and any finite subset $F \subset G$,

$$Fxy \backslash F \subseteq (Fxy \backslash Fy) \bigcup (Fy \backslash F).$$

Since $|Ay| = |A|$ for any $A \subset G$ and $y \in G$, it is simple to conclude that

$$|Fx \triangle F| \le |Fx \backslash F| + |F \backslash Fx| \le 2|x| \cdot \sup_{s \in S} |Fs \backslash F| \le 2|x| \cdot |FS \backslash F|.$$

This leads to

$$\frac{|F_n x \triangle F_n|}{|F_n|} \to 0$$

as $n \to \infty$, for any fixed $x \in G$. Thus, $(F_n)_n$ is a Følner sequence. $\qquad :) \checkmark$

Solution to Exercise 5.3 :(

Let $(F_n)_n$ be a Følner sequence; that is,

$$\lim_{n \to \infty} \frac{|F_n x \triangle F_n|}{|F_n|} = 0$$

for any $x \in G$. Set $\tilde{F}_n = (F_n)^{-1} = \{y^{-1} : y \in F_n\}$. Then, for any $x \in G$,

$$|x\tilde{F}_n \triangle \tilde{F}_n| = \left|\left(F_n x^{-1}\right)^{-1} \triangle (F_n)^{-1}\right| = \left|F_n x^{-1} \triangle F_n\right|.$$

Since $|F_n| = |\tilde{F}_n|$, this completes the equivalence of the existence of right and left Følner sequences.   :) ✓

Solution to Exercise 5.4   :(
The first assertion is immediate by definition.
   For the second assertion, let μ be a symmetric adapted probability measure on G. Let $S \subset \mathrm{supp}(\mu)$ be a finite symmetric generating set for G. Write $\varepsilon := \min_{s \in S} \mu(s) > 0$.
   Note that for $a \in A$,

$$\sum_{a \in A} \mathbb{P}_a[X_1 \notin A] = \sum_{a \in A} \sum_x \mu(x)\mathbf{1}_{\{ax \notin A\}} \geq \varepsilon|AS\backslash A|.$$

So dividing by $|A|$ and taking the infimum over finite subsets $A \subset G$, we arrive at

$$|S| \cdot \Phi_S \leq \frac{1}{\varepsilon}\Phi_\mu.$$

So if $\Phi_\mu = 0$ then G is amenable.
   For the other direction, assume that G is amenable. Fix some finite symmetric generating set S for G. Let $\varepsilon > 0$, and choose $r > 0$ large enough so that

$$\sum_{|x|>r} \mu(x) < \varepsilon.$$

Let $T = B_S(1, r) = \{x : |x| \leq r\}$. T generates G, so $\Phi_T = 0$. We now have that

$$\sum_{a \in A} \mathbb{P}_a[X_1 \notin A] = \sum_{a \in A} \sum_x \mu(x)\mathbf{1}_{\{ax \notin A\}} \leq |AT\backslash A| + |A| \cdot \varepsilon.$$

Dividing by $|A|$ and taking an infimum over finite subsets A we get that

$$\Phi_\mu \leq |T| \cdot \Phi_T + \varepsilon = \varepsilon.$$

As this holds for any $\varepsilon > 0$, we get that $\Phi_\mu = 0$.
   This implies that $\Phi_\mu = 0$ if and only if G is amenable.   :) ✓

Solution to Exercise 5.5   :(
We have using the symmetry of μ,

$$\mathbb{P}_a[X_1 \notin A] = \sum_{x \notin A} \mu\left(a^{-1}x\right) = \sum_{x \notin A} \mathbb{P}_{x^{-1}}[X_1 = a].$$

Summing over a we get

$$\sum_{a \in A} \mathbb{P}_a[X_1 \notin A] = \sum_{x \notin A^{-1}} \mathbb{P}_x[X_1 \in A].$$:) ✓

Solution to Exercise 5.8   :(
Compute:

$$||Pf||_c^2 = \sum_x c_x \left|\sum_y P(x, y)f(y)\right|^2 \leq \sum_{x,y} c_x P(x, y)|f(y)|^2 = \sum_{x,y} c(x, y)|f(y)|^2 \leq \sum_y c_y |f(y)|^2$$

$$= ||f||_c^2.$$:) ✓

Solution to Exercise 5.9   :(
Since (G, c) is connected, P is irreducible. So, for any x, y, z, w, there exist $n, k > 0$ such that $P^n(x, z)$, $P^k(w, y) > 0$. Hence, for any t,

$$P^{t+n+k}(x, y) \geq P^n(x, z)P^t(z, w)P^k(w, y),$$

which implies that

$$\limsup_{t \to \infty} (P^t(z, w))^{1/t} \le \limsup_{t \to \infty} (P^{t+n+k}(x, y))^{1/t} = \limsup_{t \to \infty} (P^t(x, y))^{1/t}. \qquad \text{:)} \checkmark$$

Solution to Exercise 5.11 :(

First, we show that $||Pf||_c \le \tilde{\rho}||f||_c$ for any $f \in \ell^0(G, c)$ (that is, including complex-valued functions). Indeed, for any $f \in \ell^0(G, c)$,

$$||Pf||_c^2 = \sum_x c_x \left| \sum_y P(x, y)(\text{Re}f(y) + i\text{Im}f(y)) \right|^2$$

$$\le \sum_{x,y} c_x P(x, y)|\text{Re}f(y)|^2 + \sum_{x,y} c_x P(x, y)|\text{Im}f(y)|^2$$

$$= ||P\text{Re}f||_c^2 + ||P\text{Im}f||_c^2 \le \tilde{\rho}^2 \cdot \left(||\text{Re}f||_c^2 + ||\text{Im}f||_c^2 \right).$$

Since

$$||f||_c^2 = \sum_x c_x |f(x)|^2 = \sum_x c_x |\text{Re}f(x)|^2 + \sum_x c_x |\text{Im}f(x)|^2 = ||\text{Re}f||_c^2 + ||\text{Im}f||_c^2,$$

we conclude that $||Pf||_c \le \tilde{\rho}||f||_c$ for all $f \in \ell^0(G, c)$.

Now, fix $\varepsilon > 0$ and $f \in \ell^2(G, c)$. Since $\sum_x c_x |f(x)|^2 < \infty$, there exists a finite subset $S_\varepsilon \subset G$ such that $\sum_{x \notin S_\varepsilon} c_x |f(x)|^2 < \varepsilon^2$. Take $g = f 1_{S_\varepsilon}$. Then g has finite support, so $||Pg||_c \le \tilde{\rho}||g||_c$. Also, $f - g = f 1_{(G \setminus S_\varepsilon)}$, so $||f - g||_c^2 < \varepsilon^2$. Thus,

$$||Pf||_c \le ||P(f - g)||_c + ||Pg||_c \le ||P|| \cdot ||f - g||_c + \tilde{\rho}||g||_c \le ||P||\varepsilon + \tilde{\rho}||f||_c,$$

where the last inequality follows from $||g||_c \le ||f||_c$. Taking $\varepsilon \to 0$ gives $||Pf||_c \le \tilde{\rho}||f||_c$. :) $\checkmark$

Solution to Exercise 5.12 :(

Let

$$\rho = \sup_{0 \neq f \in \ell^0(G, c)} \frac{\langle Pf, f \rangle_c}{\langle f, f \rangle_c} \quad \text{and} \quad \tilde{\rho} = \sup_{\substack{f : G \to \mathbb{R} \\ 0 \neq f \in \ell^0(G, c)}} \frac{\langle Pf, f \rangle_c}{\langle f, f \rangle_c}.$$

It is immediate that $\rho = ||P|| \ge \tilde{\rho}$ by Cauchy–Schwarz.

If $f, g \in \ell^0(G, c)$ are real-valued functions $f, g : G \to \mathbb{R}$, then using the parallelogram identity,

$$|\langle Pf, g \rangle_c| = \tfrac{1}{4} \cdot |\langle P(f + g), f + g \rangle_c - \langle P(f - g), f - g \rangle_c|$$

$$\le \tfrac{\tilde{\rho}}{4} \cdot (\langle f + g, f + g \rangle_c + \langle f - g, f - g \rangle_c) = \tfrac{\tilde{\rho}}{2} \cdot (\langle f, f \rangle_c + \langle g, g \rangle_c).$$

Using this with $g = \frac{||f||_c}{||Pf||_c} Pf$, we get that

$$||f||_c \cdot ||Pf||_c = \langle Pf, g \rangle_c \le \tfrac{\tilde{\rho}}{2} \cdot \left(||f||_c^2 + ||f||_c^2 \cdot ||Pf||_c^{-2} \cdot ||Pf||_c^2 \right) = \tilde{\rho} \cdot ||f||_c^2,$$

so that $||Pf||_c \le \tilde{\rho}||f||_c$ for all $f : G \to \mathbb{R}$ with finite support. Exercise 5.11 tells us that this can be extended to all $f \in \ell^2(G, \mu)$, so that $\rho = ||P|| \le \tilde{\rho}$. :) $\checkmark$

Solution to Exercise 5.13 :(

Let $\rho = \rho(P)$.

First note that by Exercises 5.11 and 5.12 we know that $|\lambda| \cdot ||h||_c = ||Ph||_c \le \rho ||h||_c$, so that $|\lambda| \le \rho$. Assume that h is nonnegative and assume that $x \in G$ is such that $h(x) \neq 0$. Then because $h \ge 0$,

$$P^t(x, x)h(x) \le \sum_y P^t(x, y)h(y) = (P^t h)(x) = \lambda^t h(x).$$

Thus, $\lambda \ge \rho$. :) $\checkmark$

Solution to Exercise 5.15 :(

The first assertion follows from integration by parts, Theorem 4.2.4, which tells us that

$$2\langle (I - P)f, f\rangle_c = \langle \text{div}\nabla f, f\rangle_c = \langle \nabla f, \nabla f\rangle_c = \sum_{x,y} c(x,y)|f(x) - f(y)|^2.$$

For the second assertion, note that

$$\begin{aligned} 2\langle (I + P)f, f\rangle_c &= 2\langle (I - P)f, f\rangle_c + 4\langle Pf, f\rangle_c \\ &= \sum_{x,y} c(x,y)|f(x) - f(y)|^2 + 4\sum_{x,y} c(x,y)f(x)f(y) \\ &= \sum_{x,y} c(x,y)|f(x) + f(y)|^2, \end{aligned}$$

using that $f \in \ell^2(G, c)$ and that $|f(x) - f(y)|^2 + 4f(x)f(y) = |f(x) + f(y)|^2$. :) ✓

Solution to Exercise 5.16 :(

Write

$$L(A) = \frac{\log c(A)}{\log \frac{c(A)}{c(A, A^c)}}$$

and set $d = \liminf_{c(A)\to\infty} L(A)$. Let $(A_n)_n$ be a sequence of finite subsets such that $c(A_n) \to \infty$ and $L(A_n) \to d$.

If $\dim^{\text{iso}}(G, c) \geq d + 2\varepsilon$ for some $\varepsilon > 0$, then we can find a constant $\kappa > 0$ such that $c(A_n, A_n^c)^{d+\varepsilon} \geq \kappa \cdot c(A_n)^{d+\varepsilon-1}$ for all n. But this gives the contradiction

$$d = \lim_{n\to\infty} L(A_n) \geq \lim_{n\to\infty} \frac{\log c(A_n)}{\log c(A_n) - \frac{d+\varepsilon-1}{d+\varepsilon} \cdot \log c(A_n) - \frac{1}{d+\varepsilon}\log\kappa} = d + \varepsilon.$$

On the other hand, if $\dim^{\text{iso}}(G, c) \leq d - \varepsilon$ for some $\varepsilon > 0$, then for any n we can find a finite set D_n such that $c(D_n, D_n^c)^{d-\varepsilon} < c(D_n)^{d-1-\varepsilon}$. This gives the contradiction

$$d \leq \liminf_{n\to\infty} \frac{\log c(D_n)}{\log c(D_n) - \frac{d-1-\varepsilon}{d-\varepsilon} \cdot \log c(D_n)} = d - \varepsilon. \qquad\text{:) ✓}$$

Solution to Exercise 5.17 :(

For any finite subset A, in any Cayley graph $\Gamma(G, S)$, considered as a network with conductance 1 on every edge, we have that

$$|\partial A| \leq |c(A, A^c)| \leq |S| \cdot |\partial A|,$$

where $\partial A = \{x \in A :: \exists\, y \sim x,\ y \notin A\}$.

Also, as in the proof of the Coulhon–Saloff-Coste inequality (Theorem 5.3.1), $|A\backslash Ay| \leq |y| \cdot |\partial A|$. Since $x \in A$, $xy \notin A$ is equivalent to $x \in A\backslash Ay^{-1}$, we have that

$$\#\{x \in A : \exists\, u \in S',\ xu \notin A\} \leq \max_{u\in S'} |u|_S \cdot \#\{x \in A : \exists\, s \in S,\ xs \notin A\}.$$

The conclusion from this is that for any two finite symmetric generating sets S, S', the ratio between the corresponding values for $c(A, A^c)$ in the two Cayley graphs is bounded between two constants depending only on S, S' and not on A.

This implies that the limit

$$\liminf_{|A|\to\infty} \frac{\log |A|}{\log \frac{|A|}{|c(A, A^c)|}}$$

will be the same in both Cayley graphs. :) ✓

Solution to Exercise 5.18 :(

If G is non-amenable, and S is some finite symmetric generating set, then $\Phi_S > 0$. Thus, there exists a constant

$\kappa > 0$ such that for any finite subset $A \subset G$ we have that

$$|\partial A| \geq \frac{|AS \setminus A|}{|S|} \geq \kappa |A| \geq \kappa |A|^{(d-1)/d},$$

for all $d > 0$.   :) ✓

Solution to Exercise 5.19   :(
A Taylor expansion gives that $r^d \leq (r-1)^d + dr^{d-1}$ for any $r \geq 1$.

For $x \in \Gamma$, let $S(x, r) = \{y \in \Gamma : \text{dist}(x, y) = r\}$.

Since $\dim^{\text{iso}}(\Gamma) \geq d$ we know that there is a constant $\varepsilon > 0$ such that for any finite subset A we have that $|\partial A| \geq \varepsilon \cdot c(A)^{(d-1)/d}$. Here, $\partial A = \{(x, y) : x \sim y, \ x \in A, \ y \notin A\}$ and $c(A) = c(A, G) = \{x \sim y : x \in A, \ y \in G\}$. Specifically, for any $r \geq 2$ and any $x \in G$, we have

$$c(B(x, r)) = \sum_{y \sim z} \mathbf{1}_{\{\text{dist}(y,x) \leq r\}} = c(B(x, r-1)) + \sum_{y \sim z} \mathbf{1}_{\{\text{dist}(y,x) = r\}}$$

$$\geq c(B(x, r-1)) + |\partial S(x, r)| \geq c(B(x, r-1)) + \varepsilon c(B(x, r-1))^{(d-1)/d}.$$

Summing this from 1 to r, we have by induction on r that

$$c(B(x, r)) \geq \sum_{k=1}^{r} c(B(x, k)) - c(B(x, k-1)) \geq \varepsilon \cdot \sum_{k=1}^{r} c(B(x, k-1))^{(d-1)/d}$$

$$\geq \varepsilon \cdot \alpha^{(d-1)/d} \cdot \sum_{k=1}^{r-1} k^{d-1} \geq \varepsilon \cdot \alpha^{(d-1)/d} \cdot \frac{1}{d} \cdot \sum_{k=1}^{r-1} \left(k^d - (k-1)^d \right)$$

$$= \varepsilon \cdot \alpha^{(d-1)/d} \cdot \frac{1}{d} \cdot (r-1)^d \geq \varepsilon \cdot \alpha^{(d-1)/d} \cdot \frac{1}{d} \cdot 2^{-d} \cdot r^d.$$

In order for this to complete the induction, α must be chosen so that $\varepsilon \cdot \alpha^{(d-1)/d} \cdot \frac{1}{d} \cdot 2^{-d} \geq \alpha$, which is equivalent to $\alpha \leq \varepsilon^d d^{-d} 2^{-d^2}$.   :) ✓

Solution to Exercise 5.20   :(
Since the ball of radius r, B_r, has volume $|B_r| \geq cr^d$ and boundary $|\partial B_r| \leq Cr^{d-1}$, we get that $\dim^{\text{iso}}\left(\mathbb{Z}^d\right) \leq d$.

Also, by the Coulhon–Saloff-Coste inequality, since $|B_r| \geq cr^d$, we get that for any finite subset $A \subset \mathbb{Z}^d$ we have $|\partial A| \geq \frac{|A|}{2r(2|A|)}$ where $r(n) \leq C'n^{1/d}$. Thus, $|\partial A| \geq c' \cdot |A|^{(d-1)/d}$, and we conclude that $\dim^{\text{iso}}\left(\mathbb{Z}^d\right) \geq d$.   :) ✓

Solution to Exercise 5.21   :(
Recall that P is self-adjoint with respect to the inner product $\langle \cdot, \cdot \rangle_c$. Using Cauchy–Schwarz,

$$c_x P^{2t}(x, y) = \left\langle P^{2t} \delta_y, \delta_x \right\rangle_c = \left\langle P^t \delta_y, P^t \delta_x \right\rangle_c$$

$$\leq ||P^t \delta_y||_2 \cdot ||P^t \delta_x||_2 \leq ||P^t||_{1 \to 2}^2 \cdot ||\delta_y||_1 \cdot ||\delta_x||_1 = ||P^t||_{1 \to 2}^2 \cdot c_y \cdot c_x.$$

Also, for any $f \in \ell^1(G, c)$ with $||f||_1 = 1$, we have

$$\left\langle P^t f, P^t f \right\rangle_c = \left\langle P^{2t} f, f \right\rangle_c = \sum_{x,y} c_x P^{2t}(x, y) f(x) f(y)$$

$$\leq \sup_{x,y} \frac{P^{2t}(x, y)}{c_y} \cdot \sum_{x,y} c_x |f(x)| c_y |f(y)| = \sup_{x,y} \frac{P^{2t}(x, y)}{c_y} \cdot ||f||_1^2 = \sup_{x,y} \frac{P^{2t}(x, y)}{c_y}.$$

Together these show that

$$\sup_{x,y} \frac{P^{2t}(x, y)}{c_y} = ||P^t||_{1 \to 2}^2.$$

Also

$$\|P^t \delta_x\|_c^2 = \left\langle P^{2t} \delta_x, \delta_x \right\rangle_c = c_x P^{2t}(x, x),$$

so by Cauchy–Schwarz,

$$c_x P^{2t}(x, y) = \left\langle P^t \delta_y, P^t \delta_x \right\rangle_c \le \sqrt{c_x P^{2t}(x, x) \cdot c_y P^{2t}(y, y)},$$

so

$$\sup_{x,y} \frac{P^{2t}(x, y)}{c_y} \le \sup_x \frac{P^{2t}(x, x)}{c_x},$$

which actually implies equality. :) ✓

Solution to Exercise 5.22 :(
Just note that since $\|Q\|_{2\to2} \le 1$,

$$\left\|Q^{t+1}f\right\|_2 \le \|Q\|_{2\to2}\|Q^t f\|_2 \le \|Q^t f\|_2. \qquad \text{:) ✓}$$

Solution to Exercise 5.23 :(
First note that

$$\|Q^t f\|_2^2 - \|Q^{t+1}f\|_2^2 = \left\langle \left(I - Q^2\right) Q^t f, Q^t f \right\rangle.$$

Compute:

$$\left\langle \left(I - Q^2\right) Q^{t+1}f, Q^{t+1}f \right\rangle - \left\langle \left(I - Q^2\right) Q^t f, Q^t f \right\rangle$$
$$= \left\langle \left(I - Q^2\right) Q^{t+1}f, Q^{t+1}f - Q^t f \right\rangle + \left\langle \left(I - Q^2\right) \left(Q^{t+1}f - Q^t f\right), Q^t f \right\rangle$$
$$= \left\langle \left(I - Q^2\right) Q^{t+1}f, (Q - I)Q^t f \right\rangle + \left\langle (Q - I)Q^t f, \left(I - Q^2\right) Q^t f \right\rangle$$
$$= \left\langle (Q - I)Q^t f, \left(I - Q^2\right) \left(Q^{t+1}f + Q^t f\right) \right\rangle = \left\langle (Q - I)Q^t f, \left(I - Q^2\right)(Q + I)Q^t f \right\rangle$$
$$= \left\langle (Q - I)(I + Q)Q^t f, \left(I - Q^2\right) Q^t f \right\rangle = -\left\|\left(I - Q^2\right) Q^t f\right\|_2^2 \le 0. \qquad \text{:) ✓}$$

Solution to Exercise 5.24 :(
Let $S \subset \text{supp}(\mu)$ be a finite symmetric generating set. By Exercise 1.75 and the assumptions, we have that $B_S(1, r) \ge \alpha' r^d$ for some $\alpha' > 0$ and all r. Thus, $r(n) := \inf\{r : |B_S(1, r)| \ge n\}$ satisfies $r(n) \le Cn^{1/d}$ for some $C = C(\alpha) > 0$. Set $\varepsilon := \min_{s \in S} \mu(s)$. Then, by the Coulhon–Saloff-Coste inequality (Theorem 5.3.1), we know that for any finite subset $A \subset G$,

$$c(A, A^c) \ge \varepsilon \cdot |\partial A| \ge \frac{\varepsilon |A|}{2C(2|A|)^{1/d}} \ge \kappa c(A)^{(d-1)/d},$$

where $c = c_\mu$ is the conductance of the network (G, μ), given by $c_\mu(x, y) = \mu\left(x^{-1}y\right)$, and $\kappa = \kappa(d, \mu, \alpha) > 0$. That is, (G, μ) satisfies a d-dimensional isoperimetric inequality. Thus, Theorem 5.5.1 guarantees the proper conclusion. :) ✓

Solution to Exercise 5.25 :(
This is a straightforward application of Exercise 1.75. :) ✓

Solution to Exercise 5.26 :(
Fix some finite generating set $S \subset \text{supp}(\mu)$ of G, and consider the Cayley graph with respect to this generating set S. Set $\alpha = \min\{\mu(s) : s \in S\} > 0$.

Let $A \subset G$ be a finite subset. For any $x \in A$, $y \notin A$, we have that $\mu\left(x^{-1}y\right) \ge 1_{\{x^{-1}y \in S\}}\alpha$, so that

$$\sum_{x \in A, y \notin A} \mu\left(x^{-1}y\right) \ge \sum_{x \in A} 1_{\{x \in \partial A\}}\alpha = |\partial A|\alpha \ge \frac{|A|}{2r(2|A|)},$$

where $r(n) = \inf\{r \geq 0 : |B_r| \geq n\}$ is the inverse growth rate, and we have used the Coulhon–Saloff-Coste inequality, Theorem 5.3.1. Because G has exponential growth, for some constant $C > 0$ we have that $r(n) \leq C \log n$, so that for some constant $\kappa > 0$ we have that

$$\sum_{x \in A, y \notin A} \mu\left(x^{-1}y\right) \geq \kappa \frac{|A|}{\log|A|}.$$:) ✓

Solution to Exercise 5.27 :(
As in the proof of Theorem 5.5.1, we set $P(x, y) = \mu\left(x^{-1}y\right)$ and $Q = \frac{1}{2}(I + P)$, the lazy version of P. So

$$\mathcal{E}(f, f) \leq 4\|f\|_2^2 - 4\|Qf\|_2^2$$

for all $f \in \ell^2(G, \mu)$.

By the previous exercise (G, μ) satisfies an $(\mathcal{I}, \kappa)$-isoperimetric inequality with $\mathcal{I}(t) = \frac{t}{\log t}$. Since $t \mapsto \log t$ is nondecreasing, we get that (G, μ) satisfies an $\mathcal{N}$-Nash inequality with $\mathcal{N}(t) = \frac{2}{\kappa^2}\log^2(4t)$. So any $f \in \ell^0(G, \mu)$ admits

$$\|f\|_2^2 \leq \frac{32}{\kappa^2} \cdot \left(\log\left(\frac{\|f\|_1}{\|f\|_2}\right)\right)^2 \cdot \left(\|f\|_2^2 - \|Qf\|_2^2\right).$$

Approximating by a finitely supported function implies that the above also holds for all $f \in \ell^2(G, \mu)$, and specifically for the function $f = Q^t \delta_x$. This function has $\|f\|_1 = 1$, so for $\xi(t) = \|f\|_2^2 = Q^{2t}(1, 1)$, we arrive at

$$\xi(t) \leq \frac{8}{\kappa^2} \cdot (\log\xi(t))^2 \cdot (\xi(t) - \xi(t+1)).$$

By Exercises 5.22 and 5.23, the functions $t \mapsto \xi(t)$ and $t \mapsto (\xi(t+1) - \xi(t))$ are nonincreasing, so we may interpolate ξ to be a function $\xi : [0, \infty) \to [0, \infty)$ that is nonincreasing and convex. Specifically, $\xi(t+1) - \xi(t) \geq \xi'(t)$ and we arrive at

$$\xi(t) \leq -\frac{8}{\kappa^2} \cdot (\log\xi(t))^2 \cdot \xi'(t).$$

Setting $\zeta(t) = \log\xi(t)$, we have

$$1 \leq -\frac{8}{\kappa^2}\zeta(t)^2\zeta'(t).$$

Integrating from 0 to t, we arrive at

$$t \leq -\frac{8}{3\kappa^2}\left(\zeta(t)^3 - \zeta(0)^3\right) = -\frac{8}{3\kappa^2}\left(\log Q^{2t}(1, 1)\right)^3,$$

which implies that

$$Q^{2t}(1, 1) \leq \exp\left(-ct^{1/3}\right), \qquad c = \left(\tfrac{3\kappa^2}{8}\right)^{1/3}.$$

Finally, as in the proof of Theorem 5.5.1, we know that $P^{2t}(1, 1) \leq 2Q^{2t}(1, 1)$ and also $P^{2t}(x, y) \leq P^{2t}(1, 1)$ for all x, y. Since for $t > 0$,

$$\mathbb{P}[X_t = 1] = P^t(1, 1) = \sum_x \mu(x)P^{t-1}(x, 1),$$

the proof is complete.   :) ✓

Solution to Exercise 5.28 :(
Let $(U_t)_{t \geq 1}$ be i.i.d. each uniform on $\{-1, 1\}$, and let $(J_t, I_t)_{t \geq 1}$ be i.i.d. Bernoulli-$\frac{1}{2}$ random variables, independent of $(U_t)_t$. Define $(X_t, \sigma_t) \in L$ inductively by $X_0 = 0$, $\sigma_0 = 0$ and for $t > 0$,

$$(X_t, \sigma_t) = (X_{t-1}, \sigma_{t-1}) \cdot \begin{cases} (U_t, 0) & \text{if } I_t = 1, \\ (0, \delta_0) & \text{if } J_t = 1 \neq I_t, \\ (0, 0) & \text{if } J_t = I_t = 0. \end{cases}$$

It is simple to verify that $((X_t, \sigma_t))_t$ is a μ-random walk, for μ uniform on $\{(1,0), (-1,0), (0, \delta_0), (0,0)\}$.

Define $V_t = U_t I_t$. It is immediate that $X_t = V_1 + \cdots + V_t$. Since $(J_t)_t$ and $(U_t, I_t)_t$ are independent, we have that $(J_t)_t$ and $(X_t)_t$ are also independent. Thus, $(J_t)_t$ and $(Q_t)_t$ are independent.

Note that $X_{k+1} = X_k$ if and only if $I_k = 0$.

For $x \in \mathbb{Z}$ define the following.

$$K(x,t) = \{0 \le k \le t-1 : X_{k+1} = X_k = x\},$$

$$L_t(x) = \sum_{k \in K(x,t)} J_{k+1} = \sum_{k=0}^{t-1} \mathbf{1}_{\{X_{k+1}=X_k=x\}} J_{k+1}.$$

So $L_t(x)$ is the number of times the lamp at x has been flipped up to time t. Thus, $\sigma_t(x) = L_t(x) \pmod 2$.

Note that $x \in Q_t$ if and only if $K(x,t) \ne \emptyset$. Also, the sets $(K(x,t))_{x \in Q_t}$ are measurable with respect to $\sigma(X_0, \ldots, X_t)$, and these sets are pairwise disjoint. We have that $L_t(x)$ is determined by $(J_k)_{k \in K(x,t)}$. Since $(J_n)_n$ are mutually independent, we have that conditioned on $(X_n)_n$, the random variables $(L_t(x))_{x \in Q_t}$ are mutually independent. Since $(J_n)_n$ and $(X_n)_n$ are independent, we conclude that: conditioned on $(X_n)_n$, the conditional distribution of $(L_t(x))_{x \in Q_t}$ is that of independent Binomial random variables, each $L_t(x)$ having Binomial-$\left(|K(x,t)|, \frac{1}{2}\right)$ distribution.

It is a simple exercise to show that

$$\mathbb{P}[L_t(x) = 1 \pmod 2 \mid x \in Q_t] = \mathbb{P}\left[\mathrm{Bin}\left(|K(x,t)|, \tfrac{1}{2}\right) = 1 \pmod 2\right] = \tfrac{1}{2}.$$

We conclude that the conditional distribution of $(\sigma_t(x))_{x \in Q_t}$, conditioned on $(X_n)_n$, is that of independent Bernoulli-$\frac{1}{2}$ random variables. :) ✓

Solution to Exercise 5.29 :(

This can be done by induction on n.

The base case where $n = 1$ is just $\mathbb{P}[X \text{ is even }] = \mathbb{P}[X = 0] = \frac{1}{2}$.

For $n > 1$, take $X = Y + Z$ for Y, Z independent and $Y \sim \mathrm{Bin}\left(n-1, \frac{1}{2}\right)$ and $Z \sim \mathrm{Ber}\left(\frac{1}{2}\right)$. Let E_X be the event that X is even and O_X the event that X is odd. Similarly, let E_Y be the event that Y is even and O_Y the event that Y is odd. Then, by induction, $\mathbb{P}[E_Y] = 1 - \mathbb{P}[O_Y] = \frac{1}{2}$.

$$\mathbb{P}[E_X] = \mathbb{P}[E_Y, \ Z = 0] + \mathbb{P}[O_Y, \ Z = 1] = \tfrac{1}{2}(\mathbb{P}[Z = 0] + \mathbb{P}[Z = 1]) = \tfrac{1}{2},$$

completing the induction step. :) ✓

Solution to Exercise 5.30 :(

Note that since the function $x \mapsto e^{\varepsilon x}$ is convex, by Jensen's inequality we have for any $|x| \le 1$ (by writing $x = \frac{x+1}{2} \cdot 1 + \frac{1-x}{2} \cdot (-1)$),

$$e^{\varepsilon x} \le \tfrac{x+1}{2} e^{\varepsilon} + \tfrac{1-x}{2} e^{-\varepsilon}.$$

Taking expectations, we have

$$\mathbb{E}\left[e^{\varepsilon X}\right] \le \tfrac{1}{2}(e^{\varepsilon} + e^{-\varepsilon}) = \tfrac{1}{2} \sum_{k=0}^{\infty} \tfrac{2\varepsilon^{2k}}{(2k)!} \le \exp\left(\tfrac{\varepsilon^2}{2}\right),$$

where we have used that $(2k)! \ge 2^k k!$. :) ✓

Solution to Exercise 5.31 :(

By shifting M_t to $\frac{M_t - M_0}{b}$, we may assume without loss of generality that $M_0 = 0$, $b = 1$.

Since $\mathbb{E}[M_{t+1} - M_t \mid \mathcal{F}_t] = 0$, we have that for any $\varepsilon > 0$,

$$\mathbb{E}\left[e^{\varepsilon(M_{t+1} - M_t)} \mid \mathcal{F}_t\right] \le \exp\left(\tfrac{\varepsilon^2}{2}\right)$$

by Hoeffding's inequality (Exercise 5.30). This leads to,

$$\mathbb{E}\left[e^{\varepsilon M_t}\right] = \mathbb{E}\left[\mathbb{E}[e^{\varepsilon(M_t - M_{t-1})} \mid \mathcal{F}_{t-1}] e^{\varepsilon M_{t-1}}\right] \le \exp\left(\tfrac{\varepsilon^2}{2}\right) \mathbb{E}\left[e^{\varepsilon M_{t-1}}\right] \le \cdots \le \exp\left(\tfrac{t\varepsilon^2}{2}\right).$$

Thus, for any $\lambda, \varepsilon > 0$,

$$\mathbb{P}[M_t \geq \lambda] = \mathbb{P}\left[e^{\varepsilon M_t} \geq e^{\varepsilon \lambda}\right] \leq \exp\left(\frac{t\varepsilon^2}{2} - \varepsilon\lambda\right),$$

by Markov's inequality. Optimizing over the choice of $\varepsilon > 0$, we choose $\varepsilon = \frac{\lambda}{t}$ to arrive at the required conclusion.   :) ✓

Solution to Exercise 5.32 :(
Since μ is symmetric, $(X_t)_t$ is easily seen to be a martingale. Since μ has finite support, there exists $b > 0$ such that $|X_{t+1} - X_t| \leq b$ a.s. for all t. Since $X_t, -X_t$ have the same distribution, by Azuma's inequality we have that

$$\mathbb{P}[X_t = x] = \mathbb{P}[X_t = |x|] \leq \mathbb{P}[X_t \geq |x|] \leq \exp\left(-\frac{|x|^2}{2b^2 t}\right).$$

:) ✓

Solution to Exercise 5.34 :(
By the Varopolous–Carne bound (Theorem 5.6.1), C may be chosen so that for all $t > 0$ and $x, y \in G$,

$$\mathbb{P}_x[X_t = y] \leq C \exp\left(-\frac{|x^{-1}y|^2}{t}\right).$$

By modifying C and by Exercise 5.27, for all $t > 0$,

$$\mathbb{P}_x[X_t = x] \leq C \exp\left(-Ct^{1/3}\right).$$

As in the proof of Theorem 5.5.1, Cauchy–Schwarz tells us that for all $x, y \in G$ and any $t \geq 3$, we have

$$\mathbb{P}_x[X_t = y] \leq \mathbb{P}\left[X_{\lfloor (t-1)/2 \rfloor} = 1\right].$$

Thus, we may compute:

$$g(1, x) = \sum_{t=0}^{m} \mathbb{P}[X_t = x] + \sum_{t=m+1}^{\infty} \mathbb{P}[X_t = x]$$

$$\leq C \sum_{t=0}^{m} \exp\left(-C\frac{|x|^2}{t}\right) + C \sum_{t=m+1}^{\infty} \exp\left(-Ct^{1/3}\right)$$

$$\leq C(m+1) \exp\left(-C\frac{|x|^2}{m}\right) + C \int_{m}^{\infty} \exp\left(-C\xi^{1/3}\right) d\xi.$$

Note that

$$\int_{m}^{\infty} \exp\left(-C\xi^{1/3}\right) d\xi = 3 \int_{m^{1/3}}^{\infty} \xi^2 \exp(-C\xi) d\xi$$

$$\leq 3 \exp\left(-\frac{C}{2}m^{1/3}\right) \cdot \frac{2}{C} \cdot \int_{0}^{\infty} \xi^2 \frac{C}{2} \exp\left(-\frac{C}{2}\xi\right) d\xi$$

$$\leq \frac{6}{C} \cdot \frac{8}{C^2} \cdot \exp\left(-\frac{C}{2}m^{1/3}\right).$$

Thus, by choosing $m = \lfloor |x|^{3/2} \rfloor$ we are done, because $g(x, y) = g\left(1, x^{-1}y\right)$.   :) ✓

Solution to Exercise 5.35 :(
Let $\rho < 1$ and $\nu > 1$ be such that for all $t > 0$ and $r > 0$,

$$\sup_{x} \mathbb{P}[X_t = x] \leq C\rho^t \quad \text{and} \quad |B(1, r)| \leq C\nu^r.$$

Let $\alpha > 0$ be small enough so that $\rho\nu^\alpha < 1$ (e.g. $\alpha < -\frac{\log \rho}{\log \nu}$). Then,

$$\mathbb{P}[|X_t| \leq \alpha t] \leq C\rho^t \cdot |B(1, \alpha t)| \leq C^2 (\rho\nu^\alpha)^t.$$

Since this is exponentially decaying in t it is summable, and by Borel–Cantelli,

$$\mathbb{P}\left[\liminf_{t\to\infty} \frac{|X_t|}{t} \le \alpha\right] = 0. \hspace{3cm} :)\checkmark$$

Solution to Exercise 5.36   :(

Fix a finite symmetric generating set S for G, and let $(X_t)_t$ be a simple random walk on G (i.e. each step is uniform on S).

Fix $\alpha \in \left(1, \frac{4}{3}\right)$. As in Exercise 4.43, we may choose an i.i.d.-ν sequence $(U_k)_{k\ge 1}$, for

$$\nu(n) = \mathbf{1}_{\{n\ge 1\}}\zeta(\alpha)^{-1}n^{-\alpha}$$

and $\zeta(\alpha) = \sum_{n=1}^{\infty} n^{-\alpha}$. Set $T_n = \sum_{k=1}^{n} U_k$. It is easy to see that $\mathbb{E}\,|U_1|^{\frac{1}{3}-\varepsilon} < \infty$ as long as $\varepsilon > \frac{4}{3} - \alpha$.
Also, just as in Exercise 4.43, we can show that

$$\mathbb{E}\left[(T_n)^{-1/2}\right] \le C(\alpha)n^{-\beta}, \qquad \beta = \frac{3}{2} - \frac{2(\alpha-1)}{\alpha}.$$

For $\alpha < \frac{4}{3}$ we have that $\beta > 1$.

By Corollary 5.5.2, we know that there exists some constant $C > 0$ such that for all $t \ge 1$ we have $\mathbb{P}[X_t = 1] \le Ct^{-1/2}$. Since $(X_t)_t$ and T_n are independent, we get that

$$\mathbb{P}[X_{T_n} = 1] = \mathbb{E}\,\mathbb{P}[X_{T_n} = 1 \mid T_n] \le C\,\mathbb{E}\left[(T_n)^{-1/2}\right] \le C \cdot C(\alpha)n^{-\beta}.$$

Since $\beta > 1$, the process $(X_{T_n})_n$ is transient.

It is easy to prove that for $\mu(x) = \mathbb{P}[X_{T_1} = x]$, the process $(X_{T_n})_n$ is a μ-random walk, similarly to Exercise 4.44. $\hspace{2cm} :)\checkmark$

6

Bounded Harmonic Functions

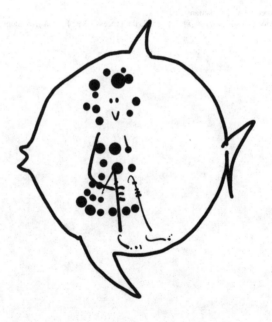

In this chapter we study the space of bounded harmonic functions. We will introduce the *Liouville property* (Definition 6.2.4), that is, the property that all bounded harmonic functions are constant, and we will consider different necessary and sufficient conditions for this property, related to speed, amenability, and entropy. Theorem 6.4.3 summarizes the main results of this chapter.

6.1 The Tail and Invariant σ-Algebras

Let G be a finitely generated group. Let μ be a symmetric and adapted probability measure on G. Let $(X_t)_t$ denote the μ-random walk on G. We consider the measurable space over which the probability measures $\mathbb{P}_x$ (for the random walk on G) are defined. Recall that this is the space $G^{\mathbb{N}}$ equipped with the σ-algebra $\mathcal{F}$ spanned by cylinder sets, see Section 1.2. As usual, we define $\mathcal{F}_t = \sigma(X_0, \ldots, X_t)$.

On the space $G^{\mathbb{N}}$ we have the *shift operator* $\theta \colon G^{\mathbb{N}} \to G^{\mathbb{N}}$ defined by $\theta(\omega)_t := \omega_{t+1}$. An event $A \in \mathcal{F}$ is called **invariant** if $\theta^{-1}(A) = A$.

Exercise 6.1 Show that the collection of all invariant events is a σ-algebra.

▷ solution ◁

Definition 6.1.1 The **tail σ-algebra** is defined as

$$\mathcal{T} = \bigcap_{t=0}^{\infty} \sigma(X_t, X_{t+1}, \ldots,).$$

An event $A \in \mathcal{T}$ is called a **tail event**.

The **invariant σ-algebra**, denoted $\mathcal{I}$, is defined to be the collection of all invariant events.

Exercise 6.2 Show that the event $A = \{\exists\, t_0,\ \forall\, t > t_0,\ X_t \neq 1\}$ is a tail event and an invariant event. (This event is just the event that the walk eventually stops returning to the origin; i.e. it is just what is known as *transience*.)

Show that the event $A' = \{\exists\, t_0,\ \forall\, t > t_0,\ X_{2t} \neq 1\}$ is a tail event, but not an invariant event.

Give an example of an event that is not a tail event.

There is another way to construct these σ-algebras.

Recall that given an equivalence relation $\sim$ on $G^{\mathbb{N}}$, we say that an event A *respects* $\sim$ if for any $\omega \sim \omega' \in G^{\mathbb{N}}$ we have that $\omega \in A \iff \omega' \in A$.

Define two equivalence relations on $G^{\mathbb{N}}$:

- $\omega \sim \omega'$ if and only if there exist $n, k \in \mathbb{N}$ such that $\theta^k(\omega) = \theta^n(\omega')$,
- $\omega \approx \omega'$ if and only if there exists n such that $\theta^n(\omega) = \theta^n(\omega')$.

Of course, $\omega \approx \omega'$ implies $\omega \sim \omega'$, but not the other way around.

Exercise 6.3 Show that $A \in \sigma(X_t, X_{t+1}, \ldots)$ if and only if A respects the relation $\sim_t$ on $G^{\mathbb{N}}$, which is defined by $\omega \sim_t \omega'$ if and only if $\theta^t(\omega) = \theta^t(\omega')$.

▷ solution ◁

Exercise 6.4 Show that an event A is a tail event if and only if for any $\omega \in A$ and $\omega' \approx \omega$ we have $\omega' \in A$ as well (i.e. A respects $\approx$).

Show that an event A is an invariant event if and only if for any $\omega \in A$ and $\omega' \sim \omega$ we have $\omega' \in A$ as well (i.e. A respects $\sim$). ▷ solution ◁

Exercise 6.5 Show that $\mathcal{I} \subset \mathcal{T}$ (any invariant event is also a tail event).

Exercise 6.6 Show that if $A \in \mathcal{T}$ then $\theta^{-1}(A) \in \mathcal{T}$ as well. ▷ solution ◁

Exercise 6.7 Assume that $A \in \mathcal{T} \setminus \mathcal{I}$. Show that there exists $\emptyset \neq B \in \mathcal{T}$ such that $B \cap \theta(B) = \emptyset$. ▷ solution ◁

Exercise 6.8 Show that a random variable Y on $\left(G^{\mathbb{N}}, \mathcal{F}\right)$ is $\mathcal{I}$-measurable if and only if $Y = Y \circ \theta$. ▷ solution ◁

Exercise 6.9 Let Y be a random variable on $\left(G^{\mathbb{N}}, \mathcal{F}\right)$. Show that the following are equivalent.

(1) Y is measurable with respect to $\sigma(X_n, X_{n+1}, \ldots)$.
(2) For any $\omega, \omega' \in G^{\mathbb{N}}$ such that $\theta^n(\omega) = \theta^n(\omega')$, we have $Y(\omega) = Y(\omega')$.
(3) There exists a random variable Z such that $Y = Z \circ \theta^n$. ▷ solution ◁

Exercise 6.10 Show that a random variable Y on $\left(G^{\mathbb{N}}, \mathcal{F}\right)$ is $\mathcal{T}$-measurable if and only if there is a sequence of random variables $(Y_n)_n$ such that $Y = Y_n \circ \theta^n$ for every n.

6.2 Parabolic and Harmonic Functions

Recall the averaging operator $P(x, y) = \mu\left(x^{-1}y\right)$; $Pf(x) = \sum_y P(x, y)f(y) = \mathbb{E}_x[f(X_1)]$.

Definition 6.2.1 A function $f: G \times \mathbb{N} \to \mathbb{C}$ is called **parabolic** if $Pf(\cdot, n+1) = f(\cdot, n)$ for all n. By this we mean

$$\text{for all } x \in G \text{ and } n \in \mathbb{N}, \quad \sum_y P(x, y) f(y, n+1) = f(x, n),$$

and all sums converge absolutely.

We may view a function $f: G \times \mathbb{N} \to \mathbb{C}$ as a sequence of functions $f_n: G \to \mathbb{C}$ by $f_n(x) = f(x, n)$. So a parabolic function is a sequence of functions admitting $Pf_{n+1} = f_n$. Recall that $f: G \to \mathbb{C}$ is harmonic if $Pf = f$. Given a harmonic function f we may define $f_n = f$ for all n, and we see that any harmonic function induces a parabolic function.

Exercise 6.11 Give an example of a parabolic function that is not harmonic.

Exercise 6.12 Show that the function $f: G \times \mathbb{N} \to \mathbb{C}$ is parabolic if and only if $(f(X_t, t))_t$ is a martingale.

Show that if $f: G \times \mathbb{N} \to \mathbb{C}$ is a parabolic function then $(f(X_t, t+n))_t$ is a martingale, for any fixed $n \geq 0$.

Exercise 6.13 Show that $f: G \times \mathbb{N} \to \mathbb{C}$ is parabolic if and only if $\mathrm{Re} f$ and $\mathrm{Im} f$ are both parabolic.

We now provide the relation between bounded parabolic functions and the tail σ-algebra, and also between bounded harmonic functions and the invariant σ-algebra.

Exercise 6.14 Let μ be a symmetric, adapted probability measure on a finitely generated group G. Let $(X_t)_t$ be the μ-random walk.

Show that if $h \in \mathrm{BHF}(G, \mu)$ is a bounded harmonic function then there exists a $\mathcal{I}$-measurable integrable random variable L such that

$$L = \lim_{t \to \infty} h(X_t) \quad \text{a.s.}$$

(In particular, the limit above exists a.s.) ▷ solution ◁

Exercise 6.15 Let μ be a symmetric, adapted probability measure on a finitely generated group G. Let $(X_t)_t$ be the μ-random walk.

Show that if $h: G \times \mathbb{N} \to \mathbb{C}$ is a bounded parabolic function then there exists a $\mathcal{T}$-measurable integrable random variable L such that

$$L = \lim_{t \to \infty} h(X_t, t) \quad \text{a.s.}$$

(In particular, the limit above exists a.s.) ▷ solution ◁

Exercise 6.16 Let μ be a symmetric, adapted probability measure on a finitely generated group G. Let $(X_t)_t$ be the μ-random walk.

Let L be an integrable random variable. Show that $\mathbb{E}_y[L] = \mathbb{E}_x[L \circ \theta^t \mid X_t = y]$.

▷ solution ◁

Proposition 6.2.2 (Poisson formula (parabolic case)) *Let G be a finitely generated group and μ an adapted probability measure on G.*

There is a correspondence between bounded $\mathcal{T}$-measurable random variables and bounded parabolic functions on G.

If L is a bounded $\mathcal{T}$-measurable (complex) random variable, then $f_L(x, t) := \mathbb{E}_x[L_t]$ is a bounded parabolic function. Here, L_t are random variables such that $L = L_t \circ \theta^t$.

Conversely, if f is a bounded parabolic function then

$$L_f := \limsup_{t \to \infty} f(X_t, t)$$

is $\mathcal{T}$-measurable (and obviously bounded).

These mappings are "inverses" of one another, in the sense that $f_{L_f} = f$ and $\mathbb{P}_x[L_{f_L} = L] = 1$ for all $x \in G$.

Proof If L is $\mathcal{T}$-measurable and bounded, by the Markov property,

$$f_L(X_t, t) = \mathbb{E}_{X_t}[L_t] = \mathbb{E}_x[L_t \circ \theta^t \mid \mathcal{F}_t] = \mathbb{E}_x[L \mid \mathcal{F}_t].$$

So $(f_L(X_t, t))_t$ is a martingale (by the tower property), implying that f_L is indeed a bounded parabolic function.

If f is a bounded parabolic function, then $L_f = \limsup_{t \to \infty} f(X_t, t)$ is $\mathcal{T}$-measurable by Exercise 6.15, and $L_f = \lim_{t \to \infty} f(X_t, t)$ a.s. by the martingale convergence theorem (Theorem 2.6.3).

Note that $L_f = \lim_{n \to \infty} f(X_{t+n}, t + n)$ a.s., so that for

$$K_t = \limsup_{n \to \infty} f(X_n, t + n),$$

we have $L_f = K_t \circ \theta^t$ a.s. Also, $f(\cdot, \cdot + t)$ is a bounded parabolic function, so that $(f(X_n, n+t))_n$ is a martingale. This implies by dominated convergence that

$$f_{L_f}(x, t) = \mathbb{E}_x[K_t] = \mathbb{E}_x[\lim_n f(X_n, t + n)] = \lim_n \mathbb{E}_x[f(X_n, t + n)] = f(x, t).$$

Also, if $L = L_t \circ \theta^t$ is a bounded $\mathcal{T}$-measurable random variable, then a.s.,

$$L_{f_L} = \lim_{t \to \infty} f_L(X_t, t) = \lim_{t \to \infty} \mathbb{E}_x[L \mid \mathcal{F}_t] = \mathbb{E}_x[L \mid \mathcal{F}] = L,$$

where we have used that $\sigma(\bigcup_t \mathcal{F}_t) = \mathcal{F}$. □

Exercise 6.17 (Poisson formula) Let G be a finitely generated group and μ an adapted probability measure on G.

Show that there is a correspondence between bounded $\mathcal{I}$-measurable random variables and bounded harmonic functions on G, as follows.

If L is a bounded $\mathcal{I}$-measurable (complex) random variable, then $h_L(x) :=$ $\mathbb{E}_x[L]$ is a bounded harmonic function.

Conversely, if h is a bounded harmonic function then $L_h := \limsup_{t\to\infty} h(X_t)$ is $\mathcal{I}$-measurable (and obviously bounded).

These maps are "inverses" of one another, in the sense that $h_{L_h} = h$ and $\mathbb{P}_x[L_{h_L} = L] = 1$ for all $x \in G$. ▷ solution ◁

Definition 6.2.3 A sub-σ-algebra $\mathcal{G} \subset \mathcal{F}$ is called **trivial** if for every $x \in G$ and any event $A \in \mathcal{G}$, we have that $\mathbb{P}_x[A] \in \{0, 1\}$.

In general, the value of $0, 1$ for the probability of a given event may depend on the specific measure $\mathbb{P}_x$. However, if the invariant σ-algebra is trivial, then this value cannot depend on the specific starting point, as the next exercise shows.

Exercise 6.18 Show that $\mathcal{I}$ is trivial if and only if for every bounded $\mathcal{I}$-measurable random variable L, there is a constant c such that $\mathbb{P}_x[L = c] = 1$ for all $x \in G$. ▷ solution ◁

Exercise 6.19 Give an example of a tail event $A \in \mathcal{T}$ such that $\mathbb{P}_x[A] \in \{0, 1\}$ for any $x \in G$ but there exist x, y such that $\mathbb{P}_x[A] \neq \mathbb{P}_y[A]$.

Exercise 6.20 (Kolmogorov 0-1 law) Let A be a tail event, and suppose that A is independent of $\mathcal{F}_t$ for all t. Prove that $\mathbb{P}_x[A] \in \{0, 1\}$ for all $x \in G$.

Conclude that if $\mathcal{T}$ is independent of $\mathcal{F}_t$ for all t then $\mathcal{T}$ is trivial. ▷ solution ◁

Recall Liouville's theorem (which was actually first proved by Cauchy), that any bounded harmonic function on $\mathbb{R}^n$ is actually constant. Here, harmonicity is with respect to the classical Laplace operator $\Delta = \sum \frac{\partial^2}{\partial x_j^2}$. Inspired by this theorem, we have the following definition.

Definition 6.2.4 Let G be a finitely generated group and μ a probability measure on G. We call (G, μ) **Liouville** if every bounded μ-harmonic function is constant.

Recall Theorem 2.7.1, which tells us that the space of bounded μ-harmonic functions, $\mathrm{BHF}(G, \mu)$, is either only the constant functions, or has infinite dimension. That is, a restatement of Theorem 2.7.1 is: (G, μ) is Liouville if and only if $\dim \mathrm{BHF}(G, \mu) < \infty$.

Our first connection is between the Liouville property and the invariant σ-algebra.

Proposition 6.2.5 *Let G be a finitely generated group and μ a symmetric and adapted probability measure on G. Then $\mathcal{I}$ is trivial if and only if every bounded harmonic function is constant (the Liouville property).*

Proof The correspondence between bounded harmonic functions and bounded $\mathcal{I}$-measurable random variables together with Exercise 6.18 prove the proposition.

If every bounded harmonic function is constant, then for any bounded $\mathcal{I}$-measurable random variable, we can write $L = \limsup_{t \to \infty} h(X_t)$ for some bounded harmonic function. Since h is constant, so is L. Thus $\mathcal{I}$ is trivial, by Exercise 6.18.

For the other direction, if $\mathcal{I}$ is trivial, then for any bounded harmonic function h, we have that $h(x) = \mathbb{E}_x[L]$ for some bounded $\mathcal{I}$-measurable random variable L. By Exercise 6.18 there is c such that $\mathbb{P}_x[L = c] = 1$ for all $x \in G$, implying that $h \equiv c$ is constant. $\square$

Exercise 6.21 Let G be a finitely generated group, and let μ be an adapted probability measure on G.

Show that if (G, μ) is recurrent, then (G, μ) is Liouville.   ▷ solution ◁

6.3 Entropic Criterion

6.3.1 Entropy

Let X be a discrete random variable, meaning that X takes only countably many possible values. For any x, let $p(x)$ be the probability that $X = x$. The (**Shannon**) **entropy** of X is defined as $H(X) = \mathbb{E}[-\log p(X)] = -\sum_x p(x) \log p(x)$ with the convention $0 \log 0 = 0$. Strictly speaking, H is not a function of the specific random variable X, rather only of the distribution of X (so a function of p in the notation above). For a probability measure ν on a countable set Ω, one may similarly define $H(\nu) = -\sum_x \nu(x) \log \nu(x)$, again with $0 \log 0 = 0$. The notation $H(X)$ is convenient, if somewhat misleading. See Section B.1 for some motivation behind this definition, and Cover and Thomas (1991) for a more in-depth discussion of *entropy*.

Let σ be a σ-algebra on Ω. For a discrete random variable X, we can define the conditional probability $p_\sigma(x) = \mathbb{E}\left[\mathbf{1}_{\{X=x\}}|\sigma\right]$. Since a.s. $\sum_x p_\sigma(x) = 1$, we have a "random" entropy

$$H_\sigma(X) = -\sum_x p_\sigma(x) \log p_\sigma(x).$$

Taking expectations, we define

$$H(X \mid \sigma) = \mathbb{E}[H_\sigma(X)].$$

If Y is another discrete random variable then we define $H(X|Y) = H(X \mid \sigma(Y))$ It is important to note that $H_\sigma(X)$ is a random variable, but $H(X \mid \sigma)$ is just some number (possibly infinite).

In the next few exercises we work out some basic properties of entropy.

Exercise 6.22 Note that if X, Y are discrete random variables, then (X, Y) is also a discrete random variable.

Show that $H(X \mid Y) = H(X, Y) - H(Y)$. ▷ solution ◁

Exercise 6.23 Show that $H(X) - H(X|Y) = H(Y) - H(Y|X)$. ▷ solution ◁

Exercise 6.24 Let X, Y be discrete random variables taking values in a set G. Let K_G denote the collection of all finite subsets of G. Assume that there exists a function $\varphi \colon G \to K_G$ such that $\mathbb{P}[X \in \varphi(Y)] = 1$.

Prove that $H(X \mid Y) \leq \mathbb{E}[\log|\varphi(Y)|]$.

Conclude that for a random variable X taking values in some finite set F, we have that $H(X) \leq \log|F|$. ▷ solution ◁

Exercise 6.25 Show that if $\sigma \subseteq \mathcal{G}$ then $H(X \mid \mathcal{G}) \leq H(X \mid \sigma)$. ▷ solution ◁

Exercise 6.26 Show that $H(X \mid \sigma) = H(X)$ if and only if X is independent of σ. ▷ solution ◁

Exercise 6.27 Show that if $\mathcal{G}$ is a σ-algebra such that for any $A \in \mathcal{G}$ we have $\mathbb{P}[A] \in \{0, 1\}$ then $H(X \mid \mathcal{G}) = H(X)$. ▷ solution ◁

Exercise 6.28 Show that $H(X, Y) \leq H(X) + H(Y)$. ▷ solution ◁

Exercise 6.29 Show that $H(X) \le H(X, Y)$. ▷ solution ◁

We will also require the following convergence theorem.

Lemma 6.3.1 *Let $(\mathcal{F}_n)$ be an increasing sequence of σ-algebras, and let $\mathcal{F}_\infty = \sigma(\bigcup_n \mathcal{F}_n)$. Let (σ_n) be a decreasing sequence of σ-algebras, and let $\sigma_\infty = \bigcap_n \sigma_n$. Then, for any random variable X with $H(X) < \infty$,*

$$\lim_{n \to \infty} H(X | \mathcal{F}_n) = H(X | \mathcal{F}_\infty),$$

$$\lim_{n \to \infty} H(X | \sigma_n) = H(X | \sigma_\infty).$$

Proof Let $Y_n = \mathbb{E}\left[\mathbf{1}_{\{X=x\}} \mid \mathcal{F}_n\right]$. The tower property shows that this is a bounded martingale. So $Y_n \to Y_\infty$ a.s. for some integrable Y_∞, by martingale convergence (Theorem 2.6.3).

For any $A \in \mathcal{F}_n$ we have that $\mathbb{E}[Y_\infty \mathbf{1}_A] = \mathbb{E}[\lim_m Y_m \mathbf{1}_A] = \lim_m \mathbb{E}[Y_m \mathbf{1}_A] = \lim_m \mathbb{E}[Y_n \mathbf{1}_A] = \mathbb{E}\left[\mathbf{1}_{\{X=x\}} \mathbf{1}_A\right]$, where we have used dominated convergence of $Y_m \mathbf{1}_A \to Y_\infty \mathbf{1}_A$ and the fact that $\mathbb{E}[Y_m \mid \mathcal{F}_n] = Y_n$ for all $m > n$ by the tower property. (This is where it is used that $(\mathcal{F}_n)_n$ are increasing.) Since $\bigcup_n \mathcal{F}_n$ is a π-system generating the σ-algebra $\mathcal{F}_\infty = \sigma(\bigcup_n \mathcal{F}_n)$, we get that $\mathbb{E}[Y_\infty \mathbf{1}_A] = \mathbb{E}[\mathbf{1}_{\{X=x\}} \mathbf{1}_A]$ for all $A \in \mathcal{F}_\infty$ (by Dynkin's lemma). Also, Y_∞ is $\mathcal{F}_\infty$ measurable, so $\mathbb{E}\left[\mathbf{1}_{\{X=x\}} \mid \mathcal{F}_\infty\right] = Y_\infty = \lim_n \mathbb{E}\left[\mathbf{1}_{\{X=x\}} \mid \mathcal{F}_n\right]$.

We conclude that $\mathbb{E}\left[\mathbf{1}_{\{X=x\}} \mid \mathcal{F}_n\right] \to \mathbb{E}\left[\mathbf{1}_{\{X=x\}} \mid \mathcal{F}_\infty\right]$ for all x.

Let $\phi(\xi) = -\xi \log \xi$. For any x the sequence $\left(\phi\left(\mathbb{E}\left[\mathbf{1}_{\{X=x\}} \mid \mathcal{F}_n\right]\right)\right)_n$ is bounded $\left(\text{above by } e^{-1} \text{ and below by } 0\right)$. Since

$$\phi\left(\mathbb{E}\left[\mathbf{1}_{\{X=x\}} \mid \mathcal{F}_n\right]\right) \to \phi\left(\mathbb{E}\left[\mathbf{1}_{\{X=x\}} \mid \mathcal{F}_\infty\right]\right) \quad \text{a.s.},$$

it also converges in expectation $\left(L^1\right)$, by dominated convergence.

Let $f_n(x) = \mathbb{E}\phi\left(\mathbb{E}\left[\mathbf{1}_{\{X=x\}} \mid \mathcal{F}_n\right]\right)$. Let $f_\infty = \mathbb{E}\phi\left(\mathbb{E}\left[\mathbf{1}_{\{X=x\}} \mid \mathcal{F}_\infty\right]\right)$. Let $g(x) = \phi(\mathbb{P}[X = x])$. Note that the above is the assertion that $f_n \to f_\infty$ pointwise. Also, by Jensen,

$$f_n(x) \le \phi\left(\mathbb{E}\mathbb{E}\left[\mathbf{1}_{\{X=x\}} \mid \mathcal{F}_n\right]\right) = \phi(\mathbb{P}[X = x]) = g(x).$$

Since $\sum_x g(x) = H(X) < \infty$, we get by dominated convergence that $\sum_x f_n(x) \to \sum_x f_\infty(x)$. Finally, note that since $f_n(x) \ge 0$, we have that

$$H(X \mid \mathcal{F}_n) = \mathbb{E} \sum_x \phi\left(\mathbb{E}\left[\mathbf{1}_{\{X=x\}} \mid \mathcal{F}_n\right]\right) = \sum_x f_n(x),$$

and similarly for $H(X \mid \mathcal{F}_\infty)$. So we conclude that $H(X \mid \mathcal{F}_n) \to H(X \mid \mathcal{F}_\infty)$.

For the second assertion, we use the backward martingale convergence theorem (Exercise 2.35) to obtain that $\mathbb{E}\left[\mathbf{1}_{\{X=x\}} \mid \sigma_n\right] \to \mathbb{E}\left[\mathbf{1}_{\{X=x\}} \mid \sigma_\infty\right]$. Similarly to the first case, we can easily deduce from this the second assertion. $\square$

Exercise 6.30 Show that there exists a universal constant $C > 0$ such that the following holds.

Let X be a random variable taking values in the natural numbers $\mathbb{N}$ ($\mathbb{P}[X \in \mathbb{N}] = 1$). Then, $H(X) \le C \, \mathbb{E}[\log(X + 1)] + C$. ▷ solution ◁

Exercise 6.31 Give an example of a discrete real-valued bounded random variable X with $H(X) = \infty$. ▷ solution ◁

6.3.2 The Entropy Criterion

Proposition 6.3.2 *Let G be a finitely generated group, and μ a probability measure on G with finite entropy. Let $(X_t)_t$ denote the μ-random walk. Denote $\mathcal{F}_t = \sigma(X_0, \ldots, X_t)$ and $\sigma_t = \sigma(X_t, X_{t+1}, \ldots)$.*

For any $k < n$, and any $m > 0$,

$$H(X_1, \ldots, X_k | X_n) = H(X_1, \ldots, X_k | X_n, \ldots, X_{n+m}) = H(X_1, \ldots, X_k | \sigma_n).$$

Also,

$$H(X_1, \ldots, X_k, X_n) = kH(X_1) + H(X_{n-k}).$$

Proof Fix some $\ell \ge 1$. Let $Y_t^{(\ell)} = (X_\ell)^{-1} X_{t+\ell}$. Note that $\left(Y_t^{(\ell)}\right)_t$ is a μ-random walk, which is independent of $X_1, \ldots, X_\ell$. Thus,

$$H(X_1, \ldots, X_k, X_n, \ldots, X_{n+m}) = H\left(X_1, \ldots, X_k, X_n, \ldots, X_{n+m-1}, Y_1^{(n+m-1)}\right)$$

$$= H(X_1, \ldots, X_k, X_n, \ldots, X_{n+m-1}) + H(X_1).$$

Similarly,

$$H(X_n, \ldots, X_{n+m}) = H(X_n, \ldots, X_{n+m-1}) + H(X_1),$$

implying that

$$H(X_1, \ldots, X_k | X_n, \ldots, X_{n+m}) = H(X_1, \ldots, X_k | X_n, \ldots, X_{n+m-1})$$

for all $m \ge 1$. Since the σ-algebras $(\sigma(X_n, \ldots, X_{n+m}))_{m \ge 0}$ increase to σ_n, by Lemma 6.3.1 we conclude that for all $m \ge 0$,

$$H(X_1, \ldots, X_k | X_n) = H(X_1, \ldots, X_k | X_n, \ldots, X_{n+m}) = H(X_1, \ldots, X_k | \sigma_n).$$

The second assertion follows inductively on k, using that

$$H(X_1, \ldots, X_k, X_n) = H\left(X_1, Y_1^{(1)}, \ldots, Y_{k-1}^{(1)}, Y_{n-1}^{(1)}\right)$$

$$= H(X_1) + H(X_1, \ldots, X_{k-1}, X_{n-1}). \qquad \square$$

Definition 6.3.3 For a probability measure μ on a finitely generated group G, define the **random walk entropy** to be

$$h(G, \mu) = \lim_{t \to \infty} \frac{H(X_t)}{t}.$$

(Note that $h(G, \mu)$ is also known as **Avez entropy**.)

Exercise 6.32 Show that the limit $h(G, \mu)$ is well defined and nonnegative. Show that $h(G, \mu) \leq H(X_1)$. ▷ solution ◁

Exercise 6.33 Show that if $\mu \in SA(G, 1)$ then $H(\mu) < \infty$. ▷ solution ◁

Exercise 6.34 Give an example of a finitely generated group G and a symmetric, adapted probability measure μ such that $\mu \in SA(G, 1 - \varepsilon)$ for any $\varepsilon > 0$, and $H(\mu) = \infty$. ▷ solution ◁

We now make another connection, this time between random walk entropy and the tail σ-algebra.

Proposition 6.3.4 (Kaimanovich–Vershik entropic criterion) *Let μ be a probability measure on a finitely generated group G. Assume that $H(\mu) < \infty$.*
Then, $h(G, \mu) = 0$ if and only if $\mathcal{T}$ is trivial.

Proof Let $(X_t)_t$ be the μ-random walk. Let $\sigma_n = \sigma(X_n, X_{n+1}, \ldots)$. Let $h = h(G, \mu)$. For any $k < n$,

$$H(X_1, \ldots, X_k | X_n) = kH(X_1) - H(X_n) + H(X_{n-k})$$

by Proposition 6.3.2. This leads to the fact that

$$H(X_{n+1}) - H(X_n) = H(X_{n+1}) - H(X_{n+1}|X_1) = H(X_1) - H(X_1|X_{n+1})$$
$$= H(X_1) - H(X_1|\sigma_{n+1})$$

is a nonincreasing nonnegative sequence. Thus, it converges to a limit given by

$$\lim_{n \to \infty} H(X_n) - H(X_{n-1}) = \lim_{n \to \infty} \frac{1}{n} \sum_{k=1}^{n} H(X_k) - H(X_{k-1}) = \lim_{n \to \infty} \frac{H(X_n)}{n} = h.$$

Thus, for any $k \geq 1$,

$$kh = \lim_{n \to \infty} \sum_{j=0}^{k-1} H(X_{n-j}) - H(X_{n-j-1}) = \lim_{n \to \infty} H(X_n) - H(X_{n-k})$$

$$= kH(X_1) - H(X_1, \ldots, X_k | \mathcal{T}) = H(X_1, \ldots, X_k) - H(X_1, \ldots, X_k | \mathcal{T}).$$

We conclude that if $h = 0$ then $\mathcal{T}$ is independent of any $(X_1, \ldots, X_k)$. This implies that $\mathcal{T}$ is trivial by the Kolmogorov 0-1 law (Exercise 6.20). On the other hand, if $\mathcal{T}$ is trivial, then $H(X_1 | \mathcal{T}) = H(X_1)$ and so $h = 0$. □

6.4 Triviality of Invariant and Tail σ-Algebras

In this section we will compare bounded parabolic and harmonic functions. We will require the following estimate regarding couplings of binomial distributions. For background on couplings, see Appendix C.

Exercise 6.35 Show that there exists a constant $c > 0$ such that for any $p \in (0, 1)$ and any $n \geq 1$, there exists a coupling of $B \sim \mathrm{Bin}(n, p)$ and $B' \sim \mathrm{Bin}(n + 1, p)$ such that $\mathbb{P}[B \neq B'] \leq cn^{-1/2}$. ▷ solution ◁

Exercise 6.36 Let ν be a probability measure on G (not necessarily symmetric or adapted). Let $p \in (0, 1)$ and let $\mu = p\delta_1 + (1 - p)\nu$ be a lazy version of ν.

Show that if $(X_t)_t$ is a ν-random walk and $B_t \sim \mathrm{Bin}(t, 1 - p)$ is independent of $(X_t)_t$, then X_{B_t} has law μ^t (the tth step of a μ-random walk). ▷ solution ◁

Lemma 6.4.1 *Let ν be a probability measure on G (not necessarily symmetric or adapted). Let $p \in (0, 1)$ and let $\mu = p\delta_1 + (1 - p)\nu$ be a lazy version of ν.*

Then, any bounded μ-parabolic function $(f_n)_n$ is actually a μ-harmonic function $f_n = f_0$.

Proof If $(X_t)_t$ is a ν-random walk and $B \sim \mathrm{Bin}(t, 1 - p)$ then X_B has the distribution of the tth step of a μ-random walk, by Exercise 6.36.

Let $B \sim \mathrm{Bin}(t, 1 - p)$, $B' \sim \mathrm{Bin}(t + 1, 1 - p)$, coupled so that $\mathbb{P}[B \neq B'] \leq ct^{-1/2}$, for some constant $c > 0$, as in Exercise 6.35. Then, if f is μ-parabolic, then $f(x, n) = \mathbb{E}_x[f(X_{B'}, n + t + 1)]$ and $f(x, n + 1) = \mathbb{E}_x[f(X_B, n + 1 + t)]$. If f is bounded by M then

$$|f(x, n) - f(x, n + 1)| \leq \mathbb{E}_x \left[|f(X_{B'}, n + t + 1) - f(X_B, n + 1 + t)| \mathbf{1}_{\{B \neq B'\}} \right]$$

$$\leq 2M \cdot \mathbb{P}[B \neq B'] \leq 2Mct^{-1/2}.$$

Taking $t \to \infty$, we have that $f(x, n) = f(x, n + 1)$ for all $x \in G$ and all n. This immediately implies that $f(\cdot, n) = f(\cdot, 0)$ is μ-harmonic. □

Proposition 6.4.2 *Let μ be an adapted probability measure on a finitely generated group G. For any tail event $A \in \mathcal{T}$, there exists an invariant event $B \in \mathcal{I}$ such that $\mathbb{P}_x[A \triangle B] = 0$ for all $x \in G$. Specifically, if $\mathcal{I}$ is trivial then $\mathcal{T}$ is also trivial.*

Proof Let $k = \inf\{t \geq 1 : \mathbb{P}[X_t = 1] > 0\}$. Since μ is adapted, $k < \infty$. Also, set $p = \mathbb{P}[X_k = 1]$, and define the probability measure

$$\nu = \frac{1}{1-p} \left(\mu^k - p\delta_1 \right)$$

$\left(\text{recalling that } \mu^k \text{ denotes the } k\text{th convolution power of } \mu\right)$.

Now, let $A \in \mathcal{T}$ be a tail event. By Proposition 6.2.2, we can find a bounded parabolic function $(f_n)_n$ such that for any $x \in G$ we have $\mathbb{P}_x[f_n(X_n) \to 1_A] = 1$.

We have that $(f_{kn+\ell})_n$ is a bounded μ^k-parabolic function, for any fixed $0 \leq \ell < k$. Since $\mu^k = p\delta_1 + (1 - p)\nu$, by Lemma 6.4.1 we conclude that $f_{kn+\ell} = f_\ell$ for all $0 \leq \ell < k$ and all $n \geq 0$, and f_ℓ is a μ^k-harmonic function. Also, the function $g = \frac{1}{k} \sum_{\ell=0}^{k-1} f_\ell$ is μ-harmonic; indeed,

$$Pg = \frac{1}{k} \sum_{\ell=0}^{k-1} Pf_\ell = \frac{1}{k} \sum_{\ell=0}^{k-1} Pf_{k+\ell} = \frac{1}{k} \sum_{\ell=0}^{k-1} f_{k+\ell-1} = \frac{1}{k} \sum_{\ell=0}^{k-1} f_\ell = g.$$

The above argument tells us that for any $0 \leq \ell < k$ and any $x \in G$,

$$\mathbb{P}_x[\lim_{n\to\infty} f_\ell(X_{kn+\ell}) = 1_A] = 1.$$

This implies that

$$\mathbb{P}_x[\lim_{n\to\infty} f_\ell(X_{kn}) = 1_A] = \sum_y \mathbb{P}_x[X_{k-\ell} = y] \cdot \mathbb{P}_y[\lim_{n\to\infty} f_\ell(X_{kn+\ell}) = 1_A] = 1.$$

Thus, defining $g = \frac{1}{k} \sum_{\ell=0}^{k-1} f_\ell$, we find that $g(X_{kn}) \to 1_A \, \mathbb{P}_x$-a.s. for any $x \in G$.

However, since g is a bounded μ-harmonic function, the random variable $L = \limsup_n g(X_n)$ is $\mathcal{I}$-measurable, and by Exercise 6.17 we know that $g(X_n) \to L \, \mathbb{P}_x$-a.s. for any $x \in G$. Thus, for any $x \in G$ we have that $\mathbb{P}_x[L = 1_A] = 1$, which implies that $\mathbb{P}_x[A \triangle B] = 0$ for $B = \{L = 1\} \in \mathcal{I}$. □

Exercise 6.37 Let G be a finitely generated group, and let μ be a symmetric and adapted probability measure on G.

Let $(f_n = f(\cdot, n))_n$ be a bounded parabolic function. Show that $h = f_0 + f_1$ is a μ-harmonic function.

We summarize this section so far with the following theorem.

Theorem 6.4.3 *Let G be a finitely generated group and μ an adapted probability measure on G with finite entropy. The following are equivalent:*

- *Every bounded μ-harmonic function is constant (i.e. (G, μ) is Liouville).*
- $\dim \mathrm{BHF}(G, \mu) < \infty$.
- $\dim \mathrm{BHF}(G, \mu) = 1$.
- *The invariant σ-algebra $\mathcal{I}$ is trivial.*
- *The tail σ-algebra $\mathcal{T}$ is trivial.*
- $h(G, \mu) = 0$.

Exercise 6.38 Let μ be an adapted probability measure on a finitely generated group G, with $H(\mu) < \infty$.

Show that (G, μ^n) is Liouville if and only if (G, μ) is Liouville.   ▷ solution ◁

6.5 An Entropy Inequality

In this section we quantify one direction of the entropic criterion for the Liouville property. We begin by introducing more notions related to entropy and information theory.

Definition 6.5.1 Let μ, ν be two probability measures on a countable set Ω. The **Kullback–Leibler divergence**, denoted $D(\mu||\nu)$, is defined as

$$D(\mu||\nu) = \sum_x \mu(x) \log \frac{\mu(x)}{\nu(x)},$$

where $p \log \frac{p}{0}$ is interpreted as ∞ for $p > 0$, and $0 \log \frac{0}{0}$ is interpreted as 0 (so that $D(\mu||\nu)$ is finite only if $\mu \ll \nu$).

If X, Y are discrete random variables with laws $\mu(x) = \mathbb{P}[X = x]$ and $\nu(y) = \mathbb{P}[Y = y]$, we define $D(X||Y) = D(\mu||\nu)$.

Be careful! $D(\mu||\nu)$ does not necessarily equal $D(\nu||\mu)$.

Definition 6.5.2 Let X, Y be discrete random variables with finite entropy. The **mutual information** of X and Y is defined as

$$I(X,Y) := H(X) + H(Y) - H(X,Y).$$

Exercise 6.39 Show that $I(X,Y) = I(Y,X)$.

Give an example where $D(X\|Y) \neq D(Y\|X)$, but both quantities are finite.

The relationship between Kullback–Leibler divergence and mutual information is given in the following exercise.

Exercise 6.40 Let X, Y be discrete random variables with finite entropy. For
every y in the support of Y, let $(X|Y = y)$ be the random variable with
distribution $\mathbb{P}[(X|Y = y) = x] = \mathbb{P}[X = x \mid Y = y]$. Show that

$$I(X,Y) = \sum_y \mathbb{P}[Y = y] \cdot D((X|Y = y)\|X). \qquad \text{▷ solution ◁}$$

Proposition 6.5.3 *Let X and Y be two discrete random variables (i.e. taking
values in some countable set). Let f be some complex-valued function defined
on the range of X and Y such that $\mathbb{E}\left[|f(X)|^2\right], \mathbb{E}\left[|f(Y)|^2\right]$ are finite.*
Then,

$$\left|\mathbb{E}[f(X)] - \mathbb{E}[f(Y)]\right|^2 \leq 2D(X\|Y) \cdot \left(\mathbb{E}[|f(X)|^2] + \mathbb{E}[|f(Y)|^2]\right).$$

Proof Let

$$J(X,Y) = \sum_z \frac{(\mathbb{P}[X = z] - \mathbb{P}[Y = z])^2}{\mathbb{P}[X = z] + \mathbb{P}[Y = z]}.$$

Let $R_X = \{z : \mathbb{P}[X = z] > 0\}$ and $R_Y = \{z : \mathbb{P}[Y = z] > 0\}$. If there exists
$z \in R_X \backslash R_Y$, then $D(X\|Y) = \infty$ and there is nothing to prove. Let us now
assume that $R_X \subseteq R_Y$. Hence we can write $p(z) := \mathbb{P}[X = z]/\mathbb{P}[Y = z]$ for
$z \in R_Y$ and

$$J(X,Y) = \sum_{z \in R_Y} \mathbb{P}[Y = z] \cdot \frac{(1 - p(z))^2}{1 + p(z)}.$$

Consider the function $\phi(x) = \xi \log \xi$ (with $\phi(0) = 0$). We have that $\phi'(\xi) = \log \xi + 1$, $\phi''(\xi) = 1/\xi$. Thus, expanding around 1 we have that for all $\xi > 0$,
$\xi \log \xi - \xi + 1 \geq \frac{(\xi-1)^2}{2(1+\xi)}$. This implies

$$J(X,Y) \leq 2\sum_z \mathbb{P}[Y = z](1 - p(z)) + 2D(X\|Y) = 2D(X\|Y).$$

Using Cauchy–Schwarz, one obtains

$$|\mathbb{E}[f(X)] - \mathbb{E}[f(Y)]| \leq \sum_{z \in R_Y} |\mathbb{P}[X = z] - \mathbb{P}[Y = z]| \cdot |f(z)|$$

$$= \sum_{z \in R_Y} \frac{|\mathbb{P}[X = z] - \mathbb{P}[Y = z]|}{\sqrt{\mathbb{P}[X = z] + \mathbb{P}[Y = z]}} \cdot \sqrt{\mathbb{P}[X = z] + \mathbb{P}[Y = z]} \cdot |f(z)|$$

$$\leq \sqrt{J(X,Y)} \cdot \sqrt{\mathbb{E}\left[|f(X)|^2\right] + \mathbb{E}\left[|f(Y)|^2\right]}$$

$$\leq \sqrt{2D(X\|Y)} \cdot \sqrt{\mathbb{E}\left[|f(X)|^2\right] + \mathbb{E}\left[|f(Y)|^2\right]},$$

as required. □

For background on total variation distance see Appendix C.

Exercise 6.41 (Pinsker's inequality) Prove Pinsker's inequality:
For any two finite entropy probability measures μ, ν, we have $\|\mu - \nu\|_{\text{TV}} \leq 2\sqrt{D(\mu\|\nu)}$. ▷ solution ◁

Exercise 6.42 Let G be a finitely generated group and μ an adapted measure on G. Let $(X_t)_t$ denote the μ-random walk. For $t \geq 0$ and $x \in G$, define

$$K_t(x) = \|\mathbb{P}_x[X_t = \cdot] - \mathbb{P}[X_{t+n(x)} = \cdot]\|_{\text{TV}},$$

where $n(x) = \inf\{n \geq 1 : \mu^n(x) > 0\}$.

Show that (G, μ) is Liouville if and only if for any $x \in G$ we have $K_t(x) \to 0$ as $t \to \infty$. ▷ solution ◁

Exercise 6.43 Give an example of a finitely generated group G and $\mu \in$ SA(G, ∞) for which (G, μ) is Liouville, but there exists some x with

$$\|\mathbb{P}_x[X_t = \cdot] - \mathbb{P}[X_t = \cdot]\|_{\text{TV}} = 1$$

for all t. ▷ solution ◁

Corollary 6.5.4 Let X, Y be discrete random variables. Let f be some complex-valued function on the range of X such that $\mathbb{E}\left[|f(X)|^2\right] < \infty$. Then,

$$\mathbb{E}|\mathbb{E}[f(X)|Y] - \mathbb{E}[f(X)]| \leq 2\sqrt{I(X,Y)}\sqrt{\text{Var}[f(X)]}.$$

Proof By subtracting the constant $\mathbb{E}[f(X)]$ from f we may assume without loss of generality that $\mathbb{E}[f(X)] = 0$.

Recall that for any y such that $\mathbb{P}[Y = y] > 0$ we have the random variable $(X|Y = y)$, which has law $\mathbb{P}[(X|Y = y) = x] = \mathbb{P}[X = x \mid Y = y]$. By

Proposition 6.5.3 applied to X and $(X|Y = y)$,

$$|\mathbb{E}[f(X|Y = y)] - \mathbb{E}[f(X)]|^2$$
$$\leq 2D((X|Y = y)||X) \cdot \left(\mathbb{E}[|f(X|Y = y)|^2] + \mathbb{E}[|f(X)|^2]\right).$$

Note that by Exercise 6.40,

$$I(X, Y) = \sum_y \mathbb{P}[Y = y]D((X|Y = y)||X),$$

and also

$$\sum_y \mathbb{P}[Y = y]\mathbb{E}\left[|f(X|Y = y)|^2\right] = \mathbb{E}[|f(X)|^2].$$

Thus, the Cauchy–Schwarz inequality implies that

$$\mathbb{E}|\mathbb{E}[f(X)|Y] - \mathbb{E}[f(X)]| \leq \sum_y \mathbb{P}[Y = y] \cdot |\mathbb{E}[f(X|Y = y)] - \mathbb{E}[f(X)]|$$

$$\leq \sum_y \mathbb{P}[Y = y]\sqrt{2D((X|Y = y)||X)\,(\mathbb{E}[|f(X|Y = y)|^2] + \mathbb{E}[|f(X)|^2])}$$

$$\leq \sqrt{2}\sqrt{I(X, Y)} \cdot \sqrt{2\,\mathbb{E}\left[|f(X)|^2\right]},$$

which can be seen to be the proper conclusion. $\qquad\square$

Theorem 6.5.5 *Let G be a finitely generated group and μ a probability measure on G, with $H(\mu) < \infty$. Let $(X_t)_{t \geq 0}$ be the μ-random walk on G. Let $h: G \to \mathbb{C}$ be a μ-harmonic function such that $\mathbb{E}\left[|h(X_t)|^2\right] < \infty$. Then,*

$$(\mathbb{E}_z\,|h(X_1) - h(z)|)^2 \leq 4\mathbb{E}_z\left[|h(X_t) - h(z)|^2\right] \cdot (H(X_t) - H(X_{t-1})).$$

Proof Since h is harmonic we have that $|h(X_1) - h(z)| = |\mathbb{E}_z[h(X_t)|X_1] - \mathbb{E}_z[h(X_t)]|$ a.s. Using Corollary 6.5.4 (with X being X_t, Y being X_1, and $f(x) = h(x) - h(z)$), we find that

$$\mathbb{E}_z\,|h(X_1) - h(z)| \leq 2 \cdot \sqrt{I(X_t, X_1)} \cdot \sqrt{\mathbb{E}_z\left[|h(X_t) - h(z)|^2\right]}.$$

Proposition 6.3.2 implies that $I(X_t, X_1) = H(X_t) - H(X_{t-1})$, which completes the proof. $\qquad\square$

The inequality in Theorem 6.5.5 actually provides a quantitative estimate on the growth of harmonic functions, which quantifies one direction of the entropic criterion for the Liouville property.

Exercise 6.44 Let μ be an adapted probability measure on a finitely generated group G, with $H(\mu) < \infty$. Let $h \colon G \to \mathbb{R}$ be a real-valued harmonic function.

Show that if for some $x \in G$,

$$\liminf_{t \to \infty} \tfrac{1}{t} \operatorname{Var}_x[h(X_t)] \cdot H(X_t) = 0,$$

then h is constant.

▷ solution ◁

Exercise 6.45 Let μ be an adapted probability measure on a finitely generated group G, with $H(\mu) < \infty$.

Show that (G, μ) is not Liouville if and only if there exists a nonconstant real-valued harmonic function $h \colon G \to \mathbb{R}$ such that $\sup_t \operatorname{Var}[h(X_t)] < \infty$.

▷ solution ◁

6.6 Coupling and Liouville

Recall the definition of a coupling and total variation distance (see Appendix C).

Definition 6.6.1 Let G be a finitely generated group and μ an adapted probability measure on G. Fix some Cayley graph of G.

For $x, y \in G$ and $r > 0$ let $D_r(x, y)$ be the total variation distance between the exit measure of $B(1, r)$ started from x and from y; that is,

$$D_r(x, y) = \| \mathbb{P}_x[X_{E_r} = \cdot] - \mathbb{P}_y[X_{E_r} = \cdot] \|_{\mathrm{TV}},$$

where $E_r = \inf\{t : |X_t| > r\}$.

Theorem 6.6.2 *Let G be a finitely generated group and μ an adapted probability measure on G.*

Then (G, μ) is Liouville if and only if $D_r(x, y) \to 0$ as $r \to \infty$ for all x, y.

In fact, for any $x, y \in G$ we have that $D_r(x, y) \to 0$ as $r \to \infty$ if and only if for all $f \in \mathrm{BHF}(G, \mu)$ we have $f(x) = f(y)$.

Proof Let $x, y \in G$, and for $r > 0$ let (X, Y) be a coupling of X_{E_r} started at x and started at y, respectively, such that $\mathbb{P}[X \neq Y] = D_r(x, y)$.

If f is a bounded harmonic function then $(f(X_t))_t$ is a bounded martingale. The optimal stopping theorem (Theorem 2.3.3) guarantees that $f(x) =$

$\mathbb{E}_x\left[f(X_{E_r})\right] = \mathbb{E}[f(X)]$ and $f(y) = \mathbb{E}_y\left[f(X_{E_r})\right] = \mathbb{E}[f(Y)]$. Thus,

$$|f(x) - f(y)| = |\mathbb{E}\left[(f(X) - f(Y))\mathbf{1}_{\{X \neq Y\}}\right]|$$
$$\leq 2\|f\|_\infty \cdot \mathbb{P}[X \neq Y] = 2\|f\|_\infty D_r(x, y).$$

Hence, if $D_r(x, y) \to 0$ then $f(x) = f(y)$ for any bounded harmonic function f.

The other direction is slightly more involved. If for some $x, y \in G$ we have $D_r(x, y) \not\to 0$, then let $d := \limsup_{r\to\infty} D_r(x, y) > 0$. Let $(r_k)_k$ be a subsequence such that $D_{r_k}(x, y) \to d$. For any $r > 0$, let A_r be a set such that $\left|\mathbb{P}_x\left[X_{E_r} \in A_r\right] - \mathbb{P}_y\left[X_{E_r} \in A_r\right]\right| = D_r(x, y)$. For any $r > 0$ we may define $f_r(z) = \mathbb{P}_z\left[X_{E_r} \in A_r\right]$. This function is harmonic in the ball of radius r around 1.

Since $(f_r)_r$ are uniformly bounded, there exists a subsequence $(r'_n = r_{k_n})_n$ for which the subsequential limit $f_{r'_n} \to f$ (pointwise convergence). It is immediate that f is a bounded harmonic function. Also,

$$|f(x) - f(y)| = \lim_{n\to\infty}\left|f_{r'_n}(x) - f_{r'_n}(y)\right| = \lim_{n\to\infty} D_{r_{k_n}}(x, y) = d > 0,$$

so $f(x) \neq f(y)$. □

Exercise 6.46 Let S be a finite symmetric generating set for a group G. Let μ be a probability measure supported on S.

Show that for any μ-harmonic function f on G, we have

$$|f(x) - f(y)| \leq 2D_r(x, y) \cdot \max_{|z|=r+1} |f(z)|. \qquad \triangleright \text{solution} \triangleleft$$

Remark 6.6.3 We see from the above proof that for fixed $x, y \in G$ we have that

$$\lim_{r\to\infty} D_r(x, y) = 0 \iff \forall f \in \text{BHF}(G, \mu),\ f(x) = f(y).$$

Tointon (2016) proved that for fixed $x, y \in G$ we have that

$$f(x) = f(y) \text{ for all } \mu\text{-harmonic functions } f$$
$$\iff \text{there exists } r_0, \text{ for all } r \geq r_0,\ D_r(x, y) = 0.$$

That is, convergence to 0 of $D_r(x, y)$ is equivalent to the fact that *bounded* harmonic functions cannot differentiate between x and y. But the (stronger) property that the sequence $(D_r(x, y))_r$ stabilizes at 0 is equivalent to the fact that *any* harmonic function cannot differentiate between x and y.

6.7 Speed and Entropy

Recall that in Exercise 6.30 we showed that $H(|X_t|) \leq C \mathbb{E} \log(|X_t| + 1) + C$ for some universal constant $C > 0$.

Exercise 6.47 Show that for a probability measure μ on a finitely generated group G, the following limit always exists: $\lim_{t \to \infty} t^{-1} \mathbb{E} |X_t|$, where $(X_t)_t$ is the μ-random walk.

 We call this limit the **speed** of the random walk.   ▷ solution ◁

Exercise 6.48 Let μ be an adapted probability measure on a finitely generated group G. Show that if μ has speed 0, then (G, μ) is Liouville.   ▷ solution ◁

Proposition 6.7.1 (Entropy and speed) *Let G be a finitely generated group and $\mu \in SA(G, \infty)$. Consider the μ-random walk $(X_t)_t$.*

 Then there exists a constant $C_\mu > 0$ such that

$$\frac{\mathbb{E} |X_t|^2}{C_\mu \cdot t} \leq H(X_t) \leq C_\mu \cdot \mathbb{E} |X_t|.$$

Proof We know that $H(X_t) \leq C \mathbb{E} |X_t|$ for some constant $C > 0$, by Exercise 6.30, as in the previous exercise.

 For the other inequality, we use the Varopolous–Carne bound (Theorem 5.6.1). Since $H(X_t) \geq H(X_{t-1}) \geq \cdots \geq H(X_1) > 0$, we only need to consider those t for which $\mathbb{E} |X_t|^2 \geq t$, otherwise the inequality is trivial.

 Now, by the Varopolous–Carne bound (Theorem 5.6.1), for some constant $c = c_\mu > 0$, using $m = c^{-1}(2t - 1)$,

$$H(X_t) = - \sum_x \mathbb{P}[X_t = x] \log \mathbb{P}[X_t = x]$$

$$\geq \sum_{2 \leq |x| \leq m+1} \mathbb{P}[X_t = x] \cdot \frac{c^2 |x|^2}{8t} + \sum_{|x| > m+1} \mathbb{P}[X_t = x] \cdot \frac{c}{4} |x|^2$$

$$= \frac{c^2}{8} \cdot \frac{\mathbb{E} |X_t|^2}{t} + \mathbb{E} \left[|X_t|^2 \mathbf{1}_{\{|X_t| > m+1\}} \right] \cdot \frac{c}{4} (1 - \frac{c}{2t}) - \mathbb{P}[|X_t| = 1] \cdot \frac{c^2}{8t}.$$

Note that $\mathbb{P}[|X_t| = 1] \leq \#\{x : |x| = 1\} \cdot \mathbb{P}[X_t = 1] \to 0$ as $t \to \infty$. Thus, we may choose C_μ appropriately so that the assertion holds.   □

Corollary 6.7.2 (Liouville and speed) Let $\mu \in SA(G, \infty)$ for a finitely generated group G.

 Then (G, μ) is Liouville if and only if μ has speed 0.

Proof If the speed is 0 then (G, μ) is Liouville even without the exponential tail assumption for μ.

If (G, μ) is Liouville and $\mu \in \mathsf{SA}(G, \infty)$, then by the Varopolous–Carne bound (Theorem 5.6.1), we have that

$$\left(\lim_{t\to\infty} \frac{\mathbb{E}\,|X_t|}{t} \right)^2 \le \limsup_{t\to\infty} \frac{\mathbb{E}\,|X_t|^2}{t^2} \le C_\mu \cdot \limsup_{t\to\infty} \frac{H(X_t)}{t} = h(G, \mu) = 0. \quad \square$$

Example 6.7.3 Consider a symmetric and adapted probability measure μ on the group $\mathbb{Z}^d$. Let $(X_t)_t$ denote the μ-random walk. The symmetry of μ implies that

$$\mathbb{E}\left[||X_{t+1}||^2 \mid \mathcal{F}_t \right] = ||X_t||^2 + \mathbb{E}\,||X_{t+1} - X_t||^2 + 2\,\mathbb{E}[\langle X_{t+1} - X_t, X_t \rangle \mid \mathcal{F}_t]$$
$$= ||X_t||^2 + \mathbb{E}\,||X_1||^2,$$

so that $M_t = ||X_t||^2 - ct$ is a martingale for $c = \mathbb{E}\,||X_1||^2$. (We have used that the increment $X_{t+1} - X_t$ is independent of $\mathcal{F}_t$.) This implies that $\mathbb{E}\,||X_t||^2 = ct$, for all t.

Note that the norm $|| \cdot ||$ is Euclidean distance in $\mathbb{R}^d$, but may be easily compared to the Cayley graph distance (with the standard generators) in $\mathbb{Z}^d$. Indeed, for any $z = (z_1, \ldots, z_d) \in \mathbb{Z}^d$, we have that

$$|z| = \sum_{k=1}^{d} |z_j| \le \sqrt{d} \cdot ||z||.$$

Thus,

$$\lim_{t\to\infty} \frac{\mathbb{E}\,|X_t|}{t} \le \sqrt{d} \cdot \limsup_{t\to\infty} \frac{\sqrt{\mathbb{E}\,||X_t||^2}}{t} = 0.$$

So a symmetric random walk on $\mathbb{Z}^d$ always has 0 speed, and thus is always Liouville.

We will see that this is part of a broader phenomenon, known as the *Choquet–Deny theorem* (Corollary 7.1.2).   △▽△

Example 6.7.4 Consider the free group $\mathbb{F}_d$ on $d \ge 2$ generators, with the standard Cayley graph; that is, the generating set is $S = \left\{ (a_1)^{\pm 1}, \ldots, (a_d)^{\pm 1} \right\}$. Let $(X_t)_t$ be the simple random walk on $\mathbb{F}_d$.

Note that for any $x \ne 1$, we have that multiplying by exactly one generator from S will bring us closer to 1, and all the rest will take us farther away. Thus,

$$\mathbb{P}[|X_{t+1}| = |X_t| + 1 \mid |X_t| > 0] = \frac{2d-1}{2d}$$
$$\text{and} \quad \mathbb{P}[|X_{t+1}| = |X_t| - 1 \mid |X_t| > 0] = \frac{1}{2d}.$$

Of course, when $X_t = 1$, then any generator will take us further from the origin.

This implies that the process $(|X_t|)_t$ is a Markov chain on $\mathbb{N}$, with the transition matrix $P(n, n+1) = 1 - \frac{1}{2d} = 1 - P(n, n-1)$ for $n > 0$, and $P(0, 1) = 1$. This Markov chain is precisely the weighted random walk induced by the network on $\mathbb{N}$ given by placing conductance $c(n, n+1) = (2d-1)^n$.

Let us provide a lower bound on the speed of the above random walk:

$$\mathbb{E}\,|X_{t+1}| = \mathbb{E}\left[|X_{t+1}|\mathbf{1}_{\{|X_t|>0\}}\right] + \mathbb{E}\left[|X_{t+1}|\mathbf{1}_{\{|X_t|=0\}}\right]$$

$$= \frac{2d-1}{2d} \cdot \mathbb{E}\left[(|X_t|+1)\mathbf{1}_{\{|X_t|>0\}}\right] + \frac{1}{2d} \cdot \mathbb{E}\left[(|X_t|-1)\mathbf{1}_{\{|X_t|>0\}}\right] + \mathbb{P}[X_t = 1]$$

$$= \mathbb{E}\,|X_t| + 1 - \frac{1}{d}\,\mathbb{P}[X_t \ne 1],$$

so that

$$\mathbb{E}\,|X_t| \ge \mathbb{E}\,|X_{t-1}| + \frac{d-1}{d} \ge \cdots \ge t \cdot \frac{d-1}{d}.$$

Hence, the speed of the simple random walk on $\mathbb{F}_d$ is at least $\frac{d-1}{d}$, which is strictly positive, implying that this walk is not Liouville. △▽△

6.7.1 The Karlsson–Ledrappier Theorem

In the above we used the Varopolous–Carne bound (Theorem 5.6.1), which required random walks with exponential tails (at least by our method of proof).

However, the relationship of the Liouville property with the speed of the random walk is much broader. A fundamental result is the following theorem from Karlsson and Ledrappier (2007).

Theorem 6.7.5 (Karlsson–Ledrappier Theorem) *Let $(X_t)_t$ be a μ-random walk for some adapted measure μ on a group G; μ does not necessarily have to be symmetric. Assume that μ has finite first moment, so $\mathbb{E}\,|X_1| < \infty$, and assume that (G, μ) is Liouville; that is, every bounded μ-harmonic function is constant.*

Then, there exists a group homomorphism $\varphi \colon G \to \mathbb{R}$ such that

$$\lim_{t\to\infty} \frac{1}{t}\,\mathbb{E}\,|X_t| = \mathbb{E}\,\varphi(X_1).$$

In particular, if μ is symmetric then $\mathbb{E}\,\varphi(X_1) = 0$, so the speed is 0.

Proof To prove this theorem we will actually be dealing with the theory of *horofunctions*, although it is not necessary to be familiar with these objects. We have already encountered them in Exercise 1.97.

Fix a finite symmetric generating set S for G. Recall the Lipschitz semi-norm:

$$\|\nabla_S f\|_\infty = \sup_{s\in S}\sup_{x\in G} |f(xs) - f(x)|.$$

Consider the set of functions

$$L = \{f \colon G \to \mathbb{R} : \|\nabla_S f\|_\infty \le 1,\ f(1) = 0\}.$$

We topologize this set with pointwise convergence. Exercise 1.97 shows that with this topology, L is a compact topological space.

G acts on L by $x.f(y) = f\left(x^{-1}y\right) - f\left(x^{-1}\right)$. An important observation here is that a fixed point for this action is a homomorphism; that is, $x.f = f$ for all $x \in G$ if and only if $f(xy) = f(x) + f(y)$ for all $x, y \in G$. The action is also continuous.

An important subset of L is given by the so-called *Busemann functions*: for $x \in G$ we have $b_x \in L$ defined by $b_x(y) := \left|y^{-1}x\right| - |x|$.

Now, let $(X_t)_t$ be the μ-random walk. Define

$$f_n(x) = \frac{1}{n} \sum_{k=0}^{n-1} \mathbb{E}\left[b_{X_k}(x)\right].$$

As a convex combination of Busemann functions, $f_n \in L$. By compactness, there is a subsequence $n_j \to \infty$ such that $f_{n_j} \to f$ for some $f \in L$.

Set $\psi(x) = \check{f}(x) := f\left(x^{-1}\right)$. Note that

$$\sum_y \mu(y)\,\mathbb{E}[|xyX_k|] = \mathbb{E}[|xX_{k+1}|],$$

which gives us that

$$\sum_y \mu(y) f_n\left((xy)^{-1}\right) = \frac{1}{n} \sum_{k=0}^{n-1} \sum_y \mu(y)\,\mathbb{E}[|xyX_k| - |X_k|]$$

$$= \frac{1}{n} \sum_{k=0}^{n-1} \mathbb{E}[|xX_{k+1}| - |X_{k+1}|] + \frac{1}{n} \sum_{k=0}^{n-1} \mathbb{E}[|X_{k+1}| - |X_k|]$$

$$= \frac{1}{n} \sum_{k=1}^{n} \mathbb{E}[|xX_k| - |X_k|] + \frac{1}{n} \mathbb{E}[|X_n|]$$

$$= f_n\left(x^{-1}\right) + \frac{1}{n}(\mathbb{E}\,|xX_n| - |x|).$$

Since $|X_n| - |x| \le |xX_n| \le |X_n| + |x|$, we have that

$$\sigma := \lim_{n \to \infty} \frac{1}{n}\mathbb{E}\,|xX_n| = \lim_{n \to \infty} \frac{1}{n}\mathbb{E}\,|X_n|.$$

Taking the limit along $n_j \to \infty$, we get that

$$\sum_y \mu(y)\psi(xy) = \psi(x) + \sigma.$$

Hence, for any fixed $x \in G$, the function

$$h_x(y) := \psi(xy) - \psi(y)$$

is a harmonic function. Note that $h_x(y) = y.f\left(x^{-1}\right)$. Since $\|\nabla_S y.f\|_\infty = \|\nabla_S f\|_\infty \leq 1$, for all $y \in G$, we get that $|h_x(y)| \leq |x|$ for all $x, y \in G$. That is, h_x are all bounded harmonic functions. By the assumption that (G, μ) is Liouville, we have that $y.f\left(x^{-1}\right) = h_x(y) = h_x(1) = f\left(x^{-1}\right)$ for all $x, y \in G$; that is, f is a fixed point for the G-action, implying that f and ψ are actually homomorphisms.

Finally, as computed above, $\mathbb{E}\psi(X_1) = \sigma$. If μ is symmetric this must be 0. $\qquad\square$

Corollary 6.7.6 Let G be a finitely generated group and $\mu \in \mathrm{SA}(G, 1)$. Then, (G, μ) is Liouville if and only if μ has 0 speed.

6.8 Amenability and Liouville

In this section we connect the notions of amenability and Liouville. In fact, the next theorem proves that any finite entropy (symmetric, adapted) random walk on a non-amenable group is non-Liouville.

Theorem 6.8.1 (Liouville implies amenable) *Let G be a non-amenable finitely generated group.*

Then, for any symmetric, adapted probability measure μ on G with finite entropy $H(\mu) < \infty$, we have that (G, μ) is not Liouville.

Proof It suffices to prove that the entropy is linearly growing. Since G is non-amenable, by Kesten's amenability criterion (Theorem 5.2.4), we may choose $\rho < 1, C > 0$ such that $\mathbb{P}_x[X_t = y] \leq C\rho^t$ for all $x, y \in G$ and $t > 0$. So

$$H(X_t) = -\sum_y P^t(x, y) \log P^t(x, y)$$

$$\geq -\sum_y P^t(x, y) \log(C\rho^t) = -\log C + t \cdot \log(1/\rho).$$

Thus, $\frac{H(X_t)}{t} \to h(G, \mu) \geq -\log \rho > 0$.

So (G, μ) is not Liouville by the entropic criterion. $\qquad\square$

Exercise 6.49 Recall Exercise 5.35, which shows that random walks on non-amenable groups have positive speed.

Give an alternative proof of this fact using the Liouville property. That is, show that if G is a non-amenable finitely generated group, then any $\mu \in \mathrm{SA}(G, 1)$ must have positive speed.

Exercise 6.50 Let G be a finitely generated group such that G has *sub-exponential growth*. That is, we have $\frac{1}{r} \log |B(1, r)| \to 0$ as $r \to \infty$.

Show that (G, μ) is Liouville for any adapted μ with finite first moment.

Conclude that G is amenable.   ▷ solution ◁

6.9 Lamplighter Groups

In this section we review a useful class of examples, known as *lamplighter groups*. These will typically be examples of amenable groups that are non-Liouville.

First, let us recall the notion of a *semi-direct product* from Section 1.5.9. Let G, H be groups and suppose that G acts on H by group automorphisms. That is, every element $g \in G$ can be thought of as an automorphism of H, and we denote by $g.h \in H$ the image of $h \in H$ under the automorphism $g \in G$. Define the group $G \ltimes H$ to be the set $\{(g, h) : g \in G, \ h \in H\}$, with the following product:

$$(g, h)(g', h') := (g \cdot g', h \cdot g.h').$$

In Exercise 1.68 we saw that this is indeed a group, that the identity element of $G \ltimes H$ is $(1_G, 1_H)$, and that inverses are given by $(g, h)^{-1} = \left(g^{-1}, g^{-1}.h^{-1}\right)$.

Some examples can be found in Exercises 1.70, 1.71, and 1.72.

The example we wish to consider here is the *lamplighter group over a group* G. Let G be a finitely generated group. Consider the group

$$\Sigma(G) = \oplus_G \{0, 1\} = \{\sigma : G \to \{0, 1\} \mid |\text{supp}(\sigma)| < \infty\}.$$

Equipped with pointwise addition modulo 2, this is an Abelian group (when G is infinite it is not finitely generated), and G acts on these functions naturally by translation $x.\sigma(y) = \sigma\left(x^{-1}y\right)$. The **lamplighter group** over G is the group $L(G) := G \ltimes \Sigma(G)$. That is, elements of $L(G)$ are pairs (x, σ) for $x \in G$ and $\sigma : G \to \{0, 1\}$. Multiplication is given by $(x, \sigma)(y, \tau) = (xy, \sigma + x.\tau)$. Inverse elements are given by $(x, \sigma)^{-1} = \left(x^{-1}, x^{-1}.\sigma\right)$. We have already encountered the group $L(\mathbb{Z})$ in Example 5.5.3.

Exercise 6.51 Consider the group $L(G)$.

Compute the conjugation $(x, \sigma)^{(y, \tau)}$. What is $(1, \sigma)^{(y, \tau)}$? Show that $\Sigma(G) \cong \{1\} \times \Sigma(G)$ is a normal subgroup of $L(G)$.

What is $L(G)/(\{1\} \times \Sigma(G))$ isomorphic to?

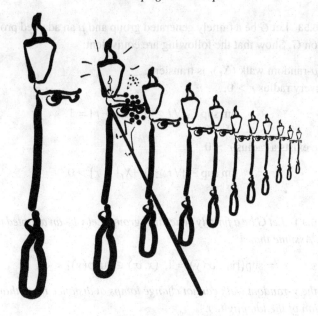

Show that if $G = \langle S \rangle$ then $L(G)$ is generated by $\{(s, 0), (1, \delta_1) : s \in S\}$, where $\delta_x \in \Sigma(G)$ is given by $\delta_x(y) = \mathbf{1}_{\{x=y\}}$.

Show that the commutator subgroup of $L(G)$ satisfies $[L(G), L(G)] \subset [G, G] \times \Sigma(G)$ (as sets).

▷ solution ◁

Let us interpret this group in a probabilistic way. We think of G as a "street" on which "lamps" are placed at each site $x \in G$. The lamps can be on or off, indicated by $0, 1$, respectively, so the configuration of lamps is σ. A "lamplighter" walks on the street G and can switch the state of the lamps. So (x, σ) indicates the position of the lamplighter by x, and the state of all lamps by σ. Multiplying (x, σ) on the right by an element $(s, 0)$ is moving the lamplighter on the street G by s. Multiplying (x, σ) on the right by $(1, \delta_z)$ is flipping the state of the lamp at xz. So under the generators $(s, 0), s \in S$, and $(1, \delta_1)$, we have a lamplighter moving around on G and flipping 0-1 lamps on G.

Exercise 6.52 Show that $L(G)$ has exponential growth; that is, show that there exists $c > 0$ such that $|B(1, r)| \geq e^{cr}$ for all $r > 0$.

Conclude that $L(G)$ is transient.

▷ solution ◁

In Exercise 6.63 we will precisely compute the entropy of certain random walks on $L(G)$. The more general phenomenon, not requiring an exact computation, is that when the "lamplighter" is walking recurrently on the "street" G, the random walk on $L(G)$ is Liouville. This is made precise in Theorem 6.9.1.

Exercise 6.53 Let G be a finitely generated group and μ an adapted probability measure on G. Show that the following are equivalent:

(1) The μ-random walk $(X_t)_t$ is transient.
(2) For every radius $r > 0$,

$$\limsup_{|x| \to \infty} \mathbb{P}_x[\forall t \geq 0, \ |X_t| > r] = 1.$$

(3) There exists a radius $r > 0$

$$\limsup_{|x| \to \infty} \mathbb{P}_x[\forall t \geq 0, \ |X_t| > r] > 0. \qquad \qquad \triangleright \text{solution} \triangleleft$$

Theorem 6.9.1 *Let G be a finitely generated group. Let ν be an adapted measure on $L(G)$. Assume that*

$$r := \sup\{|y| : \sigma(y) = 1, \ (x, \sigma) \in \mathrm{supp}(\nu)\} < \infty.$$

(That is, the ν-random walk cannot change lamps at distance more than r from the location of the lamplighter.)
    Define the measure μ on G by $\mu(x) = \sum_{\sigma \in \Sigma(G)} \nu(x, \sigma)$.
    Then $(L(G), \nu)$ is Liouville if and only if the μ-random walk on G is recurrent.

Proof Let $((X_t, \sigma_t))_t$ be the ν-random walk. Note that $(X_t)_t$ is a μ-random walk on G.
    For $(x, \sigma) \in L(G)$ define

$$h(x, \sigma) = \mathbb{P}_{(x, \sigma)}[\limsup_{t \to \infty} \sigma_t(1) = 0].$$

That is, $h(x, \sigma)$ is the probability starting at (x, σ) that the lamp at 1 is eventually off. Note that by our assumption, if $|X_t| > r$ then $\sigma_{t+1}(1) = \sigma_t(1)$.
    Consider the event $A = \{\forall t \geq 0, \ |X_t| > r\}$. That is, the walk never enters the ball of radius r (so on A the walk cannot change the lamp at 1). Note that by our assumption, $A \subset \{\forall t \geq 0, \ \sigma_t(1) = \sigma_0(1)\}$.
    If the μ-random walk is transient, then by Exercise 6.53 there exists $x \in G$ such that $\mathbb{P}_{(x, \sigma)}[A] \geq \frac{3}{4}$ for all $\sigma \in \Sigma(G)$. However, in that case, $h(x, 0) \geq \mathbb{P}_{(x, 0)}[A] \geq \frac{3}{4}$ and $h(x, \delta_1) \leq \mathbb{P}_{(x, \delta_1)}[A^c] \leq \frac{1}{4}$. So h is a nonconstant, bounded harmonic function.
    Now, for the case that the μ-random walk on G is recurrent. Let $f : L(G) \to \mathbb{C}$ be a bounded harmonic function. Fix some $\sigma \in \Sigma(G)$. Since ν is adapted, there

exists $k \geq 0$ and $\varepsilon > 0$ such that $\mathbb{P}[(X_k, \sigma_k) = (1, \sigma)] = \varepsilon$. By the Markov property we conclude that

$$\text{for all } t, \quad \mathbb{P}[(X_{t+k}, \sigma_{t+k}) = (X_t, \sigma_t)(1, \sigma) \mid \mathcal{F}_t] = \varepsilon \quad \text{a.s.}$$

Since $(X_t)_t$ is a μ-random walk on G, it is recurrent. Define $T_0 = 0$ and inductively $T_{n+1} = \inf\{t > T_n + k : X_t = 1\}$. By recurrence, all T_n are finite stopping times a.s. Also, the strong Markov property implies that for all n,

$$\mathbb{P}\left[(X_{T_n+k}, \sigma_{T_n+k}) = (1, \sigma_{T_n} + \sigma) \mid \mathcal{F}_{T_n}\right] = \varepsilon \quad \text{a.s.}$$

Thus, we have that the set

$$\Gamma = \{n : (X_{T_n+k}, \sigma_{T_n+k}) = (1, \sigma_{T_n} + \sigma)\}$$

is a.s. infinite. Specifically, since $X_{T_n} = 1$ a.s.,

$$\liminf_{t \to \infty} |f(X_{t+k}, \sigma_{t+k}) - f(X_t, \sigma_t + \sigma)| = 0 \quad \text{a.s.}$$

However, since $(f(X_{t+k}, \sigma_{t+k}) - f(X_t, \sigma_t + \sigma))_t$ is a bounded martingale, it must converge a.s. and in L^1 by the martingale convergence theorem (Theorem 2.6.3 and Exercise 2.33). Specifically, for any $x \in G$ and any $\sigma \in \Sigma(G)$, we have

$$|f(x, 0) - f(x, \sigma)| = |f(x, 0) - (1, \sigma).f(x, 0)|$$

$$\leq \mathbb{E}_{(x,0)}[|f(X_{t+k}, \sigma_{t+k}) - (1, \sigma).f(X_t, \sigma_t)|] \to 0,$$

implying that $f(x, 0) = f(x, \sigma)$ for all $x \in G$ and $\sigma \in \Sigma(G)$. This immediately leads to the fact that $h(x) := f(x, 0)$ is a μ-harmonic function on G. The assumption that μ is recurrent implies that (G, μ) is Liouville by Exercise 6.21. So h must be constant. Thus, $f(x, \sigma) = h(x) = h(1) = f(1, 0)$ for all $(x, \sigma) \in L(G)$. □

Exercise 6.54 Show that if G is a finitely generated amenable group then $L(G)$ is also amenable.

Conclude that $L\left(\mathbb{Z}^3\right)$ is an amenable group that is non-Liouville for any finitely supported, adapted random walk.   ▷ solution ◁

6.10 An Example: Infinite Permutation Group S_∞^*

In light of Conjecture 2.7.2, we may wish to point out that something similar does not hold for non-finitely generated groups, as can be seen by the following example.

Let S_∞^* denote the group of all finitely supported permutations of a countably infinite set Ω. Specifically, if $\sigma : \Omega \to \Omega$ let $\text{supp}(\sigma) = \{\omega \in \Omega : \sigma(\omega) \neq \omega\}$ and define

$$S_\infty^* = \{\sigma : \Omega \to \Omega : |\text{supp}(\sigma)| < \infty \text{ and } \sigma \text{ is a bijection}\}.$$

(Since all countably infinite sets are in bijection with one another, the precise set Ω is not important, as all versions of S_∞^* will be isomorphic.)

Exercise 6.55 Show that S_∞^* is not finitely generated.

Hint: show that S_∞^* is *locally finite*; that is, any finitely generated subgroup of S_∞^* is finite.

▷ solution ◁

Proposition 6.10.1 *There exist adapted symmetric probability measures μ, ν on S_∞^* such that $h(S_\infty^*, \mu) = 0$ and $h(S_\infty^*, \nu) > 0$.*

The specific constructions of μ, ν as above are in the following two examples.

Example 6.10.2 This is an example of a random walk on S_∞^* with 0 entropy.

Identify Ω with $\mathbb{N}$ and consider the transpositions $\pi_n = (n \ n+1)$. Let μ be the measure $\mu(\pi_n) = 2^{-n-1}$ for all $n \geq 0$. Since $\pi_n = \pi_n^{-1}$, μ is symmetric. Also, $(\pi_n)_n$ generate S_∞^*, so μ is adapted.

Let $(X_t)_t$ be the μ-random walk. Let $U_{t+1} = X_t^{-1} X_{t+1}$ be the jump at time $t+1$. So $U_t \in \{\pi_n : n \in \mathbb{N}\}$. Set $|U_t| = n$ for n such that $U_t = \pi_n$. Let

$$J_t = \max\{|U_k| : 1 \leq k \leq t\}$$

be the "maximal jump" up to time t.

Let us bound the entropy of X_t. Note that if $J_t \leq r$ then $\text{supp}(X_t) \subset \{0, 1, \ldots, r+1\}$. On this event, the number of possibilities for X_t is at most $(r+2)! \leq (r+2)^{r+2}$. Thus,

$$H(X_t) \leq H(X_t \mid J_t) + H(J_t) \leq \mathbb{E}[(J_t + 2) \log(J_t + 2)] + H(J_t).$$

Now,

$$\mathbb{P}[J_t \geq r] \leq \mathbb{P}[\exists \ 1 \leq k \leq t, \ |U_k| \geq r] \leq t 2^{-r}.$$

It is a simple computation that

$$\mathbb{E}[(J_t + 2) \log(J_t + 2)] \leq C \log t \cdot \log \log t,$$

for large enough constant $C > 0$. Moreover, since $p \mapsto -p \log p$ is maximized at $p = e^{-1}$ on $(0, 1)$, increasing on $(0, e^{-1})$, and decreasing on $(e^{-1}, 1)$, we obtain

$$H(J_t) = -\sum_{r=0}^{\infty} \mathbb{P}[J_t = r] \log \mathbb{P}[J_t = r]$$

$$\leq C \log t \cdot \tfrac{1}{e} - \sum_{r > C \log t} (t 2^{-r}) \log(t 2^{-r})$$

$$\leq C \log t \cdot \tfrac{1}{e} + O\left(t 2^{-C \log t}\right) = O(\log t).$$

Thus, altogether, for some constant $C > 0$, we have $H(X_t) \leq C \log t \cdot \log \log t$. Hence $h(S_\infty^*, \mu) = 0$.   △▽△

Example 6.10.3 This is an example of a positive-entropy random walk on S_∞^*. For every $n \geq 1$ define

$$\Omega_n^+ := \{\omega = (\omega_1, \dots, \omega_n) \in \{-1, 1\}^n : \omega_n = +1\},$$

$$\Omega_n^- := \{\omega = (\omega_1, \dots, \omega_n) \in \{-1, 1\}^n : \omega_n = -1\},$$

$$\Omega_n := \Omega_n^+ \uplus \Omega_n^-,$$

$$\Omega := \biguplus_n \Omega_n.$$

Consider S_∞^* as permutations of Ω.

For each $n \geq 1$ define

$$\tau_n^\pm(\omega) = \begin{cases} (\omega_1, \dots, \omega_n) & \text{if } \omega = (\omega_1, \dots, \omega_n, \pm 1) \in \Omega_{n+1}^\pm, \\ (\omega_1, \dots, \omega_n, \pm 1) & \text{if } \omega = (\omega_1, \dots, \omega_n) \in \Omega_n, \\ \omega & \text{otherwise.} \end{cases}$$

Note that $(\tau_n^\pm)^2 = 1$. Also, let ρ be the permutation

$$\rho(\omega) = \begin{cases} -\omega & \text{if } \omega \in \Omega_1, \\ \omega & \text{otherwise.} \end{cases}$$

Note also that $\rho^2 = 1$.

Finally, define μ by setting

$$\mu(\rho) = 1 - \alpha, \qquad \mu(\tau_n^\pm) = \tfrac{A-1}{2A^n} \cdot \alpha,$$

for some fixed $1 < A < 2$ and $\alpha \in (0, 1)$ arbitrary. So μ is symmetric and adapted. Also, A has been chosen so that for all $n > 1$,

$$\frac{\mu(\tau_n^+) + \mu(\tau_n^-)}{\mu(\tau_n^+) + \mu(\tau_n^-) + \mu\left(\tau_{n-1}^+\right)} = \frac{\mu(\tau_n^+) + \mu(\tau_n^-)}{\mu(\tau_n^+) + \mu(\tau_n^-) + \mu\left(\tau_{n-1}^-\right)} = \frac{1}{1 + \frac{A}{2}} > \frac{1}{2}.$$

Set $o = (+1) \in \Omega_1^+$. For $\sigma \in S_\infty^*$, if $\sigma^{-1}(o) = (\omega_1, \dots, \omega_n)$ define $\|\sigma\| = n$ and $\text{sgn}(\sigma) = \omega_1$.

Let $(\sigma_t)_t$ be the μ-random walk on S_∞^*, and consider the processes $X_t = ||\sigma_t||$ and $Y_t = \mathrm{sgn}(\sigma_t)$. It is an exercise (Exercise 6.56) to show that $(X_t, Y_t)_t$ forms a Markov chain with the properties:

- $\mathbb{P}[Y_{t+1} \neq Y_t \mid X_t > 1] = 0$,
- $\mathbb{P}[X_{t+1} = X_t + 1 \mid X_{t+1} \neq X_t, \ X_t > 1] \geq \dfrac{1}{1+\frac{A}{2}} > \dfrac{1}{2}$.

Let $\tau = \inf\{t : X_t = 1\}$. Comparing with a random walk on a rooted binary tree (see Exercise 6.57), we may then see that for any $n > 1$,

$$\mathbb{P}[\tau = \infty \mid X_0 = n > 1] \geq \mathbb{P}[\forall \, t, \ X_t \geq X_0 \mid X_0 = n > 1]$$

$$\geq 2 - \frac{1}{\inf_t \mathbb{P}[X_{t+1} = X_t + 1 \mid X_{t+1} \neq X_t, \ X_t > 1]}$$

$$= 2 - \left(1 + \tfrac{A}{2}\right) = 1 - \tfrac{A}{2}.$$

Since $(Y_t)_t$ can only change sign when $X_t \in \{-1, 1\}$, we conclude that starting at any $X_0 = n > 1$, with positive probability at least $1 - \frac{A}{2}$ the process $(X_t, Y_t)_t$ never changes the sign of Y_t. That is,

$$\mathbb{P}[\forall \, t, \ Y_t = Y_0 \mid X_0 = n > 1] \geq 1 - \tfrac{A}{2}.$$

Thus, the event

$$\mathcal{E} = \{\exists \, t_0, \ \forall \, t > t_0, \ Y_t = 1\}$$

is an invariant event, with probability not in $\{0, 1\}$. Hence, the invariant σ-algebra is nontrivial, and hence the tail σ-algebra as well. This can be used in Corollary 6.5.4 (as in the proof of Theorem 6.5.5) to show that the entropy $h(S_\infty^*, \mu) > 0$ is positive. $\triangle \triangledown \triangle$

Exercise 6.56 Show that $(X_t, Y_t)_t$ defined above form a Markov chain on $\{1, 2, \ldots\} \times \{-1, 1\}$ with transition probabilities given by:

$$P((1, y), (1, -y)) = \mu(\rho) = 1 - \alpha,$$

$$P((1, y), (2, y)) = \mu\left(\tau_1^+\right) + \mu\left(\tau_1^-\right) = \tfrac{A-1}{A}\alpha,$$

$$P((1, y), (1, y)) = \tfrac{\alpha}{A},$$

$$P((n, y), (n + 1, y)) = \mu\left(\tau_n^+\right) + \mu\left(\tau_n^-\right) = \tfrac{A-1}{A^n}\alpha,$$

$$P((n, y), (n - 1, y)) = \mu\left(\tau_{n-1}^+\right) = \mu\left(\tau_{n-1}^-\right) = \tfrac{A-1}{2A^{n-1}}\alpha,$$

$$P((n, y), (n, y)) = 1 - \alpha \cdot \tfrac{(A-1)(A+2)}{2A^n},$$

for all $n > 1$ and $y \in \{-1, 1\}$.

Show that $\mathbb{P}[Y_{t+1} \neq Y_t \mid X_t > 1] = 0$, and that

$$\mathbb{P}[X_{t+1} = X_t + 1 \mid X_{t+1} \neq X_t, \, X_t > 1] \geq \frac{1}{1 + \frac{4}{2}} > \frac{1}{2}.$$

Exercise 6.57 Let T be the rooted binary tree. That is, $T = \{(\omega_1, \ldots, \omega_n) \in \Omega : \omega_1 = 1\}$, for Ω as in the previous example.

For $x = (\omega_1, \ldots, \omega_n)$, $y = (\omega_1, \ldots, \omega_n, \omega_{n+1})$ we write $\hat{y} = x$. We also write $x \pm 1 = (\omega_1, \ldots, \omega_n, \pm 1)$.

Suppose that P is a transition matrix on T such that $P(x, y) > 0$ only if $\hat{y} = x$ or $\hat{x} = y$. Also, assume that $P(x, x + 1) + P(x, x - 1) \geq \alpha$, for some $\alpha > \frac{1}{2}$ independent of x.

Show that this is a transient Markov chain.

Prove that for any $x \neq (1)$ that is not the root,

$$\mathbb{P}_x[\forall \, t, \, X_t \neq \hat{x}] \geq 2 - \frac{1}{p} \geq 2 - \frac{1}{\alpha},$$

where $p = \inf_y (P(y, y + 1) + P(y, y - 1)) \geq \alpha$.

(Hint: couple the walk on the tree with a walk on $\mathbb{N}$.) ▷ solution ◁

Exercise 6.58 Define an explicit bounded nonconstant harmonic function for the non-Liouville random walk μ defined in Example 6.10.3. ▷ solution ◁

6.11 Additional Exercises

Let G be a finitely generated group. Recall the *lamplighter group* $L(G)$ from Section 6.9. In Exercises 6.59–6.63, we will precisely compute the entropy of certain random walks on $L(G)$.

We start with a generalization of Exercise 5.28.

Exercise 6.59 Let G be a finitely generated group, and let μ be a symmetric, adapted measure on G. Assume that $\mu(1) = 0$. Fix $\varepsilon, p \in (0, 1)$. Let ν be the measure on $L(G)$ given by $\nu(s, 0) = \varepsilon\mu(s)$, and $\nu(1, \delta_1) = \frac{1}{2}(1 - \varepsilon)$ and $\nu(1, 0) = \frac{1}{2}(1 - \varepsilon)$. Let $((X_t, \sigma_t))_t$ be the ν-random walk on $L(G)$. For $t \geq 1$ set

$$Q_t = \{x \in G : \exists \, 0 \leq k \leq t - 1, \, X_{k+1} = X_k = x\}.$$

Show that for any $t > 0$, the conditional distribution of $(\sigma_t(x))_{x \in Q_t}$ conditioned on $(X_n)_n$ is that of independent Bernoulli-$\frac{1}{2}$ random variables. That is, for any $\xi: Q_t \to \{0, 1\}$,

$$\mathbb{P}[\text{for all } x \in Q_t \ \sigma_t(x) = \xi(x) \mid (X_n)_n] = 2^{-|Q_t|} \qquad \text{a.s.} \qquad \triangleright \text{solution} \triangleleft$$

Exercise 6.60 Let G be a finitely generated group and μ a symmetric adapted measure on G. Assume that $\mu(1) > 0$. Let $(X_t)_t$ be μ-random walk and set $Q_t = \{x \in G : \exists \, 0 \le k \le t-1, \ X_{k+1} = X_k = x\}$.

Show that the limit $\ell := \lim_{t\to\infty} \frac{1}{t} \mathbb{E}|Q_t|$ exists. (Hint: recall Exercise 4.45.)

Show that

$$\mu(1)p_{\text{esc}} \le \ell \le p_{\text{esc}} := \mathbb{P}[\forall \, t > 0, \ X_t \ne 1].$$

Conclude that $\ell = 0$ if and only if (G, μ) is recurrent. $\qquad \triangleright \text{solution} \triangleleft$

Exercise 6.61 Let G be a finitely generated group and μ a symmetric adapted measure on G. Assume that $\mu(1) > 0$. Let $(X_t)_t$ be μ-random walk and set $Q_t = \{x \in G : \exists \, 0 \le k \le t-1, \ X_{k+1} = X_k = x\}$.

Show that

$$\ell := \lim_{t\to\infty} \frac{1}{t} \mathbb{E}|Q_t| = \frac{p_{\text{esc}}\mu(1)}{\mu(1) + p_{\text{esc}}}. \qquad \triangleright \text{solution} \triangleleft$$

Exercise 6.62 Let G be a finitely generated group and μ a symmetric adapted measure on G, with finite entropy. Let $\nu = (1 - \varepsilon)\delta_1 + \varepsilon\mu$ be a lazy version of μ. Show that

$$h(G, \nu) = \varepsilon \cdot h(G, \mu). \qquad \triangleright \text{solution} \triangleleft$$

Exercise 6.63 Let G be a finitely generated group, and let μ be a symmetric, adapted measure on G, with finite entropy. Assume that $\mu(1) = 0$. Fix $\varepsilon, p \in (0, 1)$. Let ν be the measure on $L(G)$ given by $\nu(s, 0) = \varepsilon\mu(s)$ and $\nu(1, \delta_1) = \frac{1}{2}(1 - \varepsilon)$ and $\nu(1, 0) = \frac{1}{2}(1 - \varepsilon)$.

Compute the random walk entropy $h(L(G), \nu)$, and show that

$$h(L(G), \nu) = \log 2 \cdot \frac{(1 - \varepsilon)p_{\text{esc}}}{1 - \varepsilon + p_{\text{esc}}} + \varepsilon h(G, \mu),$$

where $p_{\text{esc}} = \mathbb{P}[\forall \, t \ge 1, \ X_t \ne 1]$ and $(X_t, \sigma_t)_t$ is the ν-random walk. $\qquad \triangleright \text{solution} \triangleleft$

Exercise 6.64 Show that there exists a finitely supported adapted but non-symmetric measure μ on $\mathbb{Z}$ such that $(\mathbb{Z}, \mu)$ is transient.

Conclude that there exists a finitely supported adapted but nonsymmetric measure ν on $L(\mathbb{Z})$ such that $(L(G), \nu)$ is non-Liouville. $\qquad \triangleright \text{solution} \triangleleft$

Exercise 6.65 Let $\varepsilon > 0$.

Show that there exists $\nu \in \mathsf{SA}\left(L(\mathbb{Z}^2), 2 - \varepsilon\right)$ such that $\left(L(\mathbb{Z}^2), \nu\right)$ is non-Liouville.

▷ solution ◁

For the following exercises, let G be a finitely generated group, and μ an adapted probability measure on G. Let $(X_t)_t$ denote the μ-random walk. Define

$$\mathcal{A}(G, \mu) = \{f \in \ell^\infty(G) : (f(X_t))_t \text{ converges a.s. }\}.$$

Exercise 6.66 Show that $\mathsf{BHF}(G, \mu) \subset \mathcal{A}(G, \mu)$. ▷ solution ◁

Exercise 6.67 Show that $\mathcal{A}(G, \mu)$ is an algebra with pointwise addition and multiplication of functions (i.e. $\mathcal{A}(G, \mu)$ is a sub-algebra of $\ell^\infty(G)$). ▷ solution ◁

Exercise 6.68 Define

$$I(G, \mu) = \{f \in \mathcal{A}(G, \mu) : f(X_t) \xrightarrow{\text{a.s.}} 0\}.$$

Show that $I(G, \mu)$ is an ideal. ▷ solution ◁

Exercise 6.69 Show that for any $f \in \mathcal{A}(G, \mu)$ the random variable $L_f := \limsup_{t \to \infty} f(X_t)$ is measurable with respect to the invariant σ-algebra I.

Define $h_f(x) = \mathbb{E}_x[L_f]$. Show that $h_f \in \mathsf{BHF}(G, \mu)$.

Show that $f - h_f \in I(G, \mu)$.

Conclude that $\mathcal{A}(G, \mu) = \mathsf{BHF}(G, \mu) \oplus I(G, \mu)$. ▷ solution ◁

6.12 Remarks

In this chapter we have only just glimpsed at the rich world of bounded harmonic functions. A deep notion that we have not mentioned at all is that of the *Poisson boundary*. Given an adapted random walk μ on a group G, the *Poisson boundary* is a measure space with a G-action that is *maximal* with respect to certain equivalent properties. This deserves treatment of its own, and we refer the reader to Lyons and Peres (2016, chapter 14) and references therein. The construction and study of the Poisson boundary was initiated in Furstenberg (1963, 1971). A brief description of the construction from the latter paper is as follows:

We start with the *von Neumann algebra* $\ell^\infty(G)$. Let μ be some adapted probability measure on G. Recall $\mathcal{A}(G, \mu) = \mathsf{BHF}(G, \mu) \oplus I(G, \mu)$ from Exercise 6.69. Since $I(G, \mu)$ is an ideal (Exercise 6.68), we have that $\mathsf{BHF}(G, \mu) \cong$

$\mathcal{A}(G, \mu)/\mathcal{I}(G, \mu)$ as *commutative unital von Neumann algebras*. A classical result by von Neumann states that BHF$(G, \mu) \cong L^\infty(\Pi, \Sigma, \nu)$ for some probability space (Π, Σ, ν). It can be shown that the G action on BHF(G, μ) induces an action on (Π, Σ, ν). It turns out that the resulting action on ν by $g\nu(A) = \nu\left(g^{-1}A\right)$ is μ-*stationary*; that is, $\sum_g \mu(g)g\nu = \nu$. Furstenberg called the space (Π, Σ, ν) the *Poisson boundary*, since it is the space on which we may specify boundary values (a bounded function) to obtain some bounded harmonic function on the group. Specifically, the so-called *Furstenberg transform* is the map $L^\infty(\Pi, \sigma, \nu) \ni f \mapsto h_f \in$ BHF(G, μ) given by

$$h_f(g) = \int_\Pi f(g\xi)d\nu(\xi).$$

This defines a μ-harmonic function because ν is μ-stationary.

Blackwell (1955) first noted the *Poisson formula* (Exercise 6.17 and Proposition 6.2.2), relating the invariant σ-algebra to bounded harmonic functions. Avez (1972, 1976) proved that 0 random walk entropy implies the Liouville property (formally for finitely supported random walks). Vershik and Kaimanovich (1979) and Kaimanovich and Vershik (1983) proved the equivalence of 0 entropy and triviality of the tail σ-algebra, Proposition 6.3.4, and the proof presented is basically the one from Kaimanovich and Vershik (1983). The equivalence of triviality of the tail and invariant σ-algebras, Proposition 6.4.2, was proven independently by Derriennic *et al.* (1980), Vershik and Kaimanovich (1979), and Kaimanovich and Vershik (1983), both providing proofs of Theorem 6.4.3.

Theorem 6.8.1, stating that finite entropy symmetric adapted random walks on non-amenable groups are always non-Liouville, was proved by Furstenberg (1973). In fact, it is shown in Furstenberg (1973) that any adapted random walk on a non-amenable group is non-Liouville. The characterization of such "strongly non-Liouville" groups as non-amenable groups was shown by Rosenblatt (1981) and Kaimanovich and Vershik (1983). They prove that any amenable group admits some Liouville random walk. Chapter 7 deals with the polar case: for which groups are all adapted random walks Liouville? This is known as the *Choquet–Deny property*.

Versions of the inequality from Theorem 6.5.5 appear independently in Benjamini *et al.* (2015, 2017), Erschler and Karlsson (2010), and Ozawa (2018). We provided the exposition from Benjamini *et al.* (2017).

Regarding the Liouville property, we have already mentioned Conjecture 2.7.2, which may be restated as follows:

Conjecture 6.12.1 (Restatement of Conjecture 2.7.2) Let G be a finitely generated group. Then, for any two $\mu, \nu \in SA(G, 2)$, we have that (G, μ) is Liouville if and only if (G, ν) is Liouville.

That is, the conjecture states that the Liouville property is invariant to symmetric random walks with finite second moment. For finitely supported, symmetric and adapted μ, ν the above has been conjectured by Kaimanovich and Vershik (1983). Note that this conjecture is "tight" in the sense that the conditions of symmetry and second moment are necessary, by Exercises 6.64 and 6.65.

The examples on S_∞^* in Section 6.10 are from Kaimanovich and Vershik (1983).

6.13 Solutions to Exercises

Solution to Exercise 6.1 :(
We have that G^N is obviously invariant. Since $\theta^{-1}(A^c) = \theta^{-1}(A)^c$ and $\theta^{-1}(\cup_n A_n) = \cup_n \theta^{-1}(A_n)$, we have that the invariant events form a σ-algebra. :) ✓

Solution to Exercise 6.3 :(
In Section 1.2 we saw that $\sigma(X_t, X_{t+1}, \ldots) = \theta^{-t} \mathcal{F}$.
  We have that $\sigma(X_t, X_{t+1}, \ldots)$ is generated by the sets $\{X_{t+j}^{-1}(g) : j \geq 0, \ g \in G\}$. For any $j \geq 0, g \in G$, we have that $X_{t+j}^{-1}(g)$ respects $\sim_t$. Indeed, if $\omega \sim_t \omega'$ and $\omega \in X_{t+j}^{-1}(g)$, then $\omega_{t+j} = g$, implying that $\omega'_{t+j} = \omega_{t+j} = g$ (because $\omega \sim_t \omega'$), which in turn implies that $\omega' \in X_{t+j}^{-1}(g)$ as well. This shows that any event in $\sigma(X_t, X_{t+1}, \ldots)$ respects $\sim_t$.
  For the other direction, assume that A respects $\sim_t$. Note that

$$\theta^{-t}\left(\theta^t(A)\right) = \left\{\omega : \theta^t(\omega) \in \theta^t(A)\right\}$$
$$= \left\{\omega : \exists \omega' \in A, \ \theta^t(\omega) = \theta^t(\omega')\right\} = A.$$

Since $\theta^t(A) \in \mathcal{F}$, we get that $A = \theta^{-t}\left(\theta^t(A)\right) \in \theta^{-t}\mathcal{F} = \sigma(X_t, X_{t+1}, \ldots)$. :) ✓

Solution to Exercise 6.4 :(
Define a relation $\omega \approx_t \omega'$ if and only if $\theta^t(\omega) = \theta^t(\omega')$.
  If A respects $\approx$, then for any t, A respects $\approx_t$. So if A respects $\approx$ then $A \in \sigma(X_t, X_{t+1}, \ldots)$ for all t, implying that $A \in \mathcal{T}$.
  Conversely, if $A \in \mathcal{T}$, then for any t, it must be that A respects $\approx_t$. Now, if $\omega \approx \omega'$, and $\omega \in A$, then for some t we have $\omega \approx_t \omega'$, which implies that $\omega' \in A$ as well. This shows that any tail event A also respects $\approx$.
  Now, assume that A is an invariant event. Note that if $\omega \in A = \theta^{-1}(A)$ then $\theta(\omega) \in A$. Iterating this, if $\omega \sim \omega'$ and $\omega \in A$, then taking k, n so that $\theta^k(\omega) = \theta^n(\omega')$, we have that

$$\omega' \in \theta^{-n}(\theta^n(\omega')) = \theta^{-n}\left(\theta^k(\omega)\right) \subset \theta^{-n}(A) = A.$$

So any invariant event A respects $\sim$.
  Finally, assume that A respects $\sim$. Since $\omega \sim \theta(\omega)$ by definition, it must be that $\omega \in A$ if and only if $\theta(\omega) \in A$. This shows that $A = \theta^{-1}(A)$. :) ✓

Solution to Exercise 6.6 :(
That $A \in \mathcal{T}$ implies that A respects $\approx$.

If $\omega \approx \omega'$ and $\omega \in \theta^{-1}(A)$, then $\theta(\omega) \in A$ and $\theta(\omega) \approx \theta(\omega')$, so $\theta(\omega') \in A$, implying that $\omega' \in \theta^{-1}(A)$ as well.

Hence $\theta^{-1}(A)$ respects $\approx$, so $\theta^{-1}(A) \in \mathcal{T}$. :) ✓

Solution to Exercise 6.7 :(

Let $B = A \backslash \theta^{-1}(A)$. Note that $\theta(B) \subset A^c$, so $\theta(B) \cap B = \emptyset$. Similarly, if $B' = \theta^{-1}(A) \backslash A$, then $\theta(B') \cap B' = \emptyset$.

If $B = B' = \emptyset$, then $A = \theta^{-1}(A)$, so $A \in \mathcal{I}$. :) ✓

Solution to Exercise 6.8 :(

Just note that for any Borel subset $B \subset \mathbb{C}$, we have $(Y \circ \theta)^{-1}(B) = \theta^{-1} Y^{-1}(B)$. :) ✓

Solution to Exercise 6.9 :(

Assume Y is measurable with respect to $\sigma(X_n, X_{n+1}, \ldots) = \theta^{-n} \mathcal{F}$. For any Borel subset $B \subset \mathbb{C}$, there exists $A = A_B \in \mathcal{F}$ such that $Y^{-1}(B) = \theta^{-n}(A)$. If $B = \{c\}$ then this implies that $Y(\omega) = c$ if and only if $\theta^n(\omega) \in A_{\{c\}}$. So if $\theta^n(\omega) = \theta^n(\omega')$ then $Y(\omega) = Y(\omega')$.

Now assume that for any ω, ω' such that $\theta^n(\omega) = \theta^n(\omega')$, we have $Y(\omega) = Y(\omega')$. By this assumption, $Y(\omega) \in B$ if and only if $\theta^{-n}(\theta^n(\omega)) \subset Y^{-1}(B)$. For $\eta \in \theta^{-n}(\omega)$, define $Z(\omega) = Y(\eta)$, which is well defined by the assumption. Here, Z is a random variable, since for any Borel subset $B \subset \mathbb{C}$ we have

$$Z^{-1}(B) = \left\{ \omega : \exists \, \eta \in Y^{-1}(B), \ \omega = \theta^n(\eta) \right\} = \theta^n \left(Y^{-1}(B) \right) \in \mathcal{F}.$$

Note that we have that $Y = Z \circ \theta^n$ for this random variable Z.

Now, if $Y = Z \circ \theta^n$ for some random variable Z, then for any Borel subset $B \subset \mathbb{C}$, we get $Y^{-1}(B) = \theta^{-n} Z^{-1}(B) \in \theta^{-n} \mathcal{F}$, implying that Y is measurable with respect to $\theta^{-n} \mathcal{F} = \sigma(X_n, X_{n+1}, \ldots)$. :) ✓

Solution to Exercise 6.14 :(

We have that $(h(X_t))_t$ is a bounded martingale, so converges to an integrable random variable a.s. by the martingale convergence theorem (Theorem 2.6.3). So we just define $L = \limsup_{t \to \infty} h(X_t)$.

To see that L is $\mathcal{I}$-measurable, note that

$$L \circ \theta(\omega) = \limsup_{t \to \infty} h(\omega_{t+1}) = L(\omega).$$

So L is $\mathcal{I}$-measurable by Exercise 6.8. :) ✓

Solution to Exercise 6.15 :(

As before, the limit exists because $(h(X_t, t))_t$ is a bounded martingale, and we can take $L = \limsup_{t \to \infty} h(X_t, t)$ (by the martingale convergence theorem, Theorem 2.6.3).

To see that L is $\mathcal{T}$-measurable, let $n \in \mathbb{N}$ and assume that $\theta^n(\omega) = \theta^n(\omega')$. Then $\omega_t = \omega'_t$ for all $t \geq n$, so

$$L(\omega) = \lim_{t \to \infty} h(\omega_t, t) = \lim_{t \to \infty} h(\omega'_t, t) = L(\omega').$$

Therefore L is $\sigma(X_n, X_{n+1}, \ldots)$-measurable by Exercise 6.9, and since this holds for any n, we have that L is $\mathcal{T}$-measurable. :) ✓

Solution to Exercise 6.16 :(

Define $Y_n := X_{t+n}$ for all $n \geq 0$. The Markov property tells us that conditioned on $X_t = y$, we have that $(Y_n)_n$ has the distribution of a μ-random walk started at $Y_0 = y$. Also, $\theta^t(X_0, X_1, \ldots) = (Y_0, Y_1, \ldots)$ by definition. Thus,

$$\mathbb{E}_x \left[L \circ \theta^t \mid X_t = y \right] = \mathbb{E}_y[L]. \qquad :) ✓$$

Solution to Exercise 6.17 :(

For a bounded $\mathcal{I}$-measurable L, we know that $L = L \circ \theta$. Note that by the Markov property, $\mathbb{P}_x$-a.s.,

$$\mathbb{E}_{X_t}[L] = \mathbb{E}_x \left[L \circ \theta^t \mid \mathcal{F}_t \right].$$

So

$$\mathbb{E}_x[h_L(X_1)] = \mathbb{E}_x \, \mathbb{E}_{X_1}[L] = \mathbb{E}_x[L \circ \theta] = \mathbb{E}_x[L] = h_L(x).$$

Thus $h_L \in \mathrm{BHF}(G, \mu)$.

The inverse mapping is well defined because if $h \in \mathrm{BHF}(G, \mu)$ then $(h(X_t))_t$ is a bounded martingale and thus converges a.s. In Exercise 6.14 we saw that $L_h = \limsup_{t \to \infty} h(X_t)$ is $\mathcal{I}$-measurable, so that $\lim_{t \to \infty} h(X_t) = L_h$ a.s.

To show this is indeed the inverse mapping, we show that $\mathbb{E}_x[L_h] = h(x)$. Indeed, since $(h(X_t))_t$ converges a.s.,

$$\mathbb{E}_x[L_h] = \mathbb{E}_x[\limsup_{t \to \infty} h(X_t)] = \mathbb{E}_x[\lim_{t \to \infty} h(X_t)] = \lim_{t \to \infty} \mathbb{E}_x[h(X_t)] = h(x).$$

The exchange of limit and expectation is justified by dominated convergence since $(h(X_t))_t$ is uniformly bounded.

Also, for a bounded $\mathcal{I}$-measurable L, we have that $\mathbb{P}_x$-a.s.,

$$\limsup_{t \to \infty} h_L(X_t) = \limsup_{t \to \infty} \mathbb{E}_{X_t}[L]$$

$$= \limsup_{t \to \infty} \mathbb{E}_x \left[L \circ \theta^t \mid \mathcal{F}_t \right]$$

$$= \limsup_{t \to \infty} \mathbb{E}_x[L \mid \mathcal{F}_t] = \mathbb{E}_x[L \mid \mathcal{F}] = L.$$

We have used that $\sigma(\bigcup_t \mathcal{F}_t) = \mathcal{F}$.   :) ✓

Solution to Exercise 6.18 :(

If every bounded $\mathcal{I}$-measurable random variable is constant $\mathbb{P}_x$-a.s. for all $x \in G$, then applying this to indicators completes one direction.

For the other direction, assume $\mathcal{I}$ is trivial. If $L \geq 0$ is a bounded $\mathcal{I}$-measurable, then for any x, y such that $\mathbb{P}_x[X_1 = y] > 0$, we have

$$\mathbb{E}_x[L] = \mathbb{E}_x[L \circ \theta] \geq \mathbb{P}_x[X_1 = y] \cdot \mathbb{E}_y[L].$$

Thus, $\mathbb{E}_x L = 0$ implies $\mathbb{E}_y L = 0$. Since this holds for any pair x, y with $P(x, y) = \mu\left(x^{-1}y\right) > 0$, and since μ is adapted, this implies that $\mathbb{E}_x[L]$ is either always 0 or always nonzero, for all $x \in G$.

Now, if $L = \mathbf{1}_A$ for $A \in \mathcal{I}$, this argument implies that $\mathbb{P}_x[A] = \mathbb{P}_y[A] \in \{0, 1\}$ for all $x, y \in G$ (as we assumed that $\mathcal{I}$ is trivial).

Thus, this holds for finite linear combinations of indicators as well.

If Y is a nonnegative $\mathcal{I}$-measurable random variable, then we can approximate Y by $Y_n \nearrow Y$, which converges $\mathbb{P}_x$-a.s. for all x, and Y_n is a finite linear combination of indicators (e.g. take $Y_n = 2^{-n} \lfloor 2^n Y \rfloor \wedge n$). Thus, $\mathbb{P}_x[Y = c] = 1$ for some $c \geq 0$ and all $x \in G$. If Y is a general (complex) $\mathcal{I}$-measurable random variable, we may write $Y = (Y_1 - Y_2) + i(Y_3 - Y_4)$ where Y_j are all $\mathcal{I}$-measurable and nonnegative. Thus, $\mathbb{P}_x[Y_j = c_j] = 1$ for all $x \in G$ and some constants $c_j \geq 0$, which implies that $\mathbb{P}_x[Y = (c_1 - c_2) + i(c_3 - c_4)] = 1$ for all $x \in G$.   :) ✓

Solution to Exercise 6.20 :(

A tail event is measurable with respect to $\sigma(X_t, X_{t+1}, X_{t+2}, \dots)$ for all t. Since A is independent of $\mathcal{F}_t$ we have that $\mathbb{E}_x[\mathbf{1}_A] = \mathbb{E}_x[\mathbf{1}_A \mid \mathcal{F}_t]$. Now, $M_t = \mathbb{E}_x[\mathbf{1}_A \mid \mathcal{F}_t]$ is a bounded martingale. Thus, it converges a.s. and in L^1 to the integrable random variable $M_\infty = \limsup_{t \to \infty} M_t$. Now, since $M_\infty = \limsup_{t \to \infty} M_t$, and M_t is measurable with respect to $\mathcal{F}_t$ for all t, we have that M_∞ is measurable with respect to $\mathcal{T}$. Also, for any $B \in \mathcal{T}$ we have $M_t \mathbf{1}_B \to M_\infty \mathbf{1}_B$ in L^1, so

$$\mathbb{E}_x[\mathbf{1}_A \mathbf{1}_B] = \mathbb{E}_x[\mathbb{E}_x[\mathbf{1}_A \mid \mathcal{F}_t] \mathbf{1}_B] \to \mathbb{E}_x[M_\infty \mathbf{1}_B].$$

Thus, $M_\infty = \mathbb{E}_x[\mathbf{1}_A \mid \mathcal{T}]$.

Because $A \in \mathcal{T}$, we have that $\mathbb{E}_x[\mathbf{1}_A] = \mathbb{E}_x[\mathbf{1}_A \mid \mathcal{T}] = \mathbf{1}_A \in \{0, 1\}$ a.s. So $\mathbf{1}_A$ is a.s. constant in $\{0, 1\}$.   :) ✓

Solution to Exercise 6.21 :(

If $h : G \to \mathbb{C}$ is a bounded harmonic function, then $(h(X_t))_t$ is a bounded martingale. Since (G, μ) is recurrent, $T_1 = \inf\{t \geq 1 : X_t = 1\}$ is a.s. finite. Thus, the optional stopping theorem (Theorem 2.3.3) tells us that $h(x) = \mathbb{E}_x\left[h(X_{T_1})\right] = h(1)$, for any $x \in G$, implying that h is constant.   :) ✓

Solution to Exercise 6.22 :(
Using that the random variables are discrete, since

$$\mathbb{E}\left[1_{\{X=x\}} | \sigma(Y)\right] = \sum_y 1_{\{Y=y\}} \mathbb{P}[X=x, Y=y] / \mathbb{P}[Y=y],$$

we get that

$$H(X|Y) = -\sum_{x,y} \mathbb{P}[X=x, Y=y] \log \mathbb{P}[X=x|Y=y].$$

So

$$H(X,Y) = -\sum_{x,y} \mathbb{P}[X=x, Y=y] \log \mathbb{P}[X=x, Y=y]$$

$$= -\sum_{x,y} \mathbb{P}[X=x, Y=y] \log \mathbb{P}[X=x|Y=y] - \sum_{x,y} \mathbb{P}[X=x, Y=y] \log \mathbb{P}[Y=y]$$

$$= H(X|Y) + H(Y). \qquad \qquad :)\checkmark$$

Solution to Exercise 6.23 :(
Compute:

$$H(X) - H(X|Y) = H(X) + H(Y) - H(X,Y) = H(Y) - H(Y|X). \qquad :)\checkmark$$

Solution to Exercise 6.24 :(
Let $R_Y = \{y \in G : \mathbb{P}[Y=y] > 0\}$. In the solution to Exercise 6.22 we have seen that

$$H(X \mid Y) = -\sum_{y \in R_Y} \sum_x \mathbb{P}[X=x, Y=y] \log \mathbb{P}[X=x|Y=y].$$

So using this computation, and the assumption $\mathbb{P}[X \in \varphi(Y)] = 1$, we find using Jensen's inequality that

$$H(X \mid Y) = \sum_{y \in R_Y} \mathbb{P}[Y=y] \sum_{x \in \varphi(y)} \mathbb{P}[X=x|Y=y] \log \frac{1}{\mathbb{P}[X=x|Y=y]}$$

$$\leq \sum_{y \in R_Y} \mathbb{P}[Y=y] \cdot \log |\varphi(y)| = \mathbb{E}[\log |\varphi(Y)|]. \qquad :)\checkmark$$

Solution to Exercise 6.25 :(
The function $\phi(x) = -x \log x$ is a concave function on $(0, 1)$. Thus, Jensen's inequality tells us that

$$H(X|\mathcal{G}) = \sum_x \mathbb{E}\left[\phi\left(\mathbb{E}\left[1_{\{X=x\}} | \mathcal{G}\right]\right)\right] = \sum_x \mathbb{E}\,\mathbb{E}\left[\phi\left(\mathbb{E}\left[1_{\{X=x\}} | \mathcal{G}\right]\right) | \sigma\right]$$

$$\leq \sum_x \mathbb{E}\,\phi\left(\mathbb{E}\left[1_{\{X=x\}} | \sigma\right]\right) = H(X|\sigma). \qquad :)\checkmark$$

Solution to Exercise 6.26 :(
We use the "equality version" of Jensen's inequality. The function $\phi(x) = -x \log x$ is strictly concave on $(0, 1)$. Since $H(X|\sigma) \leq H(X)$, we have that equality holds if and only if $\mathbb{E}\left[\phi(\mathbb{E}[1_{\{X=x\}} | \sigma])\right] = \phi(\mathbb{P}[X=x])$ for every x. Thus, with $Z = \mathbb{E}[1_{\{X=x\}} | \sigma]$, we have that this holds if and only if $\mathbb{E}\left[1_{\{X=x\}} | \sigma\right] = \mathbb{P}[X=x]$ for every x a.s., which is if and only if X is independent of σ. $\qquad :)\checkmark$

Solution to Exercise 6.27 :(
Since $\mathbb{P}[A] \in \{0, 1\}$ for all $A \in \mathcal{G}$, any X is independent of $\mathcal{G}$, so $H(X \mid \mathcal{G}) = H(X)$. $\qquad :)\checkmark$

Solution to Exercise 6.28 :(
This is equivalent to $H(X|Y) \leq H(X)$, which follows from the fact that $\{\emptyset, \Omega\} \subset \sigma(Y)$, so that $H(X|Y) \leq H(X|\{\emptyset, \Omega\}) = H(X)$. $\qquad :)\checkmark$

Solution to Exercise 6.29 :(
Just note that $H(X, Y) - H(X) = H(Y|X) \geq 0$. :) ✓

Solution to Exercise 6.30 :(
Let $p_n = \mathbb{P}[X = n]$. Define $A = \left\{ n \in \mathbb{N} : p_n > (n+1)^{-2} \right\}$.

The function $\varphi(\xi) = -\xi^{1/4} \log \xi$ is maximized on the interval $[0, 1]$ at the point $\xi = e^{-4}$, obtaining the value $\varphi\left(e^{-4}\right) = \frac{4}{e}$. Thus, for all $n \in \mathbb{N}$,

$$-p_n \log p_n = (p_n)^{3/4} \cdot \varphi(p_n) \leq \frac{4}{e} (p_n)^{3/4}.$$

We conclude that

$$H(X) = -\sum_{n=0}^{\infty} p_n \log p_n \leq -\sum_{n \in A} p_n \log p_n - \sum_{n \notin A} p_n \log p_n$$

$$\leq -\sum_{n \in A} p_n \log \left((n+1)^{-2}\right) + \frac{4}{e} \sum_{n \notin A} (p_n)^{3/4}$$

$$\leq 2 \sum_{n=0}^{\infty} p_n \log(n+1) + \frac{4}{e} \sum_{n=0}^{\infty} (n+1)^{-3/2},$$

as required. :) ✓

Solution to Exercise 6.31 :(
For integers $n > 1$, let $\mathbb{P}\left[X = \frac{1}{n}\right] = p_n$ where $p_n = \frac{C}{n(\log n)^2}$, with $C > 0$ chosen so that $\sum_{n=2}^{\infty} p_n = 1$.
Then,

$$H(X) = -\sum_{n=2}^{\infty} p_n \log p_n = C \cdot \sum_{n=2}^{\infty} \frac{\log n + 2 \log \log n - \log C}{n(\log n)^2},$$

which is infinite.

We have used the fact that $\sum_{n=2}^{\infty} \frac{1}{n(\log n)^2} < \infty$ but $\sum_{n=2}^{\infty} \frac{1}{n \log n} = \infty$. :) ✓

Solution to Exercise 6.32 :(
By Proposition 6.3.2, $H(X_{t+s} \mid X_t) = H(X_s)$. This implies that

$$H(X_{t+s}) \leq H(X_{t+s}, X_t) = H(X_{t+s} \mid X_t) + H(X_t) \leq H(X_s) + H(X_t),$$

so that the sequence $(H(X_t))_t$ is subadditive. Hence, the limit

$$h(G, \mu) = \inf_t \frac{H(X_t)}{t} = \lim_{t \to \infty} \frac{H(X_t)}{t}$$

exists by Fekete's lemma.

Finally, iterating subadditivity, $H(X_t) \leq tH(X_1)$ showing that $h(G, \mu) \leq H(X_1)$. :) ✓

Solution to Exercise 6.33 :(
Since G has at most exponential growth, there exists $D > 0$ such that for any $r > 0$ we have $\log |B(1, r)| \leq Dr$. (Here $B(1, r)$ is the ball of radius r in some fixed Cayley graph of G.)

Let U have law μ. Since $U \in B(1, |U|)$, by Exercise 6.24,

$$H(U) \leq H(|U|) + H(U \mid |U|) \leq H(|U|) + \mathbb{E} \log |B(1, |U|)| \leq H(|U|) + D \mathbb{E} |U|,$$

so it suffices to prove that $H(|U|) < \infty$.

We have that $|U|$ is a random variable with nonnegative integer values such that $\mathbb{E}[|U|] < \infty$ (because μ has finite 1st moment). So $H(|U|) < \infty$ by Exercise 6.30. :) ✓

Solution to Exercise 6.34 :(
Let $G = \mathbb{F}_d$ be the free group on d generators. Consider the Cayley graph with respect to the standard generators. Let $S_r = \{x \in G : |x| = r\}$. So $|S_r| \geq e^{cr}$ for some constant $c = c(d) > 0$ and all $r \geq 0$.

Define a probability measure μ on G by

$$\mu(x) = \alpha \cdot |x|^{-2} \cdot \left|S_{|x|}\right|^{-1} \mathbf{1}_{\{|x|>0\}},$$

for $\alpha^{-1} = \sum_{r=1}^{\infty} r^{-2}$. Let U be a random element of G with law μ.

For any $\varepsilon > 0$ we have that

$$\mathbb{E}\left[|U|^{1-\varepsilon}\right] = \sum_{r=1}^{\infty} r^{1-\varepsilon} \cdot \alpha r^{-2} \sum_{|x|=r} |S_r|^{-1} = \alpha \cdot \sum_{r=1}^{\infty} r^{-1-\varepsilon} < \infty.$$

So $\mu \in SA(G, 1 - \varepsilon)$. However,

$$H(\mu) = \sum_{r=1}^{\infty} \sum_{|x|=r} \alpha r^{-2} |S_r|^{-1} \log\left(\alpha^{-1} r^2 |S_r|\right) \geq \alpha \sum_{r=1}^{\infty} r^{-2} \log\left(\alpha r^2 e^{cr}\right) \geq c\alpha \sum_{r=1}^{\infty} r^{-1} = \infty. \qquad :) \checkmark$$

Solution to Exercise 6.35 :(

Let $p(k) = \binom{n}{k}p^k(1-p)^{n-k}$ and $q(k) = \binom{n+1}{k}p^k(1-p)^{n+1-k}$ be the laws of B, B', respectively. It suffices to prove that $||p - q||_{TV} \leq cn^{-1/2}$ (see Appendix C), where $||p - q||_{TV}$ is the total variation distance; that is,

$$2||p - q||_{TV} = \sum_{k=0}^{n} \left|\binom{n}{k}p^k(1-p)^{n-k} - \binom{n+1}{k}p^k(1-p)^{n+1-k}\right| + p^{n+1}$$

$$= \sum_{k=0}^{n+1} \binom{n+1}{k}p^k(1-p)^{n-k} \cdot \left|\tfrac{n+1-k}{n+1} - (1-p)\right|$$

$$= \tfrac{1}{n+1} \sum_{k=0}^{n+1} \binom{n+1}{k}p^k(1-p)^{n-k} \cdot \left|k - (n+1)p\right|$$

$$\leq \frac{\sqrt{\mathrm{Var}[\mathrm{Bin}(n+1, p)]}}{n+1} = \frac{\sqrt{(n+1)p(1-p)}}{n+1} \leq \frac{1}{2\sqrt{n+1}},$$

using $p(1-p) \leq \tfrac{1}{4}$. $\qquad :) \checkmark$

Solution to Exercise 6.36 :(

We prove this by induction on t. If $t = 1$ then

$$\mathbb{P}\left[X_{B_1} = x\right] = p\,\mathbb{P}[X_0 = x] + (1-p)\,\mathbb{P}[X_1 = x] = p\delta_1(x) + (1-p)\nu(x) = \mu(x).$$

Now, for $t \geq 1$, let $B_t \sim \mathrm{Bin}(t, 1-p)$ and let $M \sim \mathrm{Ber}(1-p)$ be independent of B_t. So $B_t + M \sim \mathrm{Bin}(t+1, 1-p)$. By induction we have that

$$\mathbb{P}\left[X_{B_t+M} = x\right] = p\,\mathbb{P}\left[X_{B_t} = x\right] + (1-p)\,\mathbb{P}\left[X_{B_t+1} = x\right]$$

$$= p\mu^t(x) + (1-p)\sum_y \mathbb{P}\left[X_{B_t} = y\right]\nu\left(y^{-1}x\right)$$

$$= \sum_y \mu^t(y)\left(p\delta_1(y^{-1}x) + (1-p)\nu(y^{-1}x)\right) = \left(\mu^t * \mu\right)(x) = \mu^{t+1}(x),$$

which completes the induction. $\qquad :) \checkmark$

Solution to Exercise 6.38 :(

We have that $\left(\frac{H(X_{tn})}{tn}\right)_t$ is a subsequence of $\left(\frac{H(X_t)}{t}\right)_t$, so

$$h(G, \mu^n) = \lim_{t\to\infty} \frac{H(X_{tn})}{t} = n \cdot \lim_{t\to\infty} \frac{H(X_t)}{t} = n \cdot h(G, \mu). \qquad :) \checkmark$$

Solution to Exercise 6.40 :(

Just a straightforward computation:

$$\sum_y \mathbb{P}[Y = y] \cdot D((X|Y = y)||X) = \sum_y \mathbb{P}[Y = y] \sum_x \mathbb{P}[X = x \mid Y = y] \log \frac{\mathbb{P}[X = x, Y = y]}{\mathbb{P}[Y = y]\mathbb{P}[X = x]}$$

$$= \sum_{x,y} \mathbb{P}[X = x, Y = y] \log \mathbb{P}[X = x, Y = y]$$

$$- \sum_{x,y} \mathbb{P}[X = x, Y = y] \log \mathbb{P}[Y = y] - \sum_{x,y} \mathbb{P}[X = x, Y = y] \log \mathbb{P}[X = x]$$

$$= -H(X, Y) + H(Y) + H(X) = I(X, Y). \qquad :) \checkmark$$

Solution to Exercise 6.41 :(

Let X have law μ and Y have law ν. Fix an event A. Let $f = \mathbf{1}_A$. So $\mathbb{E}[f(X)] = \mathbb{P}[X \in A]$ and $\mathbb{E}[f(Y)] = \mathbb{P}[Y \in A]$. Also, $\mathbb{E}\left[f(X)^2\right] + \mathbb{E}\left[f(Y)^2\right] \le 2$. By Proposition 6.5.3,

$$|\mathbb{P}[X \in A] - \mathbb{P}[Y \in A]| \le \sqrt{2D(X||Y) \cdot 2} = 2 \cdot \sqrt{D(X||Y)}.$$

Taking the supremum over events A we obtain the inequality. $\qquad :) \checkmark$

Solution to Exercise 6.42 :(

Let $x \in G$ and let $n = n(x)$. So $\mu^n(x) > 0$. Fix $t > 0$. Recall that $\mu^t(z) = \mathbb{P}[X_t = z]$. So

$$\mathbb{P}[X_{t+n} = z \mid X_n = y] = \mathbb{P}[yX_t = z] = \mu^t\left(y^{-1}z\right) = y.\mu^t(z).$$

Using Pinsker's inequality (Exercise 6.41), Jensen's inequality, Proposition 6.3.2, and Exercise 6.40, we find that

$$\mu^n(x) \cdot K_t(x) = \mu^n(x) \cdot ||x.\mu^t - \mu^{t+n}||_{\mathrm{TV}} \le \sum_y \mu^n(y)||y.\mu^t - \mu^{t+n}||_{\mathrm{TV}}$$

$$\le 2 \sum_y \mu^n(y)\sqrt{D(y.\mu^t \, || \, \mu^{t+n})} \le 2\left(\sum_y \mu^n(y)D((X_{t+n}|X_n = y) \, || \, X_{t+n})\right)^{1/2}$$

$$= 2\sqrt{I(X_{t+n}, X_n)} = 2\sqrt{H(X_{t+n}) - H(X_{t+n} \mid X_n)}$$

$$= 2\left(\sum_{k=0}^{n-1} H(X_{t+k+1}) - H(X_{t+k})\right)^{1/2}.$$

We have seen in the proof of Proposition 6.3.4 that $(H(X_{t+1}) - H(X_t))_t$ is a nonincreasing sequence converging to $h(G, \mu)$. Thus, for any $x \in G$ we have that

$$\limsup_{t \to \infty} K_t(x) \le \frac{2}{\mu^{n(x)}(x)} \cdot \sqrt{n(x)h(G, \mu)}.$$

Hence, if (G, μ) is Liouville then $K_t(x) \to 0$ for any $x \in G$.

For the other direction, assume that $K_t(x) \to 0$ as $t \to \infty$ for all $x \in G$. Let $f \in \mathrm{BHF}(G, \mu)$. For $t > 0$, let (X_t, Y_{t+1}) be a coupling of two μ-random walks at times t and $t + 1$ such that $\mathbb{P}[xX_t \ne Y_{t+1}] = K_t(x)$. Then,

$$|f(x) - f(1)| = |\mathbb{E}[f(xX_t) - f(Y_{t+1})]| \le 2||f||_\infty \mathbb{P}[xX_t \ne Y_{t+1}] = 2||f||_\infty K_t(x).$$

Taking $t \to \infty$ shows that $f(x) = f(1)$, for all $x \in G$. So any $f \in \mathrm{BHF}(G, \mu)$ is constant. $\qquad :) \checkmark$

Solution to Exercise 6.43 :(

Let $G = \mathbb{Z}$ and μ uniform on $\{-1, 1\}$. Let $x = 1$. Since

$$\mathbb{P}[X_{2t} \in 2\mathbb{Z}] = \mathbb{P}[X_{2t+1} \in 2\mathbb{Z} + 1] = 1$$

for all t, and since $\mathbb{P}_1[X_t \in A] = \mathbb{P}[X_t \in A - 1]$ for all $A \subset \mathbb{Z}$, we have that

$$|| \mathbb{P}_x[X_{2t} = \cdot] - \mathbb{P}[X_{2t} = \cdot]||_{\mathrm{TV}} \ge |\mathbb{P}_1[X_{2t} \in 2\mathbb{Z}] - \mathbb{P}[X_{2t} \in 2\mathbb{Z}]| = 1,$$

$$|| \mathbb{P}_x[X_{2t+1} = \cdot] - \mathbb{P}[X_{2t+1} = \cdot]||_{\mathrm{TV}} \ge |\mathbb{P}_1[X_{2t+1} \in 2\mathbb{Z}] - \mathbb{P}[X_{2t+1} \in 2\mathbb{Z}]| = 1. \qquad :) \checkmark$$

Solution to Exercise 6.44   :(
Assume that $h\colon G \to \mathbb{R}$ is a harmonic function and $x \in G$ is such that

$$\liminf_{t\to\infty} \tfrac{1}{t}\, \mathrm{Var}_x[h(X_t)] \cdot H(X_t) = 0.$$

Note that since $(h(X_t))_t$ is a martingale, we have that

$$
\begin{aligned}
\mathrm{Var}_x[h(X_t)] &= \mathbb{E}_x\left[|h(X_t) - h(X_{t-1})|^2\right] + \mathbb{E}_x\left[|h(X_{t-1}) - h(x)|^2\right] \\
&\quad + 2\,\mathbb{E}_x\left[(h(X_{t-1}) - h(x))\,\mathbb{E}[(h(X_t) - h(X_{t-1}))\mid \mathcal{F}_{t-1}]\right] \\
&= \mathbb{E}_x\left[|h(X_t) - h(X_{t-1})|^2\right] + \mathrm{Var}_x[h(X_{t-1})],
\end{aligned}
$$

which implies that $(\mathrm{Var}_x[h(X_t)])_t$ is a nondecreasing sequence. Thus,

$$\limsup_{t\to\infty} \frac{1}{t}\, \mathrm{Var}_x[h(X_t)] \cdot H(X_t) = 0.$$

Now, let $y \in G$ and consider $m \geq 1$ large enough so that $\mathbb{P}_x[X_m = y] > 0$ (using that μ is adapted). Note that h is μ^m-harmonic as well. Also, $(Y_t = X_{mt})_t$ is a μ^m-random walk, with

$$\tfrac{1}{t}\, \mathrm{Var}_x[h(Y_t)] \cdot H(Y_t) = m \cdot \tfrac{1}{mt}\, \mathrm{Var}_x[h(X_{mt})] \cdot H(X_{mt}) \to 0.$$

Thus, by Theorem 6.5.5, we then have that $\mathbb{E}_x|h(Y_1) - h(x)| = 0$, implying that $h(Y_1) = h(x)$ $\mathbb{P}_x$-a.s. Since $\mathbb{P}_x[Y_1 = y] > 0$, this implies that $h(y) = h(x)$.
This holds for arbitrary $y \in G$, implying that h is constant.   :) ✓

Solution to Exercise 6.45   :(
If (G, μ) is not Liouville then there exists a bounded harmonic function $h\colon G \to \mathbb{C}$. It is immediate that $\mathrm{Re}\,h$ is also a bounded harmonic function, and must therefore have bounded variance.
For the other direction, let $h\colon G \to \mathbb{R}$ be a nonconstant harmonic function with $\sup_t \mathrm{Var}[h(X_t)] \leq M < \infty$. Since h is nonconstant and since μ is adapted, there exist $x \in G$ and $u \in \mathrm{supp}(\mu)$ such that $h(xu) \neq h(x)$. By replacing h with $x.h$ we may assume that $h(u) \neq h(1)$. So $\mathbb{E}|h(X_1) - h(1)| \geq \mu(u)|h(u) - h(1)| > 0$.
By Theorem 6.5.5, we have

$$(\mathbb{E}|h(X_1) - h(1)|)^2 \leq 4\mathbb{E}|h(X_t) - h(1)|^2 \cdot (H(X_t) - H(X_{t-1})) \leq 4M\tfrac{1}{t}H(X_t).$$

Taking $t \to \infty$ we have that $h(G, \mu) \geq \tfrac{C}{4M}$ where $C = (\mathbb{E}|h(X_1) - h(1)|)^2 > 0$, which is positive. This implies that (G, μ) is non-Liouville.   :) ✓

Solution to Exercise 6.46   :(
Let (X, Y) be a coupling of X_{E_r} started at x and at y such that $\mathbb{P}[X \neq Y] = D_r(x, y)$. Because μ is supported on S, we have that $|X| = |Y| = r + 1$. Also, $|X_t| \leq r$ for all $t < E_r$.
Let f be a μ-harmonic function. Note that $M_t = f(X_{E_r \wedge t})$ is a martingale, uniformly bounded by $\max_{|z|\leq r+1}|f(z)|$. Thus, the optional stopping theorem (Theorem 2.3.3) guarantees that

$$\mathbb{E}[f(X)] = \mathbb{E}_x[f(X_{E_r})] = f(x) \qquad \text{and} \qquad \mathbb{E}[f(Y)] = \mathbb{E}_y[f(X_{E_r})] = f(y).$$

So,

$$|f(x) - f(y)| \leq \mathbb{E}\left[|f(X) - f(Y)|\mathbf{1}_{\{X\neq Y\}}\right] \leq 2\max_{|z|=r+1}|f(z)| \cdot D_r(x, y). \qquad :) ✓$$

Solution to Exercise 6.47   :(
Note that for all integers $t, s > 0$,

$$\mathbb{E}|X_{t+s}| \leq \mathbb{E}|X_t| + \mathbb{E}\left|(X_t)^{-1}X_{t+s}\right| = \mathbb{E}|X_t| + \mathbb{E}|X_s|.$$

Thus, the limit exists by subadditivity (Fekete's lemma).   :) ✓

Solution to Exercise 6.48   :(
By Exercise 6.30,

$$H(|X_t|) \leq C\,\mathbb{E}\log(|X_t| + 1) + C \leq C'\,\mathbb{E}|X_t|$$

for some universal constants $C, C' > 0$.

Also, there exists a constant $C > 0$ such that for all $r \geq 0$ we have $|B(1, r)| \leq \exp(Cr)$, so that by Exercise 6.24,

$$H(X_t) \leq H(|X_t|) + H(X_t \mid |X_t|) \leq H(|X_t|) + \mathbb{E} \log |B(1, |X_t|)|$$
$$\leq C' \, \mathbb{E} \, |X_t| + C \, \mathbb{E} \, |X_t|,$$

so that

$$h(G, \mu) \leq (C' + C) \cdot \lim_{t \to \infty} \frac{\mathbb{E} \, |X_t|}{t} = 0. \qquad :) \checkmark$$

Solution to Exercise 6.50 :(

Let $\mu \in SA(G, 1)$. Since μ has finite first moment we know that $\mathbb{E} \, |X_t| \leq Ct$ for some constant $C > 0$ and all $t \geq 0$.

Let $\varepsilon > 0$, and let $R > 0$ be such that for all $r > R$ we have $\log |B(1, r)| \leq \varepsilon r$. As in Exercise 6.48, using Exercises 6.24 and 6.30, we have that for some constant $C' > 0$,

$$H(X_t \mid |X_t|) \leq \mathbb{E}[\log |B(1, |X_t|)|] \leq \varepsilon \, \mathbb{E} \, |X_t| + \log |B(1, R)|,$$
$$H(|X_t|) \leq C' \, \mathbb{E}[\log(|X_t| + 1)] + C' \leq C' \varepsilon \, \mathbb{E} \, |X_t| + C' \log(R + 1) + C',$$

which implies that

$$h(G, \mu) = \lim_{t \to \infty} \frac{H(X_t)}{t} \leq (C' + 1)C\varepsilon.$$

Since $\varepsilon > 0$ was arbitrary, this proves that $h(G, \mu) = 0$. $\qquad :) \checkmark$

Solution to Exercise 6.51 :(

We have that

$$(x, \sigma)^{(y, \tau)} = \left(y^{-1}, y^{-1}.\tau\right)(x, \sigma)(y, \tau) = \left(x^y, y^{-1}.(\tau + \sigma) + \left(y^{-1}x\right).\tau\right),$$

and since $y^{-1}.\tau + y^{-1}.\tau = 0$,

$$(1, \sigma)^{(y, \tau)} = \left(1, y^{-1}.\sigma\right).$$

Note that this implies that $\{1\} \times \Sigma(G) \subset L(G)$ is a normal subgroup.

Also, the map $(x, \sigma) \mapsto x$ is a homomorphism from $L(G)$ onto G, with kernel exactly $\{1\} \times \Sigma(G)$, so $L(G)/(\{1\} \times \Sigma(G)) \cong G$.

Note that for any $y \in \text{supp}(\sigma)$, we have that $(1, \delta_1)^{(y^{-1}, 0)} = (1, \delta_y)$. Thus, if $\text{supp}(\sigma) = \{y_1, \ldots, y_k\}$ then

$$(1, \sigma) = (1, \delta_1)^{(y_1^{-1}, 0)} \cdots (1, \delta_1)^{(y_k^{-1}, 0)}.$$

So we only have to show that $\{(s, 0) : s \in S\}$ generates any element of the form $(y, 0)$ for $y \in G$. Indeed, if $y \in G$ then $y = s_1 \cdots s_n$ for some $s_j \in S$, so

$$(y, 0) = (s_1, 0) \cdots (s_n, 0).$$

Using the above computation of conjugation in $L(G)$, we can compute

$$[(x, \sigma), (y, \tau)] = \left(x^{-1}, x^{-1}.\sigma\right)(x, \sigma)^{(y, \tau)} = \left([x, y], x^{-1}.(\sigma + y^{-1}.\sigma + y^{-1}.\tau + y^{-1}x.\tau)\right).$$

So $[L(G), L(G)] \subset [G, G] \times \Sigma(G)$ as sets. $\qquad :) \checkmark$

Solution to Exercise 6.52 :(

Transience follows from exponential growth by the Nash inequality, see Exercise 5.27.

For exponential growth, consider a finite symmetric generating set S for G, and for $L(G)$ use the generating set $S' = \{(s, 0), (1, \delta_1) : s \in S\}$. Let $(s_n)_n$ be a sequence of generators in S and denote $x_n = s_1 \cdots s_n$ for all n. Choose $(s_n)_n$ so that $|x_n| = n$ for all n, and set $x_0 = 1$. Then, consider

$$F_n = \{(x_n, \sigma) : \text{supp}(\sigma) \subset \{x_0, x_1, \ldots, x_n\}\}.$$

It is immediate that $|F_n| = 2^{n+1}$. It is also easy to show that for any $(x_n, \sigma) \in F_n$ one has $|(x_n, \sigma)| \le 2(n+1)$. Thus, $B_{S'}((1, 0), 2r) \ge 2^r$ for all r.   :)✓

Solution to Exercise 6.53 :(
If the random walk is recurrent then

$$\mathbb{P}_x[\forall t \ge 0, \ |X_t| > r] \le \mathbb{P}_x[\forall t \ge 0, \ X_t \ne 1] = \mathbb{P}_x[T_1 = \infty] = 0.$$

This proves (3) $\Rightarrow$ (1).
   (2) $\Rightarrow$ (3) is trivial.
   To prove (1) $\Rightarrow$ (2), let $Z_r = G \backslash B(1, r)$ be the complement of the ball of radius r. If the walk is transient then

$$0 < \mathbb{P}\left[T_1^+ = \infty\right] \le \mathbb{P}\left[T_{Z_r} < T_1^+\right] \cdot \sup_{x \in Z_r} \mathbb{P}_x\,[T_1 = \infty].$$

For every $r > 0$ let $z_r \in Z_r$ be such that

$$\mathbb{P}_{z_r}[T_1 = \infty] \ge \sup_{x \in Z_r} \mathbb{P}_x[T_1 = \infty]\left(1 - r^{-1}\right).$$

Then, as $r \to \infty$,

$$\mathbb{P}_{z_r}[T_1 = \infty] \ge \left(1 - r^{-1}\right) \frac{\mathbb{P}\left[T_1^+ = \infty\right]}{\mathbb{P}\left[T_{Z_r} < T_1^+\right]} \to 1.$$

Since for any $r > 0$ the set $B(1, r)$ is finite, and since μ is adapted, we have $\sup_{|x| \le r} \mathbb{P}_x[T_1 = \infty] < \infty$. Thus, it must be that $|z_r| \to \infty$ and we are done.   :)✓

Solution to Exercise 6.54 :(
Let $F \subset G$ be a finite set. Define $\tilde{F} \subset L(G)$ by

$$\tilde{F} = \{(x, \sigma) : x \in F, \ \text{supp}(\sigma) \subset F\}.$$

It is easy to compute that

$$|\tilde{F}| = 2^{|F|} \cdot |F|$$

$$|\partial \tilde{F}| = \#\{(x, \sigma) : \text{supp}(\sigma) \subset F, \ x \in \partial F\} = 2^{|F|} \cdot |\partial F|,$$

so that $\frac{|\partial \tilde{F}|}{|\tilde{F}|} = \frac{|\partial F|}{|F|}$. Thus, a Følner sequence in G gives rise to a Følner sequence in $L(G)$.   :)✓

Solution to Exercise 6.55 :(
Let $\sigma_1, \ldots, \sigma_n$ be elements in S_∞^* and let G be the subgroup generated by these elements. Let $M \subset \Omega$ be defined by

$$M = \bigcup_{j=1}^{n} \text{supp}(\sigma_j).$$

Note that for any j we have that $\sigma_j|_M$ is a bijection on M. Thus, for any element $g \in G$ also $g|_M$ is a bijection on M (because composing bijections is a bijection). Thus, the map $g \mapsto g|_M$ is a homomorphism from G to the group of permutations on M. Also, the kernel of this homomorphism is precisely those $g \in G$ such that $g|_M$ is the identity on M. However, since $\text{supp}(g) \subset M$ for any $g \in G$, it must be that the kernel is trivial. So G admits an injective homomorphism into a finite group (the group of permutations on the finite set M), implying that G is finite.
   Now, S_∞^* is not finite, because, for example, if we identify Ω with $\mathbb{N}$, and consider the transpositions $\pi_n = (n \ n + 1)$ then there are infinitely many different permutations with size 2 support.   :)✓

Solution to Exercise 6.57 :(
For $x = (\omega_1, \ldots, \omega_n) \in T$ denote $|x| = n$ and consider the process $Y_t = |X_t|$. Note that for any t we have a.s.

$$\mathbb{P}[Y_{t+1} = Y_t + 1 \mid X_t] \ge P(X_t, X_t + 1) + P(X_t, X_t - 1) \ge \alpha.$$

Thus, we may couple $(Y_t)_t$ with a process $(Z_t)_t$ on $\mathbb{N} \setminus \{0\}$ such that $Z_t \leq Y_t$ a.s. for all t, and $Z_0 = Y_0$ and such that $(Z_t)_t$ is a Markov chain with transition probabilities $P(z, z+1) = 1 - P(z, z-1) = \alpha$ for all $z > 1$.

Under such a coupling we have the inclusion of events

$$\{\forall\, t,\ Z_t \geq Z_0\} \subset \{\forall\, t,\ Y_t \geq Y_0\} \subset \{\forall\, t,\ X_t \neq \mathring{X}_0\}.$$

Thus we only need to prove a lower bound on $\mathbb{P}[\forall\, t,\ Z_t \geq Z_0 \mid Z_0 = n > 1]$.

Denote $\varphi(n) = \mathbb{P}[\forall\, t,\ Z_t \geq Z_0 \mid Z_0 = n]$. Since $(Z_t)_t$ moves only by ± 1 values, we have that $\varphi(n) = \varphi(n+1) = \varphi$ for all $n > 1$. We have:

$$\varphi(n) = \mathbb{P}[Z_1 = n+1 \mid Z_0 = n] \cdot \mathbb{P}[\forall\, t,\ Z_t \geq n-1 \mid Z_0 = n+1]$$

$$= \alpha \cdot \mathbb{P}[\forall\, t,\ Z_t \geq n+1 \mid Z_0 = n+1] + \alpha \cdot \mathbb{P}[\exists\, t,\ Z_t = n \mid Z_0 = n+1] \cdot \mathbb{P}[\forall\, t,\ Z_t \geq n \mid Z_0 = n]$$

$$= \alpha \cdot \varphi + \alpha \cdot (1 - \varphi) \cdot \varphi.$$

We thus arrive at the equation $1 = \alpha(2 - \varphi)$, which gives us for any $|x| > 1$,

$$\mathbb{P}_x[\forall\, t,\ X_t \neq \mathring{x}] \geq \varphi = 2 - \tfrac{1}{\alpha}. \qquad\qquad\qquad\qquad \text{:)}\ \checkmark$$

Solution to Exercise 6.58 :(

Set

$$h(\sigma) = \mathbb{P}_\sigma[\limsup_t Y_t = 1],$$

where X_t, Y_t are defined in Example 6.10.3.

Exercise 6.57 shows that

$$\mathbb{P}[\forall\, t,\ X_t \geq X_0 \mid X_0 > 1] \geq \varepsilon := 1 - \tfrac{A}{2} > 0.$$

Let $T_n = \inf\{t : X_t = n\}$. Since $X_{t+1} - X_t \in \{-1, 0, 1\}$ always, conditioned on $X_0 \geq n$ we have that $T_n < T_{n-1} < T_{n-2} < \cdots < T_2 < T_1$. Since $\mathbb{P}[T_{n-1} = \infty \mid X_0 = n] \geq \varepsilon$, we have by repeated use of the strong Markov property,

$$\mathbb{P}[T_1 < \infty \mid X_0 = n] = \mathbb{P}[T_{n-1} < \infty \mid X_0 = n] \cdot \mathbb{P}[T_1 < \infty \mid X_0 = n-1] \leq \cdots \leq (1 - \varepsilon)^{n-1}.$$

Since $\mathbb{P}[Y_{t+1} \neq Y_t \mid X_t > 1] = 0$, we conclude that

$$\mathbb{P}[\forall\, t,\ Y_t = Y_0 \mid X_0 = n] \geq \mathbb{P}[T_1 = \infty \mid X_0 = n] \geq 1 - (1 - \varepsilon)^{n-1}.$$

Thus, if $\sigma_0 = \sigma$ is such that $X_0 = n, Y_0 = -1$ then $h(\sigma) \leq (1 - \varepsilon)^{n-1}$ and if $\sigma_0 = \sigma'$ is such that $X_0 = n, Y_0 = 1$ then $h(\sigma') \geq 1 - (1 - \varepsilon)^{n-1}$. Taking n large enough so that $(1 - \varepsilon)^{n-1} < \tfrac{1}{2}$, we obtain $\sigma \neq \sigma'$ for which $h(\sigma) < \tfrac{1}{2} < h(\sigma')$. $\qquad\qquad\qquad\qquad \text{:)}\ \checkmark$

Solution to Exercise 6.59 :(

Let $(U_t, J_t, I_t)_{t \geq 1}$ be mutually independent random variables such that $U_t \in G$ has distribution μ, J_t is a Bernoulli-$\tfrac{1}{2}$ random variable, and I_t is a Bernoulli-ε random variable.

Define $(X_t, \sigma_t) \in L(G)$ inductively by $X_0 = 1$, $\sigma_0 = 0$, and for $t > 0$,

$$(X_t, \sigma_t) = (X_{t-1}, \sigma_{t-1}) \cdot \begin{cases} (U_t, 0) & \text{if } I_t = 1, \\ (1, \delta_1) & \text{if } J_t = 1 \neq I_t, \\ (1, 0) & \text{if } J_t = I_t = 0. \end{cases}$$

It is simple to verify that $((X_t, \sigma_t))_t$ is a ν-random walk, for ν given by $\nu(s, 0) = \varepsilon\mu(s)$ and $\nu(1, \delta_1) = \tfrac{1}{2}(1 - \varepsilon)$ and $\nu(1, 1) = \tfrac{1}{2}(1 - \varepsilon)$.

Define

$$V_t = \begin{cases} U_t & \text{if } I_t = 1, \\ 1 & \text{if } I_t = 0. \end{cases}$$

It is immediate that $X_t = V_1 \cdots V_t$. Since $(J_t)_t$ and $(U_t, I_t)_t$ are independent, we have that $(J_t)_t$ and $(X_t)_t$ are also independent. Thus, $(J_t)_t$ and $(Q_t)_t$ are independent.

Note that $X_{k+1} = X_k$ if and only if $I_k = 0$.

For $x \in G$ define the following.

$$K(x,t) = \{0 \leq k \leq t-1 : X_{k+1} = X_k = x\},$$

$$L_t(x) = \sum_{k \in K(x,t)} J_{k+1} = \sum_{k=0}^{t-1} \mathbf{1}_{\{X_{k+1}=X_k=x\}} J_{k+1}.$$

So $L_t(x)$ is the number of times the lamp at x has been flipped up to time t. Thus, $\sigma_t(x) = L_t(x) \pmod{2}$.

Note that $x \in Q_t$ if and only if $K(x,t) \neq \emptyset$. Also, the sets $(K(x,t))_{x \in Q_t}$ are measurable with respect to $\sigma(X_0, \ldots, X_t)$, and these sets are pairwise disjoint, and $L_t(x)$ is determined by $(J_k)_{k \in K(x,t)}$. Since $(J_n)_n$ are mutually independent, we have that conditioned on $(X_n)_n$, the random variables $(L_t(x))_{x \in Q_t}$ are mutually independent. Since $(J_n)_n$ and $(X_n)_n$ are independent, we conclude that: conditioned on $(X_n)_n$, the conditional distribution of $(L_t(x))_{x \in Q_t}$ is that of independent Binomial random variables, each $L_t(x)$ having Binomial-$\left(|K(x,t)|, \frac{1}{2}\right)$ distribution.

It is a simple exercise to show that

$$\mathbb{P}[\sigma_t(x) = 1 \mid x \in Q_t] = \mathbb{P}\left[\mathsf{Bin}\left(|K(x,t)|, \tfrac{1}{2}\right) = 1 \pmod 2\right] = \frac{1}{2},$$

see Exercise 5.29.

Since $(L_t(x))_{x \in Q_t}$ are conditionally independent conditioned on $(X_n)_n$, it is also true that $(\sigma_t(x))_{x \in Q_t}$ are conditionally independent. We conclude that the conditional distribution of $(\sigma_t(x))_{x \in Q_t}$, conditioned on $(X_n)_n$, is that of independent Bernoulli-$\frac{1}{2}$ random variables. :) ✓

Solution to Exercise 6.60 :(

For $t, n \geq 1$ define

$$Q_{t,n} = \{x \in G : \exists\, t \leq k \leq t+n-1,\ X_{k+1} = X_k = x\}.$$

Thus, $Q_{t+n} \subset Q_t \cup Q_{t,n}$.

Also, note that $\mathbb{P}_x[y \in Q_n] = \mathbb{P}\left[x^{-1}y \in Q_n\right]$, so that $\mathbb{E}_x|Q_n| = \mathbb{E}|Q_n|$.

The Markov property tells us that

$$\mathbb{E}|Q_{t,n}| = \sum_x \mathbb{P}[X_t = x]\,\mathbb{E}_x|Q_n| = \mathbb{E}|Q_n|,$$

so that $(\mathbb{E}|Q_t|)_t$ is a subadditive sequence. Thus, the limit $\ell := \lim_{t \to \infty} \frac{1}{t}\mathbb{E}|Q_t|$ exists by Fekete's lemma.

Note that $Q_t \subset R_{t-1}$ implying that $\mathbb{E}|Q_t| \leq \mathbb{E}|R_{t-1}|$.

The strong Markov property at time $T_x \wedge t$ implies that

$$\mathbb{P}[x \in Q_t] \geq \mathbb{P}[T_x < t] \cdot \mu(1) = \mathbb{P}[x \in R_{t-1}] \cdot \mu(1),$$

so that $\mathbb{E}|Q_t| \geq \mu(1) \cdot \mathbb{E}|R_{t-1}|$.

Thus, by Exercise 4.45,

$$\mu(1) \cdot p_{\mathrm{esc}} \leq \lim_{t \to \infty} \frac{1}{t}\mathbb{E}|Q_t| \leq p_{\mathrm{esc}}.$$

Since $p_{\mathrm{esc}} = 0$ if and only if (G, μ) is recurrent, and since $\mu(1) > 0$, this completes the proof. :) ✓

Solution to Exercise 6.61 :(

If (G, μ) is recurrent, then as we have seen in Exercise 6.60 that $\ell = p_{\mathrm{esc}} = 0$.

So assume that (G, μ) is transient. Let $\mathsf{g}(x, y) = \sum_{t=0}^{\infty} \mathbb{P}_x[X_t = y]$ be the Green function.

Let $T_0 = 0$ and $T_{n+1} = \inf\{t > T_n : X_t = 1\}$, with the convention that $\inf \emptyset = \infty$. These are the successive return times to $1 \in G$.

Let $N = \max\{n \geq 1 : T_n < \infty\}$, which is a.s. finite because of transience. We have seen in Example 3.5.1 that the random variables $(T_{n+1} - T_n)_{n=0}^{N-1}$ are independent.

Also, note that $1 \in Q_t$ if and only if there exists $n < N$ such that $T_n + 1 = T_{n+1} \leq t$. Thus, the event that $T_{n+1} > T_n + 1$ for all $n < N$ implies that $1 \notin Q_t$. Note that the strong Markov property at time T_1 gives that

$$\mathbb{P}[\forall\, n < N,\ T_{n+1} > T_n + 1] = \mathbb{P}[1 < T_1 < \infty] \cdot \mathbb{P}[\forall\, n < N,\ T_{n+1} > T_n + 1] + \mathbb{P}[T_1 = \infty]$$

$$= (\mathbb{P}[T_1 < \infty] - \mu(1)) \cdot \mathbb{P}[\forall\, n < N,\ T_{n+1} > T_n + 1] + \mathbb{P}[T_1 = \infty],$$

which implies that

$$\mathbb{P}[\forall\, n < N,\, T_{n+1} > T_n + 1] = \frac{\mathbb{P}[T_1 = \infty]}{\mu(1) + \mathbb{P}[T_1 = \infty]} = \frac{p_{\mathrm{esc}}}{\mu(1) + p_{\mathrm{esc}}}.$$

Denote $r := \frac{p_{\mathrm{esc}}}{\mu(1) + p_{\mathrm{esc}}}$. Thus,

$$\mathbb{P}[1 \notin Q_t] \geq \mathbb{P}[\forall\, n < N,\, T_{n+1} > T_n + 1] = r.$$

On the other hand, as $t \to \infty$,

$$\mathbb{P}[1 \notin Q_t] \leq \mathbb{P}[\forall\, n < N,\, T_{n+1} > t \max(T_n + 1)] \nearrow \mathbb{P}[\forall\, n < N,\, T_{n+1} > T_n + 1].$$

Now, fix $\varepsilon > 0$ and let t_0 be large enough so that for all $t \geq t_0$ we have $r \leq \mathbb{P}[1 \notin Q_t] \leq (1 + \varepsilon) \cdot r$.

Now, for any $x \in G$, using the strong Markov property at time $T_x \wedge n$, for $t \geq n$ we have that

$$\mathbb{P}[x \in R_{n-1} \backslash Q_t] \geq \mathbb{P}[T_x < n] \cdot \mathbb{P}_x[x \notin Q_t],$$
$$\mathbb{P}[x \in R_{n-1} \backslash Q_t] \leq \mathbb{P}[T_x < n] \cdot \mathbb{P}_x[x \notin Q_{t-n+1}].$$

Since $\mathbb{P}_x[x \notin Q_t] = \mathbb{P}[1 \notin Q_t]$, as long as $t - n \geq t_0$ we have that

$$1 \leq \frac{\mathbb{P}[x \in R_{n-1} \backslash Q_t]}{r\, \mathbb{P}[T_x < n]} \leq 1 + \varepsilon.$$

Summing over x we have that for any $t \geq t_0 + n$,

$$\mathbb{E}\,|R_{n-1}| \cdot r \leq \mathbb{E}\,|R_{n-1} \backslash Q_t| \leq \mathbb{E}\,|R_{n-1}| \cdot r(1 + \varepsilon).$$

Recalling that $Q_t \subset R_{t-1}$ and that $|R_{t-1} \backslash R_{n-1}| \leq t - n$, this implies that

$$(\mathbb{E}\,|R_{t-1}| - (t - n)) \cdot r \leq \mathbb{E}\,|R_{t-1} \backslash Q_t| \leq \mathbb{E}\,|R_{n-1} \backslash Q_t| \leq \mathbb{E}\,|R_{t-1}| \cdot r(1 + \varepsilon).$$

Fixing $t = n + t_0$ and taking $t \to \infty$, we have

$$p_{\mathrm{esc}} \cdot r \leq p_{\mathrm{esc}} - \ell \leq p_{\mathrm{esc}} \cdot r(1 + \varepsilon).$$

Since this holds for all $\varepsilon > 0$, we obtain $\ell = p_{\mathrm{esc}}(1 - r)$.   :) ✓

Solution to Exercise 6.62 :(

Let $(U_t, I_t)_{t \geq 1}$ be mutually independent, with each U_t having the distribution of μ and each I_t a Bernoulli-ε random variable. Set $B_t = \sum_{k=1}^{t} I_k$, and note that B_t is a Binomial-(t, ε) random variable. Let $X_t = U_1 \cdots U_t$ and $Y_t = X_{B_t}$.

We have that

$$
\begin{aligned}
\mathbb{P}[Y_{t+1} = Y_t x \mid Y_t] &= \mathbb{P}[X_{B_t + I_{t+1}} = X_{B_t} x \mid X_{B_t}] \\
&= \mathbb{P}[I_{t+1} = 0,\, x = 1 \mid X_{B_t}] + \mathbb{P}[I_{t+1} = 1,\, U_{B_t + 1} = x \mid X_{B_t}] \\
&= (1 - \varepsilon)\delta_1(x) + \varepsilon\mu(x) = \nu(x).
\end{aligned}
$$

Thus, $(Y_t)_t$ is a ν-random walk.

Finally, note that

$$H(Y_t) = H(X_{B_t}) \geq H(X_{B_t} \mid B_t) = \sum_{k=0}^{t} \mathbb{P}[B_t = k]H(X_k),$$

$$H(Y_t) = H(X_{B_t}) \leq H(X_{B_t} \mid B_t) + H(B_t) \leq \sum_{k=0}^{t} \mathbb{P}[B_t = k]H(X_k) + \log(t + 1),$$

$$h(G, \nu) = \lim_{t \to \infty} \frac{H(Y_t)}{t} = \lim_{t \to \infty} \frac{1}{t}\sum_{k=0}^{t} \mathbb{P}[B_t = k]k \cdot \frac{1}{k}H(X_k).$$

Now, let $\eta > 0$ and choose K large enough so that $\left| \frac{1}{k} H(X_k) - h(G, \mu) \right| < \eta$ for all $k \geq K$. Then,

$$h(G, \nu) \geq (h(G, \mu) - \eta) \cdot \lim_{t \to \infty} \frac{1}{t} \sum_{k=K}^{t} \mathbb{P}[B_t = k] k$$

$$\geq (h(G, \mu) - \eta) \cdot \varepsilon - \lim_{t \to \infty} \frac{K(K+1)}{t} = (h(G, \mu) - \eta) \cdot \varepsilon,$$

$$h(G, \nu) \leq (h(G, \mu) + \eta) \cdot \varepsilon + (h(G, \mu) + \eta) \cdot \lim_{t \to \infty} \sum_{k=0}^{K} \mathbb{P}[B_t = k] k$$

$$\leq (h(G, \mu) + \eta) \cdot \varepsilon + \lim_{t \to \infty} \frac{K(K+1)}{t} = (h(G, \mu) + \eta) \cdot \varepsilon.$$

Taking $\eta \to 0$ completes the proof. :) ✓

Solution to Exercise 6.63 :(

Let $(X_t, \sigma_t)_t$ be the ν-random walk on $L(G)$. Let $Q_t = \{x \in G : \exists\, 0 \leq k \leq t - 1, \ X_{k+1} = X_k = x\}$.
We have

$$H(\sigma_t \mid X_t) + H(X_t) = H(X_t, \sigma_t) \leq H(\sigma_t) + H(X_t).$$

Also, since $|Q_t|$ independent Bernoulli-$\frac{1}{2}$ have entropy $|Q_t| \cdot \log 2$, by Exercise 6.59,

$$H(\sigma_t \mid |Q_t|) = \log 2 \cdot \mathbb{E} |Q_t|,$$

and similarly,

$$H(\sigma_t \mid X_t) \geq H(\sigma_t \mid |Q_t|, X_t) = \log 2 \cdot \mathbb{E} |Q_t|.$$

Note that $(X_t)_t$ is a $\tilde{\mu}$-random walk with $\tilde{\mu} = (1 - \varepsilon)\delta_1 + \varepsilon\mu$, and specifically $\tilde{\mu}(1) = 1 - \varepsilon$. So, by Exercises 6.61 and 6.62,

$$h(L(G), \nu) = \log 2 \cdot \lim_{t \to \infty} \frac{1}{t} \mathbb{E} |Q_t| + \lim_{t \to \infty} \frac{1}{t} H(X_t) = \log 2 \cdot \frac{(1 - \varepsilon) p_{\mathrm{esc}}}{1 - \varepsilon + p_{\mathrm{esc}}} + \varepsilon h(G, \mu).$$

Here $p_{\mathrm{esc}} = \mathbb{P}[\forall\, t \geq 1, \ X_t \neq 1]$. :) ✓

Solution to Exercise 6.64 :(

As in Exercise 4.27, if we consider the random walk measure $\mu(1) = \frac{1}{2}p$, $\mu(-1) = \frac{1}{2}(1 - p)$, and $\mu(0) = \frac{1}{2}$ for some $\frac{1}{2} < p < 1$, then the deterministic path $\gamma = (0, 1, 2, \ldots)$ is a SIT in a network inducing the μ-random walk. Thus $(\mathbb{Z}, \mu)$ is transient.
Define ν on $L(\mathbb{Z})$ by

$$\nu(1, 0) = \tfrac{1}{2}p, \qquad \nu(-1, 0) = \tfrac{1}{2}(1 - p), \qquad \nu(0, \delta_1) = \tfrac{1}{2}.$$

So

$$r := \sup\{|y| : \sigma(y) = 1, \ (x, \sigma) \in \mathrm{supp}(\nu)\} = 0,$$

and

$$\mu(x) = \sum_{\sigma} \nu(x, \sigma).$$

Thus, by Theorem 6.9.1, we know that $(L(\mathbb{Z}), \nu)$ is non-Liouville. :) ✓

Solution to Exercise 6.65 :(

Exercise 4.44 provides us with $\mu \in \mathrm{SA}\left(\mathbb{Z}^2, 2 - \varepsilon\right)$ such that $\left(\mathbb{Z}^2, \mu\right)$ is transient. Define ν on $L\left(\mathbb{Z}^2\right)$ by $\nu(z, 0) = \frac{1}{2}\mu(z)$ and $\nu(0, \delta_1) = \frac{1}{2}$. Note that for $\tilde{\mu}(z) = \sum_{\sigma} \nu(z, \sigma)$, we have that $\tilde{\mu}(z) = \frac{1}{2}\mu(z) + \frac{1}{2}\mathbf{1}_{\{z=0\}}$. So the $\tilde{\mu}$-random walk is just a lazy version of the μ-random walk. Exercise 3.13 tells us that $\left(\mathbb{Z}^2, \tilde{\mu}\right)$ is transient if and only if $\left(\mathbb{Z}^2, \mu\right)$ is transient. Since $\max\{|y| : \sigma(y) = 1, \ (x, \sigma) \in \mathrm{supp}(\nu)\} = 0$, by Theorem 6.9.1 we get that $\left(L(\mathbb{Z}^2), \nu\right)$ is non-Liouville. :) ✓

Solution to Exercise 6.66 :(
This is just the martingale convergence theorem (Theorem 2.6.3 and Exercise 2.32).   :) ✓

Solution to Exercise 6.67 :(
This follows since pointwise sums and products of converging sequences are convergent.   :) ✓

Solution to Exercise 6.68 :(
If $f \in \mathcal{I}(G,\mu)$ and $h \in \mathcal{A}(G,\mu)$, then $h(X_t)f(X_t) \to 0$ a.s., because $f(X_t) \to 0$ a.s.   :) ✓

Solution to Exercise 6.69 :(
We have that L_f is $\mathcal{I}$-measurable just as in Exercise 6.14, and $h_f \in \mathrm{BHF}(G,\mu)$ just as in Exercise 6.17.

By Exercise 6.17, $h_f(X_t) \to L_f$ a.s. Thus, $(f - h_f)(X_t) \to L_f - L_f = 0$ a.s., implying that $f - h_f \in \mathcal{I}(G,\mu)$.

If $h \in \mathrm{BHF}(G,\mu) \cap \mathcal{I}(G,\mu)$, then by Exercise 6.17 we see that $h \equiv 0$. So $\mathcal{A}(G,\mu) = \mathrm{BHF}(G,\mu) \oplus \mathcal{I}(G,\mu)$.   :) ✓

7

Choquet–Deny Groups

7.1 The Choquet–Deny Theorem

In this chapter we depart from the usual framework of this book and work with *general* probability measures on a finitely generated group G; that is, the measures considered in this section are not necessarily assumed to be symmetric or have moment conditions.

We have already seen in Theorem 6.8.1 that if G is a finitely generated non-amenable group, then every symmetric adapted random walk on G with finite entropy is non-Liouville. (In fact, on a non-amenable group any adapted random walk is non-Liouville; see Kaimanovich and Vershik, 1983; Rosenblatt, 1981.) The main objective of this section is to understand the extreme opposite case: what are the finitely generated groups G for which (G, μ) is Liouville for *all* adapted probability measures μ. For this section it is wise to be familiar with the basic notions of virtually nilpotent groups, see Section 1.5.4.

In 1960, Choquet and Deny proved that Abelian groups never admit non-constant harmonic functions. We will basically present their result using a probabilistic proof.

Theorem 7.1.1 *Let G be a group and let μ be an adapted measure on G. Let K be the kernel of the canonical left action of G on $\mathrm{BHF}(G, \mu)$. Let $Z = Z(G)$ be the center of G (i.e. $Z = \{x \in G \; : \; \forall \, y \in G \, [x, y] = 1\}$.)*

Then, $Z \lhd K$. That is, the center of G acts trivially on $\mathrm{BHF}(G, \mu)$.

Proof Let $x \in Z$. Let $(X_t)_t$ be the μ-random walk on G. Since μ is adapted, there exist $k = k(x) > 0$ and $\alpha > 0$ such that for any t we have for all $y \in G$,

$$\mathbb{P}_y \left[X_{t+k} = X_t x^{-1} \mid X_t \right] = \mathbb{P} \left[X_k = x^{-1} \right] =: \alpha > 0.$$

Specifically, this implies that for any $f \in \mathrm{BHF}(G, \mu)$,

$$\mathbb{E}_y[|f(X_{t+k}) - f(X_t)|] \geq \alpha \, \mathbb{E}_y \left[|f(X_t x^{-1}) - f(X_t)| \right]$$
$$= \alpha \, \mathbb{E}_y[|x.f(X_t) - f(X_t)|] \geq \alpha |x.f(y) - f(y)|.$$

We have used that $f\left(X_t x^{-1}\right) = f\left(x^{-1} X_t\right) = x.f(X_t)$ since $x \in Z$.

On the other hand, since $f \in \mathrm{BHF}(G, \mu)$, the sequence $(f(X_t))_t$ is a bounded martingale, and thus converges a.s. and in L^1 by the martingale convergence theorem (Theorem 2.6.3 and Exercise 2.33). Thus, $\mathbb{E}_y[|f(X_{t+k}) - f(X_t)|] \to 0$.

We obtain that for any $y \in G$, any $f \in \mathrm{BHF}(G, \mu)$, and any $x \in Z$, we have $x.f(y) = f(y)$. This implies that Z acts trivially on $\mathrm{BHF}(G, \mu)$. $\qquad \square$

Corollary 7.1.2 (Choquet–Deny theorem) If G is an Abelian group then (G, μ) is Liouville for any adapted μ.

Proof When G is Abelian then $Z(G) = G$.   □

Exercise 7.1 Let G be a finitely generated group and μ an adapted probability measure on G.

Show that if $K \lhd G$ is a normal subgroup acting trivially on $\mathrm{BHF}(G, \mu)$, then $\mathrm{BHF}(G, \mu) \cong \mathrm{BHF}(G/K, \bar{\mu})$, where $\bar{\mu}$ is the (projected) measure on G/K given by $\bar{\mu}(Kx) = \sum_{y \in K} \mu(yx)$.   ▷ solution ◁

In light of the Choquet–Deny theorem for Abelian groups, we have the following definition.

Definition 7.1.3 We say that a group G is **Choquet–Deny** if for any adapted measure μ on G we have that (G, μ) is Liouville.

Corollary 7.1.4 Let G be a virtually nilpotent group. Then, G is Choquet–Deny; that is, (G, μ) is Liouville for any adapted μ.

Proof By Theorem 3.9.7, it suffices to prove that any nilpotent group is Choquet–Deny, because passing to a finite index results in an isomorphic space of bounded harmonic functions. Also, the Abelian case (or 1-step nilpotent) is Corollary 7.1.2.

Now, if G is a n-step nilpotent group, then $Z(G)$ acts trivially on $\mathrm{BHF}(G, \mu)$, by Theorem 7.1.1. So Exercise 7.1 tells us that $\mathrm{BHF}(G, \mu) \cong \mathrm{BHF}(G/Z(G), \bar{\mu})$. Also $G/Z(G)$ is $(n-1)$-step nilpotent. Thus, inductively, we have that $\mathrm{BHF}(G, \mu)$ is just the space of constant functions.   □

Naturally the question arises whether virtually nilpotent groups are the only examples of Choquet–Deny groups. Although the Choquet–Deny theorem has basically been known since 1960, the converse was only shown in 2018.

Theorem 7.1.5 (Frisch, Hartman, Tamuz, Vahidi-Ferdowsi) *Let G be a finitely generated group. The following are equivalent:*

(1) G is virtually nilpotent,
(2) G is Choquet–Deny,
(3) (G, μ) is Liouville for every symmetric, adapted measure μ on G with finite entropy $H(\mu) < \infty$.

We have seen that every virtually nilpotent group is Choquet–Deny, so to prove Theorem 7.1.5, it suffices to show the following.

Theorem 7.1.6 *Let G be a finitely generated group that is not virtually nilpotent. Then there exists a symmetric, adapted measure μ on G, with finite entropy $H(\mu) < \infty$, such that $h(G, \mu) > 0$.*

The proof of Theorem 7.1.6 is carried out in a few steps over the remainder of this chapter.

7.2 Centralizers

Let G be a group, and let $N \triangleleft G$. Define

$$C_G^N(x) = \{y \in G \ : \ [x, y] \in N\}.$$

So $C_G(x) := C_G^{\{1\}}(x)$ is the *centralizer* of x; that is, all elements of G that commute with x.

Exercise 7.2 Show that $C_G^N(x)$ is a subgroup. ▷ solution ◁

Exercise 7.3 Show that $N \triangleleft C_G^N(x)$. ▷ solution ◁

Exercise 7.4 Let G be a group, and let $N \triangleleft M \triangleleft G$ be normal subgroups of G.
Show that $C_G^N(x) \leq C_G^M(x)$.
Show that $C_G^M(x)/N = C_{G/N}^{M/N}(Nx)$. ▷ solution ◁

Exercise 7.5 Show that $C_G^N(x^y) = \left(C_G^N(x)\right)^y$. ▷ solution ◁

Definition 7.2.1 Let G be a group. Let $\mathcal{N}(G)$ denote the collection of all normal subgroups of G.
Define $\tau = \tau_G \colon \mathcal{N}(G) \to \mathcal{N}(G)$ by

$$\tau(N) = \tau_G(N) = \left\{x \in G \mid \left[G : C_G^N(x)\right] < \infty\right\}.$$

Exercise 7.6 Show that τ is well defined. That is, show that $\tau(N)$ is always a normal subgroup. ▷ solution ◁

Exercise 7.7 Show that $N \triangleleft \tau(N)$. ▷ solution ◁

Exercise 7.8 Show that if $N \lhd G$ is of finite index $[G : N] < \infty$, then $\tau(N) = G$.

▷ solution ◁

Exercise 7.9 Show that if $N \lhd M \lhd G$ are normal subgroups of G then $\tau(N) \lhd \tau(M)$.

▷ solution ◁

Definition 7.2.2 Let G be a group. Define $\tau^0 = \tau_G^0 = \{1\}$, and inductively $\tau^{n+1} = \tau_G^{n+1} = \tau_G(\tau^n)$.

Define

$$\tau^\infty = \tau_G^\infty = \bigcup_n \tau^n.$$

Exercise 7.10 Let $(N_n)_n$ be a sequence of normal subgroups of G such that $N_n \subset N_{n+1}$ for all n.

Show that $N_\infty = \bigcup_n N_n$ is also a normal subgroup of G.

Conclude that τ_G^∞ is a normal subgroup of G.

▷ solution ◁

Exercise 7.11 Let G be a group, and let $N \lhd M \lhd G$ be normal subgroups of G.

Show that $\tau_G(M)/N = \tau_{G/N}(M/N)$.

Conclude that for $n > k$, $\tau_G^n / \tau_G^k = \tau_{G/\tau_G^k}^{n-k}$.

▷ solution ◁

Exercise 7.12 Show that $Z_n(G) \lhd \tau_G^n$.

▷ solution ◁

Exercise 7.13 Show that if $H \leq G$ has finite index $[G : H] < \infty$, then $\tau_H^n = \tau_G^n \cap H$.

▷ solution ◁

The connection of τ^∞ and nilpotence is given by the following.

Theorem 7.2.3 *Let G be a finitely generated group. Then G is virtually nilpotent if and only if $\tau^\infty = G$.*

Proof Assume that $\tau^\infty = G$. If S is a finite generating set for G, then there exists n such that $S \subset \tau^n$. So $\tau^n = G$ for some n.

We will show by induction on n that G admits a finite index subgroup $H \leq G$, $[G : H] < \infty$ such that $\gamma_n(H) = \{1\}$.

If $n = 1$, then we claim that $[G : Z] < \infty$, for $Z = Z_1(G)$. (This suffices since Z is Abelian, so $\gamma_1(Z) = \{1\}$.) Indeed, if $s_1, \ldots, s_d$ generate G, then $Z = \bigcap_{j=1}^d C_G(s_j)$. Since $[G : C_G(s_j)] < \infty$ for all j $\left(\text{because } \tau^1 = G\right)$, we get that $[G : Z] < \infty$ (by Proposition 1.3.3), so G is virtually Abelian. This proves the base step of the induction.

Now assume that $n > 1$ and $\tau^n = G$. Let $N = \tau^1$. Since $G/N = \tau^n/\tau^1 = \tau^{n-1}_{G/N}$, by induction we have that there exists a finite index subgroup $N \le H \le G$, $[G : H] < \infty$, such that $\gamma_{n-1}(H/N) = \{1\}$. Thus, $\gamma_{n-1}(H) \le N = \tau_G(\{1\})$. This implies that for any $x \in \gamma_{n-1}(H)$ we have that $[G : C_G(x)] < \infty$. Since $C_H(x) = C_G(x) \cap H$ and since $[G : H] < \infty$, we conclude (by Proposition 1.3.3) that $[H : C_H(x)] \le [G : C_G(x) \cap H] < \infty$, for any $x \in \gamma_{n-1}(H)$.

Since $[G : H] < \infty$ and G is finitely generated, H is also finitely generated (Exercise 1.61). Now, if $H = \langle S \rangle$ for some finite set S, then by Exercise 1.34,

$$\gamma_{n-1}(H) = \langle [s_1, \ldots, s_{n-1}]^x : s_j \in S, x \in H \rangle.$$

For any $s_1, \ldots, s_{n-1} \in S$, we have that $[H : C_H(y)] < \infty$ for $y = [s_1, \ldots, s_{n-1}]$. This implies that the set $y^H = \{y^x : x \in H\}$ is finite (by the orbit-stablizer theorem, Exercise 1.17). Thus, the generating set

$$\{[s_1, \ldots, s_{n-1}]^x : s_j \in S, x \in H\}$$

is a finite set, implying that $\gamma_{n-1}(H)$ is finitely generated.

So let $U = \{u_1, \ldots, u_d\}$ be a finite generating set for $\gamma_{n-1}(H)$. Let $K = \bigcap_{j=1}^d C_H(u_j)$. Since $[H : C_H(u_j)] < \infty$ for all j, we have that $[H : K] < \infty$ (Proposition 1.3.3), and so $[G : K] < \infty$. Also, $\gamma_{n-1}(K) \le \gamma_{n-1}(H)$. Since K was chosen such that its elements commute with those of $\gamma_{n-1}(H)$, we get that $\gamma_n(K) = [K, \gamma_{n-1}(K)] = \{1\}$, completing the induction step.

Thus, we have proved that if $\tau^n = G$ then G is virtually nilpotent, which is one direction of the theorem.

For the other direction, assume that G is virtually nilpotent. So there exists $H \le G$, $[G : H] < \infty$ such that $Z_n(H) = H$ for some n. By Exercise 7.12 we find that $\tau^n_H = H$. By Exercise 7.13, we get that $\tau^n_G \supset \tau^n_H = H$, so $[G : \tau^n_G] < \infty$, which implies that $\tau^{n+1}_G = \tau_G(\tau^n_G) = G$ by Exercise 7.8, completing the proof. □

7.3 ICC Groups

Definition 7.3.1 Let G be a group. Consider $T = \{N \lhd G : \tau(N) = N\}$, the set of all fixed points for $\tau = \tau_G$. Define

$$\tau^* = \tau^* = \bigcap_{N \in T} N.$$

Since $\tau(G) = G$, this is well defined. Note that if $\tau(N) = N$ for some $N \lhd G$, then $\tau^* \lhd N$.

Exercise 7.14 Show that for any $n \geq 1$ we have that $\tau^n \lhd \tau^*$.
Conclude that $\tau^\infty \lhd \tau^*$.   ▷ solution ◁

Exercise 7.15 Show that $\tau(\tau^*) = \tau^*$.   ▷ solution ◁

That is, τ^* is the smallest fixed point of τ, and it contains all the finite applications of τ to the trivial subgroup.

Definition 7.3.2 A group G has the **infinite conjugacy class** (ICC) property if $[G : C_G(x)] = \infty$ for all $1 \neq x \in G$.

This name comes from the fact that $[G : C_G(x)] = \left| x^G \right|$ (where $x^G = \{x^y : y \in G\}$). This fact is just the orbit-stabilizer theorem (Exercise 1.17), since $C_G(x)$ is the stabilizer of x for the G-action by conjugation.

It is immediate that an infinite group G is ICC if and only if $\tau_G(\{1\}) = \{1\}$.

Exercise 7.16 Show that G/τ^* is either trivial or ICC.   ▷ solution ◁

Exercise 7.17 Let G be a finitely generated group.
Show that if $\tau^\infty = G$ then there exists n such that $\tau^* = \tau^n = G$.   ▷ solution ◁

The relationship of τ^* to bounded harmonic functions is given by the following theorem.

Theorem 7.3.3 *Let G be a finitely generated group and μ an adapted measure on G. Let $K = \{x \in G : \forall h \in \mathrm{BHF}(G, \mu), x.h = h\}$ be the kernel of the G-action on $\mathrm{BHF}(G, \mu)$.*
Then $\tau(K) = K$. Consequently, $\tau^* \lhd K$.

Proof Since $K \lhd \tau(K)$ we only need to show that $\tau(K) \lhd K$. To this end, assume that $x \in \tau(K)$. Then, $\left[G : C_G^K(x) \right] < \infty$. Set $C = C_G^K(x)$.

Let $(X_t)_t$ be a μ-random walk, let $y \in G$, and consider the Markov chain $(CyX_t)_t$. This is a Markov chain on a finite network, so must be recurrent.

Let $\varepsilon > 0$ and k be such that $\mathbb{P}[X_k = x] = \varepsilon > 0$, which is possible as μ is adapted. Then, for any t we have that $\mathbb{P}[X_{t+k} = X_t x \mid X_t] = \varepsilon$.

Define inductively $T_0 = 0$ and

$$T_{n+1} = \inf\{t > T_n + k : yX_t \in C\}.$$

Recurrence of the process $(CyX_t)_t$ ensures that $T_n < \infty$ for all n a.s. Also, the strong Markov property tells us that $\mathbb{P}[yX_{T_n+k} = yX_{T_n}x \mid \mathcal{F}_{T_n}] = \varepsilon$ for all n. Since $yX_{T_n} \in C = C_G^K(x)$ by definition, we have that $yX_{T_n}x \equiv xyX_{T_n} \pmod{K}$.

Because K acts trivially on $\mathrm{BHF}(G, \mu)$, this implies that for any $h \in \mathrm{BHF}(G, \mu)$ we have

$$x^{-1}.h(yX_{T_n}) = h(xyX_{T_n}) = h(yX_{T_n}x),$$

so that

$$\mathbb{P}\left[x^{-1}.h(yX_{T_n}) \neq h(yX_{T_n+k}) \mid \mathcal{F}_{T_n}\right] \leq 1 - \varepsilon.$$

Thus, $x^{-1}.h(yX_{T_n}) = h(yX_{T_n+k})$ for infinitely many n, a.s.

Now, for $h \in \mathrm{BHF}(G, \mu)$, the process $\left(M_t := x^{-1}.h(yX_t) - h(yX_{t+k})\right)_t$ is a bounded martingale. So it converges a.s. and in L^1 by the martingale convergence theorem (Theorem 2.6.3 and Exercise 2.33). Since

$$\liminf_{t \to \infty} |M_t| \leq \liminf_{n \to \infty} |x^{-1}.h(yX_{T_n}) - h(yX_{T_n+k})| = 0 \qquad \text{a.s.,}$$

it must be that $\mathbb{E}|M_t| \to 0$. Thus,

$$\left|x^{-1}.h(y) - h(y)\right| = \left|\mathbb{E}\left[x^{-1}.h(yX_t) - h(yX_{t+k})\right]\right| \leq \mathbb{E}|M_t| \to 0,$$

implying that $x^{-1}.h(y) = h(y)$.

This holds for any $y \in G$, any $h \in \mathrm{BHF}(G, \mu)$, and all $x \in \tau(K)$.

We conclude that $\tau(K)$ acts trivially on $\mathrm{BHF}(G, \mu)$, so $\tau(K) \triangleleft K$. □

The key property of ICC groups is the following.

Definition 7.3.4 Let G be a finitely generated group, and fix some Cayley graph of G. Let $F \subset G$ be a finite subset.

An element $x \in G$ is said to **shatter** F, if for all $\eta \in \{-1, 1\}$ we have $x^\eta Fx \cap F \subset \{1\}$.

Note that if x shatters F, then

$$F \cap \left(xFx \cup x^{-1}Fx \cup xFx^{-1} \cup x^{-1}Fx^{-1}\right) \subset \{1\}.$$

Lemma 7.3.5 *Let G be a finitely generated ICC group. For any finite subset $F \subset G$ and any $r > 0$ there exists $x \in G$ with $|x| > r$ such that x shatters F.*

Proof Fix some finite, symmetric generating set S for G, and let $(X_t)_t$ be a simple random walk on G; that is, the jumps $X_{t-1}^{-1}X_t$ are i.i.d. uniform on S.

Fix a finite subset $F \subset G$. We denote by F^* the set of all $x \in G$ such that x shatters F. Define

$$F_1 = \left\{ x \in G \ : \ x^{-1}Fx \cap F \not\subset \{1\} \right\},$$
$$F_2 = \{ x \in G \ : \ xFx \cap F \not\subset \{1\} \}.$$

So $G \backslash F^* \subset F_1 \cup F_2$.

Let $I = \left\{ x \in G \ : \ x^2 = 1 \right\}$. Note that $I \cap (G \backslash F_1) \subset F^*$, so $I \subset F^* \cup F_1$.

Assume that $H \leq G$ is an infinite index subgroup $[G : H] = \infty$. We have seen in Corollary 5.5.2, as a consequence of the Nash inequality, that there exists a universal constant $C > 0$ such that for all $t \geq 0$ we have

$$\mathbb{P}[X_t \in xHy] = \mathbb{P}\left[x^{-1}X_t \in Hy \right] \leq \mathbb{P}\left[Hx^{-1}X_t = Hy \right] \leq C(t+1)^{-1/2},$$

where we have crucially used here that $[G : H] = \infty$. This fact will be used repeatedly in what follows, with different subgroups.

Recall that $y^G = \{y^x \ : \ x \in G\}$. For $1 \neq y \in G$ and $z \in y^G$ fix some $x_{y,z} \in G$ such that $y^{x_{y,z}} = z$. Note that if $y^x = z$ then $x \in C_G(y)x_{y,z}$. This implies that

$$F_1 \subset \bigcup_{y,z \in F \backslash \{1\}} C_G(y)x_{y,z}. \tag{7.1}$$

Now, let $K = \left\{ (y,z) \in F^2 \ : \ \exists\, x \in G, \ xyx = z \right\}$. For any $(y,z) \in K$ fix some $w_{y,z}$ such that $w_{y,z}yw_{y,z} = z$. Note that for any $x \in G$ and $w = w_{y,z}$, if $wxywx = z$ as well, then $wxywx = wyw$, so $ywx = x^{-1}yw$, implying that $(x^{yw})^{-1} = \left(x^{-1}\right)^{yw} = x$. Thus, $x^{(yw)^2} = x$, implying that $x \in C_G\left((yw)^2\right)$. In conclusion, for any $(y,z) \in K$ we have that

$$w_{y,z}^{-1}\{v \ : \ vyv = z\} \subset C_G\left((yw_{y,z})^2\right).$$

If $(yw_{y,z})^2 = 1$, then $z = w_{y,z}yw_{y,z} = y^{-1}$, so $(yx)^2 = 1$ for any $x \in G$ such that $xyx = z$. That is, $(yw_{y,z})^2 = 1$ implies that $\{x \ : \ xyx = z\} = \left\{ x \ : \ (yx)^2 = 1 \right\} = y^{-1}I$.

Set $K' = \left\{ (y,z) \in K \ : \ (yw_{y,z})^2 \neq 1 \right\}$. We then have that

$$F_2 \subset \bigcup_{(y,z) \in K'} w_{y,z} C_G\left((yw_{y,z})^2\right) \bigcup_{y \in F} y^{-1}I. \tag{7.2}$$

Denote by B_r the ball of radius r around 1 and $F_r^* = F^* \backslash B_r$. Then, combining (7.1) and (7.2), and using that $[G : C_G(y)] = \infty$ for all $y \neq 1$, and that

$\left[G : C_G((yw_{y,z})^2) \right] = \infty$ for all $(y,z) \in K'$, we arrive at

$$\mathbb{P}[X_t \notin F_r^*] \leq \mathbb{P}[X_t \in F_1] + \mathbb{P}[X_t \in F_2 \backslash B_r] + \mathbb{P}[|X_t| \leq r]$$

$$\leq \mathbb{P}[X_t \in F_1] + \sum_{(y,z) \in K'} \mathbb{P}\left[X_t \in w_{y,z} C_G((yw_{y,z})^2) \right]$$

$$+ \sum_{v \in F} \mathbb{P}[vX_t \in I \backslash B_r] + \mathbb{P}[|X_t| \leq r]$$

$$\leq (|F| + 1) \cdot \max_{v \in F \cup \{1\}} \sum_{y,z \in F \backslash \{1\}} \mathbb{P}\left[X_t \in v^{-1} C_G(y) x_{y,z} \right]$$

$$+ \sum_{(y,z) \in K'} \mathbb{P}\left[X_t \in w_{y,z} C_G((yw_{y,z})^2) \right]$$

$$+ |F| \cdot \max_{v \in F} \mathbb{P}[vX_t \in F^* \backslash B_r] + \sum_{|v| \leq r} \mathbb{P}[X_t = v]$$

$$\leq \frac{C \cdot \left((|F| + 1)|F|^2 + |K'| + |B_r| \right)}{\sqrt{t + 1}} + |F| \cdot \max_{v \in F} \mathbb{P}\left[vX_t \in F_r^* \right].$$

For

$$p = \max_{y \in F \cup \{1\}} \mathbb{P}\left[yX_t \in F_r^* \right]$$

this implies that

$$1 - p \leq \mathbb{P}\left[X_t \notin F_r^* \right] \leq \frac{C \cdot \left((|F| + 1)|F|^2 + |K'| + |B_r| \right)}{\sqrt{t + 1}} + |F| p,$$

or

$$p \geq \left(1 - \frac{C \cdot \left((|F| + 1)|F|^2 + |K'| + |B_r| \right)}{\sqrt{t + 1}} \right) \cdot \frac{1}{|F| + 1}.$$

So when t is large enough (only as a function of $|F|$ and $|B_r|$), we have that $p > 0$, implying that $F_r^* = F^* \backslash B_r$ cannot be empty. $\qquad \square$

7.4 JNVN Groups

Definition 7.4.1 A group G is called **just not virtually nilpotent**, or **JNVN**, if G is not virtually nilpotent and for every nontrivial normal subgroup $\{1\} \neq N \triangleleft G$, the quotient G/N is virtually nilpotent.

For example, since nilpotent groups are never simple, an infinite simple group is always JNVN, because every nontrivial quotient of a simple group is the trivial group.

The following lemma tells us that every finitely generated group that is not virtually nilpotent has a JNVN quotient.

Lemma 7.4.2 *Let G be an infinite finitely generated group. If G is not virtually nilpotent, then there exists $N \lhd G$ such that G/N is JNVN.*

Proof We will use two properties of finitely generated virtually nilpotent groups.

Fact I. Quotients of virtually nilpotent groups are virtually nilpotent. See Exercise 1.43.

Fact II. Every finitely generated virtually nilpotent group is finitely presented. See Exercise 1.67.

Now to continue the proof of Lemma 7.4.2.

Consider the collection

$$\mathcal{N} = \{N \lhd G \ : \ G/N \text{ is not virtually nilpotent}\}.$$

If $(N_k)_k$ is an increasing chain in $\mathcal{N}$, then $N_\infty := \bigcup_k N_k$ is a normal subgroup (Exercise 7.10). We will show that N_∞ is in $\mathcal{N}$.

Indeed, if $N_\infty \notin \mathcal{N}$, then G/N_∞ is virtually nilpotent. So G/N_∞ is finitely presented (Fact II). If G is generated by the finite set S, then there exists a free group $\mathbb{F} = \mathbb{F}_S$, and finitely many words $r_1, \ldots, r_m \in \mathbb{F}$, such that if we let $R \lhd \mathbb{F}$ be the smallest normal subgroup containing $r_1, \ldots, r_m$, then under the canonical projection $\varphi \colon \mathbb{F} \to G$, we have that $\varphi(R) = N_\infty$. Since $N_\infty = \bigcup_k N_k$, there must exist some k for which $\varphi(r_j) \in N_k$ for all $1 \le j \le m$. But then $\varphi^{-1}(N_k) \lhd \mathbb{F}$ is a normal subgroup containing $r_1, \ldots, r_m$, implying that $R \lhd \varphi^{-1}(N_k)$, so that $N_\infty = N_k$, a contradiction!

We conclude that any nondecreasing chain in $\mathcal{N}$ has an upper bound in $\mathcal{N}$, so by Zorn's lemma $\mathcal{N}$ contains a maximal element. That is, there exists a maximal element $N \in \mathcal{N}$. So G/N is not virtually nilpotent.

Now, assume that G/N has a non-virtually nilpotent quotient. That implies the existence of some $N \lhd M \lhd G$ such that G/M is not virtually nilpotent. By the maximality of N, we have that $N = M$. So the only non-virtually nilpotent quotient of G/N is by the trivial group. That is, G/N is JNVN. $\qquad\square$

Theorem 7.4.3 *If G is a finitely generated JNVN group, then G is ICC.*

Proof Let $N = \tau_G(\{1\}) \lhd G$. If $N \ne \{1\}$ then since G is JNVN we have that G/N is virtually nilpotent. So by Theorem 7.2.3, we know that for some n,

$$\tau_G^{n+1}/N = \tau_{G/N}^n = G/N.$$

But this implies that $\tau_G^{n+1} = G$, so G is virtually nilpotent by Theorem 7.2.3 again, a contradiction!

Hence we deduce that $\tau_G(\{1\}) = \{1\}$, which is to say that G is ICC. $\qquad\square$

Exercise 7.18 Let $\mathcal{P}$ be a group property. Assume that any group that is $\mathcal{P}$ is also finitely presented. Assume that any quotient of a group that is $\mathcal{P}$ is also $\mathcal{P}$. (That is, $\mathcal{P}$ is closed under quotients.)

Show that for any finitely generated group G that is not $\mathcal{P}$, there exists a normal subgroup $N \lhd G$ such that G/N is not $\mathcal{P}$, but all nontrivial quotients of G/N are $\mathcal{P}$. (That is, G/N is **just not** $\mathcal{P}$.) ▷ solution ◁

Exercise 7.19 Show that the group $\mathbb{Z}$ is just-not-finite (i.e. just-infinite); that is, $\mathbb{Z}$ is an infinite group such that any nontrivial quotient of $\mathbb{Z}$ is finite. ▷ solution ◁

Exercise 7.20 Show that if G is a finitely generated infinite group, then there exists $N \lhd G$ such that $[G : N] = \infty$, but for any $N \lhd M \lhd G$ such that $N \neq M$, we have that $[G : M] < \infty$. ▷ solution ◁

7.5 Choquet–Deny Groups Are Virtually Nilpotent

Exercise 7.21 Let G be a finitely generated group and let $N \lhd G$. Show that $\dim \mathrm{BHF}(G, \mu) \geq \dim \mathrm{BHF}(G/N, \bar{\mu})$ where $\bar{\mu}$ is the projection of μ onto the group G/N.

Conclude that if $(G/N, \bar{\mu})$ is not Liouville, then neither is (G, μ). ▷ solution ◁

Exercise 7.22 Let G be a group and $N \lhd G$. Let ν be an adapted probability measure on G/N. Show that there exists an adapted probability measure μ on G such that the projection $\bar{\mu}$ on G/N satisfies $\bar{\mu} = \nu$.

Show that if ν is symmetric, then we can choose μ to be symmetric.

Show that if $H(\nu) < \infty$ then μ can be chosen so that $H(\mu) < \infty$. ▷ solution ◁

In order to prove Theorem 7.1.6, we will need to construct some probability measure on a finitely generated ICC group G. The construction will utilize a suitable sequence of group elements $(g_n)_n$, choosing g_n with probability roughly $n^{-\alpha}$ for some $\alpha \in (1, 2)$. Thus, there is some underlying probability measure on $\mathbb{N}$ governing which element in the sequence is sampled.

We begin with the construction of the sequence $(g_n)_n$. Fix some parameter $\alpha \in (1, 2)$.

Start the construction by fixing a finite, symmetric generating set S for G, and consider the Cayley graph with respect to S. We assume $1 \notin S$. Let B_r denote the ball of radius r around $1 \in G$.

Set $G_0 = \{1\}$ and $G_1 = S$. Let $r_0 = 0$. For $n > 1$, given $G_1, \ldots, G_{n-1}$, we let

$$r_{n-1} = 4n^{2(\alpha-1)} \max \left\{ |x| \; : \; x \in \bigcup_{k=1}^{n-1} G_k \right\}$$

and choose g_n such that $|g_n| > r_{n-1}$ and g_n shatters $B_{r_{n-1}}$. We have required that G is ICC, to be able to find such a g_n using Lemma 7.3.5. Set $G_n = \left\{ g_n, g_n^{-1} \right\}$. Let $G_* = \bigcup_{n=0}^{\infty} G_n$. (The exact constant 4 above is not important, but it needs to be large enough for what follows.)

For $n \geq 0$, we write $\|x\| = n$ for any $x \in G_n$. Note that if $x = u_1 \cdots u_t$ where $u_k \in G_*$ for all k, then writing $m = \max_{k \leq n} \|u_k\|$, we have that $|x| \leq t \max_{k \leq m} |g_k|$.

For $u_1, \ldots, u_t \in G_*$, let $M(u_1, \ldots, u_t) = \max\{ \|u_j\| \; : \; 1 \leq j \leq t \}$. Denote

$$J(u_1, \ldots, u_t) = \sum_{j=1}^{t} \mathbf{1}_{\{ \|u_j\| = M(u_1, \ldots, u_t) \}},$$

and note that $J(u_1, \ldots, u_t) \geq 1$. Define

$$\mathcal{U}_t = \left\{ (u_1, \ldots, u_t) \in G_*^t \; : \; J(u_1, \ldots, u_t) = 1, \; M(u_1, \ldots, u_t)^{2(\alpha-1)} > t \right\}.$$

Lemma 7.5.1 *Let $t, s \geq 2$. Let $(u_1, \ldots, u_t) \in \mathcal{U}_t$ and $(v_1, \ldots, v_s) \in \mathcal{U}_s$. Let $a \in S$ and assume that $au_1 \cdots u_t = v_1 \cdots v_s$.*

Then, there exist $k < t$ and $\ell < s$ such that $au_1 \cdots u_k = v_1 \cdots v_\ell$.

Proof We can write $u_1 \cdots u_t = xgz$ where:

- $x = u_1 \cdots u_k$ for some $1 \leq k < t$,
- $g = u_{k+1}$ admits $\|g\| = M(u_1, \ldots, u_t) > 0$,
- $z = u_{k+2} \cdots u_t$ (and $z = 1$ if $k + 1 = t$),
- for all $j \neq k + 1$ we have $\|u_j\| < \|g\|$.

Similarly, we write $v_1 \cdots v_t = x'g'z'$ where:

- $x' = v_1 \cdots v_\ell$ for some $1 \leq \ell < s$,
- $g' = v_{\ell+1}$ admits $\|g'\| = M(v_1, \ldots, v_t) > 0$,
- $z' = v_{\ell+2} \cdots v_s$ (and $z' = 1$ if $\ell + 1 = s$),
- for all $j \neq \ell + 1$ we have $\|v_j\| < \|g'\|$.

If $\|g\| > \|g'\|$ then $axgz = x'g'z'$ implies that for $n = \|g\|$,

$$4n^{2(\alpha-1)} \max_{k \leq n-1} |g_k| = r_{n-1} < |g| \leq |a| + |x| + |z| + |x'| + |z'| + |g'|$$

$$\leq 1 + (t - 1 + s) \cdot \max_{k \leq n-1} |g_k| \leq (t + s) \cdot \max_{k \leq n-1} |g_k|,$$

which implies that

$$t + s < M(u_1, \ldots, u_t)^{2(\alpha-1)} + M(v_1, \ldots, v_s)^{2(\alpha-1)} \le 2n^{2(\alpha-1)} < t + s,$$

a contradiction! Similarly, by interchanging the roles of g, g', we also get a contradiction if we assume that $\|g\| < \|g'\|$.

So it must be that $\|g\| = \|g'\|$, and that for $n = M(u_1, \ldots, u_t) = M(v_1, \ldots, v_s)$ we have $a x g_n^\eta z = x' g_n^\xi z'$ for some $\eta, \xi \in \{-1, 1\}$. Since

$$\left|(x')^{-1} a x\right| \le (t + s - 1) \max_{k \le n-1} |g_k| < 2n^{2(\alpha-1)} \max_{k \le n-1} |g_k| \le r_{n-1},$$

and similarly,

$$\left|z' z^{-1}\right| \le (t + s - 2) \max_{k \le n-1} |g_k| < r_{n-1},$$

and since g_n shatters $B_{r_{n-1}}$, the identity $g_n^{-\xi}(x')^{-1} a x g_n^\eta = z' z^{-1}$ actually implies that $a u_1 \cdots u_k = a x = x' = v_1 \cdots v_\ell$. $\qquad\square$

We now fix an additional integer parameter $K > 0$, which will later be taken to be large enough. Define $\beta = \frac{\alpha-1}{2}$ and

$$\mathcal{E}_t(K) =$$

$$\{(u_1, \ldots, u_t) \in G_*^t \mid \forall\, 1 \le j < K^\beta,\ u_j = 1,\ \forall\, K^\beta \le k \le t,\ (u_1, \ldots, u_k) \in \mathcal{U}_k\}.$$

Lemma 7.5.2 *Let $t, s \ge 1$ and let $(u_1, \ldots, u_t) \in \mathcal{E}_t(K)$ and $(v_1, \ldots, v_s) \in \mathcal{E}_s(K)$. Let $a \in S$.*
Then, $a u_1 \cdots u_t \ne v_1 \cdots v_s$.

Proof Let $N = \lceil K^\beta \rceil$.

Assume for a contradiction that $a u_1 \cdots u_t = v_1 \cdots v_s$. Define

$$k = \min\{0 \le j \le t : \exists\, 1 \le \ell \le s,\ a u_1 \cdots u_j = v_1 \cdots v_\ell\},$$

where $k = 0$ implies that $v_1 \cdots v_\ell = a$ for some $1 \le \ell \le s$. After fixing k, let

$$\ell = \min\{1 \le j \le s : a u_1 \cdots u_k = v_1 \cdots v_j\}$$

(which satisfies $\ell \ge 1$ since $a \ne 1$).

If $k, \ell \ge N$, then $(u_1, \ldots, u_k) \in \mathcal{U}_k$ and $(v_1, \ldots, v_\ell) \in \mathcal{U}_\ell$, so by Lemma 7.5.1 there exist $k' < k$ and $\ell' < \ell$ such that $a u_1 \cdots u_{k'} = v_1 \cdots v_{\ell'}$, contradicting the minimality of k, ℓ.

We conclude that it must be that either $k < N$ or $\ell < N$.

If $k < N \le \ell$, then since $u_1 = \cdots = u_{N-1} = 1$, we have $v_1 \cdots v_\ell = a$. As in the proof of Lemma 7.5.1, we may find $\ell' < \ell$ such that $\|v_{\ell'+1}\| = n :=$

$M(v_1, \ldots, v_\ell)$ and for all $1 \leq j \leq \ell$ with $j \neq \ell' + 1$ we have $||v_j|| < n$. But this would imply that

$$4\ell \max_{j \leq n-1} |g_j| < 4n^{2(\alpha-1)} \max_{j \leq n-1} |g_j| = r_{n-1} < |v_{\ell'+1}| \leq 1 + (\ell - 1) \max_{j \leq n-1} |g_j|,$$

a contradiction! If $\ell < N \leq k$ we would arrive at a similar contradiction.

Hence, it must be that $k < N$ and $\ell < N$. But then $a = au_1 \cdots u_k = v_1 \cdots v_\ell = 1$, a contradiction!

Therefore, we have shown that it is impossible that $au_1 \cdots u_t = v_1 \cdots v_s$ for $(u_1, \ldots, u_t) \in \mathcal{E}_t(K)$ and $(v_1, \ldots, v_s) \in \mathcal{E}_s(K)$.   □

We now define the required probability measure on G. Using the above lemmas, we will prove it has the necessary properties, completing a proof of Theorem 7.1.6.

Let $\kappa = \frac{1}{\zeta(\alpha)}$ for $\zeta(\alpha) = \sum_{n=1}^{\infty} n^{-\alpha}$. For $n \geq 1$ and $x \in G_n$ define

$$\mu(x) = \mathbf{1}_{\{x \in G_n\}} \cdot \frac{\kappa}{|G_n|} \cdot (n + K)^{-\alpha},$$

and define

$$\mu(1) = 1 - \sum_{n=K+1}^{\infty} n^{-\alpha}.$$

Since $G_1 = S$, we see that μ is adapted. Since $(G_n)^{-1} = G_n$ for all n, we see that μ is symmetric.

The crucial property of this μ is given by the following lemma.
Define

$$G_t(K) := \{u_1 \cdots u_t \in G \ : \ (u_1, \ldots, u_t) \in \mathcal{E}_t(K)\}.$$

Lemma 7.5.3 *Let $(X_t)_t$ be the μ-random walk. For any $\varepsilon > 0$, there exists K_0 such that for all $K \geq K_0$ and all $t \geq 1$ we have that $\mathbb{P}[X_t \notin G_t(K)] < \varepsilon$.*

Proof Let $(U_t)_{t=1}^{\infty}$ by i.i.d.-μ elements, and $X_t = U_1 \cdots U_t$. We bound:

$$\mathbb{P}[X_t \notin G_t(K)] \leq \mathbb{P}[(U_1, \ldots, U_t) \notin \mathcal{E}_t(K)]$$

$$\leq \sum_{j \leq K^\beta} \mathbb{P}[U_j \neq 1] + \sum_{j > K^\beta} \mathbb{P}[(U_1, \ldots, U_j) \notin \mathcal{U}_t]$$

$$\leq K^\beta (1 - \mu(1)) + \sum_{j > K^\beta} \mathbb{P}[J(U_1, \ldots, U_j) \geq 2]$$

$$+ \sum_{j > K^\beta} \mathbb{P}\left[M(U_1, \ldots, U_j)^{2(\alpha-1)} \leq j\right]. \qquad (7.3)$$

If we can choose K large enough to make each of the three terms above smaller than $\frac{1}{3}\varepsilon$, we are done.

We will make repeated use of the following estimate: for any $r \geq 1$,

$$\sum_{n=r}^{\infty} n^{-\alpha} \geq \int_r^{\infty} \xi^{-\alpha} d\xi = \tfrac{1}{\alpha-1} r^{1-\alpha},$$

$$\sum_{n=r}^{\infty} n^{-\alpha} \leq r^{-\alpha} + \int_r^{\infty} \xi^{-\alpha} d\xi \leq \tfrac{\alpha}{\alpha-1} r^{1-\alpha}.$$

(This was also used in the solution to Exercise 4.43.)

For the first term in (7.3), note that by our choice of $\beta = \frac{\alpha-1}{2}$,

$$K^\beta(1 - \mu(1)) < K^\beta \sum_{n=K+1}^{\infty} n^{-\alpha} \leq \tfrac{\alpha}{\alpha-1} K^{-\beta},$$

which can be made as small as required by taking K large enough.

For the second term in (7.3), we compute for any $t \geq 2$ and $n \geq 1$, using the independence of $(U_j)_j$,

$$\mathbb{P}[J(U_1,\ldots,U_t) \geq 2, M(U_1,\ldots,U_t) = n \geq 1]$$

$$\leq \binom{t}{2} (\mathbb{P}[\|U_1\| = n])^2 \cdot (1 - \mathbb{P}[\|U_1\| \geq n])^{t-2}$$

$$\leq \tfrac{1}{2} t^2 \cdot \left(\kappa^2(n+K)^{-2\alpha} \wedge (1 - \tfrac{1}{\alpha-1}(n+K)^{1-\alpha})^{t-2} \right).$$

Also, for any $t \geq 1$,

$$\mathbb{P}[M(U_1,\ldots,U_t) = 0] \leq \mu(1)^t \leq \left(1 - \tfrac{1}{\alpha-1}(1+K)^{1-\alpha}\right)^t.$$

Hence, for $t \geq 2$ and any $r \geq 1$, by summing over $n \leq r$ and $n > r$ separately, we have that

$$\mathbb{P}[J(U_1,\ldots,U_t) \geq 2] \leq \left(1 - \tfrac{1}{\alpha-1}(1+K)^{1-\alpha}\right)^t$$
$$+ \tfrac{1}{2} t^2 \kappa^2 \cdot \tfrac{2\alpha}{2\alpha-1} (K+r)^{1-2\alpha}$$
$$+ \tfrac{1}{2} t^2 \cdot r \exp\left(-\tfrac{1}{\alpha-1}(K+r)^{1-\alpha}(t-2)\right).$$

By choosing r so that $r + K \geq \left(\frac{t}{C_1 \log t}\right)^{1/(\alpha-1)}$ for a large enough constant $C_1 = C_1(\alpha)$, we have that for some constants $C_2 = C_2(\alpha) > 0$ and $C_3 = C_3(\alpha) > 0$,

$$\mathbb{P}[J(U_1,\ldots,U_t) \geq 2] \leq C_2 \cdot t^{-\gamma}(\log t)^{(2\alpha-1)/(\alpha-1)} \leq C_3 t^{-\gamma+\eta},$$

where $\gamma = \frac{2\alpha-1}{\alpha-1} - 2 = \frac{1}{\alpha-1}$ and $\eta = \frac{2-\alpha}{2(\alpha-1)}$. Since $\gamma - \eta = \frac{\alpha}{2(\alpha-1)}$, summing over $t > K^\beta$, we have that for some constant $C_4 = C_4(\alpha) > 0$,

$$\sum_{j>K^\beta} \mathbb{P}[J(U_1,\ldots,U_j) \geq 2] \leq C_4 K^{-\beta(2-\alpha)/(\alpha-1)},$$

which takes care of the second term in (7.3).

Finally, for the third term in (7.3), we compute for $r^{2(\alpha-1)} = t$,

$$\mathbb{P}[M(U_1, \ldots, U_t) \le r] \le (1 - \mathbb{P}[\|U_1\| > r])^t \le \exp(-C_5 \sqrt{t}),$$

where $C_5 = C_5(\alpha) > 0$ is some constant. Thus, for some constants $C_6 = C_6(\alpha)$ and $C_7 = C_7(\alpha) > 0$ we have

$$\sum_{t > K^\beta} \mathbb{P}\left[M(U_1, \ldots, U_t)^{2(\alpha-1)} \le t\right] \le C_6 \exp\left(-C_7 K^{\beta/2}\right).$$

This takes care of the third term in (7.3), completing the proof.   □

We are now ready to prove that $\mathsf{BHF}(G, \mu)$ contains a nonconstant function.

Proof of Theorem 7.1.6   Fix some $a \in S$. Let $K_t(a)$ denote the total variation distance

$$K_t(a) = \| \mathbb{P}_a[X_t = \cdot] - \mathbb{P}[X_{t+1} = \cdot]\|_{\mathsf{TV}}.$$

By Exercise 6.42, it suffices to prove that $K_t(a)$ is bounded away from 0, as $t \to \infty$.

Fix some $t \ge 1$. We know that there is a coupling (X_t, Y_{t+1}) of two μ-random walks at time t and $t + 1$ such that $\mathbb{P}[aX_t \ne Y_t] = K_t(a)$. Choose K in Lemma 7.5.3 such that $\mathbb{P}[X_t \notin \mathcal{G}_t(K)] < \frac{1}{4}$ and $\mathbb{P}[Y_{t+1} \in \mathcal{G}_{t+1}(K)] < \frac{1}{4}$. Note that if $X_t \in \mathcal{G}_t(K)$ and $Y_{t+1} \in \mathcal{G}_{t+1}(K)$ then by Lemma 7.5.2 it must be that $aX_t \ne Y_t$. Thus,

$$1 - K_t(a) = \mathbb{P}[aX_t = Y_{t+1}] \le \mathbb{P}[X_t \in \mathcal{G}_t(K)] + \mathbb{P}[Y_{t+1} \in \mathcal{G}_{t+1}(K)] < \frac{1}{2}.$$

So $K_t(a) > \frac{1}{2}$ for all $t \ge 1$, and we are done.   □

7.6 Additional Exercises

Exercise 7.23 Let G be a finitely generated group and μ an adapted symmetric probability measure on G. Let h be a nonnegative μ-harmonic function. Show that there exists $c > 0$ such that for all $x \in G$ we have $h(x) \le e^{c|x|} h(1)$.

Conclude that if h is a nonnegative μ-harmonic function, and that $h(x) = 0$ for some $x \in G$, then $h \equiv 0$.   ▷ solution ◁

Exercise 7.24 Let G be a finitely generated group and μ an adapted probability measure on G.

Show that the center $Z = Z(G)$ acts trivially on any positive μ-harmonic function.   ▷ solution ◁

Exercise 7.25 Recall the definitions of τ, τ^n, τ^* (Definitions 7.2.1 and 7.3.1).

Let G be a finitely generated group and μ an adapted probability measure on G. Let $\mathcal{P}$ be the collection of all positive μ-harmonic functions. Let $K = \{x \in G : \forall h \in \mathcal{P}, x.h = h\}$.

Show that $K \triangleleft G$.

Show that $\tau(K) = K$.

Conclude that $\tau^\infty \triangleleft \tau^* \triangleleft K$. ▷ solution ◁

Exercise 7.26 Let G be a finitely generated virtually nilpotent group. Let μ be an adapted probability measure on G.

Show that any positive μ-harmonic function is constant. ▷ solution ◁

Recall Definition 6.6.1 and Theorem 6.6.2.

Let G be an infinite finitely generated group, and μ be an adapted symmetric probability measure on G. Let $(X_t)_t$ be the μ-random walk. For a finite subset $A \subset G$ let

$$E_A = \inf\{t \geq 0 : X_t \notin A\}$$

be the exit time from A. Define

$$M_{A,x}(z) = \mathbb{P}_x[X_{E_A} = z]$$

and

$$D_A(x \mid y) = \sup_{z \, : \, M_{A,x}(z) > 0} \left| 1 - \frac{M_{A,y}(z)}{M_{A,x}(z)} \right|.$$

Exercise 7.27 Show that if $A \subset A' \subset G$ are finite subsets, then for any $x, y \in G$ it holds that $D_{A'}(x \mid y) \leq D_A(x \mid y)$. ▷ solution ◁

Exercise 7.28 For this exercise assume that μ has finite support.

Let $x, y \in G$. Assume that there exists a nondecreasing sequence of finite subsets $A_n \subset A_{n+1}$ such that $\bigcup_n A_n = G$ and such that

$$\lim_{n \to \infty} D_{A_n}(x \mid y) = 0.$$

Show that for any positive μ-harmonic function h it holds that $h(x) = h(y)$. ▷ solution ◁

Exercise 7.29 Let $x, y \in G$. Assume that there exists a nondecreasing sequence of finite subsets $A_n \subset A_{n+1}$ such that $\bigcup_n A_n = G$ and such that

$$\limsup_{n\to\infty} D_{A_n}(x \mid y) > 0.$$

Prove that there exists a positive μ-harmonic function h such that $h(x) \neq h(y)$.

▷ solution ◁

Exercise 7.30 Assume that any positive μ-harmonic function on G is constant. Let $S \subset \mathrm{supp}(\mu)$ be a finite symmetric generating set and let $B_r = B_S(1, r)$ denote the ball of radius r around $1 \in G$ in the Cayley graph with respect to S.

Show that for any $\varepsilon > 0$ there exists $c > 0$ such that for all $r > 0$ and any $|z| = r + 1$,

$$M_{B_r, 1}(z) = \mathbb{P}[X_{E_{B_r}} = z] \geq c e^{-\varepsilon r}.$$

Conclude that G has sub-exponential growth; that is,

$$\lim_{r\to\infty} \frac{1}{r} \log |B_r| = 0.$$

▷ solution ◁

7.7 Remarks

The Choquet–Deny theorem (Corollary 7.1.2) was first shown by Blackwell (1955) for $\mathbb{Z}^d$, by Choquet and Deny (1960) for Abelian groups, and by Dynkin and Malyutov (1961) for finitely generated nilpotent groups.

It took quite a few decades before the converse was obtained by Frisch, Hartman, Tamuz, and Vahidi-Ferdowsi (2019). They prove that for any non-virtually nilpotent group there exists a finite entropy symmetric adapted random walk that is non-Liouville, Thoerem 7.1.5. This characterizes the Choquet–Deny groups as the virtually nilpotent groups, tying together the analytic phenomenon and algebraic property.

The equivalent condition for the existence of nonconstant positive harmonic functions (in Exercises 7.28 and 7.29), as well as the proof that any group of exponential growth admits a nonconstant positive harmonic function (Exercise 7.30), was shown in Amir and Kozma (2017).

Note that by Exercise 7.26, finitely generated virtually nilpotent groups (which by Gromov's theorem, Theorem 9.0.1, are the same as polynomial growth groups) do not admit nonconstant positive harmonic functions. In contrast, Exercise 7.30 tells us that any group of exponential growth admits some nonconstant positive harmonic function. The following question was raised in Amir and Kozma (2017), and is still open at the time of writing.

Question 7.7.1 Do groups of intermediate growth admit nonconstant positive harmonic functions?

Specifically, does the Grigorchuk Group (from Grigorchuk, 1980, 1984) admit nonconstant positive harmonic functions?

In Perl and Yadin (2023) it is shown that if $\mathsf{HF}_k(G, \mu)$ contains a nonconstant positive harmonic function then $\dim \mathsf{HF}_k(G, \mu) = \infty$. See Exercise 2.38 for the case where $\mathsf{LHF}(G, \mu)$ contains a nonconstant positive harmonic function.

Question 7.7.2 Let G be a finitely generated group and $\mu \in \mathsf{SA}(G, \infty)$. Assume that $\dim \mathsf{HF}_k(G, \mu) = \infty$.

Does $\mathsf{HF}_k(G, \mu)$ contain a nonconstant positive harmonic function?

7.8 Solutions to Exercises

Solution to Exercise 7.1 :(

If $f \in \mathsf{BHF}(G, \mu)$ then $\bar{f}(Kx) := f(x)$ is a well-defined function on G/K (because K acts trivially on f). It is also simple to verify that $\bar{f}$ is $\bar{\mu}$-harmonic.

Finally, this is a linear map, and its kernel is trivial. Indeed, if $\bar{f} \equiv 0$, then for any $x \in G$ we have that $f(x) = \bar{f}(Kx) = 0$, so $f \equiv 0$. :) ✓

Solution to Exercise 7.2 :(

If $y, z \in C_G(x)$, then since $N \lhd G$,

$$[x, yz] = x^{-1}z^{-1}y^{-1}xyz = [x, z] \cdot ([x, y])^z \in N,$$

$$\left[x, y^{-1}\right] = x^{-1}yxy^{-1} = ([y, x])^{y^{-1}} \in N,$$

because $[y, x] = ([x, y])^{-1}$. :) ✓

Solution to Exercise 7.3 :(

If $y \in N$ then $[x, y] = \left(y^{-1}\right)^x y \in N$, so $y \in C_G^N(x)$. :) ✓

Solution to Exercise 7.4 :(

The first assertion is immediate.

For the second assertion, let $\pi \colon G \to G/N$ be the canonical projection, and let $\varphi = \pi|_{C_G^M(x)}$ be the restriction of π to $C_G^M(x)$.

If $y \in C_G^M(x)$ then $[\pi(x), \pi(y)] = \pi([x, y]) \in M/N$, because $[x, y] \in M$. Thus φ is a homomorphism from $C_G^M(x)$ into $C_{G/N}^{M/N}(\pi(x))$.

Now, if $Ny \in C_{G/N}^{M/N}(\pi(x))$, then $\pi([x, y]) = [\pi(x), \pi(y)] \in M/N$, so $N[x, y] = Nm$ for some $m \in M$. Since $N \lhd M$, this implies that $[x, y] \in M$.

Thus, $\varphi \colon C_G^M(x) \to C_{G/N}^{M/N}(Nx)$ is surjective, completing the proof. :) ✓

Solution to Exercise 7.5 :(
We have that

$$z \in C_G^N(x^y) \iff [x^y, z] \in N \iff \left[x, z^{y^{-1}}\right] \in N \iff z \in \left(C_G^N(x)\right)^y . \qquad :)\checkmark$$

Solution to Exercise 7.6 :(
Since

$$\left[x^{-1}, y\right] = xy^{-1}x^{-1}y = ([y, x])^{x^{-1}} = \left([x, y]^{-1}\right)^{x^{-1}},$$

we have that $C_G^N\left(x^{-1}\right) = C_G^N(x)$.
   Also, because

$$[xy, z] = y^{-1}x^{-1}z^{-1}xyz = ([x, z])^y [y, z],$$

we know that $C_G^N(xy) \supset C_G^N(x) \cap C_G^N(y)$, so $x, y \in \tau(N)$ implies that $xy \in \tau(N)$ (by Proposition 1.3.3).
   This shows that $\tau(N)$ is indeed a subgroup.
   Now let $y \in G$. The automorphism $z \mapsto z^y$ of G maps $C_G^N(x)$ onto $C_G^N(x^y)$. Thus, $\left[G : C_G^N(x)\right] = \left[G : C_G^N(x^y)\right]$, implying that $x \in \tau(N)$ if and only if $x^y \in \tau(N)$. This shows that $\tau(N) \lhd G$. $\qquad :)\checkmark$

Solution to Exercise 7.7 :(
If $x \in N$, then $[x, y] \in N$ for all $y \in G$. That is, $C_G^N(x) = G$, which implies that $x \in \tau(N)$. $\qquad :)\checkmark$

Solution to Exercise 7.8 :(
Since $N \lhd C_G^N(x)$ for any x, we have that $\left[G : C_G^N(x)\right] \leq [G : N] < \infty$ for all x, implying that $\tau(N) = G$.
$\qquad :)\checkmark$

Solution to Exercise 7.9 :(
Just note that since $N \lhd M$, we have that $C_G^N(x) \leq C_G^M(x)$. Thus, $\left[G : C_G^M(x)\right] \leq \left[G : C_G^N(x)\right]$. This easily implies the assertion. $\qquad :)\checkmark$

Solution to Exercise 7.10 :(
Let $x, y \in N_\infty$ and $z \in G$. Since $(N_n)_n$ is a nondecreasing sequence, there exists n such that $x, y \in N_n$. So $x^{-1}y \in N_n \subset N_\infty$, and also $x^z \in N_n \subset N_\infty$. $\qquad :)\checkmark$

Solution to Exercise 7.11 :(
Since $C_G^M(x)/N = C_{G/N}^{M/N}(Nx)$, we know that $\left[G : C_G^M(x)\right] = \left[G/N : C_{G/N}^{M/N}(Nx)\right]$. So we have that $x \in \tau_G(M)$ if and only if $Nx \in \tau_{G/N}(M/N)$, which is to say that $\tau_G(M)/N = \tau_{G/N}(M/N)$.
   The second assertion follows inductively on $n - k$ by noting that by induction, for $N = \tau^k$,

$$\tau^{n+1}/N = \tau(\tau^n)/N = \tau_{G/N}(\tau^n/N) = \tau_{G/N}\left(\tau_{G/N}^{n-k}\right) = \tau_{G/N}^{n+1-k}. \qquad :)\checkmark$$

Solution to Exercise 7.12 :(
This is done by induction on n.
   The base case of $n = 1$ is just the observation that if $z \in Z_1$ then $C_G(z) = G$.
   For $n > 1$, we know that $Z_n/Z_{n-1} = Z_1(G/Z_{n-1})$. So

$$Z_n/Z_{n-1} \subset \tau_{G/Z_{n-1}}^1 = \tau_G(Z_{n-1})/Z_{n-1}.$$

Since $Z_{n-1} \lhd Z_n \cap \tau_G(Z_{n-1})$ we get that $Z_n \lhd \tau_G(Z_{n-1})$. By induction we now have that $Z_n \lhd \tau_G(Z_{n-1}) \lhd \tau_G\left(\tau_G^{n-1}\right) = \tau_G^n$, completing the proof. $\qquad :)\checkmark$

Solution to Exercise 7.13 :(

If $N \lhd G$, then for any $x \in H$ we have that $C_H^{H \cap N}(x) = C_G^N(x) \cap H$. As $[G : H] < \infty$ we conclude that $\tau_H(H \cap N) = \tau_G(N) \cap H$. Thus, inductively,

$$\tau_H^n = \tau_H\left(\tau_H^{n-1}\right) = \tau_H\left(\tau_G^{n-1} \cap H\right) = \tau_G\left(\tau_G^{n-1}\right) \cap H = \tau_G^n \cap H. \qquad :) \checkmark$$

Solution to Exercise 7.14 :(

We prove this by induction on n.

For $n = 1$ this follows since $\{1\} \lhd N$ for any $N \lhd G$. So if $\tau(N) = N$ then $\tau^1 \lhd \tau(N) = N$. Thus, $\tau^1 \lhd \tau^*$.

For the induction step, let $n > 1$. Note that by induction, $\tau^{n-1} \lhd \tau^*$. So for any $N \lhd G$ such that $\tau(N) = N$, since $\tau^* \lhd N$, we have that $\tau^n \lhd \tau(\tau^*) \lhd \tau(N) = N$. Intersecting over all such N completes the proof. :) $\checkmark$

Solution to Exercise 7.15 :(

For any $N \lhd G$ such that $\tau(N) = N$ we have that $\tau^* \lhd N$, hence $\tau(\tau^*) \lhd \tau(N) = N$. Intersecting over all such N gives that $\tau(\tau^*) \lhd \tau^*$. The other inclusion is true for any normal subgroup. :) $\checkmark$

Solution to Exercise 7.16 :(

If $[G : \tau^*] < \infty$, then $\tau^* = \tau(\tau^*) = G$, so we may assume that $[G : \tau^*] = \infty$.

Since

$$\tau_{G/\tau^*}(\{1\}) = \tau_G(\tau^*)/\tau^* = \tau^*/\tau^*,$$

we conclude that G/τ^* is ICC as long as it is infinite. :) $\checkmark$

Solution to Exercise 7.17 :(

Let $s_1, \ldots, s_d$ be generators of G. Since $G = \tau^\infty = \bigcup_n \tau^n$, there exists n such that $s_1, \ldots, s_d \in \tau^n$, which implies that $G = \tau^n \lhd \tau^*$. :) $\checkmark$

Solution to Exercise 7.18 :(

This is almost identical to the proof of Lemma 7.4.2.

Consider the collection

$$\mathcal{N} = \{N \lhd G : G/N \text{ is not } \mathcal{P}\}.$$

If $(N_k)_k$ is an increasing chain in $\mathcal{N}$, then $N_\infty := \bigcup_k N_k$ is a normal subgroup.

If $N_\infty \notin \mathcal{N}$, then G/N_∞ is $\mathcal{P}$. So G/N_∞ is finitely presented. Say

$$G/N_\infty = \langle s_1, \ldots, s_d \mid r_1, \ldots, r_m \rangle.$$

Just as in the proof of Lemma 7.4.2, there exists k such that all relations r_j would be mapped into N_k by the canonical projection onto G. This would imply that $N_\infty = N_k$, meaning that G/N_∞ is not $\mathcal{P}$, a contradiction!

We conclude that any nondecreasing chain in $\mathcal{N}$ has an upper bound in $\mathcal{N}$, so by Zorn's lemma $\mathcal{N}$ contains a maximal element. That is, there exists a maximal element $N \in \mathcal{N}$. So G/N is not $\mathcal{P}$.

Now, assume that G/N has a non-$\mathcal{P}$ quotient. That implies the existence of some $N \lhd M \lhd G$ such that G/M is not $\mathcal{P}$. By the maximality of N, we have that $N = M$. So the only non-$\mathcal{P}$ quotient of G/N is by the trivial group, as required. :) $\checkmark$

Solution to Exercise 7.19 :(

If $0 \neq z \in N \lhd \mathbb{Z}$, it must be that $\{nz : n \in \mathbb{N}\} \subset N$, so $[\mathbb{Z} : N] \leq |z|$. :) $\checkmark$

Solution to Exercise 7.20 :(

All finite groups are finitely presented, and quotients of finite groups are also finite. So by Exercise 7.18, there exists $N \lhd G$ such that G/N is just-infinite. That is, G/N is infinite, but any nontrivial quotient of G/N is finite. So if $N \neq M$ for $N \lhd M \lhd G$, then $[G/N : M/N] = [G : M] < \infty$. :) $\checkmark$

Solution to Exercise 7.21 :(

Let $f_1, \ldots, f_n \in \mathrm{BHF}(G/N, \bar{\mu})$ be linearly independent bounded harmonic functions. For each $1 \leq k \leq n$, define $h_k : G \to \mathbb{C}$ by $h_k(x) = f_k(Nx)$. It is easy to check that h_k is μ-harmonic because f_k is $\bar{\mu}$-harmonic. It is also straightforward that $h_1, \ldots, h_n$ are linearly independent. :) $\checkmark$

Solution to Exercise 7.22 :(

Fix some Cayley graph of G. Let ρ be the probability measure on N given by

$$\rho(x) = \mathbf{1}_{\{x \in N\}} \cdot C \cdot \frac{1}{(|x|+1)^3 \, |S_{|x|} \cap N|}$$

for an appropriate constant $C > 0$, where $S_r = \{x \in G \; : \; |x| = r\}$.

Fix a set of representatives $R \subset G$ so that $G = \biguplus_{r \in R} Nr$. For any $x \in G$, there are unique elements $n_x \in N$ and $r_x \in R$ such that $x = n_x r_x$. Denote $\bar{n}_x = n_{x^{-1}}$. Thus, the following is well defined:

$$\mu(x) := \tfrac{1}{2}(\rho(n_x) + \rho(\bar{n}_x))\nu(Nx).$$

Since $r^{-1}N = Nr^{-1}$, we have that for any $r \in R$,

$$\sum_{n \in N} \rho\left(n_{r^{-1}n^{-1}}\right) = \sum_{x \in r^{-1}N} \rho(n_x) = \sum_{n \in N} \rho(n) = 1.$$

Thus,

$$\begin{aligned}
\bar{\mu}(Nr) &= \sum_{n \in N} \tfrac{1}{2}\left(\rho(n) + \rho(n_{r^{-1}n^{-1}})\right)\nu(Nr) \\
&= \frac{1}{2}\sum_{n \in N}\rho(n)\nu(Nr) + \frac{1}{2}\sum_{x \in r^{-1}N}\rho(n_x)\nu(Nr) \\
&= \frac{1}{2}\sum_{n \in N}\rho(n)\nu(Nr) + \frac{1}{2}\sum_{n \in N}\rho(n)\nu(Nr) = \nu(Nr),
\end{aligned}$$

which implies that $\bar{\mu} = \nu$ and that μ is indeed a probability measure.

If ν is symmetric, then $\nu(Nx) = \nu\left(Nx^{-1}\right)$, so that μ is also symmetric.

For any $r \in R$, let $\rho_r(n) := \rho\left(n_{r^{-1}n^{-1}}\right)$, which we have seen above is a probability measure. If X is a random element of law μ, then one easily checks that r_X has law ν, and $(X \mid r_X = r)$ has law $\tfrac{1}{2}(\rho + \rho_r)$.

Note that $|S_k| \leq D^k$ for some $D > 1$ and all $k \geq 0$. Thus,

$$\begin{aligned}
H(\rho) &\leq \sum_{k=0}^{\infty} \sum_{n \in S_k \cap N} \frac{C \log |S_k| + 3C \log(k+1) - C \log C}{(k+1)^3 |S_k \cap N|} \\
&\leq \sum_{k=0}^{\infty} \frac{C \log D \cdot k + 3C \log k - C \log C}{(k+1)^3} < \infty.
\end{aligned}$$

Also,

$$H(\rho_r) = -\sum_{x \in r^{-1}N} \rho(n_x) \log \rho(n_x) = -\sum_{x \in Nr^{-1}} \rho(n_x) \log \rho(n_x) = H(\rho).$$

So for any $r \in R$ we have that $H\left(\tfrac{1}{2}(\rho + \rho_r)\right) \leq \log 2 + H(\rho)$, implying that if X is a random element of G with law μ, then

$$H(X) \leq H(X \mid r_X) + H(r_X) \leq \log 2 + H(\rho) + H(\nu).$$

Finally, to show that μ is adapted, let $n \in N$ and $r \in R$. Write $Nr = Np_1 \cdots p_k$ for $Np_j \in \operatorname{supp}(\nu)$ and $p_j \in R$ for all j. Let $m \in N$ be such that $mp_1 \cdots p_k = nr$. Note that $\rho(m) > 0$ (since ρ is supported on all of N). Since $\mu(mp_1) = \rho(m)\nu(Np_1) > 0$ and $\mu(p_j) = \nu(Np_j) > 0$ for all j, we have that $mp_1, p_2, \ldots, p_k \in \operatorname{supp}(\mu)$. Since $nr = mp_1p_2 \cdots p_k$, and this holds for arbitrary $n \in N, r \in R$, we find that μ is adapted.

:) ✓

Solution to Exercise 7.23 :(

Let S be a finite symmetric generating set for G such that $S \subset \operatorname{supp}(\mu)$. Consider the Cayley graph of G with respect to S. Set $\mu_* = \min_{s \in S} \mu(s) > 0$.

Let $x \in G$ and write $x = s_1 \cdots s_n$ for $s_j \in S$ and $n = |x|$. Then, because h is nonnegative,

$$h(1) = \sum_y \mu(y)h(y) \geq \mu_* h(s_1) = \mu_* \sum_y \mu(y)h(s_1 y) \geq (\mu_*)^2 h(s_1 s_2)$$

$$\geq \cdots \geq (\mu_*)^n h(s_1 \cdots s_n) = (\mu_*)^{|x|} h(x).$$

This proves the first assertion.

Now, assume that h is a nonnegative μ-harmonic function and that $h(x) = 0$. Set $f = x.h$. So f is a nonnegative μ-harmonic function with $f(1) = 0$. By the first assertion, $0 \leq f(y) \leq e^{c|y|} \cdot f(1) = 0$ for some $c > 0$ and all $y \in G$. This implies that $h \equiv 0$. :)✓

Solution to Exercise 7.24 :(
This is similar to the proof of Theorem 7.1.1.

Let f be any positive μ-harmonic function.

Let $x \in Z$. Let $(X_t)_t$ be the μ-random walk on G. Since μ is adapted, there exist $k = k(x) > 0$ and $\alpha > 0$ such that for any t we have for all $y \in G$,

$$\mathbb{P}_y \left[X_{t+k} = X_t x^{-1} \mid X_t \right] = \mathbb{P} \left[X_k = x^{-1} \right] =: \alpha > 0.$$

Specifically,

$$\mathbb{E}_y[|f(X_{t+k}) - f(X_t)|] \geq \alpha \, \mathbb{E}_y \left[|f(X_t x^{-1}) - f(X_t)| \right]$$
$$= \alpha \, \mathbb{E}_y[|x.f(X_t) - f(X_t)|] \geq \alpha |x.f(y) - f(y)|.$$

We have used that $f\left(X_t x^{-1}\right) = f\left(x^{-1} X_t\right) = x.f(X_t)$ since $x \in Z$.

On the other hand, since the sequence $(f(X_t))_t$ is a nonnegative martingale, it converges a.s. and in L^1 by the martingale convergence theorem (Theorem 2.6.3). This implies that $(f(X_{t+k}) - f(X_t))_t$ converges to 0 a.s. and in L^1. Thus, $\mathbb{E}_y[|f(X_{t+k}) - f(X_t)|] \to 0$.

We obtain that for any $y \in G$ and any $x \in Z$ we have $x.f(y) = f(y)$. This implies that Z acts trivially on f. :)✓

Solution to Exercise 7.25 :(
It is easily shown that $K \lhd G$.

For the rest of the proof, since $K \lhd \tau(K)$ we only need to show that $\tau(K) \lhd K$. This is exactly as in the proof of Theorem 7.3.3, with $\mathcal{P}$ replacing $\mathrm{BHF}(G, \mu)$, and using the fact that positive martingales converge a.s. and in L^1. :)✓

Solution to Exercise 7.26 :(
By Theorem 7.2.3, since G is virtually nilpotent, $\tau^\infty = G$, so also $G = \tau^*$. By Exercise 7.25, G acts trivially on any positive μ-harmonic function, so any such function must be constant. :)✓

Solution to Exercise 7.27 :(
Note that since $A \subset A'$, we have that $E_A \leq E_{A'}$. Thus, by the strong Markov property,

$$\mathbb{P}_a \left[X_{E_{A'}} = b \right] = \sum_w \mathbb{P}_a \left[X_{E_A} = w \right] \cdot \mathbb{P}_w \left[X_{E_{A'}} = b \right].$$

This leads to

$$|M_{A',x}(z) - M_{A',y}(z)| = \sum_w M_{A',w}(z) \cdot |M_{A,x}(w) - M_{A,y}(w)|$$

$$\leq D_A(x \mid y) \cdot \sum_w M_{A',w}(z) \cdot M_{A,x}(w) = D_A(x \mid y) \cdot M_{A',x}(z).$$

Maximizing over z proves the claim. :)✓

Solution to Exercise 7.28 :(
Let A be a finite subset, and consider the exit time E_A. If h is a positive μ-harmonic function and $(X_t)_t$ is a μ-random walk, then $(M_t := h(X_{t \wedge E_A}))_t$ is a bounded martingale. Indeed,

$$0 < M_t \leq \max_{\substack{x \in A \\ s \in \text{supp}(\mu)}} h(xs) < \infty.$$

Thus, by the optional stopping theorem (Theorem 2.3.3), $h(x) := \mathbb{E}_x \left[h(X_{E_A}) \right]$ for any $x \in G$.

Now, let $n(x, y)$ be large enough so that $x, y \in A_n$ for all $n \geq n(x, y)$ $\left($which is possible because $(A_n)_n$ are nondecreasing and $\bigcup_n A_n = G\right)$. Using that h is positive we can now compute,

$$|h(x) - h(y)| = \sum_z h(z) \cdot |\mathbb{P}_x \left[X_{E_{A_n}} = z \right] - \mathbb{P}_y \left[X_{E_{A_n}} = z \right]|$$

$$\leq \sum_z h(z) M_{A,x}(z) \cdot D_{A_n}(x \mid y) = h(x) \cdot D_{A_n}(x \mid y) \to 0.$$

This implies that $h(x) = h(y)$. :) ✓

Solution to Exercise 7.29 :(

Let $\varepsilon > 0$ and $(z_n)_n$ be such that for all n we have

$$|M_{A_n, x}(z_n) - M_{A_n, y}(z_n)| > \varepsilon M_{A_n, x}(z_n).$$

For each n define

$$h_n(w) = \frac{M_{A_n, w}(z_n)}{M_{A_n, x}(z_n)}.$$

Let S be a finite symmetric generating set for G such that $S \subset \text{supp}(\mu)$. Consider the Cayley graph of G with respect to S. Set $\mu_* = \min_{s \in S} \mu(s) > 0$.

Let $w \in G$. Because $\bigcup_n A_n = G$ and $A_n \subset A_{n+1}$ for all n, there exists $n(w)$ such that $S \cup \{w\} \subset A_n$ for any $n \geq n(w)$. For any $s \in S$ and $n \geq n(w)$, we have that

$$\mathbb{P}_{ws} \left[X_{E_{A_n}} = z_n, \ E_{A_n} > 1, \ X_1 = w \right] \geq \mu_* \cdot \mathbb{P}_w \left[X_{E_{A_n}} = z_n \right].$$

Thus, for any $n \geq n(w)$,

$$h_n(ws) \geq \mu_* \cdot h_n(w).$$

Since S generates G, there exist $s_1, \ldots, s_{|w|} \in S$ such that $w s_1 \cdots s_{|w|} = 1$. We can choose $m(w)$ large enough so that

$$\left\{ w, w s_1, w s_1 s_1, \ldots, w s_1 \cdots s_{|w|} = 1 \right\} \subset A_n$$

for all $n \geq m(w)$. Thus, for all $n \geq m(w)$ we have that

$$h_n(1) \geq (\mu_*)^{|w|} \cdot h_n(w).$$

Since this holds for any $w \in G$, an Arzelà–Ascoli type argument (as in the solution to Exercise 1.97) proves that there is a subsequence $(h_{n_k})_k$ such that the limit

$$h(w) := \lim_{k \to \infty} h_{n_k}(w)$$

exists for each $w \in G$.

We now have that

$$h(x) = \lim_{k \to \infty} h_{n_k}(x) = 1,$$

$$|1 - h(y)| = \lim_{k \to \infty} \left| 1 - h_{n_k}(y) \right| > \varepsilon.$$

So $h(x) \neq h(y)$.

We are left with showing that h is μ-harmonic. Let $w \in G$. Fix $\delta > 0$ small, and let $F = F_\delta$ be a finite subset such that $\mathbb{P}[X_1 \notin F] < \delta$. Since F is finite, there exists $n_{w,F}$ such that for all $n \geq n_F$ we have $wF \cup \{w\} \subset A_n$. Thus, for any $n \geq n_{w,F}$,

$$\left| \mathbb{P}_w \left[X_{E_{A_n}} = z_n \right] - \sum_{z \in F} \mathbb{P}_w \left[X_{E_{A_n}} = z_n, \, X_1 = wz \right] \right|$$

$$\leq \sum_{z \notin F} \mathbb{P}_w[X_1 = wz] \leq \mathbb{P}[X_1 \notin F] < \delta.$$

For $z \in F$ and $n \geq n_{w,F}$, we have

$$\mathbb{P}_w \left[X_{E_{A_n}} = z_n, \, X_1 = wz \right] = \mu(z) \, \mathbb{P}_{wz} \left[X_{E_{A_n}} = z_n \right]$$

by the Markov property. We conclude that for all $n \geq n_{w,F}$,

$$\left| \mathbb{P}_w \left[X_{E_{A_n}} = z_n \right] - \sum_z \mu(z) \, \mathbb{P}_{wz} \left[X_{E_{A_n}} = z_n \right] \right|$$

$$\leq \delta + \sum_{z \notin F} \mu(z) = \delta + \mathbb{P}[X_1 \notin F] < 2\delta.$$

We conclude that for any $\delta > 0$ and any $w \in G$, there exists n_δ such that for all $n \geq n_\delta$,

$$\left| h_n(w) - \sum_z \mu(z) h_n(wz) \right| < 2\delta.$$

Taking a limit along the subsequence $\left(h_{n_k} \right)_k$ shows that h is harmonic at w, for any $w \in G$. :) ✓

Solution to Exercise 7.30 :(
An important observation is that since $S \subset \text{supp}(\mu)$, any path in the Cayley graph has positive probability of being traversed by the μ-random walk. Specifically, if we set $\mu_* = \min_{s \in S} \mu(s) > 0$, for any $s_1, \dots, s_n \in S$ and any $x \in G$, we have that

$$\mathbb{P}_x[\forall \, 1 \leq k \leq n, \, X_k = x s_1 \cdots s_k] \geq (\mu_*)^n.$$

For any $|x| \leq r$ and any $|z| = r + 1$, there is a path in the Cayley graph from x to z of length at most $|x| + |z| \leq 2r + 1$, which stays inside B_r until hitting z (e.g. follow a geodesic from x to 1, and then a geodesic from 1 to z). We thus have that $M_{B_r, x}(z) \geq (\mu_*)^{2r+1} > 0$.

Fix $\varepsilon > 0$.

Since any positive harmonic function is constant, by Exercise 7.29 we can choose R be large enough so that for any $r \geq R$ we have

$$D_r := \max_{s \in S} (D_{B_r}(1 \mid s) \vee D_{B_r}(s \mid 1)) < \varepsilon.$$

Note that the $x \mapsto yx$ is an automorphism of the Cayley graph, and since $M_{A,x}(z) = M_{yA, yx}(yz)$, we get that

$$D_{B(x,r)}(x \mid xs) \vee D_{B(x,r)}(xs \mid x) \leq D_r.$$

Now, let $s \in S$ and $x \in G$. For any $r \geq R + |x|$, since $B(x, R) \subset B_r$, by Exercise 7.27 we have that

$$D_{B_r}(x \mid xs) \vee D_{B(x,r)}(xs \mid x) \leq D_R < \varepsilon.$$

Now, let $r \geq R + |x|$ and choose $s_1, \dots, s_{|x|} \in S$ such that $x = s_1 \cdots s_{|x|}$. Write $y_k = s_1 \cdots s_k$ for $1 \leq k \leq |x|$, and $y_0 = 1$. By the above, for any $|z| = r + 1$,

$$M_{B_r, x}(z) = M_{B_r, y_{|x|-1} s_{|x|}}(z) \leq (1 + \varepsilon) M_{B_r, y_{|x|-1}}(z)$$

$$\leq \cdots \leq (1 + \varepsilon)^{|x|} \cdot M_{B_r, 1}(z) \leq e^{\varepsilon |x|} \cdot M_{B_r, 1}(z).$$

Now, let $r > R$ and let $z \in G$ be such that $|z| = r + 1$. Write $z = s_1 \cdots s_{r+1}$ for $s_j \in S$. Set $y_0 = 1$ and $y_j = s_1 \cdots s_j$ for all $1 \leq j \leq r + 1$. Note that $|y_j| = j \leq r$. Let $x = y_{r-R}$. We have that

$$M_{B_r, x}(z) \geq \mathbb{P}_x[\forall \, 1 \leq t \leq R + 1, \, X_j = y_{r-R+t}] \geq (\mu_*)^{R+1}.$$

Thus we have that

$$M_{B_r,1}(z) \geq e^{-\varepsilon(r-R)} \cdot (\mu_*)^{R+1} = \mu_* \cdot (e^\varepsilon \mu_*)^R \cdot e^{-\varepsilon r}.$$

This holds for all $r > R$ and all $|z| = r + 1$. Choosing an appropriate constant $c > 0$ gives that there is a constant $c > 0$ such that for all $r > 0$ and all $|z| = r + 1$ it holds that $M_{B_r,1}(z) > ce^{-\varepsilon r}$.

Summing over z, we find

$$1 \geq \#\{z \ : \ |z| = r + 1\} \cdot ce^{-\varepsilon r}.$$

Summing over r we conclude that

$$|B_r| = 1 + \sum_{k=0}^{r-1} \#\{z \ : \ |z| = k + 1\} \leq 1 + \frac{1}{c} \sum_{k=0}^{r-1} e^{\varepsilon k}$$

$$\leq 1 + \frac{1}{c(e^\varepsilon - 1)} \cdot e^{\varepsilon r}.$$

Thus, for some constant $C > 0$ and all $r > 0$ we have that $\log |B_r| \leq C + \varepsilon r$. Dividing by r and taking $r \to \infty$ completes the proof.   :) ✓

8

The Milnor–Wolf Theorem

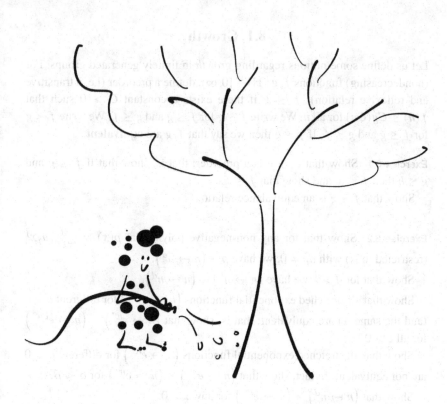

This chapter is devoted to proving a special dichotomy that holds in the case of solvable groups: any finitely generated solvable group is either of *exponential growth* or of *polynomial growth*. So solvable groups do not exhibit *intermediate growth*. The result, Theorem 8.2.4, is due to Milnor (1968a) and Wolf (1968).

This result is a very nice example of the algebra of the group (e.g. the property of solvability) constraining the geometry of the group (e.g. growth).

Wolf also proved in his 1968 paper that finitely generated nilpotent groups have polynomial growth (Theorem 8.2.1). In Chapter 9 we will prove Gromov's theorem (Theorem 9.0.1), which states that a finitely generated group is virtually nilpotent if and only if it has polynomial growth.

8.1 Growth

Let us define some notions regarding *growth* in finitely generated groups. For (nondecreasing) functions $f, g : \mathbb{N} \to [0, \infty)$, define a preorder (i.e. a transitive and reflexive relation): $f \preceq g$ if there exists a constant $C > 0$ such that $f(n) \leq Cg(Cn)$ for all n. We write $f \sim g$ for $f \preceq g$ and $g \preceq f$. We write $f \prec g$ for $f \preceq g$ and $g \not\preceq f$. If $f \sim g$ then we say that f, g are **equivalent**.

Exercise 8.1 Show that $f \preceq g$ is a preorder; that is, show that if $f \preceq g$ and $g \preceq h$ then $f \preceq h$, and show that $f \preceq f$.

Show that $f \sim g$ is an equivalence relation.

Exercise 8.2 Show that for any non-negative polynomial $p(x) = \sum_{j=0}^{d} a_j x^j$ (restricted to $\mathbb{N}$) with $a_d \neq 0$, we have $p \sim \left(n \mapsto n^d \right)$.

Show that for $d > k$ we have $\left(n \mapsto n^k \right) \prec \left(n \mapsto n^d \right)$.

Show that all stretched exponential functions $\left(n \mapsto e^{cn^{\alpha}} \right)$ for different $c > 0$ (and the same α) are equivalent; that is, show that $\left(n \mapsto e^{n^{\alpha}} \right) \sim \left(n \mapsto e^{cn^{\alpha}} \right)$ for all $c > 0$.

Show that all stretched exponential functions $\left(n \mapsto e^{n^{\alpha}} \right)$ for different $\alpha > 0$ are not equivalent. In fact, show that $\left(n \mapsto e^{n^{\alpha}} \right) \prec \left(n \mapsto e^{n^{\beta}} \right)$ for $\alpha < \beta$.

Show that $\left(n \mapsto n^d \right) \prec \left(n \mapsto e^{n^{\alpha}} \right)$ for any $\alpha > 0$.

Definition 8.1.1 Let G be a finitely generated group, and fix some Cayley graph of G. Let $f(r) = |B_r|$ where $B_r = \{ x : |x| \leq r \}$ is the ball of radius r around 1.

We say that G has **polynomial growth** if $f(r) \preceq \left(r \mapsto r^d \right)$ for some $d > 0$.

We say that G has **exponential growth** if $(r \mapsto e^r) \preceq f(r)$.

Otherwise we say that G has **intermediate growth**.

If $f(r) \prec (r \mapsto e^r)$ we say that G has **sub-exponential** growth.

By the **growth** of a finitely generated group G, we refer to the equivalence class of $f(r)$.

Note that when we say that a group has some growth we implicitly assume it is finitely generated, and we are assuming some implicit Cayley graph (or finite symmetric generating set) in the background. The following exercise shows that the specific Cayley graph is not important.

Exercise 8.3 Show that the growth of a group does not depend on the choice of finite symmetric generating set.

Show that the growth of a group cannot exceed exponential. $\quad$ ▷ solution ◁

Exercise 8.4 Let G be a finitely generated group.

Show that a finitely generated subgroup of G has growth not more than that of G (under the preorder on growth functions).

Show that a quotient group of G has growth not more than that of G.

▷ solution ◁

Exercise 8.5 Show that if H is a finite-index subgroup of a finitely generated group G, then the growth of H is the same as that of G. $\quad$ ▷ solution ◁

Exercise 8.6 Show that a finitely generated Abelian group has polynomial growth. $\quad$ ▷ solution ◁

8.2 Growth of Nilpotent Groups

In this section we will prove that any finitely generated nilpotent group has polynomial growth, generalizing Exercise 8.6. Recall from Section 1.5.4 the definition of a nilpotent group: G is nilpotent of step n if for some n the lower central series terminates at step n. That is, $\gamma_n(G) = \{1\}$ where $\gamma_0(G) = G$ and $\gamma_{k+1}(G) = [G, \gamma_k(G)]$.

Theorem 8.2.1 *Let G be a finitely generated nilpotent group. Then G has polynomial growth.*

We prove Theorem 8.2.1 via the following steps.

Exercise 8.7 Let G be a finitely generated nilpotent group. Show that for any n, the subgroup $\gamma_n(G)$ is finitely generated. (Hint: use Exercises 1.39 and 1.62).

▷ solution ◁

Exercise 8.8 Let $S_0 = S$ be a finite generating set for an n-step nilpotent group G (i.e. $\gamma_n(G) = \{1\}$). For $1 \le k \le n - 1$, define

$$S_k = \{[s_0, \ldots, s_k] : s_1, \ldots, s_k \in S\}.$$

Define $T_{n-1} = S_{n-1}$, and for $k < n - 1$ define inductively $T_k = S_k \cup T_{k+1}$.
Show that $\gamma_k(G) = \langle T_k \rangle$ for all $0 \le k \le n - 1$.

▷ solution ◁

Lemma 8.2.2 *Let S be a finite symmetric generating set for an infinite nilpotent group G.*

For any finite symmetric generating set T for $[G, G]$, there exists a constant $\kappa > 0$ such that for any $x \in [G, G]$, we have $|x|_T \le \kappa(|x|_S)^2$.

That is, Lemma 8.2.2 states that the internal metric on $[G, G]$ does not *distort* much the external metric inherited from the one on G. For more on the phenomena of distortion, see Exercise 8.35.

Proof We have that $G/[G, G]$ is an infinite finitely generated Abelian group, so isomorphic to $\mathbb{Z}^d \times F$ for some $d \ge 1$ and some finite Abelian group F, by Theorem 1.5.2.

Thus there exists a subgroup $[G, G] \lhd H \lhd G$ such that $[G : H] = |F|$ and $H/[G, G] \cong \mathbb{Z}^d$. Since $[G : H] < \infty$, we know that H is finitely generated (Exercise 1.61) and nilpotent. It is also simple to see that $[H, H] = \mathrm{Ker}\pi = [G, G]$ where $\pi \colon H \to \mathbb{Z}^d$ is the canonical projection. By adjusting the constant κ, if we replace G by H we may assume without loss of generality that $G/[G, G] \cong \mathbb{Z}^d$.

Let n be the nilpotent step of G. Let

$$T = \{[s_0, \ldots, s_k] : s_0, \ldots, s_k \in S, \ k \ge 1\} \backslash \{1\},$$

which is a finite set since $\gamma_n(G) = \{1\}$. Exercise 8.8 tells us that T generates $[G, G]$. Let $\pi \colon G \to \mathbb{Z}^d$ be the canonical projection (with $\mathrm{Ker}\pi = [G, G]$). Let $a_1, \ldots, a_d \in G$ be such that $\pi(a_j) = \vec{e}_j$, where $\vec{e}_1, \ldots, \vec{e}_d$ is the standard basis for $\mathbb{R}^d$ $\big($so they generate $\mathbb{Z}^d\big)$. Let $\tilde{S} = \big\{a_1^{\pm 1}, \ldots, a_d^{\pm 1}\big\}$. Note that $\tilde{S} \cap T \subseteq \tilde{S} \cap \mathrm{Ker}\pi = \emptyset$. By adjusting the constant κ, we may modify S so that $\tilde{S} \subseteq S$. Exercise 1.62 tells us that $\tilde{S} \cup T$ generates G. So it suffices to prove that there exists $\kappa > 0$ such that for all $x \in [G, G]$ we have $|x|_T \le \kappa \left(|x|_{\tilde{S} \cup T}\right)^2$.

To this end, fix some $x \in [G, G]$.

For $k \geq 1$, define

$$T_k = \{[s_0, \ldots, s_k] : s_0, \ldots, s_k \in S\} \setminus \{1\},$$

noting that $T_k = \emptyset$ for $k \geq n$ and that $T = \bigcup_{k=1}^{n-1} T_k$. Define

$$W_x = \{(u_1, \ldots, u_m) : m \geq 1, \; u_j \in \check{S} \cup T, \; u_1 \cdots u_m = x\}.$$

For $\omega = (u_1, \ldots, u_m)$ and $1 \leq j \leq d$ and $1 \leq k \leq n-1$, we define

$$|\omega| = m,$$

$$||\omega||_{0,j} = \sum_{i=1}^{m} \mathbf{1}_{\{u_i \in \{a_j, a_j^{-1}\}\}},$$

$$||\omega||_0 = \sum_{j=1}^{d} ||\omega||_{0,j} = \sum_{i=1}^{m} \mathbf{1}_{\{u_I \in \check{S}\}},$$

$$||\omega||_k = \sum_{i=1}^{m} \mathbf{1}_{\{u_i \in T_k\}},$$

$$||\omega||_T = \sum_{k=1}^{n-1} ||\omega||_k = \sum_{i=1}^{m} \mathbf{1}_{\{u_i \in T\}}.$$

Note that $|\omega| = ||\omega||_0 + ||\omega||_T$. Moreover, if $||\omega||_0 = 0$ then $|x|_T \leq |\omega| = ||\omega||_T$.

Let $\omega = (u_1, \ldots, u_m) \in W_x$. Assume that $||\omega||_{0,j} > 0$. Then, since $x \in [G, G] = \mathrm{Ker}\pi$,

$$\sum_{i=1}^{m} \pi(u_j) = \pi(u_1 \cdots u_m) = \pi(x) = 0,$$

so there must exist $1 \leq \ell' < \ell \leq k$ such that $u_\ell = u_{\ell'}^{-1} \in \{a_j, a_j^{-1}\}$. Consider

$$\omega' = (u_1, \ldots, u_{\ell'-1}, u_{\ell'+1}, [u_{\ell'+1}, u_\ell], \ldots, u_{\ell-1}, [u_{\ell-1}, u_\ell], u_{\ell+1}, \ldots, u_k).$$

Since $yz = zy[y, z]$ and since $u_{\ell'} u_\ell = 1$, we see that $\omega' \in W_x$. Also, we have that

$$||\omega'||_{0,j} \leq ||\omega||_{0,j} - 2,$$

$$||\omega'||_{0,i} \leq ||\omega||_{0,i}, \quad \text{for all } 1 \leq i \leq d,$$

$$||\omega'||_k \leq ||\omega||_k + ||\omega||_{k-1}, \quad \text{for all } k \geq 1.$$

This implies that

$$|\omega'| \leq |\omega| + ||\omega||_0 - 2,$$

$$||\omega'||_0 \leq ||\omega||_0 - 2.$$

Applying this procedure r times, we obtain $\omega'' \in W_x$ with

$$||\omega''||_{0,j} \leq ||\omega||_{0,j} - 2r,$$
$$||\omega''||_{0,i} \leq ||\omega||_{0,i}, \quad \text{for all } 1 \leq i \leq d,$$
$$|\omega''| \leq |\omega| + (||\omega||_0 - 2)r.$$

(Hence it must be that $2r \leq ||\omega||_{0,j}$.)

We find that for any $\omega \in W_x$ with $||\omega||_{0,j} > 0$, there exists $\omega' \in W_x$ with

$$||\omega'||_{0,j} = 0, \quad ||\omega'||_0 \leq ||\omega||_0, \quad \text{and} \quad |\omega'| < |\omega| + \tfrac{1}{2}(||\omega||_0)^2.$$

Repeating this at most d times, once for every $1 \leq j \leq d$, we conclude with the following fact.

For any $\omega \in W_x$, there exists $\omega' \in W_x$ with

$$||\omega'||_0 = 0 \quad \text{and} \quad |\omega'| < |\omega| + \tfrac{d}{2}(||\omega||_0)^2. \tag{8.1}$$

Finally, note that since $\tilde{S} \cup T$ generates G, we know that there exists $\omega \in W_x$ with $|\omega| = |x|_{\tilde{S} \cup T}$. Hence, using (8.1), there exists $\omega' \in W_x$ with

$$|x|_T \leq ||\omega||_T = |\omega'| < |\omega| + \tfrac{d}{2}(||\omega||_0)^2 \leq \left(\tfrac{d}{2} + 1\right)\left(|x|_{\tilde{S} \cup T}\right)^2. \qquad \square$$

Proof of Theorem 8.2.1   We prove the theorem by induction on the nilpotent step n.

If G is 1-step nilpotent, then G is Abelian, which is done in Exercise 8.6.

Assume that G is n-step nilpotent for $n > 1$.

Let Q be a set of representatives for the cosets of $[G, G]$; that is, $G = \biguplus_{q \in Q}[G,G]q$. So, for any $x \in G$ there exist unique $q_x \in Q$ and $g_x \in [G,G]$ such that $x = g_x q_x$. We may choose Q so that for any $x \in G$ we have that $|x|_S \geq |q_x|_S$. Thus, $|g_x|_S \leq 2|x|_S$. By Lemma 8.2.2, there exists a finite symmetric generating set T for $[G,G]$ and a constant $\kappa > 0$ such that $|g_x|_T \leq \kappa(|g_x|_S)^2 \leq 4\kappa(|x|_S)^2$. This implies that the map $\psi_r : B_S(1, r) \to B_T\left(1, 4\kappa r^2\right) \times (Q \cap B_S(1, r))$ given by $\psi_r(x) = (g_x, q_x)$ is injective, so we have that $|B_S(1, r)| \leq \left|B_T\left(1, 4\kappa r^2\right)\right| \cdot |Q \cap B_S(1, r)|$.

Let $\pi : G \to G/[G, G]$ be the canonical projection, and let $\tilde{S} = \pi(S)$. So $\tilde{S}$ is a finite symmetric generating set for $G/[G, G]$. Also, π maps $Q \cap B_S(1, r)$ injectively into $B_{\tilde{S}}([G, G], r) \subset G/[G, G]$.

So it suffices to show that both $[G, G]$ and $G/[G, G]$ are of polynomial growth.

Since $\gamma_k([G, G]) \leq \gamma_{k+1}(G)$, we have that $[G, G]$ is a finitely generated $(n - 1)$-step nilpotent group. Thus, $[G, G]$ has polynomial growth by induction. Also, $G/[G, G]$ is a finitely generated Abelian group, so is also of polynomial growth. $\qquad \square$

For the sake of completeness, we present the correct order of growth for nilpotent groups, although we will not prove the precise bounds. These are due to Bass (1972) and Guivarc'h (1973).

Theorem 8.2.3 (Bass–Guivarc'h Formula) *Let G be a finitely generated nilpotent group. Recall the lower central series $\gamma_0(G) = G$ and $\gamma_{k+1}(G) = [\gamma_k(G), G]$. For every $1 \leq k \leq n$ we have that $\gamma_k(G)/\gamma_{k+1}(G)$ is finitely generated and Abelian, so we may write d_k for an integer such that $\gamma_k(G)/\gamma_{k+1}(G) \cong \mathbb{Z}^{d_k} \times F_k$ and F_k finite.*

The growth of G is then given by: there exists a constant $C > 0$ such that for all $r > 0$ we have $C^{-1}r^D \leq |B(1,r)| \leq Cr^D$, where $D = \sum_{k=0}^{n-1}(k+1)d_k$ and $\gamma_n(G) = \{1\}$.

Specifically, the growth exponent is an integer.

Recall that any non-amenable group must have exponential growth (see e.g. Exercise 6.50). We have already seen examples of polynomial growth groups (e.g. nilpotent groups) and of exponential growth groups (e.g. free groups). In 1968, Milnor asked if groups of intermediate growth are in fact possible (Milnor, 1968b). This was answered affirmatively only in 1980 by Grigorchuk (see Grigorchuk, 1980, 1984). Prior to Grigorchuk's famous construction, various families of groups were shown to not contain any intermediate growth group. The Milnor–Wolf theorem tells us that in the class of solvable groups there are no groups of intermediate growth.

Theorem 8.2.4 (Milnor–Wolf theorem) *If G is a finitely generated solvable group then G either has exponential growth or polynomial growth.*

In fact, G is a finitely generated solvable group of sub-exponential growth if and only if G is virtually nilpotent.

Recall that a group G is solvable if the *derived series* terminates after finitely many steps: $G = G^{(0)} \triangleright G^{(1)} \triangleright \cdots \triangleright G^{(n)} = \{1\}$, where $G^{(j+1)} = [G^{(j)}, G^{(j)}]$. A group G is nilpotent if the *lower central series* terminates after finitely many steps: $G = \gamma_0(G) \triangleright \gamma_1(G) \triangleright \cdots \triangleright \gamma_n(G) = \{1\}$, where $\gamma_{j+1}(G) = [G, \gamma_j(G)]$. This is a good place for the reader to recall these notions, see Section 1.5.

The Milnor–Wolf theorem is related to Gromov's celebrated theorem, Theorem 9.0.1, to which Chapter 9 is devoted. Gromov's theorem states that a finitely generated group has polynomial growth if and only if it is virtually nilpotent. So the second assertion in Theorem 8.2.4 is a special case of Gromov's theorem for solvable groups.

We now proceed to develop the required tools for the proof of the Milnor–Wolf theorem.

8.3 The Milnor Trick

Proposition 8.3.1 (Milnor's trick) *Let G be a finitely generated group of sub-exponential growth. Then, for any $x, y \in G$ the subgroup $\langle x^{-n} y x^n : n \in \mathbb{Z} \rangle$ is finitely generated.*

Proof Fix $x, y \in G$ and a finite symmetric generating set for G containing x, y. Consider the set

$$A = \left\{ x y^{j_1} x y^{j_2} \cdots x y^{j_n} : j_1, \ldots, j_n \in \{0, 1\} \right\}.$$

One sees that the map $(j_1, \ldots, j_n) \mapsto x y^{j_1} x y^{j_2} \cdots x y^{j_n}$ maps $\{0, 1\}^n$ into A, and that $A \subset B(1, 2n)$. Since $|B(1, 2n)|$ grows sub-exponentially, this implies that there must exist n and $(j_1, \ldots, j_n) \neq (i_1, \ldots, i_n) \in \{0, 1\}^n$ such that

$$x y^{j_1} \cdots x y^{j_n} = x y^{i_1} \cdots x y^{i_n}.$$

We may assume without loss of generality that $j_n \neq i_n$ (by cancelling out the terms on the right until arriving at $j_n \neq i_n$). Using the notation $y_k = x^k y x^{-k}$, we get that

$$(y_1)^{j_1} \cdots (y_n)^{j_n} = (y_1)^{i_1} \cdots (y_n)^{i_n}.$$

We obtain that

$$(y_n)^{j_n - i_n} = (y_{n-1})^{-j_{n-1}} \cdots (y_1)^{-j_1} \cdot (y_1)^{i_1} \cdots (y_{n-1})^{i_{n-1}}.$$

But since $j_n \neq i_n$, this tells us that $y_n \in \langle y_1, \ldots, y_{n-1} \rangle$. Conjugating by x^{-1} we have that $y_{n+1} \in \langle y_2, \ldots, y_n \rangle \leq \langle y_1, y_2, \ldots, y_{n-1} \rangle$. Continuing inductively we have that $y_{n+k} \in \langle y_1, \ldots, y_n \rangle$ for all k.

Repeating this argument with x^{-1} instead of x we get that for some n we have $y_m \in \langle y_{-n}, \ldots, y, \ldots, y_n \rangle$ for all $m \in \mathbb{Z}$. □

Exercise 8.9 Let G be a finitely generated group. Let $H \lhd G$ such that $G/H \cong \mathbb{Z}$. Show that there exists a generating set $S = \{s_1, \ldots, s_n\}$ of G such that s_1 maps to $1 \in \mathbb{Z}$ under the canonical projection, and such that $s_j \in H$ for all $j > 1$.

▷ solution ◁

Exercise 8.10 Let G be a finitely generated group of sub-exponential growth. Let $H \lhd G$ such that $G/H \cong \mathbb{Z}$. Show that H is finitely generated. (Hint: use Exercise 1.60.)

▷ solution ◁

Proposition 8.3.2 *If G is a finitely generated group of sub-exponential growth, and $H \lhd G$ such that G/H is Abelian, then H is finitely generated.*

Proof Since G/H is finitely generated and Abelian, then $G/H \cong \mathbb{Z}^d \times F$ for a finite group F. For $d = 0$ we have that $[G : H] < \infty$, and so H is finitely generated. The proof is by induction on $d \geq 1$.

Let $\pi: G \to \mathbb{Z}^d \times F$ be the canonical projection. Let $K = \pi^{-1}(\{0\} \times F)$, which is easily seen to be a normal subgroup of G. Also, $K/H \cong \{0\} \times F \cong F$, so $[K : H] < \infty$. Thus, it suffices to prove that K is finitely generated.

Note that $G/K \cong (G/H)/(K/H) \cong \mathbb{Z}^d$. So we have a surjective homomorphism $\psi: G \to \mathbb{Z}^d$ with $K = \text{Ker}\psi$.

The base case, $d = 1$, is Exercise 8.10.

For $d > 1$ we proceed by induction as follows. We note that $\mathbb{Z}^d = \mathbb{Z}^{d-1} \times \mathbb{Z}$. Let $\varphi: \mathbb{Z}^d \to \mathbb{Z}^{d-1}$ be the projection onto the first $d-1$ coordinates, $\varphi(z_1, \ldots, z_d) = (z_1, \ldots, z_{d-1})$. Let $N = \text{Ker}(\varphi \circ \psi)$. So $K \lhd N \lhd G$ and $G/N \cong \mathbb{Z}^{d-1}$.

By induction, N is finitely generated. Thus, N has sub-exponential growth, as a finitely generated subgroup of G, by Exercise 8.4.

Now, $\psi(N) = \{\vec{z} \in \mathbb{Z}^d : z_1 = \cdots = z_{d-1} = 0\} \cong \mathbb{Z}$, so $N/K \cong \mathbb{Z}$. This implies that K is finitely generated as well, again by Exercise 8.10, completing the induction step. $\qquad\square$

The above proposition will suffice for our purposes, but let us give an easy corollary of it.

Theorem 8.3.3 (Rosset's theorem) *If G is a finitely generated group of sub-exponential growth, and $H \lhd G$ is such that G/H is solvable, then H is finitely generated.*

Proof The proof is by induction on the length of the derived series of G/H. (Recall from Section 1.5.5 that the derived series of a group Γ is $\Gamma^{(0)} = \Gamma$ and $\Gamma^{(k+1)} = \left[\Gamma^{(k)}, \Gamma^{(k)}\right]$. Γ is n-step solvable if $\Gamma^{(n)} = \{1\}$ and $\Gamma^{(n-1)} \neq \{1\}$.)

If G/H is 1-step solvable, then G/H is Abelian, which is exactly Proposition 8.3.2.

Now assume that G/H is n-step solvable for some $n > 1$. If $\pi: G \to N = G/H$ is the canonical projection, let $M = \pi^{-1}([N, N])$. So $H \lhd M$ and $M/H \cong [N, N]$ is $(n - 1)$-step solvable. Since $G/M \cong N/[N, N]$ is Abelian, we know that M is finitely generated by Proposition 8.3.2. So M has sub-exponential growth as a subgroup of G by Exercise 8.4. Since M/H is $(n - 1)$-step solvable, H is finitely generated by induction. $\qquad\square$

Exercise 8.11 A group G is called **torsion** if every element of G has finite order.

Show that if G is a finitely generated solvable group of sub-exponential growth and G is torsion, then G is finite.   ▷ solution ◁

Now, since finitely generated nilpotent groups have polynomial growth, we can use Rosset's theorem (Theorem 8.3.3) to show that all their subgroups are finitely generated.

Theorem 8.3.4 *Let G be a finitely generated nilpotent group. Then, any subgroup of G is finitely generated.*

Proof This is by induction on the nilpotent step n of G.

The base case, where $\gamma_1(G) = \{1\}$, is the case where G is finitely generated and Abelian. In this case any subgroup is normal, so if $H \leq G$, then since G/H is a finitely generated Abelian group of sub-exponential growth, H is finitely generated by Rosset's theorem (Theorem 8.3.3, or in this case even Proposition 8.3.2 suffices).

For the induction step, when $n > 1$, let H be any subgroup. Consider $H \cap [G, G]$. By Rosset's theorem (Theorem 8.3.3), since G has polynomial growth and $G/[G, G]$ is Abelian, we know that $[G, G]$ is finitely generated. Also, $\gamma_{n-1}([G, G]) \leq \gamma_n(G) = \{1\}$, so $[G, G]$ is finitely generated and nilpotent of step at most $(n - 1)$. By induction we know that $H \cap [G, G]$ is finitely generated.

Also, if $\pi\colon G \to G/[G, G]$ is the canonical projection, then $\pi(H) \cong H/(H \cap [G, G])$ is a subgroup of the finitely generated Abelian group $G/[G, G]$. Thus, $H/(H \cap [G, G])$ is finitely generated. By Exercise 1.62, H is finitely generated, completing the induction step.   □

8.4 Characteristic Subgroups

Recall that a subgroup $H \leq G$ is normal if $H^g = H$ for all $g \in G$. That is, the *inner automorphisms* of G by conjugation preserve the subgroup H. Another notion is that of a *characteristic subgroup*.

Definition 8.4.1 A subgroup $H \leq G$ is **characteristic** in G if any automorphism $\varphi \in \mathrm{Aut}(G)$ preserves H; that is, $\varphi(H) = H$.

Since conjugation by $g \in G$ is an automorphism, any characteristic subgroup is also normal. However, this is a stronger notion than normality.

Exercise 8.12 Show that if H is a characteristic subgroup of N and $N \lhd G$ is a normal subgroup of G, then H is normal in G. ▷ solution ◁

Theorem 8.4.2 *Let G be a finitely generated group. Let $H \le G$ be a finite index subgroup, $[G : H] < \infty$. Then, there exists a subgroup $N \le H$ such that $[G : N] < \infty$ and N is characteristic in G.*

Proof Assume that $[G : H] = n$. By Theorem 1.5.12, the set $X = \{K \le G : [G : K] = n\}$ is a finite set (in fact $|X| \le (n!)^d$, where d is the number of generators of G). Let $N = \cap_{K \in X} K$. Since $[G : A \cap B] \le [G : A] \cdot [G : B]$ (Proposition 1.3.3), we have that $[G : N] \le n^{|X|} < \infty$.

Moreover, for any automorphism $\varphi \in \text{Aut}(G)$, and any $K \in X$, we have that $[G : \varphi(K)] = [G : K] = n$, so that $\varphi(K) \in X$ as well. Thus, $\varphi(N) = N$, proving that N is a characteristic subgroup of G. □

Example 8.4.3 We have already seen in Exercise 1.35 that $\gamma_n(G)$ and $Z_n(G)$ are characteristic subgroups of G. △▽△

Exercise 8.13 Recall that for two subsets $A, B \subset G$, we define

$$[A, B] = \langle [a, b] : a \in A, \ b \in B \rangle.$$

Show that if $N, K \le G$ are two characteristic subgroups of G, then $[N, K]$ is also a characteristic subgroup of G. ▷ solution ◁

For a group G and $n > 0$ define

$$\bar{\gamma}_n(G) = \{x \in \gamma_{n-1}(G) \mid \exists \, 0 \ne k \in \mathbb{Z}, \ x^k \in \gamma_n(G)\}.$$

Exercise 8.14 Show that $\bar{\gamma}_n(G)$ is a characteristic subgroup of G. ▷ solution ◁

Exercise 8.15 Show that if G is finitely generated, then $\gamma_n(G)/\bar{\gamma}_{n+1}(G) \cong \mathbb{Z}^{d_n}$ for some integer $d_n = d_n(G) \ge 0$. ▷ solution ◁

Exercise 8.16 Show that if $d_n(G) = 0$ then $\gamma_{n+k}(G) = \bar{\gamma}_{n+k+1}(G) = \gamma_{n+k+1}(G)$ for all $k \ge 1$.
Conclude that if $d_n(G) = 0$ then $d_{n+1}(G) = 0$. ▷ solution ◁

Exercise 8.17 Show that if $\varphi \in \text{Aut}(G)$ then it induces an automorphism

$$\varphi_n \in \text{Aut}(\gamma_n(G)/\bar{\gamma}_{n+1}(G)) \quad \text{given by} \quad \varphi_n(\bar{\gamma}_{n+1}(G)x) := \bar{\gamma}_{n+1}(G)\varphi(x).$$

▷ solution ◁

8.5 $\mathbb{Z}$-extensions and Nilpotent Groups

Recall the notion of a *semi-direct product of groups* from Section 1.5.9 and Exercise 1.68. We have already seen examples in Exercises 1.70, 1.71, and 1.72, as well as in the *lamplighter groups* in Section 6.9. One particular case is known as a $\mathbb{Z}$-*extension* of a group, which we now explain in more detail.

If $\varphi \in \mathrm{Aut}(G)$ then $\mathbb{Z}$ acts on G via φ (specifically $a \in \mathbb{Z}$ acts on $x \in G$ by $\varphi^a(x)$). This gives a new group $H = \mathbb{Z} \ltimes_\varphi G$, which is defined as follows. Elements of H are pairs (a, x) for $a \in \mathbb{Z}, x \in G$. Multiplication is given by $(a, x)(b, y) = (a + b, x\varphi^a(y))$.

Exercise 8.18 Show that in $\mathbb{Z} \ltimes_\varphi G$ for any $x \in G$ and $a, b \in \mathbb{Z}$ we have $(a, x)^{(b,1)} = \left(a, \varphi^{-b}(x)\right)$.

Show that if $\varphi \in \mathrm{Aut}(G)$ and $\psi = \varphi^m$ for $m \in \mathbb{Z}\backslash\{0\}$, then there is a normal subgroup $H \lhd \mathbb{Z} \ltimes_\varphi G$ such that $(\mathbb{Z} \ltimes_\psi G) \cong H$ and $[(\mathbb{Z} \ltimes_\varphi G) : H] = |m|$.

Show that if $G = \langle S \rangle$ then $\mathbb{Z} \ltimes_\varphi G$ is generated by $\tilde{S} = \{(0, s), (\pm 1, 1) : s \in S\}$. Show that with this generating set, if S is a finite symmetric generating set for G, we have $|(a, x)|_{\tilde{S}} \leq |a| + |x|_S$.   ▷ solution ◁

Exercise 8.19 Let G be a finitely generated group. Let $K \lhd G$ be such that $G/K \cong \mathbb{Z}$.

Show that $G \cong \mathbb{Z} \ltimes_\varphi K$ for some $\varphi \in \mathrm{Aut}(K)$. Show that $\{0\} \times K \subset \mathbb{Z} \ltimes_\varphi K$ is a subgroup isomorphic to K.

Show that if G has growth $\sim \left(r \mapsto r^d\right)$ then K is finitely generated with growth $\leq \left(r \mapsto r^{d-1}\right)$.   ▷ solution ◁

Exercise 8.20 Let G be a group and let $\varphi \in \mathrm{Aut}(G)$. Let $H \lhd G$ be a subgroup such that $\varphi(H) = H$.

Show that the map $\psi(Hx) = H\varphi(x)$ is a well-defined automorphism of G/H.

Let $N = \{0\} \times H \subset \mathbb{Z} \ltimes_\varphi G$. Show that N is a normal subgroup of $\mathbb{Z} \ltimes_\varphi G$, and that $N \cong H$.

Show that $(\mathbb{Z} \ltimes_\varphi G)/N = \mathbb{Z} \ltimes_\psi (G/H)$.   ▷ solution ◁

Exercise 8.21 Show that if $F \lhd G$ is a finite normal subgroup such that $G/F \cong \mathbb{Z}$, then there exists a finite index subgroup $H \leq G, [G : H] < \infty$ such that $H \cong \mathbb{Z}$.

                              ▷ solution ◁

We require some results regarding matrices with integer entries. Recall that $GL_n(\mathbb{Z})$ is the group of $n \times n$ matrices with integer entries and determinant in $\{-1, 1\}$.

Exercise 8.22 Let $M \in M_n(\mathbb{Z})$ be an $n \times n$ matrix with *integer* entries. Show that if all (potentially, complex) eigenvalues of M have modulus 1 then all eigenvalues of M are roots of unity. ▷ solution ◁

Exercise 8.23 Show that if φ is an automorphism of $\mathbb{Z}^d$, then there exists a matrix $M \in GL_d(\mathbb{Z})$ such that $\varphi(x) = Mx$ for all $x \in \mathbb{Z}^d$. ▷ solution ◁

Since any $M \in GL_d(\mathbb{Z})$ is an automorphism of $\mathbb{Z}^d$, we conclude that $\mathrm{Aut}\left(\mathbb{Z}^d\right) \cong GL_d(\mathbb{Z})$.

Exercise 8.24 Let G be a group isomorphic to $\mathbb{Z}^d$. Let $i, j\colon G \to \mathbb{Z}^d$ be group isomorphisms. Let $\varphi \in \mathrm{Aut}(G)$ and consider $\varphi_i = i \circ \varphi \circ i^{-1}$ and $\varphi_j = j \circ \varphi \circ j^{-1}$.

Note that φ_i, φ_j are automorphisms of $\mathbb{Z}^d$.

Let $M_i, M_j \in GL_d(\mathbb{Z})$ be matrices such that $\varphi_i(\vec{z}) = M_i\vec{z}$ and $\varphi_j(\vec{z}) = M_j\vec{z}$ for all $\vec{z} \in \mathbb{Z}^d$.

Prove that $M_i = U^{-1}M_j U$ for some invertible matrix U. ▷ solution ◁

Using the above, we now proceed to define the notion of *eigenvalues of a group automorphism*.

Definition 8.5.1 Let G be a finitely generated group. Let $\varphi \in \mathrm{Aut}(G)$. Set $\gamma_n = \gamma_n(G)$ and $\bar{\gamma}_n = \bar{\gamma}_n(G)$.

Consider the induced automorphisms guaranteed by Exercise 8.17, $\varphi_n \in \mathrm{Aut}(\gamma_n/\bar{\gamma}_{n+1})$; that is, satisfying $\varphi_n(\gamma_{n+1}x) = \gamma_n\varphi(x)$ for all $x \in G$.

By Exercise 8.15, there exists a group isomorphism $i_n\colon \gamma_n/\bar{\gamma}_{n+1} \to \mathbb{Z}^{d_n}$, where $d_n > 0$ if and only if $\gamma_n \neq \bar{\gamma}_{n+1}$. Thus, $\psi_n := i_n \circ \varphi_n \circ (i_n)^{-1}$ is an automorphism of $\mathbb{Z}^{d_n}$. If $d_n > 0$, then there exists $M_n \in GL_{d_n}(\mathbb{Z})$ such that $\psi_n(\vec{z}) = M_n x$ for all $\vec{z} \in \mathbb{Z}^{d_n}$ (Exercise 8.23). If $d_n > 0$, then let $\lambda_{n,1}, \ldots, \lambda_{n,d_n}$ denote the (possibly complex) eigenvalues of M_n, with multiplicities. Note that these eigenvalues are independent of the specific choice of isomorphism i_n, by Exercise 8.24.

Let $J = \inf\{j : d_j = 0\}$ where $J = \infty$ if $d_j > 0$ for all j. Exercise 8.16 tells us that if $J < \infty$ then $d_j = 0$ for all $j \geq J$.

The (possibly infinite list of) numbers $(\lambda_{j,1}, \ldots, \lambda_{j,d_j})_{j=0}^{J-1}$ are called the **eigenvalues** of the automorphism φ. (If $d_0 = 0$ then φ has no eigenvalues.)

We will now study $\mathbb{Z}$-extensions of nilpotent groups and see that the eigenvalues of the implicit automorphism provide a lot of information on the structure of the $\mathbb{Z}$-extension.

Lemma 8.5.2 *Let G be a finitely generated nilpotent group. Let $\varphi \in \mathrm{Aut}(G)$. If φ has an eigenvalue λ with $|\lambda| \neq 1$ then $\mathbb{Z} \ltimes_\varphi G$ has exponential growth.*

Proof By Exercise 8.18, for an appropriate $m \in \mathbb{Z}$ we have with $\psi = \varphi^m$ that $H \cong \mathbb{Z} \ltimes_\psi G$ is finite index in $\mathbb{Z} \ltimes_\varphi G$ and ψ has an eigenvalue $|\lambda| < 1/2$. It suffices to prove that $\mathbb{Z} \ltimes_\psi G$ has exponential growth. So we may assume without loss of generality that φ has an eigenvalue λ with $0 < |\lambda| < \frac{1}{2}$.

The eigenvalue λ must be $\lambda = \lambda_{j,i}$ for some $0 \leq j < n$ and $1 \leq i \leq d_j = d_j(G)$, which is an eigenvalue of the corresponding matrix $M_j \in \mathrm{GL}_{d_j}(\mathbb{Z})$ corresponding to the induced automorphism φ_j on $\gamma_j(G)/\bar{\gamma}_{j+1}(G) \cong \mathbb{Z}^{d_j}$; that is, for the canonical projection $\pi : \gamma_j(G) \to \gamma_j(G)/\bar{\gamma}_{j+1}(G)$, there is some isomorphism $\alpha : \gamma_j(G)/\bar{\gamma}_{j+1}(G) \to \mathbb{Z}^{d_j}$ such that for any $y \in \gamma_j(G)$, we have that

$$\varphi_j(\bar{\gamma}_{j+1}(G)y) = \varphi_j(\pi(y)) = \alpha^{-1} M_j \alpha(\pi(y)).$$

Set $d = d_j$, $M = M_j$, and $\psi = \varphi_j$. So $\psi = \alpha^{-1} \circ M \circ \alpha$. Recall that by Exercise 8.17, for any $y \in \gamma_j(G)$ we have that $\psi(\pi(y)) = \pi(\varphi(y))$.

Let $v \in \mathbb{C}^d$ be an eigenvector of M^* with eigenvalue $\bar{\lambda}$ (recall that eigenvalues of M^* are complex conjugates of the eigenvalues of M). Scaling v by an appropriate scalar we may assume that $\langle e_\ell, v \rangle = 1$ for one of the standard basis vectors $e_\ell \in \mathbb{Z}^d$. Let $x \in \gamma_j(G)$ be such that $\alpha \circ \pi(x) = e_\ell$. So for all $a \in \mathbb{Z}$,

$$M^a e_\ell = \alpha \circ \psi^a \circ \pi(x) = \alpha \circ \pi(\varphi^a(x)).$$

Thus,

$$\langle \alpha \circ \pi(\varphi^a(x)), v \rangle = \langle M^a e_\ell, v \rangle = \langle e_\ell, (M^*)^a v \rangle = \lambda^a.$$

Now, assume that $(0, x), (0, 1), (-1, 1), (1, 1)$ are all in the generating set of $H = \mathbb{Z} \ltimes_\varphi G$. Then, in the group H, for any $\varepsilon = (\varepsilon_0, \ldots, \varepsilon_k) \in \{0, 1\}^{k+1}$, consider the elements

$$g_\varepsilon = (0, x^{\varepsilon_0})(-1, 1)(0, x^{\varepsilon_1}) \cdots (-1, 1)(0, x^{\varepsilon_k}) \cdot (k, 1).$$

Note that $|g_\varepsilon| \le 2(k+1) - 1 + k + 1 < 3(k+1)$. Also, by Exercise 8.18,

$$g_\varepsilon = (0, x^{\varepsilon_0})^{(1,1)^0} \cdot (0, x^{\varepsilon_1})^{(1,1)^1} \cdots (0, x^{\varepsilon_k})^{(1,1)^k}$$
$$= \left(0, \varphi^0(x^{\varepsilon_0}) \cdot \varphi^{-1}(x^{\varepsilon_1}) \cdots \varphi^{-k}(x^{\varepsilon_k})\right).$$

Assume that for some $\varepsilon, \varepsilon' \in \{0,1\}^{k+1}$, we have $g_\varepsilon = g_{\varepsilon'}$. Then,

$$\varphi^0(x^{\varepsilon_0}) \cdots \varphi^{-k}(x^{\varepsilon_k}) = \varphi^0(x^{\varepsilon'_0}) \cdots \varphi^{-k}(x^{\varepsilon'_k}).$$

Using the fact that $\alpha \circ \pi$ is a homomorphism, we obtain

$$\sum_{i=0}^{k} \varepsilon_i \lambda^{-i} = \sum_{i=0}^{k} \left\langle \alpha \circ \pi(\varphi^{-i}(x^{\varepsilon_i})), v \right\rangle$$
$$= \left\langle \alpha \circ \pi \left(\varphi^0(x^{\varepsilon_0}) \cdots \varphi^{-k}(x^{\varepsilon_k})\right), v \right\rangle = \left\langle \alpha \circ \pi \left(\varphi^0(x^{\varepsilon'_0}) \cdots \varphi^{-k}(x^{\varepsilon'_k})\right), v \right\rangle$$
$$= \sum_{i=0}^{k} \left\langle \alpha \circ \pi \left(\varphi^{-i}(x^{\varepsilon'_i})\right), v \right\rangle = \sum_{i=0}^{k} \varepsilon'_i \lambda^{-i}.$$

Since $0 < |\lambda| < \frac{1}{2}$, this is only possible if $\varepsilon = \varepsilon'$.

We conclude that the map $\varepsilon \mapsto g_\varepsilon$ from $\{0,1\}^{k+1}$ into $B_H(1, 3(k+1))$ is injective, and hence $|B_H(1, 3(k+1))| \ge 2^{k+1}$. This implies exponential growth of $H = \mathbb{Z} \ltimes_\varphi G$. □

Exercise 8.25 Show that if $N \triangleleft G$ is such that G/N is virtually nilpotent and $[G, N] = \{1\}$, then G is virtually nilpotent. ▷ solution ◁

Exercise 8.26 Show that if $F \triangleleft G$ is a finite normal subgroup and G/F is virtually nilpotent, then G is virtually nilpotent. ▷ solution ◁

Exercise 8.27 Let $\Gamma \cong \mathbb{Z}^d$ and let $\varphi \in \operatorname{Aut}(\Gamma)$. Assume all eigenvalues of φ are exactly 1.

Show that there exists $y \in \Gamma$ such that $\varphi(y) = y$. ▷ solution ◁

Exercise 8.28 Let $K \le \mathbb{Z}^d$. Show that if K is nontrivial, then $\mathbb{Z}^d/K \cong \mathbb{Z}^m \times F$ for some $m < d$ and a finite Abelian group F. ▷ solution ◁

Exercise 8.29 Let G be a finitely generated Abelian group. Let $\varphi \in \operatorname{Aut}(G)$. Assume that $K \le G$ is a subgroup such that $\varphi(K) = K$ and $\bar{\gamma}_1(G) \le K$. Let $\psi \in \operatorname{Aut}(G/K)$ be the automorphism given by $\psi(Kx) = K\varphi(x)$.

Show that any eigenvalue of ψ is an eigenvalue of φ. ▷ solution ◁

Exercise 8.30 Let G be an infinite finitely generated $(n + 1)$-step nilpotent group. So $\gamma_n(G) \neq \{1\} = \gamma_{n+1}(G)$. Let N be a subgroup such that $N \leq \gamma_n(G)$.
Show that $N \lhd G$.
Let $\pi: G \to G/N$ be the canonical projection. Show that for all $k \geq 0$ we have $\gamma_k(\pi(G)) = \pi(\gamma_k(G))$.
Show that $\bar{\gamma}_{k+1}(\pi(G)) = \pi(\bar{\gamma}_{k+1}(G))$ for all $0 \leq k \leq n$.
Show that if $k > n$ then $\bar{\gamma}_{k+1}(\pi(G)) = \{1\}$.   ▷ solution ◁

Lemma 8.5.3 *Let G be an infinite finitely generated nilpotent group. Let $\varphi \in$ Aut(G). Assume that all eigenvalues of φ are 1.*
Then $\mathbb{Z} \ltimes_\varphi G$ is virtually nilpotent.

Proof Let n be such that $\gamma_n(G) \neq \{1\} = \gamma_{n+1}(G)$ (i.e. G is $(n + 1)$-step nilpotent).

Recall that $d_k(G) \in \mathbb{N}$ is the nonnegative integer for which $\gamma_k(G)/\bar{\gamma}_{k+1}(G) \cong \mathbb{Z}^{d_k(G)}$ (Exercise 8.15). Set $D(G) = \sum_{k=0}^\infty d_k(G)$. Since $\gamma_k(G) = \{1\}$ for all $k \geq n + 1$, we know that $D(G) = \sum_{k=0}^n d_k(G)$.

We will prove the lemma by induction on $D(G)$.

Denote $H = \mathbb{Z} \ltimes_\varphi G$, $\Gamma = \gamma_n(G)$, and $\bar{\Gamma} = \bar{\gamma}_{n+1}(G)$. Let $\pi: G \to G/\bar{\Gamma}$ be the canonical projection.

We have that $\bar{\Gamma}$ is a finitely generated Abelian group such that all elements have finite order. By Exercise 8.7, Γ is a finitely generated group. Since $[\Gamma, \Gamma] \leq \gamma_{n+1}(G) = \{1\}$, we have that Γ is an Abelian group. By Theorem 1.5.2, we know that $\Gamma \cong \mathbb{Z}^d \times F$ for some finite Abelian group F. Since any element of $\bar{\Gamma}$ has finite order, if $i: \Gamma \to \mathbb{Z}^d \times F$ is an isomorphism, then $i(\bar{\Gamma}) \subset \{\vec{0}\} \times F$. We conclude that $\bar{\Gamma}$ is a finite Abelian group.

Since $\bar{\Gamma}$ is a characteristic subgroup of G (Exercise 8.14), Exercise 8.20 gives that $\{0\} \times \bar{\Gamma} \lhd H$ is a finite normal subgroup with $H/(\{0\} \times \bar{\Gamma}) \cong \mathbb{Z} \ltimes_\psi \pi(G)$, where $\psi \in$ Aut$(\pi(G))$ is given by $\psi(\pi(x)) = \pi(\varphi(x))$. By Exercise 8.26 it suffices to prove that $\mathbb{Z} \ltimes_\psi \pi(G)$ is virtually nilpotent.

By Exercise 8.29, ψ has all eigenvalues equal to 1. Since $\pi(\Gamma) \cong \mathbb{Z}^{d_n(G)}$, by Exercise 8.27 there exists $y \in \Gamma \backslash \bar{\Gamma}$ such that $\psi(\pi(y)) = \pi(y)$. By the definition of $\bar{\Gamma}$, if $y^m \in \bar{\Gamma}$ for some $0 \neq m \in \mathbb{Z}$, then $(y^m)^\ell = 1$ for some $\ell \in \mathbb{Z}$, so $y \in \bar{\Gamma}$. Hence, $\pi(y) \in \pi(\Gamma)$ is an element of infinite order. Specifically, if we define $N = \langle \bar{\Gamma}, y \rangle$, we have that $\pi(N) = \langle \pi(y) \rangle \cong \mathbb{Z}$.

For $a, m \in \mathbb{Z}$ and $x \in G$ we can compute commutators in $\mathbb{Z} \ltimes_\psi \pi(G)$:

$$[(a, \pi(x)), (0, \pi(y^m))] = \left(-a, \psi^{-a}\left(\pi(x^{-1})\right)\psi^{-a}(\pi(y^{-m}))\right)(a, \pi(x)\psi^a(\pi(y^m)))$$
$$= \left(0, \psi^{-a}(\pi(x^{-1}))\psi^{-a}(\pi(x))\right) = 1_{\mathbb{Z} \ltimes_\psi \pi(G)}.$$

That is, $[\mathbb{Z} \ltimes_\psi \pi(G), \{0\} \times \pi(N)] = \{1\}$. Thus, by Exercises 8.20 and 8.25, it suffices to prove that

$$(\mathbb{Z} \ltimes_\psi \pi(G))/(\{0\} \times \pi(N)) \cong \mathbb{Z} \ltimes_\phi (\pi(G)/\pi(N))$$

is virtually nilpotent, where $\phi(\pi(N)\pi(x)) = \pi(N)\psi(\pi(x))$.

Since

$$\pi(G)/\pi(N) \cong (G/\bar{\Gamma})/(N/\bar{\Gamma}) \cong G/N,$$

by Exercise 8.30 we may conclude the following facts:

$$\gamma_k(\pi(G)/\pi(N))/\bar{\gamma}_{k+1}(\pi(G)/\pi(N)) \cong \gamma_k(G/N)/\bar{\gamma}_{k+1}(G/N)$$
$$\cong \pi(\gamma_k(G))/\pi(\bar{\gamma}_{k+1}(G))$$
$$\cong \gamma_k(G)/\bar{\gamma}_{k+1}(G)$$

for all $0 \le k < n$. So $d_k(\pi(G)/\pi(N)) = d_k(G)$ for all $0 \le k < n$.

If $k > n$ then $\gamma_k(\pi(G)) = \pi(\gamma_k(G)) = \{1\}$, so $d_k(\pi(G)) = 0$.

Finally, for $k = n$ we have that since $\bar{\Gamma} = \bar{\gamma}_{n+1}(G) \le N$, $\bar{\gamma}_{n+1}(\pi(G)/\pi(N)) \cong \bar{\gamma}_{n+1}(G/N) = \{1\}$. Also,

$$\gamma_n(G/\bar{\Gamma}) \cong \gamma_n(G)/\bar{\Gamma} = \Gamma/\bar{\Gamma},$$

so

$$\gamma_n(G/N) \cong \gamma_n(G)/N \cong \gamma_n(G/\bar{\Gamma})/(N/\bar{\Gamma}) \cong \Gamma/N.$$

Note that $\pi(\Gamma) \cong \Gamma/\bar{\Gamma} \cong \mathbb{Z}^{d_n(G)}$. Since $\bar{\Gamma}$ is a finite subgroup of the infinite group N, $\pi(N)$ cannot be trivial. Since $\pi(N) \le \pi(\Gamma) \cong \mathbb{Z}^{d_n(G)}$ is a nontrivial subgroup, by Exercise 8.28 we have that

$$\Gamma/N \cong \pi(\Gamma)/\pi(N) \cong \mathbb{Z}^m \times F$$

for a finite Abelian group F and some $m < d_n(G)$. Specifically, since

$$\gamma_n(\pi(G)/\pi(N)) \cong \gamma_n(G/N) \cong \Gamma/N,$$

we find that

$$d_n(\pi(G)/\pi(N)) = d_n(G/N) < d_n(G).$$

In conclusion, we have that $D(\pi(G)/\pi(N)) < D(G)$. Thus, $\mathbb{Z} \ltimes_\phi (\pi(G)/\pi(N))$ is virtually nilpotent by induction, completing the induction step. □

Lemma 8.5.4 *Let G be an infinite finitely generated nilpotent group. Let $\varphi \in$ Aut(G). If all eigenvalues of φ have modulus 1 then $\mathbb{Z} \ltimes_\varphi G$ is virtually nilpotent.*

Proof   Since G is nilpotent, φ has finitely many eigenvalues (Exercise 8.16). By Exercise 8.22, the eigenvalues of φ are all roots of unity. Thus, there exists $k > 0$ such that all eigenvalues of $\psi = \varphi^k \in \mathsf{Aut}(G)$ are exactly 1. So Lemma 8.5.3 tells us that $\mathbb{Z} \ltimes_\psi G$ is virtually nilpotent.

By Exercise 8.18, $\mathbb{Z} \ltimes_\psi G$ is isomorphic to a finite index subgroup of $\mathbb{Z} \ltimes_\varphi G$. So $\mathbb{Z} \ltimes_\varphi G$ is virtually nilpotent as well. $\square$

Let us summarize Lemmas 8.5.2 and 8.5.4. A $\mathbb{Z}$-extension of a finitely generated nilpotent group G can either be virtually nilpotent (and thus have polynomial growth), or must have exponential growth. The former case is exactly when all eigenvalues of the implicit automorphism of G have modulus 1. Note that this is a special case of the Milnor–Wolf theorem (Theorem 8.2.4) for groups that are $\mathbb{Z}$-extensions of finitely generated nilpotent groups.

8.6 Proof of the Milnor–Wolf Theorem

One consequence of Lemmas 8.5.2 and 8.5.4 is the following.

Corollary 8.6.1   Let G be a finitely generated group of sub-exponential growth. Assume that the commutator subgroup $[G, G]$ is virtually nilpotent.

Then G is virtually nilpotent.

Proof   Since $G/[G, G]$ is Abelian and since G has sub-exponential growth, we know that $[G, G]$ is finitely generated by Rosset's theorem (Theorem 8.3.3). Denote

$$\bar{\Gamma} = \bar{\gamma}_1(G) = \{x \in G : \exists\, 0 \neq m \in \mathbb{Z},\ x^m \in [G, G]\}.$$

Since $\bar{\Gamma}/[G, G]$ is an Abelian group whose elements are all of finite order, we know that $[\bar{\Gamma} : [G, G]] < \infty$. Specifically, $\bar{\Gamma}$ is virtually nilpotent by our assumption on $[G, G]$. By Exercises 8.15 and 1.50, we know that $G/\bar{\Gamma} \cong \mathbb{Z}^d$ for some $d > 0$.

We may find normal subgroups $\bar{\Gamma} = K_d \lhd \cdots \lhd K_1 \lhd K_0 = G$ such that $K_j/K_{j+1} \cong \mathbb{Z}$ for all $0 \leq j < d$. Using Rosset's theorem (Theorem 8.3.3) on these we get that $K_1, \ldots, K_d$ are all finitely generated. Exercise 8.19 gives that $K_j \cong \mathbb{Z} \ltimes_{\psi_j} K_{j+1}$ for all $0 \leq j < d$ for some $\psi_j \in \mathsf{Aut}(K_{j+1})$.

Assume that G is not virtually nilpotent. Then there exists some $0 \leq j < d$ such that K_j is not virtually nilpotent, but K_{j+1} is virtually nilpotent (recall that $\bar{\Gamma}$ is virtually nilpotent). Then, since $K_j = \mathbb{Z} \ltimes_{\psi_j} \mathsf{Aut}(K_{j+1})$ has sub-exponential growth (Exercise 8.4), Lemma 8.5.2 tells us that it must be that all eigenvalues of ψ_j are of modulus 1. Thus, by Lemma 8.5.4 we have that K_j is virtually nilpotent, a contradiction! □

We are now ready to prove the Milnor–Wolf theorem, Theorem 8.2.4, which basically states that if G is a finitely generated solvable group of sub-exponential growth, then G is virtually nilpotent.

Proof of Theorem 8.2.4 Recall the derived series of G, which is $G^{(0)} = G$, $G^{(j+1)} = \left[G^{(j)}, G^{(j)} \right]$. Let n be such that $G^{(n+1)} = \{1\}$ and $G^{(n)} \neq \{1\}$. The proof is by induction on n.

Base case, $n = 0$. In this case G is Abelian, which has polynomial growth by Exercise 8.6.

For the induction step, assume $n > 0$. Since G has sub-exponential growth, and since $G/G^{(1)}$ is Abelian, we know that $G^{(1)}$ is finitely generated by Rosset's theorem (Theorem 8.3.3). So $G^{(1)}$ is finitely generated of sub-exponential growth (Exercise 8.4). Since $\left(G^{(1)} \right)^{(n)} = G^{(n+1)} = \{1\}$, by induction $G^{(1)} = [G, G]$ is virtually nilpotent. By Corollary 8.6.1, this implies that G is virtually nilpotent as well. □

8.7 Additional Exercises

Exercise 8.31 Recall the groups $H_n(\mathbb{Z})$ from Exercise 1.46.

Consider the case $n = 3$, which is the so-called *Heisenberg Group*,

$$H_3(\mathbb{Z}) = \{((c, b, a)) : a, b, c \in \mathbb{Z}\}, \qquad \text{where} \qquad ((c, b, a)) = \begin{bmatrix} 1 & a & c \\ 0 & 1 & b \\ 0 & 0 & 1 \end{bmatrix}.$$

Let $A = ((0, 0, 1))$, $B = ((0, 1, 0))$, and $C = ((1, 0, 0))$.

Show that $((c, b, a)) = C^c B^b A^a$.

Show that $[A, B] = C$.

Conclude that A, B generate $H_3(\mathbb{Z})$. ▷ solution ◁

Exercise 8.32 Show that $Z(H_3(\mathbb{Z})) = \langle C \rangle = \{((c, 0, 0)) : c \in \mathbb{Z}\}$. ▷ solution ◁

Exercise 8.33 Let G be a group. Let $x, y \in G$ be such that $[x, y] \in Z(G)$. Show that for all integers n, m we have that $[x^n, y^m] = [x, y]^{nm}$. ▷ solution ◁

Exercise 8.34 Consider the Cayley graph of $H_3(\mathbb{Z})$ with respect to the symmetric generating set $\{A, -A, B, -B\}$, for A, B given in Exercise 8.31. Show that there exists a constant $\kappa > 0$ such that

$$|((c, b, a))| \le \kappa \left(\sqrt{|c| + 1} + |b| + |a|\right).$$ ▷ solution ◁

Exercise 8.35 Let G be a finitely generated group with finite symmetric generating sets S, S'. Let $H \lhd G$ be a finitely generated normal subgroup, with finite symmetric generating sets T, T'.

The **distortion** of H in G is defined to be

$$\delta_{(G,S),(H,T)}(r) = \max\{|x|_T : x \in H, \ |x|_S \le r\},$$

which is easily seen to be a nondecreasing function.

Show that $\delta_{(G,S),(H,T)} \sim \delta_{(G,S'),(H,T')}$ (where $\sim$ is the equivalence relation on growth functions).

Show that if the growth of H is at most ψ, the distortion of H in G is at most δ, and the growth of G/H is at most ϕ, then the growth of G is at most $\phi \cdot (\delta \circ \psi)$.

Conclude that if H has polynomial distortion in G and G/H has polynomial growth, then G has polynomial growth.   ▷ solution ◁

Exercise 8.36 Show that the growth of $H_3(\mathbb{Z})$ is $r \mapsto r^4$.   ▷ solution ◁

Exercise 8.37 Let G be a group and let H, K be subgroups of G such that:

- $\langle K, H \rangle = G$,
- $H \lhd G$,
- $H \cap K = \{1_G\}$.

Show that K acts on H by conjugation and that $K \ltimes H \cong G$.   ▷ solution ◁

Exercise 8.38 Let $G = K \ltimes H$ for some K acting on H. Let $\tilde{K} = K \times \{1_H\} \subset G$ and $\tilde{H} = \{1_K\} \times H \subset G$.

Show that $\tilde{H} \cap \tilde{K} = \{1_G\}$.

Show that $\langle \tilde{H}, \tilde{K} \rangle = G$.

Show that $\tilde{H} \lhd G$ and $\tilde{H} \cong H$.

8.8 Remarks

Nilpotent groups were shown to have polynomial growth by Wolf (1968). Later, Bass (1972) and Guivarc'h (1973) provided the precise degree of the polynomial growth rate (and showed that it is in fact an integer).

The Milnor–Wolf theorem (Theorem 8.2.4) was published in Milnor (1968a) and Wolf (1968), and is a special case of Gromov's theorem (Theorem 9.0.1) for the class of solvable groups.

Our treatment of the results in this chapter is based on Druțu and Kapovich (2018). According to Druțu and Kapovich (2018), the provided proof of Lemmas 8.5.3 and 8.5.4 is due to B. Plotkin.

Related to the Milnor–Wolf theorem, but outside the scope if this book, is another dichotomy known as the *Tits alternative*, proved by Jacques Tits (1972).

Theorem 8.8.1 (Tits alternative) *Let V be a finite-dimensional vector space over some field F. Let G be a finitely generated subgroup of $\mathsf{GL}(V)$.*

Then, either G contains a subgroup isomorphic to $\mathbb{F}_2$ (the free group generated by two elements), or G is virtually solvable.

The Tits alternative is a special case of Gromov's theorem for finitely generated linear groups (i.e. subgroups of $\mathsf{GL}(V)$ for finite-dimensional V). Note that it implies that any finitely generated amenable linear group is virtually solvable.

8.9 Solutions to Exercises

Solution to Exercise 8.3 :(

We showed in Section 1.6.1 (Exercise 1.75) that if S, T are two finite symmetric generating sets of G, then there exists a constant $\kappa = \kappa_{S,T} > 0$ such that for all $x, y \in G$,

$$\kappa^{-1} \cdot \mathrm{dist}_T(x, y) \leq \mathrm{dist}_S(x, y) \leq \kappa \cdot \mathrm{dist}_T(x, y)$$

(i.e. the metrics are bi-Lipschitz). Thus, $B_S(1, r) \subset B_T(1, \kappa r)$ for all $r \geq 0$.

So all Cayley graphs (of the same group) have equivalent growth functions.

Since $|B_S(1, r)| \leq |S|^r$, it is immediate the exponential growth is the maximal possible.   :) ✓

Solution to Exercise 8.4 :(

If $H \leq G$ is finitely generated, then we may choose $T \subset S$ such that S is a finite symmetric generating set of G and T is a finite symmetric generating set of H. It is then immediate that for any $x \in H$ we have $|x|_S \leq |x|_T$ (because any word in T is also a word in S). Thus, $B_T(1, r) \subset B_S(1, r) \cap H$, which immediately implies the first assertion.

For the second assertion, let $N \triangleleft G$ and let $\pi \colon G \to {}_N\backslash^G$ be the canonical projection. If S is a finite symmetric generating set for G, then $\pi(S)$ is a finite symmetric generating set for ${}_N\backslash^G$. Now, if $|Nx|_{\pi(S)} = r$, then there exist $s_1, \ldots, s_r \in S$ such that $Nx = Ns_1 \cdots s_r$. Set $x_j = s_1 \cdots s_j$ for $1 \leq j \leq r$. Note that $|Ny|_{\pi(S)} \leq |y|_S$ for all $y \in G$, and hence

$$r \geq |x_r|_S \geq |Nx_r|_{\pi(S)} = |Nx|_{\pi(S)} = r,$$

implying that $|x_r|_S = r$. We conclude that $\pi|_{B_r}$ is surjective onto $B_{\pi(S)}(N, r)$. Thus, $|B_{\pi(S)}(N, r)| \leq |B_S(1, r)|$, implying the second assertion.   :) ✓

Solution to Exercise 8.5   :(

The fact that H is finitely generated was shown in Exercise 1.61. So the growth of H is at most the growth of G, by Exercise 8.4.

Now, consider some finite symmetric generating set $\tilde{S}$ for H. Write $G = \biguplus_{t \in T} Ht$, for some finite set $|T| = [G : H] < \infty$. For any $x \in G$ there exist $h_x \in H$ and $t_x \in T$ such that $x = h_x t_x$. Thus, $S := \tilde{S} \cup T \cup T^{-1}$ is a finite symmetric generating set for G.

Moreover, since $\tilde{S} \subset S$ we have that $|h_x|_S \leq |h_x|_{\tilde{S}} \leq |x|_S + 1$. Thus,

$$|B_S(1, r)| \leq \#\{(h_x, t_x) : |x| \leq r\} \leq |B_{\tilde{S}}(1, r+1)| \cdot |T|,$$

which implies that the growth of G is at most the growth of H.   :) ✓

Solution to Exercise 8.6   :(

If A is a finitely generated Abelian group then $A \cong \mathbb{Z}^d \times F$ for some $d \geq 0$ and $|F| < \infty$ a finite group, by Theorem 1.5.2. Since $\mathbb{Z}^d$ has finite index in $\mathbb{Z}^d \times F$, by Exercise 8.5 it suffices to show that $\mathbb{Z}^d$ has polynomial growth for any $d \geq 0$.

This is immediate when $d = 0$ (as the group is finite) and for $d > 0$, we can take the standard basis elements to generate $\mathbb{Z}^d$, namely $\mathbb{Z}^d = \langle \pm e_1, \ldots, \pm e_d \rangle$, where e_j is the d-dimensional vector with all coordinates 0 except the jth coordinate equal to 1. With these generators, it is very easy to see that

$$B(0, r) = \left\{ (x_1, \ldots, x_d) \in \mathbb{Z}^d : \sum_{j=1}^{d} |x_j| \leq r \right\}$$

$\left(\text{recalling that } 0 \in \mathbb{Z}^d \text{ is the identity element}\right)$. This can easily be bounded by

$$B(0, r) \subset \{ (x_1, \ldots, x_d) \in \mathbb{Z}^d : \max_j |x_j| \leq r \},$$

so that $|B(0, r)| \leq (2r + 1)^d$, and by

$$\{ (x_1, \ldots, x_d) \in \mathbb{Z}^d : \max_j |x_j| \leq r/d \} \subset B(0, r),$$

so that $|B(0, r)| \geq \left(\frac{2r}{d}\right)^d$.   :) ✓

Solution to Exercise 8.7   :(

Denote $\gamma_k = \gamma_k(G)$. Let n be the nilpotent step of G; so $\gamma_n = \{1\}$.

Assume for a contradiction that there exists $1 \leq k \leq n - 1$ such that γ_k is not finitely generated. Choose a maximal such k. So γ_k is not finitely generated, but γ_{k+j} is finitely generated for all $j \geq 1$.

By Exercise 1.39 we know that γ_k / γ_{k+1} is finitely generated. Since γ_{k+1} is finitely generated by the choice of k, by Exercise 1.62 we have that γ_k is also finitely generated, a contradiction!   :) ✓

Solution to Exercise 8.8   :(

Denote $\gamma_k = \gamma_k(G)$. We prove the claim by induction on $n - k$.

The base case is $k = n - 1$. By Exercise 1.34, we know that

$$\gamma_{n-1}(G) = \langle [s_0, \ldots, s_{n-1}]^x : s_1, \ldots, s_{n-1} \in S, \ x \in G \rangle.$$

Since $\gamma_{n-1}(G) \lhd Z(G)$, we have that $[s_0, \ldots, s_{n-1}]^x = [s_0, \ldots, s_{n-1}]$. Thus, $\gamma_{n-1} = \langle S_{n-1} \rangle$.

Now, for $k < n - 1$, by Exercise 1.34,

$$\gamma_k(G) = \langle [s_0, \ldots, s_k]^x : s_1, \ldots, s_k \in S, \ x \in G \rangle.$$

Since,

$$[s_0, \ldots, s_k]^x = [s_0, \ldots, s_k] \cdot [s_0, \ldots, s_k, x],$$

and since $[s_0, \ldots, s_k, x] \in \gamma_{k+1}$, we have that if T generates γ_{k+1} then $S_k \cup T$ generates γ_k. By induction we have that $T_k = S_k \cup T_{k+1}$ generates γ_k.   :) ✓

Solution to Exercise 8.9 :(

Let $\pi : G \to \mathbb{Z}$ be a surjective homomophism with $\text{Ker}\pi = H$. Choose some element $s \in G$ such that $\pi(s) = 1 \in \mathbb{Z}$.

Let $\tilde{S}$ be any finite generating set for G. For $u \in \tilde{S}$ there exist $s_u \in H$ and $z_u \in \mathbb{Z}$ such that $\pi(u) = z_u$ and $u = s_u s^{z_u}$. Note that the set $S = \{s, s_u : u \in \tilde{S}\}$ generates $\tilde{S}$ and thus generates all of G, so S satisfies the necessary requirements. :) ✓

Solution to Exercise 8.10 :(

Using Exercise 8.9, let $S = \{s_1, \ldots, s_d\}$ be a finite generating set for G such that s_1 is mapped to $1 \in \mathbb{Z}$ under the canonical projection (and $s_j \in H$ for all $j > 1$). Note that $\langle s_1 \rangle = \{(s_1)^n : n \in \mathbb{Z}\}$ is a right-traversal for H; that is, $1 \in \langle s_1 \rangle$ and $G = \uplus_{t \in \langle s_1 \rangle} Ht$. Exercise 1.60 tells us that H is generated by $\langle s_1 \rangle S \langle s_1 \rangle^{-1} \cap H$. Note that $\langle s_1 \rangle S \langle s_1 \rangle^{-1} \cap H = \{(s_1)^{-n} s_j (s_1)^n : 1 < j \le d, \, n \in \mathbb{Z}\}$. This last set is finitely generated by Milnor's trick (Proposition 8.3.1). :) ✓

Solution to Exercise 8.11 :(

The proof is by induction on the length of the derived series.

If G is 1-step solvable, then G is Abelian, and being finitely generated is of the form $\mathbb{Z}^d \times F$ for a finite F. Since $\mathbb{Z}^d$ only has elements of infinite order, it must be that $d = 0$. This is the base case.

Now assume that G is n-step solvable for some $n > 1$. If G is torsion, then $[G, G]$ is torsion as well, and it is easy to see that $G/[G, G]$ is also torsion. Thus, by induction $G/[G, G]$ is finite. Also, $[G, G]$ is finitely generated by Rosset's theorem (Theorem 8.3.3). We have that $[G, G]$ is $(n - 1)$-step solvable, so $[G, G]$ must be finite by induction. The induction is completed via $|G| \le [G : [G, G]] \cdot |[G, G]|$. :) ✓

Solution to Exercise 8.12 :(

Let $g \in G$. The map $x \mapsto x^g$ is an automorphism of G, and thus, restricted to N is an automorphism of N since $N \triangleleft G$. Thus, it preserves H. So $H^g = H$ for any $g \in G$, implying that $H \triangleleft G$. :) ✓

Solution to Exercise 8.13 :(

Let φ be any automorphism of G. Note that $\varphi([x, y]) = [\varphi(x), \varphi(y)]$ for all $x, y \in G$. Now, any $\varphi(z) \in \varphi([N, K])$ can be written as $z = [x_1, y_1] \cdots [x_n, y_n]$ for some $x_1, \ldots, x_n \in N$ and $y_1, \ldots, y_n \in K$. Thus, $\varphi(z) \in [N, K]$.

This proves that $\varphi([N, K]) \subset [N, K]$ for any automorphism φ of G. But then, we also have that $[N, K] = \varphi\left(\varphi^{-1}([N, K])\right) \subset \varphi([N, K]) \subset [N, K]$, so equality holds. :) ✓

Solution to Exercise 8.14 :(

For any automorphism $\varphi \in \text{Aut}(G)$, we have that

$$\varphi(x)^k \in \gamma_n(G) \iff x^k \in \gamma_n(G)$$

because $\gamma_n(G)$ is a characteristic subgroup.

So we only need to show that $\bar{\gamma}_n(G)$ is indeed a subgroup. To this end, for any $x, y \in \bar{\gamma}_n(G)$, we have $x^k \in \gamma_n(G)$ and $y^m \in \gamma_n(G)$ for some $k, m \in \mathbb{Z}$. Also, since $x, y \in \gamma_{n-1}(G)$, we know that $[x, y] \in \gamma_n(G)$, so there exists $g \in \gamma_n(G)$ such that

$$\left(x^{-1} y\right)^{km} = g x^{-km} y^{km} = g \left(x^{-k}\right)^m (y^m)^k \in \gamma_n(G).$$

This implies that $x^{-1} y \in \bar{\gamma}_n(G)$, proving that $\bar{\gamma}_n(G)$ is a subgroup. :) ✓

Solution to Exercise 8.15 :(

Since $[\gamma_n(G), \gamma_n(G)] \le [\gamma_n(G), G] = \gamma_{n+1}(G) \le \bar{\gamma}_{n+1}(G)$, we have that $\gamma_n(G)/\bar{\gamma}_{n+1}(G)$ is Abelian. We know that $\gamma_n(G)/\gamma_{n+1}(G)$ is finitely generated (Exercise 1.39). Note that $\gamma_{n+1}(G)$ is a normal subgroup of both $\bar{\gamma}_{n+1}(G)$ and $\gamma_n(G)$. Thus,

$$\gamma_n(G)/\bar{\gamma}_{n+1}(G) \cong (\gamma_n(G)/\gamma_{n+1}(G))/(\bar{\gamma}_{n+1}(G)/\gamma_{n+1}(G)),$$

implying that $\gamma_n(G)/\bar{\gamma}_{n+1}(G)$ is a finitely generated Abelian group. By Theorem 1.5.2, $\gamma_n(G)/\bar{\gamma}_{n+1}(G) \cong \mathbb{Z}^{d_n} \times F$ for some finite Abelian group F and some integer $d_n \ge 0$.

For any $f \in F$ we have that $(0, f) \in \mathbb{Z}^{d_n} \times F$ has finite order in $\mathbb{Z}^{d_n} \times F$, namely $(0, f)^{|F|} = \left(0, f^{|F|}\right) = (0, 1_F)$. We want to show that $F = \{1\}$. So it suffices to prove that no nontrivial element of $\gamma_n(G)/\bar{\gamma}_{n+1}(G)$ has finite order.

Indeed, assume that $x \in \gamma_n(G)$. If $\bar{\gamma}_{n+1}(G)x$ has finite order, then $x^k \in \bar{\gamma}_{n+1}(G)$ for some k. But then $x^{km} = \left(x^k\right)^m \in \gamma_{n+1}(G)$ for some m, implying that $x \in \gamma_{n+1}(G)$. So the only coset $\bar{\gamma}_{n+1}(G)x$ of finite order in $\gamma_n(G)/\bar{\gamma}_{n+1}(G)$ is the trivial coset. :) ✓

Solution to Exercise 8.16 :(

Let $\gamma_k = \gamma_k(G)$, $\bar{\gamma}_k = \bar{\gamma}_k(G)$, and $d_k = d_k(G)$ for all k.

Let $x \in \gamma_n = \bar{\gamma}_{n+1}$ and $y \in G$. There exists $k \geq 1$ such that $x^k \in \gamma_{n+1}$. We have that

$$\left[x^k, y\right] = x^{-1}\left[x^{k-1}, y\right]y^{-1}xy = \left[x^{k-1}, y\right]\left[[x^{k-1}, y], x\right][x, y]$$
$$= \cdots = [[x, y], x] \cdots \left[[x^{k-1}, y], x\right] \cdot [x, y].$$

Since $\left[x^k, y\right] \in \gamma_{n+2}$ and $[[x^m, y], x] \in \gamma_{n+2}$ for all $m \in \mathbb{N}$, we have that $[x, y] \in \gamma_{n+2}$. This holds for all $x \in \gamma_n$ and all $y \in G$. Thus,

$$\gamma_{n+1} = \langle [x, y] : x \in \gamma_n, \ y \in G \rangle \leq \gamma_{n+2},$$

implying that $\gamma_{n+1} = \gamma_{n+2} = \bar{\gamma}_{n+2}$.

Thus, $d_{n+1} = 0$, and we can repeat the argument inductively for $n + k$ for all $k \geq 1$. :) ✓

Solution to Exercise 8.17 :(

We have that φ_n is well defined since if $\bar{\gamma}_{n+1}(G)x = \bar{\gamma}_{n+1}(G)y$, then $x = gy$ for some $g \in \bar{\gamma}_{n+1}(G)$. Hence $\varphi(x) = \varphi(g)\varphi(y)$, and since $\bar{\gamma}_{n+1}(G)$ is a characteristic subgroup, we get that $\varphi(g) \in \bar{\gamma}_{n+1}(G)$, so that $\bar{\gamma}_{n+1}(G)\varphi(x) = \bar{\gamma}_{n+1}(G)\varphi(y)$.

It is immediate to see that φ_n is an automorphism of $\gamma_n(G)/\bar{\gamma}_{n+1}(G)$. :) ✓

Solution to Exercise 8.18 :(

First, conjugation in $\mathbb{Z} \ltimes_\varphi G$ is given by

$$(a, x)^{(b, y)} = \left(-b, \varphi^{-b}(y^{-1})\right)(a + b, x\varphi^a(y)) = \left(a, \varphi^{-b}(y^{-1}x)\varphi^{a-b}(y)\right).$$

This proves the first assertion.

For the second assertion, consider the map $\lambda : \mathbb{Z} \ltimes_\psi G \to \mathbb{Z} \ltimes_\varphi G$ given by $\lambda(a, x) = (ma, x)$. Note that

$$\lambda\left(a + b, x\psi^a(y)\right) = (m(a + b), x\varphi^{ma}(y)) = (ma, x) \cdot (mb, y) = \lambda(a, x) \cdot \lambda(b, y).$$

Also, if $\lambda(a, x) = (0, 1)$ then $a = 0$ and $x = 1$. Thus, λ is an injective homomorphism. The homomorphism $\pi : \mathbb{Z} \ltimes_\varphi G \to \mathbb{Z}/|m|\mathbb{Z}$ given by $\pi(a, x) = a \pmod{|m|}$ has kernel $\ker \pi = H = \{(ma, x) : a \in \mathbb{Z}, x \in G\}$. Thus, $[\mathbb{Z} \ltimes_\varphi G : H] = |m|$. Note that $H = \lambda(\mathbb{Z} \ltimes_\psi G)$.

The last assertion is easy, since $(a, 1) = (1, 1)^a$ and $(0, xy) = (0, x)(0, y)$ and $(a, x) = (0, x)(a, 1)$. :) ✓

Solution to Exercise 8.19 :(

Let $g \in G$ be such that g is mapped to $1 \in \mathbb{Z} \cong G/K$ under the canonical projection $G \mapsto \mathbb{Z} \cong G/K$.

Define $\varphi \in \text{Aut}(K)$ by $\varphi(k) = gkg^{-1}$. It is simple to check that this induces an isomorphism of groups $\mathbb{Z} \ltimes_\varphi K \cong G$ by mapping $\lambda(a, k) = kg^a$. Indeed, for $a, b \in \mathbb{Z}$ and $k, n \in K$,

$$\lambda((a, k)(b, n)) = \varphi(a + b, kg^a n g^{-a}) = kg^a n g^{-a} g^{a+b} = \lambda(a, k)\lambda(b, n).$$

The kernel of λ is trivial because $kg^a = 1$ if and only if $a = 0$ and $k = 1$. Also, λ is surjective because given $x \in G$, there exists $a \in \mathbb{Z}$ such that $Kx = Kg^a$. So choosing $k = xg^{-a}$, we have that $\lambda(a, k) = kg^a = x$.

Note that this isomorphism maps $\{0\} \times K$ onto K, so these are isomorphic subgroups.

Now, if G has growth $\leq r^d$, then K is finitely generated by Rosset's theorem (Theorem 8.3.3). Let S be a finite symmetric generating set for K, and consider the generating set $\{(0, s), (\pm 1, 1) : s \in S\}$ for $\mathbb{Z} \ltimes K$. Note that $|(a, k)| \leq |a| + |k|$ (by Exercise 8.18). Consider the identity map of $\mathbb{Z} \times K$ into $\mathbb{Z} \ltimes K$ (as sets). This is an injective function, which takes $\{1, \ldots, r\} \times \{k \in K : |k| \leq r\}$ into the ball of radius $2r$ in $\mathbb{Z} \ltimes K$. Thus, $r \cdot \#\{k \in K : |k| \leq r\} \leq Cr^d$ for some $C > 0$, implying that K has growth $\leq r^{d-1}$. :) ✓

Solution to Exercise 8.20 :(

If $Hx = Hy$ then $x^{-1}y \in H$ so also $\varphi(x)^{-1}\varphi(y) = \varphi\left(x^{-1}y\right) \in H$, implying that ψ is well defined. It is easy to see that ψ is a homomorphism, with trivial kernel (because $x \in H \iff \varphi(x) \in H$), and is surjective because φ is surjective. So $\psi \in \text{Aut}(G/H)$.

Consider the map $\pi : \mathbb{Z} \ltimes_\varphi G \to \mathbb{Z} \ltimes_\psi (G/H)$ given by $\pi(a, x) = (a, Hx)$. We have

$$\pi(a + b, x\varphi^a(y)) = (a + b, Hx\varphi^a(y)) = (a + b, Hx\psi^a(Hy)) = (a, Hx)(b, Hy),$$

so π is a homomorphism. It is simple to see that π is surjective, and

$$\mathrm{Ker}\pi = \{(a, x) : a = 0, \; x \in H\} = N.$$

So $N \lhd \mathbb{Z} \ltimes_\varphi G$ and $(\mathbb{Z} \ltimes_\varphi G)/N \cong \mathbb{Z} \ltimes_\psi (G/H)$.
    The map $(0, x) \mapsto x$ is an isomorphism of N with H.   :) ✓

Solution to Exercise 8.21   :(
Since $G/F \cong \mathbb{Z}$, we know that $G \cong \mathbb{Z} \ltimes_\varphi F$ for some $\varphi \in \mathrm{Aut}(F)$ (Exercise 8.19). So assume without loss of generality that $G = \mathbb{Z} \ltimes_\varphi F$. Let $H = \{(z, 1) : z \in \mathbb{Z}\} \subset G$. It is easy to see that H is a (not necessarily normal) subgroup. Also, since $(a, f) = (0, f)(a, 1)$ for all $f \in F$, $a \in \mathbb{Z}$, we see that $[G : H] \le |F| < \infty$.
                                                                                :) ✓

Solution to Exercise 8.22   :(
Let $\lambda_1, \ldots, \lambda_n$ be the eigenvalues of M (with multiplicities). Let $v_k := \left(\lambda_1^k, \ldots, \lambda_n^k\right) \in \left(S^1\right)^n$ $\left(\text{here } S^1 \subset \mathbb{C}\right.$ is the unit circle$)$. So trace $\left(M^k\right) = \sum_{j=1}^n \lambda_j^k = \langle v_k, 1\rangle$, where 1 is the all ones vector. Since $\left(S^1\right)^n$ is compact, there is a convergent subsequence $(w_j := v_{k_j})_j$, which implies that for $m_j := k_j - k_{j-1}$ we have that $\lambda_i^{m_j} \to 1$ for all i, as $j \to \infty$. Thus, trace $\left(M^{m_j}\right) \to n$ as $j \to \infty$. However, trace $\left(M^{m_j}\right)$ is an integer for any m_j (this is where it is used that M has integer values). Hence, it must be that there exists j_0 such that for all $j > j_0$ we have $\sum_{i=1}^n \lambda_i^{m_j} = $ trace $\left(M^{m_j}\right) = n$. Since $|\lambda_i| = 1$ for all i, the only way this can happen is if $\lambda_i^{m_j} = 1$ for all i and $j > j_0$.   :) ✓

Solution to Exercise 8.23   :(
Let $e_1, \ldots, e_d$ denote the standard basis of $\mathbb{Z}^d$. So e_j is the d-dimensional vector with all 0's except at the jth coordinate, which is 1. Then any $x \in \mathbb{Z}^d$ can be written as $x = \sum_{j=1}^d \langle x, e_j\rangle e_j$.
    For an automorphism $\varphi \in \mathrm{Aut}\left(\mathbb{Z}^d\right)$ (recalling that the group operation is vector addition), we have that $\varphi(ne_j) = n\varphi(e_j)$ for any $n \in \mathbb{Z}$. Also, $\varphi(x + y) = \varphi(x) + \varphi(y)$ for any $x, y \in \mathbb{Z}^d$.
    Define $M_{i,j} = \langle\varphi(e_j), e_i\rangle \in \mathbb{Z}$ for $1 \le i, j \le d$. If $x \in \mathbb{Z}^d$, note that $\langle x, e_j\rangle \in \mathbb{Z}$ for all j, so that

$$\langle\varphi(x), e_i\rangle = \sum_{j=1}^d \langle x, e_j\rangle\langle\varphi(e_j), e_i\rangle = \sum_{j=1}^d M_{i,j}x_j = (Mx)_i.$$

Thus, $\varphi(x) = Mx$.
    So we are left with showing that $|\det(M)| = 1$. Since φ is an automorphism, φ^{-1} is also an automorphism. It is easy to see that M is invertible, and $\varphi^{-1}(x) = M^{-1}x$ for all $x \in \mathbb{Z}^d$. So $\det(M) \cdot \det\left(M^{-1}\right) = 1$. Since both these determinants must be integers, the only possibilities are $\det(M) = \det\left(M^{-1}\right) \in \{-1, 1\}$.   :) ✓

Solution to Exercise 8.24   :(
Consider $\psi = j \circ i^{-1} \in \mathrm{Aut}\left(\mathbb{Z}^d\right)$. By Exercise 8.23, there is a matrix $U \in \mathrm{GL}_d(\mathbb{Z})$ such that $\psi(\vec{z}) = U\vec{z}$ for all $\vec{z} \in \mathbb{Z}^d$. Note that for any $\vec{z} \in \mathbb{Z}^d$ we have

$$U^{-1}M_j U\vec{z} = \psi^{-1} \circ \varphi_j \circ \psi(\vec{z}) = i \circ j^{-1} \circ j \circ \varphi \circ j^{-1} \circ j \circ i^{-1}(\vec{z}) = \varphi_i(\vec{z}) = M_i\vec{z}. \qquad :) ✓$$

Solution to Exercise 8.25   :(
Note that if $\gamma_n(G/N) = \{1\}$ then $\gamma_n(G) \le N$. But then $\gamma_{n+1}(G) = [G, \gamma_n(G)] \le [G, N] = \{1\}$, so G is nilpotent.
    If G/N is virtually nilpotent, then there is a finite index subgroup $H \lhd G$ such that $N \lhd H$ and H/N is nilpotent. So H is nilpotent and thus G is virtually nilpotent.   :) ✓

Solution to Exercise 8.26   :(
$F \lhd G$ implies that G acts on F by conjugation. The kernel of this action is the subgroup

$$C_F = \{x \in G : \forall f \in F, \ f^x = f\}.$$

So G/C_F is isomorphic to a subgroup of the permutations of the finite set F, and thus C_F has finite index in G. So it suffices to show that C_F is virtually nilpotent.

Let $H = FC_F = \{fx : f \in F, \ x \in C_F\}$, which is a subgroup of G. The second isomorphism theorem tells us that $C_F/(F \cap C_F) \cong H/F \le G/F$, which is virtually nilpotent as a (finite index) subgroup of a virtually nilpotent group. Since $[C_F, F] = \{1\}$ by definition, we can use Exercise 8.25 to conclude that C_F is virtually nilpotent. :) ✓

Solution to Exercise 8.27 :(

By passing to an isomorphic group, we can assume that $\varphi = M \in \mathsf{GL}_d(\mathbb{Z})$, and that all eigenvalues of M are 1 (Exercise 8.23). Thus, there necessarily exists a vector $v \in \mathbb{Q}^d$ such that $Mv = v$. By multiplying v by an appropriate constant, we obtain an eigenvector in $\mathbb{Z}^d$. :) ✓

Solution to Exercise 8.28 :(

Since $\mathbb{Z}^d/K$ is a finitely generated Abelian group, $\mathbb{Z}^d/K \cong \mathbb{Z}^m \times F$ for some $m \le d$ and a finite Abelian group F (Theorem 1.5.2). There exists $M \le \mathbb{Z}^d$ with $K \le M$ such that $\mathbb{Z}^d/M \cong \mathbb{Z}^m$ and $M/K \cong F$. Let $\pi : \mathbb{Z}^d \to \mathbb{Z}^m$ be the canonical projection with $\mathrm{Ker}\,\pi = M$. Using the standard basis $e_1, \ldots, e_d \in \mathbb{Z}^d$, π can be extended to a surjective homomorphism of vector spaces $\pi : \mathbb{R}^d \to \mathbb{R}^m$. If K is nontrivial, then $\mathrm{Ker}\,\pi$ is nontrivial, so $m = \dim \pi\left(\mathbb{R}^d\right) < \dim\left(\mathbb{R}^d\right) = d$. :) ✓

Solution to Exercise 8.29 :(

Since $\bar{\gamma}_1(G) \le K$, by Exercise 8.30 we have that $\bar{\gamma}_1(G/K) = \{1\}$. Thus, $G/K \cong \mathbb{Z}^d$ for some d. Let $\pi : G \to G/K$ be the canonical projection with $\mathrm{Ker}\,\pi = K$, and let $\eta : G/K \to \mathbb{Z}^d$ be an isomorphism. Let $\phi \in \mathrm{Aut}(G/\bar{\gamma}_1(G))$ be defined by $\phi(\bar{\gamma}_1(G)x) = \bar{\gamma}_1(G)\varphi(x)$ (this is well defined because $\bar{\gamma}_1(G)$ is a characteristic subgroup). Since $(G/\bar{\gamma}_1(G))/(K/\bar{\gamma}_1(G)) \cong G/K \cong \mathbb{Z}^d$, there exist surjective homomorphisms $\alpha : G \to G/\bar{\gamma}_1(G)$ and $\beta : G/\bar{\gamma}_1(G) \to \mathbb{Z}^d$ with $\mathrm{Ker}\,\alpha = \bar{\gamma}_1(G)$ and $\mathrm{Ker}\,\beta = \alpha(K)$ such that $\beta \circ \alpha = \eta \circ \pi$. Note that this implies that $\eta(Kx) = \beta(\bar{\gamma}_1(G)x)$ for all $x \in G$.

Now, let λ be an eigenvalue of ψ, so we can choose $x \in G$ such that $\eta(\psi(Kx)) = \lambda\eta(Kx)$. Then,

$$\lambda\beta(\bar{\gamma}_1(G)x) = \lambda\eta(Kx) = \eta(\psi(Kx)) = \eta \circ \pi(\varphi(x)) = \beta(\bar{\gamma}_1(G)\varphi(x)).$$

Thus, λ is also an eigenvalue of φ. :) ✓

Solution to Exercise 8.30 :(

Since $[G, \gamma_n(G)] = \{1\}$, we know that $[G, N] = \{1\}$ so $N \vartriangleleft G$.

It is obvious that $\gamma_0(\pi(G)) = \pi(G) = \pi(\gamma_0(G))$.

Note that for any homomorphism $\pi([x, y]) = [\pi(x), \pi(y)]$, so $[\pi(H), \pi(K)] = \pi([H, K])$. By induction, for $k > 0$ we have that

$$\gamma_k(\pi(G)) = [\pi(G), \gamma_{k-1}(\pi(G))] = \pi([G, \gamma_{k-1}(G)]) = \pi(\gamma_k(G)).$$

Also, for $0 \le k \le n$, since $N \le \gamma_n(G)$, we have that $\pi(x) \in \pi(\gamma_k(G))$ if and only if $x \in \gamma_k(G)$. Thus, for $0 \le k \le n$,

$$\bar{\gamma}_{k+1}(\pi(G)) = \{\pi(x) \in \pi(\gamma_k(G)) : \exists m \in \mathbb{Z}, \ \pi(x^m) \in \pi(\gamma_{k+1}(G))\} = \pi(\bar{\gamma}_{k+1}(G)).$$

For $k \ge n + 1$, note that $\gamma_k(G) = \{1\}$, so $\bar{\gamma}_{k+1}(G) = \pi(N) = \{1\}$. :) ✓

Solution to Exercise 8.31 :(

The product in $H_3(\mathbb{Z})$ is readily computed to be

$$((c, b, a))((z, y, x)) = ((c + z + ay, y + b, x + a)).$$

Thus

$$((c, b, a))^{-1} = ((-c + ab, -b, -a)),$$

and thus,

$$[A, B] = ((0, 0, -1))((0, -1, 0))((0, 0, 1))((0, 1, 0))$$
$$= ((1, -1, -1))((1, 1, 1)) = ((1, 0, 0)) = C.$$

Finally, it is easy to show that

$$C^c B^b A^a = ((c, 0, 0))((0, b, 0))((0, 0, a)) = ((c, 0, 0))((0, b, a)) = ((c, b, a)).$$:)✓

Solution to Exercise 8.32 :(
One easily verifies that

$$((z, 0, 0))((c, b, a)) = ((c, b, a))((z, 0, 0)).$$

Moreover, for a general element $((c, b, a)) \in H_3(\mathbb{Z})$:

- $((c, b, a))((0, 0, 1)) = ((c, b, a + 1))$ and $((0, 0, 1))((c, b, a)) = ((c + b, b, a + 1))$. So $((c, b, a)) \notin Z(H_3(\mathbb{Z}))$ if $b \neq 0$.
- $((c, b, a))((0, 1, 0)) = ((c + a, b + 1, a))$ and $((0, 1, 0))((c, b, a)) = ((c, b + 1, a))$. So $((c, b, a)) \notin Z(H_3(\mathbb{Z}))$ if $a \neq 0$. :)✓

Solution to Exercise 8.33 :(
Let $z = [x, y] \in Z(G)$. Then, for any positive integers n, m,

$$[x^n, y] = x^{-n} y^{-1} x^n y = x^{1-n} z y^{-1} x^{n-1} y = z \left[x^{n-1}, y \right],$$

so inductively on n we have that $[x^n, y] = z^n$.

Since $z^n \in Z(G)$, it is also true that $[y, x^n] \in Z(G)$, so the same argument applied to y and m tells us that $[y^m, x^n] = [y, x^n]^m = z^{-nm}$, so $[x^n, y^m] = z^{nm}$ for all positive integers n, m.

For negative integers, note that if $[x, y] \in Z(G)$, then

$$\left[x^{-1}, y \right] = xy^{-1}x^{-1}y = x[y, x]x^{-1} \in Z(G),$$

so we can obtain the result for negative n, and negative m as well. :)✓

Solution to Exercise 8.34 :(
Since $C = [A, B] \in Z(H_3(\mathbb{Z}))$, by Exercise 8.33 we have that $C^{ab} = \left[A^a, B^b \right]$, implying that $\left| C^{ab} \right| \leq 2(|a| + |b|)$ for all integers a, b. Specifically,

$$|((c, b, a))| = \left| C^c B^b A^a \right| \leq \kappa \left(\sqrt{|c| + 1} + |b| + |a| \right),$$

for some constant $\kappa > 0$. :)✓

Solution to Exercise 8.35 :(
That $\delta_{(G,S),(H,T)} \sim \delta_{(G,S'),(H,T')}$ is immediate from Exercise 1.75.

Let T be a finite symmetric generating set for H for which $|B_T(1, r)| \leq \psi(r)$. Let $\pi : G \to G/H$ be the canonical projection, and let S be a finite symmetric subset of G such that $\pi(S)$ generates G/H and such that $|B_{\pi(S)}(H, r)| \leq \phi(r)$.

Exercise 1.62 tells us that $S \cup T$ generates G. We write $|x| = |x|_{S \cup T}$.

The definition of distortion tells us that for some constant $\kappa > 0$ we have for all $x \in H$ that $|x|_T \leq \kappa \delta(\kappa |x|)$.

Let Q be a set of representatives of the cosets of H, so that $G = \biguplus_{q \in Q} Hq$. So for any $x \in G$ there exist unique $q_x \in Q$ and $h_x \in H$ such that $x = h_x q_x$. Moreover, we may choose Q so that $|q_x| \leq |x|$ for all x. Thus, for any x we have that $|Hq_x|_{\pi(S)} \leq |q_x| \leq |x|$. Also, $|h_x| \leq |q_x| + |x| \leq 2|x|$, so $|h_x|_T \leq \kappa \delta(2\kappa |x|)$.

We conclude that the map $x \mapsto (h_x, Hq_x)$ maps $B_{S \cup T}(1, r)$ injectively into $B_T(1, \kappa \delta(2\kappa r)) \times B_{\pi(S)}(H, r)$. This immediately implies that

$$|B_{S \cup T}(1, r)| \leq \phi(r) \cdot \psi(\kappa \delta(2\kappa r)),$$

which proves that the growth of G is at most $\phi \cdot (\psi \circ \delta)$, as required.

The last assertion is just the fact that products and composition of polynomials is still a polynomial. :)✓

Solution to Exercise 8.36 :(
Exercise 8.34 implies that

$$\{((c, b, a)) : |c| \leq r^2 - 1, \ |a|, |b| \leq r\} \subset B(1, 3\kappa r),$$

implying that the growth of $H_3(\mathbb{Z})$ is at least $r \mapsto \left(2(r^2 - 1) + 1\right)(2r + 1)^2$, which is equivalent to $r \mapsto r^4$. This provides a lower bound on the growth.

For the upper bound, denote $H = H_3(\mathbb{Z})$ and $Z = Z(H)$.

Exercise 8.32 tells us that $H/Z \cong \mathbb{Z}^2$ (which has quadratic growth) and that $Z \cong \mathbb{Z}$ (which has linear growth). Exercise 8.34 implies that the distortion of Z in H is at most quadratic.

By Exercise 8.35 with $\psi(r) = r$ and $\delta(r) = \phi(r) = r^2$, we have that the growth of H is at most $r \mapsto r^4 = \phi(r) \cdot (\psi \circ \delta)(r)$.   :) ✓

Solution to Exercise 8.37   :(

Because $H \lhd G$, we know that $h^k \in H$ for all $h \in H$ and $k \in K$. Thus, K acts on H by $k.h = khk^{-1}$.

Also, consider $HK = \{hk : h \in H,\ k \in K\}$. Since $\langle K, H \rangle = G$, for any $g \in G$ we can write g as a finite product of elements from K and H. Moving all such elements from K to the right, replacing $kh = khk^{-1}k$, we may write any $g \in G$ as $g = hk$ for some $h \in H$ and $k \in K$. Thus, $HK = G$.

Also, if $hk = h'k'$ for $h, h' \in H$, and $k, k' \in K$, then $h^{-1}h' = k(k')^{-1} \in H \cap K = \{1\}$, so $h = h'$ and $k = k'$. Thus, for $h \in H$ and $k \in K$, the map $\varphi(hk) = (k, h)$ is a well-defined surjective map onto the set $K \times H$. Since $h, h' \in H$ and $k, k' \in K$, we have $hkh'k' = hkh'k^{-1}kk' = h(k.h')kk'$, we get that φ is a surjective homomorphism from $G = HK$ onto $K \ltimes H$.

We have that φ is an isomorphism since $(k, h) = 1_{K \ltimes H}$ if and only if $k = h = 1_G$.   :) ✓

9

Gromov's Theorem

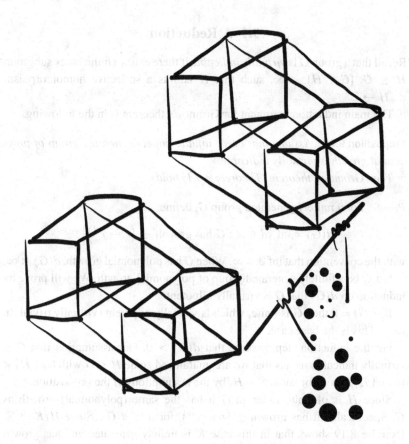

This chapter is dedicated to the following famous theorem of Gromov, connecting the geometric property of polynomial growth to the algebraic property of (virtual) nilpotence.

Theorem 9.0.1 (Gromov's theorem) *A finitely generated group G has polynomial growth if and only if G is virtually nilpotent.*

We have already seen the "easy" direction in Theorem 8.2.1, that every finitely generated virtually nilpotent group has polynomial growth. We will now work to prove the other direction of Gromov's theorem. We begin with some reductions.

9.1 A Reduction

Recall that a group G is *virtually indicable* if there exists a finite index subgroup $H \leq G$, $[G : H] < \infty$, such that H admits a surjective homomorphism $\varphi \colon H \to \mathbb{Z}$.

The main induction argument for Gromov's theorem is in the following.

Proposition 9.1.1 *Assume that every infinite finitely generated group of polynomial growth is virtually indicable.*

Then Gromov's theorem (Theorem 9.0.1) holds.

Proof   For a finitely generated group G, define

$$d(G) = \inf \left\{ d \in \mathbb{N} : G \text{ has growth } \preceq \left(r \mapsto r^d \right) \right\},$$

with the convention that $\inf \emptyset = \infty$. When G has polynomial growth, $d(G) < \infty$.

Let G be a finitely generated group of polynomial growth. We will prove by induction on $d(G)$ that G is virtually nilpotent.

If $d(G) = 0$ then G is finite, which is virtually nilpotent (virtually trivial, in fact). This is the base case.

For the induction step, assume that $d(G) > 0$. Our assumption that G is virtually indicable means that we are guaranteed some $H \leq G$ with $[G : H] < \infty$ and $H/K \cong \mathbb{Z}$ for some $K \triangleleft H$, by the assumptions of the proposition.

Since H is of finite index in G it has the same (polynomial) growth as G. Specifically, it has growth $\preceq \left(r \mapsto r^d \right)$, for $d = d(G)$. Since $H/K \cong \mathbb{Z}$, Exercise 8.19 shows that in this case K is finitely generated and has growth $\preceq \left(r \mapsto r^{d-1} \right)$, so that $d(K) \leq d(G) - 1$. Hence, by induction, K is virtually nilpotent. So we can find a finite index subgroup $N \leq K$, $[K : N] < \infty$, such that N is nilpotent. By Theorem 8.4.2, we may assume that N is characteristic in K, and thus $N \triangleleft H$. Since $(H/N)/(K/N) \cong H/K \cong \mathbb{Z}$ and K/N is finite, by Exercise 8.21 there exists a finite index subgroup $N \triangleleft M \le H$, $[H : M] < \infty$, such that $M/N \cong \mathbb{Z}$. Since N is nilpotent and $[M, M] \le N$, we have that M is solvable. Since M is finite index in G, we find that M has polynomial growth. By the Milnor–Wolf theorem (Theorem 8.2.4) M is virtually nilpotent, implying that G is virtually nilpotent, completing the induction step.   □

9.2 Unitary Actions

Recall the operator norm for a linear operator $a: V \to V$, where V is a complex normed vector space:

$$\|a\|_{op} = \sup\{\|av\| : \|v\| = 1\}.$$

If $\|a\|_{op} < \infty$, we say that a is a **bounded operator**. It is easy to verify that $\|ab\|_{op} \le \|a\|_{op} \cdot \|b\|_{op}$, and that if a is invertible with $\|a\|_{op} < \infty$, then also $\left\|a^{-1}\right\|_{op} < \infty$. So the collection of bounded invertible operators on V forms a group (with composition as the group operation). We denote this group by $GL(V)$.

The most basic example is when $V = \mathbb{C}^n$, for which the invertible linear operators are just the invertible matrices $GL_n(\mathbb{C}) = GL(\mathbb{C}^n)$. In this case all linear operators are bounded because the dimension is finite.

Exercise 9.1 Let $\mathbb{H}$ be a complex Hilbert space. Let a be a bounded linear operator. Show that the following are equivalent:

(1) $\|av\| = \|v\|$ for all $v \in \mathbb{H}$.
(2) $\langle au, av \rangle = \langle u, v \rangle$ for all $u, v \in \mathbb{H}$.
(3) $a^*a = 1$.

(Here, 1 denotes the identity operator.)   ▷ solution ◁

Let $\mathbb{H}$ be a Hilbert space. An operator $a \in GL(\mathbb{H})$ is called **unitary** if $a^*a = aa^* = 1$, where a^* denotes the adjoint of a. A unitary operator is necessarily bounded with operator norm 1 (by Exercise 9.1). We denote by $U(\mathbb{H})$ the collection of unitary operators on $\mathbb{H}$. Since $a^* = a^{-1}$ for a unitary a, and since $(ab)^* = b^*a^*$, we see that $U(\mathbb{H})$ is a group.

We use the notation $U_n(\mathbb{C}) = U(\mathbb{C}^n)$.

Exercise 9.2 Prove that $\|xa\|_{op} = \|ax\|_{op} = \|a\|_{op}$ for all $a \in GL(\mathbb{H})$, $x \in U(\mathbb{H})$.

▷ solution ◁

It is a basic theorem from linear algebra that any unitary matrix $x \in \mathsf{U}_n(\mathbb{C})$ is unitarily diagonalizable. That is, there exists $u \in \mathsf{U}_n(\mathbb{C})$ such that u^*xu is a diagonal matrix. The eigenvalues of x are precisely the elements on this diagonal. Also, the corresponding orthonormal basis of eigenvectors are the columns of u. (This follows from the fact that x is *normal*, that is, $x^*x = xx^*$, which is a more general property than being unitary. Diagonalization of normal operators also generalizes to operators over infinite-dimensional spaces, via the *spectral theorem*. We will only utilize the finite-dimensional case, which may be found in any linear algebra textbook.)

Exercise 9.3 Let $x \in \mathsf{U}_d(\mathbb{C})$. Suppose that $\lambda_1, \lambda_2, \ldots, \lambda_d$ are the eigenvalues of x.

Show that $|\lambda_j| = 1$ for all $1 \le j \le d$.

Prove that $\|x - 1\|_{\mathrm{op}} = \max_j |\lambda_j - 1|$.   ▷ solution ◁

Exercise 9.4 Let $x \in \mathsf{U}_d(\mathbb{C})$. Show that $\|x^n - 1\|_{\mathrm{op}} \le n \cdot \|x - 1\|_{\mathrm{op}}$ for any integer $n > 0$.   ▷ solution ◁

Lemma 9.2.1 *Set $c = 6^{-1/2}$. Let $1 \ne x \in \mathsf{U}_d(\mathbb{C})$.*

Then, for all $1 \le n \le \frac{c\pi}{\|x-1\|_{\mathrm{op}}}$ we have that $\|x^n - 1\|_{\mathrm{op}} \ge cn\|x - 1\|_{\mathrm{op}}$. Specifically, $x^n \ne 1$.

Proof Let $\theta \in [0, \pi]$ and $\lambda = e^{i\theta}$. A simple calculation reveals that

$$|\lambda - 1| = \sqrt{2} \cdot \sqrt{1 - \cos(\theta)}.$$

Also, a Taylor expansion of cos provides the inequality

$$c\theta \le \theta \cdot \sqrt{1 - \frac{\theta^2}{12}} \le \sqrt{2} \cdot \sqrt{1 - \cos(\theta)} \le \theta,$$

because $c^2 = \frac{1}{6} < 1 - \frac{\pi^2}{12}$.

For any $\lambda \in \mathbb{C}$ with $|\lambda| = 1$, we have $|\lambda - 1| = |\bar{\lambda} - 1|$. Thus, for any $\theta \in [-\pi, \pi]$, we get that $c|\theta| \le |e^{i\theta} - 1| \le |\theta|$.

Now, let $\lambda_1, \ldots, \lambda_d$ be the eigenvalues of $x \in \mathsf{U}_n(\mathbb{C})$. For every j write $\lambda_j = e^{i\theta_j}$ for $\theta_j \in [-\pi, \pi]$. So

$$c \max_j |\theta_j| \le \|x - 1\|_{\mathrm{op}} = \max_j |\lambda_j - 1| \le \max_j |\theta_j|.$$

Since the eigenvalues of x^n are precisely $\lambda_1^n, \ldots, \lambda_d^n$, we get that, if $n\|x-1\|_{\mathrm{op}} \le c\pi$, then $n|\theta_j| \le \pi$ for all j, so that

$$\|x^n - 1\|_{\mathrm{op}} = \max_j |\lambda_j^n - 1| \ge cn \max_j |\theta_j| \ge cn\|x - 1\|_{\mathrm{op}}. \qquad \square$$

Lemma 9.2.2 *Let $G \leq U_d(\mathbb{C})$ be a finitely generated infinite group of unitary matrices. If G has polynomial growth, then there exists a finite index normal subgroup $H \lhd G, [G : H] < \infty$ such that the following holds. Either*

- *H is Abelian, or*
- *there exists an element $a \in Z(H)$ (the center of H) such that $a \neq \lambda \cdot 1$ (a is nonscalar).*

Proof We will take advantage of the *ring* structure of linear operators (or matrices in this case). Namely, we have the ability to add them.

For any $x, y \in U_d(\mathbb{C})$,

$$
\begin{aligned}
||[x, y] - 1||_{\mathrm{op}} &= ||xy - yx||_{\mathrm{op}} \\
&= ||(x - 1)(y - 1) - (y - 1)(x - 1)||_{\mathrm{op}} \\
&\leq 2||x - 1||_{\mathrm{op}} \cdot ||y - 1||_{\mathrm{op}}.
\end{aligned}
\tag{9.1}
$$

(This "distortion" tells us that commutators are roughly closer to the identity than their original components.)

Another useful observation is as follows. Suppose that $x = [a, b] = \lambda \cdot 1$ is some unitary scalar matrix that is also a commutator. Since $\lambda^d = \det(x) = \det[a, b] = 1$ it must be that $\lambda = e^{i2\pi k/d}$ for some $0 \leq k \leq d - 1$. If $k \geq 1$ then $||x - 1||_{\mathrm{op}} = |\lambda - 1| \geq \xi_d := c\frac{2\pi}{d}$ for $c = 6^{-1/2}$ (as shown in the proof of Lemma 9.2.1). We conclude that if $x = [a, b] = \lambda \cdot 1$ and $||x - 1||_{\mathrm{op}} < \xi_d$ then $x = 1$.

Lemma 9.2.1 tells us that if $x \neq 1$, then for $1 \leq n \leq \frac{c\pi}{||x-1||_{\mathrm{op}}}$, we have $||x^n - 1||_{\mathrm{op}} \geq cn||x - 1||_{\mathrm{op}}$.

Now, we assume that G has polynomial growth. Fix some Cayley graph of G, and let $B_G(1, r)$ denote the ball of radius r in this Cayley graph. Let $D > 0$ be such that $|B_G(1, r)| \leq r^D$ for all $r \geq 1$. Fix a large enough integer n so that $n + 1 > 2^D$, and such that $\varepsilon := \frac{c}{8n} < \min\left\{\frac{1}{4}, \xi_d\right\}$.

Let $G_\varepsilon = \left\langle x \in G : ||x - 1||_{\mathrm{op}} \leq \varepsilon \right\rangle$. Since $||x^y - 1||_{\mathrm{op}} = ||(x - 1)^y||_{\mathrm{op}} = ||x - 1||_{\mathrm{op}}$ for any $y \in U_d(\mathbb{C})$, we have that $G_\varepsilon \lhd G$. Also, if $||x - y||_{\mathrm{op}} \leq \varepsilon$ then $x^{-1}y \in G_\varepsilon$. It is an exercise below to show that this implies that $[G : G_\varepsilon] < \infty$.

We have that G_ε is finitely generated (as a finite index subgroup of the finitely generated group G) and we may write the finite number of generators of G_ε as finite products of elements $x \in G$ with $||x - 1||_{\mathrm{op}} \leq \varepsilon$. Taking all elements (and their inverses) appearing in these products, there exists a finite symmetric generating set S_ε for $G_\varepsilon = \langle S_\varepsilon \rangle$ such that $||s - 1||_{\mathrm{op}} \leq \varepsilon$ for all $s \in S_\varepsilon$ and such that $1 \notin S_\varepsilon$.

We now assume for a contradiction the negation of the conclusion of the lemma for $H = G_\varepsilon$. That is, we assume that G_ε is not Abelian, and for any

$z \in G_\varepsilon$ such that $[z, x] = 1$ for all $x \in G_\varepsilon$ we have $z = \lambda \cdot 1$ for some $\lambda \in \mathbb{C}$. If we reach a contradiction, then taking $H = G_\varepsilon$ will prove the lemma.

Consider the following inductive construction.

First step.   Start by considering all $s \in S_\varepsilon$. If all such s are scalar, then $G_\varepsilon = \langle S_\varepsilon \rangle$ contains only scalars, and thus is Abelian, contradicting our assumption. Hence, we may assume that there exists some $x_1 := y_1 \in S_\varepsilon$ such that x_1 is nonscalar.

Inductive step.   Suppose we have defined a sequence $y_1, \ldots, y_k \in S_\varepsilon$ and $y_1 = x_1, x_2, \ldots, x_k$ with the following properties for all $1 \le j \le k - 1$:

- $x_{j+1} = [y_{j+1}, x_j]$.

- x_j, x_{j+1} are nonscalar.

- $\|x_{j+1} - 1\|_{\mathrm{op}} \le 2\varepsilon \cdot \|x_j - 1\|_{\mathrm{op}}$.

Then, since x_k is nonscalar, it is not in the center, by our assumption on G_ε. Hence, there must exist $y_{k+1} \in S_\varepsilon$ such that $x_{k+1} := [y_{k+1}, x_k] \ne 1$. Note that

$$\|x_{k+1}-1\|_{\mathrm{op}} \le 2 \cdot \|y_{k+1}-1\|_{\mathrm{op}} \cdot \|x_k-1\|_{\mathrm{op}} \le 2\varepsilon\|x_k-1\|_{\mathrm{op}} \le \tfrac{1}{2}(2\varepsilon)^{k+1} \le \varepsilon < \xi_d.$$

Since $\det(x_{k+1}) = 1$ and $x_{k+1} \ne 1$, we have that x_{k+1} cannot be scalar. This completes the inductive construction, with the above properties.

Now, note that for all k, we have $x_k \ne 1$ and

$$\|x_k - 1\|_{\mathrm{op}} \le \tfrac{1}{2}(2\varepsilon)^k \le \varepsilon < \tfrac{c\pi}{n}.$$

Thus, for all $1 \le j \le n$ we have $\left\|x_k^j - 1\right\|_{\mathrm{op}} \ge cj\|x_k - 1\|_{\mathrm{op}}$, and specifically $x_k^j \ne 1$.

Claim   *For each $k \ge 1$, the map $\varphi \colon \{0, \ldots, n\}^k \to G_\varepsilon$ given by $\varphi(j_1, \ldots, j_k) = (x_1)^{j_1} \cdots (x_k)^{j_k}$ is injective, with image contained in the ball of radius $3n2^k$ (with respect to the generating set S_ε).*

Proof of Claim   Note that for $r > \ell$ and $j \ne i \in \{0, \ldots, n\}$,

$$\left\|x_r^j - x_r^i\right\|_{\mathrm{op}} = \left\|x_r^{|j-i|} - 1\right\|_{\mathrm{op}} \le |j - i| \cdot \|x_r - 1\|_{\mathrm{op}} \le n \cdot \|x_r - 1\|_{\mathrm{op}},$$

$$\|x_r - 1\|_{\mathrm{op}} \le (2\varepsilon)^{r-\ell} \cdot \|x_\ell - 1\|_{\mathrm{op}},$$

$$\left\|x_\ell^j - x_\ell^i\right\|_{\mathrm{op}} = \left\|x_\ell^{|j-i|} - 1\right\|_{\mathrm{op}} \ge c|j - i| \cdot \|x_\ell - 1\|_{\mathrm{op}}.$$

So,

$$\left\| x_\ell^{j_\ell} \cdots x_k^{j_k} - x_\ell^{i_\ell} \cdots x_k^{i_k} \right\|_{\mathrm{op}} \geq \left\| x_\ell^{j_\ell} \cdots x_{k-1}^{j_{k-1}} - x_\ell^{i_\ell} \cdots x_{k-1}^{i_{k-1}} \right\|_{\mathrm{op}} - \left| x_k^{j_k} - x_k^{i_k} \right\|_{\mathrm{op}}$$

$$\geq \cdots \geq \| x_\ell^{j_\ell} - x_\ell^{i_\ell} \|_{\mathrm{op}} - n \cdot \sum_{r=\ell+1}^{k} \| x_r - 1 \|_{\mathrm{op}}$$

$$\geq c \cdot |j_\ell - i_\ell| \cdot \| x_\ell - 1 \|_{\mathrm{op}} - n \cdot \| x_\ell - 1 \|_{\mathrm{op}} \cdot \sum_{r=\ell+1}^{\infty} (2\varepsilon)^{r-\ell}.$$

Since $n \cdot \sum_{r=1}^{\infty} (2\varepsilon)^r \leq \frac{c}{2}$ (by our choice of ε), we get that if $x_\ell^{j_\ell} \cdots x_k^{j_k} = x_\ell^{i_\ell} \cdots x_k^{i_k}$ then $j_\ell = i_\ell$. Since this holds for any $1 \leq \ell \leq k$, this proves the injectivity of φ.

Now, note that in the Cayley graph of G_ε with respect to the generating set S_ε, we have that $|x_1| = 1$ and $|x_{k+1}| \leq 2(|x_k| + |y_{k+1}|) = 2|x_k| + 2$. So $|x_k| \leq 3 \cdot 2^{k-1} - 2$ by induction. Thus, for $0 \leq j_1, \ldots, j_k \leq n$ we have

$$\left| (x_1)^{j_1} \cdots (x_k)^{j_k} \right| \leq \sum_{\ell=1}^{k} j_\ell \left(3 \cdot 2^{\ell-1} - 2 \right) \leq 3n2^k.$$

This proves the claim. □

From this claim we deduce that in the Cayley graph of G_ε with respect to the generating set S_ε we have $\left| B_{S_\varepsilon}(1, 3n2^k) \right| \geq (n+1)^k$ for all k (n is large but fixed). Since $[G : G_\varepsilon] < \infty$, there exists a constant $C = C(\varepsilon) > 0$ such that $B_{S_\varepsilon}(1, r) \subseteq B_G(1, Cr)$ for all $r \geq 0$.

We conclude that for all $k \geq 1$ we have

$$(n+1)^k \leq \left| B_G(C3n2^k) \right| \leq \left(C3n2^k \right)^D = (C3n)^D \cdot \left(2^D \right)^k.$$

Since $n + 1 > 2^D$ we arrive at a contradiction for large enough k.

Thus, the proof of the lemma is concluded by taking $H = G_\varepsilon$. □

Exercise 9.5 Show that $\mathsf{U}_d(\mathbb{C})$ is compact with the metric induced by the norm $\| \cdot \|_{\mathrm{op}}$.

Show that in the proof of Lemma 9.2.2, $[G : G_\varepsilon] \leq C$ for some $C = C(d, \varepsilon) > 0$.

▷ solution ◁

Our goal in this section is to prove the following result, basically due to Jordan. It is Gromov's theorem for the special case of unitary matrix groups.

Theorem 9.2.3 *Let* $G \leq \mathsf{U}_d(\mathbb{C})$ *be a finitely generated group of unitary matrices. Then* G *has polynomial growth if and only if* G *is virtually Abelian.*

Proof If G is virtually Abelian then G has a finite index subgroup isomorphic to $\mathbb{Z}^d$ (by Theorem 1.5.2), so G has polynomial growth.

So we are left with proving the other direction. Assume $G \leq \mathsf{U}_d(\mathbb{C})$ is a finitely generated group of polynomial growth. We prove by induction on the dimension d that G is virtually Abelian.

The base case $d = 1$ is where $G \leq \mathbb{C}$ is just an Abelian group.

For the induction step, assume $d > 1$.

Let $H \lhd G$ be a finite index subgroup $[G : H] < \infty$ guaranteed by Lemma 9.2.2. If H is Abelian, we are done. Otherwise, there is an element $a \in H$ such that $a \neq \lambda \cdot 1$ and $[x, a] = 1$ for all $x \in H$.

By conjugating H (and thus moving to an isomorphic group) we may assume that a is a diagonal matrix. (Indeed, a is unitary and thus diagonalizable over $\mathbb{C}$. Conjugate all of H by the unitary matrix that conjugates a to a diagonal matrix.)

That is, $\mathbb{C}^d = W_1 \oplus \cdots \oplus W_m$ for some $m \leq d$, where $W_j = \mathsf{Ker}(a - \lambda_j \cdot 1)$ for some $\lambda_1, \ldots, \lambda_m \in \mathbb{C}$. Note that since $a - \lambda \cdot 1$ commutes with any $x \in H$ we get that $HW_j \subset W_j$. For every j set $N_j = \left\{ x \in H : xw = w, \forall\, w \in W_j \right\} \lhd H$. Note that if $x \in N_1 \cap \cdots \cap N_m$ then $xv = v$ for all $v \in \mathbb{C}^d$, implying that $x = 1$. Thus, $N_1 \cap \cdots \cap N_m = \{1\}$. It is then an easy exercise to show that the map $H \mapsto H/N_1 \times \cdots \times H/N_m$ by $x \mapsto (N_1 x, \ldots, N_m x)$ is an injective homomorphism.

Since a is not scalar, there exists j such that $\lambda_j \neq \lambda_1$. Thus, $m > 1$. Thus, $\dim W_j < d$ for all $1 \leq j \leq m$.

For any $1 \leq j \leq m$, note that H/N_j is isomorphic to a subgroup of $\mathsf{U}(W_j)$, the unitary operators on W_j. We have that H/N_j is finitely generated and of polynomial growth (as a quotient of the finite index subgroup $H \leq G$, $[G : H] < \infty$), so H/N_j is virtually Abelian by induction.

Hence, also the direct product $H/N_1 \times \cdots \times H/N_m$ is virtually Abelian. Thus, H is isomorphic to a subgroup of a virtually Abelian group, and hence H is virtually Abelian, which implies that G is virtually Abelian. $\square$

Exercise 9.6 Show that if $N_1, \ldots, N_m \lhd H$ are normal subgroups with $N_1 \cap \cdots \cap N_m = \{1\}$, then the map $x \mapsto (N_1 x, \ldots, N_m x)$ is an injective homomorphism.

▷ solution ◁

Exercise 9.7 Show that the direct product of virtually Abelian groups is virtually Abelian.

Show that any subgroup of a virtually Abelian group is virtually Abelian.

▷ solution ◁

From Theorem 9.2.3 and Exercise 1.32 we immediately obtain the following corollary.

Corollary 9.2.4 Let V be a nontrivial finite-dimensional normed vector space over $\mathbb{C}$. Let $G \leq U(V)$ be a finitely generated infinite group of unitary operators on V.

If G has polynomial growth, then there exists a finite index normal subgroup $H \lhd G$, $[G : H] < \infty$ such that H admits a surjective homomorphism onto $\mathbb{Z}$.

9.3 Harmonic Cocycles

Suppose that a group G acts on a Hilbert space $\mathbb{H}$. Suppose that for any $x \in G$ the map $v \mapsto x.v$ is a unitary operator. We say that *G acts on $\mathbb{H}$ by unitary operators*. This is equivalent to the existence of a homomorphism $U \colon G \to U(\mathbb{H})$, and the action defined as $x.v := U(x)v$ for $v \in \mathbb{H}$ and $x \in G$.

Definition 9.3.1 Let G be a group acting on a Hilbert space $\mathbb{H}$ by unitary operators. That is, $U \colon G \to U(\mathbb{H})$ a homomorphism and $x.v = U(x)v$ for $v \in \mathbb{H}$ and $x \in G$.

A map $c \colon G \to \mathbb{H}$ is called a **cocycle** (with respect to U) if

$$c(xy) = c(x) + x.c(y),$$

for all $x, y \in G$.

As in the complex-valued case, a map $h \colon G \to \mathbb{H}$ is μ-harmonic if $\sum_y \mu(y)h(xy) = h(x)$ for all $x \in G$, and the above sum converges absolutely.

Exercise 9.8 Let G be a finitely generated group acting on a Hilbert space $\mathbb{H}$ by unitary operators.

Show that if $c \colon G \to \mathbb{H}$ is a cocycle and if S is a finite symmetric generating set for G, then for any $x \in G$ we have $\|c(x)\| \leq |x| \cdot \max_{s \in S} \|c(s)\|$. ▷ solution ◁

Exercise 9.9 Let G be a finitely generated group acting on a Hilbert space $\mathbb{H}$ by unitary operators. Let $c \colon G \to \mathbb{H}$ be a cocycle.

Show that $c(1) = 0$.

Show that $c\left(x^{-1}\right) = -x^{-1}.c(x)$. ▷ solution ◁

Recall that $\mathsf{SA}(G, 1)$ is the collection of symmetric, adapted probability measures on G with finite first moment.

Exercise 9.10 Let G be a finitely generated group acting on a Hilbert space $\mathbb{H}$ by unitary operators. Let $c: G \to \mathbb{H}$ be a cocycle.

Show that for $\mu \in \mathrm{SA}(G, 1)$, c is μ-harmonic if and only if $\sum_y \mu(y)c(y) = 0$.

▷ solution ◁

Exercise 9.11 Let G be a finitely generated group acting on a Hilbert space $\mathbb{H}$ by unitary operators.

Fix $v \in \mathbb{H}$ and define $c_v(x) := x.v - v$. Show that c_v is a cocycle. (These cocycles are sometimes called **co-boundaries**.)

▷ solution ◁

Exercise 9.12 Let G be a finitely generated group acting on a Hilbert space $\mathbb{H}$ by unitary operators. Let $c: G \to \mathbb{H}$ be a cocycle. Let $v \in \mathbb{H}$. Define $h(x) := \langle c(x), v \rangle$. Let $\mu \in \mathrm{SA}(G, 1)$.

Show that $h \in \mathrm{LHF}(G, \mu)$. (Recall that LHF is the space of Lipschitz harmonic functions.)

9.3.1 Construction

This section is devoted to proving the following fundamental result. (Recall that a Hilbert space is *separable* if it has a countable orthonormal basis.)

Theorem 9.3.2 *Let G be a finitely generated infinite amenable group, and $\mu \in \mathrm{SA}(G, 1)$.*

Then, there exists a separable Hilbert space $\mathbb{H}$ such that G acts on $\mathbb{H}$ by unitary operators, and such that there exists a nonzero μ-harmonic cocycle $c: G \to \mathbb{H}$.

Remark 9.3.3 In fact, the proof of Theorem 9.3.2 will show that we may take $\mathbb{H} = \ell^2(G, \mu)$. However, the G action on $\ell^2(G, \mu)$ will not necessarily be the regular action, but some other homomorphism $U: G \to \mathsf{U}\left(\ell^2(G, \mu)\right)$.

Recall that $\ell^0(G)$ denotes the space of finitely supported complex-valued functions on G. For a probability measure μ on G, we define the *transition matrix*, *Laplacian*, and *Green function*:

$$P(x, y) = \mu\left(x^{-1}y\right), \qquad \Delta = I - P, \qquad \mathsf{g}(x, y) = \sum_{t=0}^{\infty} P^t(x, y),$$

where the latter is only defined if (G, μ) is transient. These operators are always defined on $\ell^0(G)$, but in some cases can be extended to larger spaces. Recall also the space $\ell^2(G, \mu)$, which consists of functions $f: G \to \mathbb{C}$ with

$\|f\|^2 = \sum_x |f(x)|^2 < \infty$. Also, $\ell^2(G, \mu)$ is a Hilbert space with the inner product $\langle f, h \rangle = \sum_x f(x)\overline{h(x)}$. See Chapter 4.

Exercise 9.13 Let G be an infinite finitely generated group and let μ be a symmetric adapted probability measure on G. Assume that the μ-random walk on G is transient.

Fix $f \in \ell^0(G)$. Show that $\varphi := \sum_{k=0}^{\infty} P^k f = \mathfrak{g}f$ is well defined.

Show that $\Delta\varphi = f$.

Show that if f is a nonnegative function, then for any n,

$$\langle \Delta\varphi, \varphi \rangle \geq \sum_{k=0}^{n} \langle P^k f, f \rangle. \qquad \text{▷ solution ◁}$$

Exercise 9.14 Let G be a finitely generated group and let μ be a symmetric adapted probability measure on G.

We say that (G, μ) satisfies the *bubble condition* if (G, μ) is transient and the Green function $\mathfrak{g}$ satisfies $\sum_y |\mathfrak{g}(1, y)|^2 < \infty$.

Show that if (G, μ) satisfies the bubble condition, then there exists a linear operator $\Delta^{-1} : \ell^0(G) \to \ell^2(G, \mu)$ such that $\Delta\Delta^{-1} f = f$ for any $f \in \ell^0(G)$.

▷ solution ◁

Now, consider some finite subset $A \subset G$. Note that for $f = |A|^{-1/2} \mathbf{1}_A$, we have $\|f\| = 1$ and

$$\langle P^k f, f \rangle = \frac{1}{|A|} \sum_{a \in A} \mathbb{P}_a[X_k \in A] = 1 - \frac{1}{|A|} \sum_{a \in A} \mathbb{P}_a[X_k \notin A].$$

If G is amenable, then for any n we may find a finite set F_n such that with $f_n := |F_n|^{-1/2} \mathbf{1}_{F_n}$, we have $\langle P^k f_n, f_n \rangle \geq \frac{1}{2}$ for all $0 \leq k < n$. Also, since f_n is a nonnegative function, if we could define $\varphi_n := \sum_{k=0}^{\infty} P^k f_n$, then the above exercises would give that

$$\|\Delta\varphi_n\| = 1 \qquad \text{and} \qquad \langle \Delta\varphi_n, \varphi_n \rangle \geq \frac{1}{2}n.$$

Specifically,

$$\frac{\|\Delta\varphi_n\|^2}{\langle \Delta\varphi_n, \varphi_n \rangle} \to 0.$$

The only obstacles are the convergence of the series defining φ_n, for example, when (G, μ) is recurrent, and also ensuring that $\varphi_n \in \ell^2(G, \mu)$, for example, the bubble condition. This is the content of the next lemma, where we replace the infinite sum by a truncated finite sum.

Lemma 9.3.4 *Let G be a finitely generated infinite group, and let μ be a symmetric adapted probability measure on G.*

Let $f \in \ell^2(G, \mu)$ be a real-valued function such that $\|f\| = 1$, $\|\Delta f\| \le \frac{1}{2}$, and

$$\lim_{k \to \infty} \frac{1}{k} \sum_{j=0}^{k-1} \langle P^j f, f \rangle = 0.$$

Then there exists $\varphi \in \ell^2(G, \mu)$ such that

$$\frac{\|\Delta\varphi\|^2}{\langle \Delta\varphi, \varphi \rangle} \le 16 \|\Delta f\|.$$

Proof Define $\varphi_k = \sum_{j=0}^{k-1} P^j f$. Since P is a contraction,

$$\|\Delta\varphi_k\| = \left\|\left(I - P^k\right) f\right\| \le 2\|f\| = 2.$$

Also, because μ is symmetric, P is self-adjoint, and so is $I - P^k$, and thus,

$$\langle \Delta\varphi_k, \varphi_k \rangle = \left\langle \left(I - P^k\right) f, \varphi_k \right\rangle = \left\langle f, \left(I - P^k\right) \varphi_k \right\rangle = \langle f, 2\varphi_k - \varphi_{2k} \rangle, \quad (9.2)$$

where we have used that

$$P^k \varphi_k = \sum_{j=0}^{k-1} P^{j+k} f = \varphi_{2k} - \varphi_k.$$

For any $m \ge 1$ we have that since P is a contraction,

$$\|(I - P^m)f\| \le \sum_{j=0}^{m-1} \left\|\left(P^j - P^{j+1}\right) f\right\| \le m\|\Delta f\|,$$

so by Cauchy–Schwarz,

$$1 - \langle P^m f, f \rangle = \langle (I - P^m)f, f \rangle \le m\|\Delta f\|.$$

This implies that for any m, if we set $a_m := \langle \varphi_{2^m}, f \rangle$ then

$$a_m = \sum_{j=0}^{2^m-1} \langle P^j f, f \rangle \ge \sum_{j=0}^{2^m-1} (1 - j\|\Delta f\|) \ge 2^m \left(1 - 2^{m-1}\|\Delta f\|\right).$$

By the assumptions of the lemma, $2^{-m} a_m \to 0$, so we may write

$$2^{-m} a_m = \sum_{k \ge m} (2^{-k} a_k - 2^{-(k+1)} a_{k+1}), \quad \text{implying} \quad a_m = \sum_{k \ge m} \frac{2a_k - a_{k+1}}{2^{k-m+1}}.$$

Since $\sum_{k \geq m} 2^{m-k-1} = 1$, and since $a_k \in \mathbb{R}$ (f is assumed to be real-valued), there must exist $k \geq m$ such that $2a_k - a_{k+1} \geq a_m$. Choose $m \geq 1$ such that $2^{m-1} \leq \frac{1}{2\|\Delta f\|} \leq 2^m$ (recall that we assume that $2\|\Delta f\| \leq 1$). So

$$a_m = \langle \varphi_{2^m}, f \rangle \geq 2^m \left(1 - 2^{m-1}\|\Delta f\|\right) \geq (4\|\Delta f\|)^{-1}.$$

Thus, we conclude that there exists $k \geq m$ such that $2a_k - a_{k+1} \geq (4\|\Delta f\|)^{-1}$, so that by (9.2),

$$\langle \Delta \varphi_{2^k}, \varphi_{2^k} \rangle = 2a_k - a_{k+1} \geq (4\|\Delta f\|)^{-1},$$

and consequently, for $\varphi := \varphi_{2^k}$,

$$\frac{\|\Delta \varphi\|^2}{\langle \Delta \varphi, \varphi \rangle} \leq 4 \cdot 4\|\Delta f\|. \qquad \square$$

The next lemma tells us that in order to find a harmonic cocycle for a unitary action of G on a Hilbert space $\mathbb{H}$, it suffices to find a sequence of "almost harmonic" cocycles, and take some appropriate limit. The price to pay, however, is that the unitary action for the harmonic cocycle may be different from the original action.

Lemma 9.3.5 *Let G be a finitely generated group acting on a separable Hilbert space by unitary operators. Let $\mu \in \mathrm{SA}(G, 1)$.*

Let $c_n \colon G \to \mathbb{H}$ be a sequence of cocycles such that

- *$(c_n)_n$ are uniformly Lipschitz, that is, $\sup_n \max_{s \in S} \|c_n(s)\| < \infty$, for some finite symmetric generating set S, and*
- *$(c_n)_n$ are almost harmonic, that is,*

$$\lim_{n \to \infty} \left\| \sum_x \mu(x) c_n(x) \right\| = 0.$$

Then, there exists a homomorphism $U \colon G \to \mathrm{U}(\mathbb{H})$ and a map $c \colon G \to \mathbb{H}$ such that c is a μ-harmonic cocycle with respect to U; that is, for all $x, y \in G$,

$$c(xy) = c(x) + U(x)c(y) \qquad \text{and} \qquad \sum_x \mu(x) c(x) = 0.$$

Moreover, for any fixed $x \in G$ we may choose c such that

$$\|c(x)\| \geq \limsup_{n \to \infty} \|c_n(x)\|.$$

Proof Fix some enumeration $G = \{x_1, x_2, \ldots\}$. Let $\kappa := \sup_n \max_{s \in S} \|c_n(s)\|$. Note that $\|c_n(x)\| \leq \kappa |x|$ for all n and all $x \in G$ (see Exercise 9.8). There exists an infinite set $J_0 \subset \mathbb{N}$ such that

$$\lim_{J_0 \ni n \to \infty} \|c_n(x_1)\| = \limsup_{n \to \infty} \|c_n(x_1)\|.$$

Let $(b_n)_n$ be an orthonormal basis for $\mathbb{H}$. For each n let $V_n = \text{span}\{b_1, \ldots, b_n\}$. So $V_n \leq V_{n+1}$ for all n.

Fix n, and consider the subspaces $W_{n,k} = \text{span}\{c_n(x_j) : j \leq k\}$. So $W_{n,k} \leq W_{n,k+1}$. Since $\dim W_{n,k} \leq k$, we may find a unitary operator $U_n : \mathbb{H} \to \mathbb{H}$ such that $U_n(W_{n,k}) \leq V_k$ for all $k \leq n$.

Define $\tilde{c}_n(x) = U_n c_n(x)$.

Note that for any $k \leq n$ we have that $\tilde{c}_n(x_k) \in V_k$, which is a finite-dimensional space. Also, we have that $\|\tilde{c}_n(x_k)\| = \|c_n(x_k)\| \leq \kappa |x_k|$. Thus the sequence $(\tilde{c}_n(x_k))_n$ is inside the set $R_k = \{v \in V_k : \|v\| \leq \kappa |x_k|\}$. Since $\dim V_k < \infty$, we know that R_k is compact. This implies that for any infinite set $J \subset \mathbb{N}$, there exists another infinite subset $I \subset J$ such that the subsequence $(\tilde{c}_n(x_k))_{n \in I}$ converges in $\mathbb{H}$.

By taking further and further subsequences, we obtain by induction that there exists a sequence $J_k \supset J_{k+1}$ of infinite subsets of $\mathbb{N}$ such that for any $\ell \leq k$ the subsequence $(\tilde{c}_n(x_\ell))_{n \in J_k}$ converges in $\mathbb{H}$.

Write $J_k = \{m_{k,1} < m_{k,2} < \cdots\}$. For any n let $m(n) = m_{n,n}$, and consider the subsequence $(\tilde{c}_{m(n)})_n$. Since this is a subsequence of $(\tilde{c}_n)_{n \in J_k}$ for any k, we obtain that the limit

$$c(x_k) := \lim_{n \to \infty} \tilde{c}_{m(n)}(x_k) = \lim_{J_k \ni n \to \infty} \tilde{c}_n(x_k)$$

exists in $\mathbb{H}$ for all $k \geq 1$. (This was basically the Arzelà–Ascoli theorem, see also Exercise 1.97.) Note that for any $k \geq 1$,

$$\|c(x_k)\| = \lim_{J_k \ni n \to \infty} \|c_n(x_k)\|,$$

so that

$$\|c(x_1)\| = \limsup_{n \to \infty} \|c_n(x_1)\|.$$

We now show that c is μ-harmonic at $1 \in G$. Fix some small $\varepsilon > 0$. Recalling that $\mu \in \text{SA}(G, 1)$, we can choose $A \subset G$ a large enough but finite subset such that $\sum_{x \notin A} \mu(x)|x| < \varepsilon$. Note that

$$\|c(x)\| \leq \limsup_{n \to \infty} \|\tilde{c}_n(x)\| \leq \kappa |x|,$$

so that

$$\left\| \sum_{x \notin A} \mu(x) c(x) \right\| \leq \sum_{x \notin A} \mu(x) \|c(x)\| \leq \kappa \cdot \varepsilon.$$

Since A is finite, for some large enough k we have, using the μ-harmonicity of $\tilde{c}_n$,

$$\left\| \sum_{x \in A} \mu(x)c(x) \right\| = \lim_{J_k \ni n \to \infty} \left\| \sum_{x \in A} \mu(x)\tilde{c}_n(x) \right\| = \lim_{J_k \ni n \to \infty} \left\| \sum_{x \notin A} \mu(x)\tilde{c}_n(x) \right\|$$

$$\leq \limsup_{n \to \infty} \sum_{x \notin A} \mu(x) \| \tilde{c}_n(x) \| \leq \kappa \cdot \varepsilon.$$

This implies that

$$\left\| \sum_x \mu(x)c(x) \right\| \leq 2\kappa \cdot \varepsilon,$$

and by taking $\varepsilon \to 0$ we find that $\sum_x \mu(x)c(x) = 0 \in \mathbb{H}$. Also, for some large enough $k \geq 1$ for which $x_k = 1$,

$$c(1) = \lim_{J_k \ni n \to \infty} U_n c_n(1) = 0 \in \mathbb{H}$$

(because $c_n(1) = 0$ for any cocycle c_n).

This function $c \colon G \to \mathbb{H}$ will be our μ-harmonic cocyle, but we are still left with finding the appropriate homomorphism $U \colon G \to \mathsf{U}(\mathbb{H})$ so that $c(xy) = c(x) + U(x)c(y)$.

To this end, let $V = \mathrm{cl}(\mathrm{span}(c(G)))$, which is the smallest closed subspace of $\mathbb{H}$ containing the image $c(G)$. For $x \in G$ define $U(x) \in \mathsf{U}(\mathbb{H})$ by setting $U(x)c(y) = c(xy) - c(x)$ for all $y \in G$, and extending $U(x)$ linearly and continuously to all of V. For $v \in V^{\perp}$ define $U(x)v = v$. It is an exercise below to verify that $U(x)$ is a well-defined unitary operator. Note that $U(x)V \leq V$ and $U(x)|_{V^{\perp}}$ is the identity operator. Also,

$$U(xz)c(y) = c(xzy) - c(xz) = U(x)c(zy) + c(x) - c(xz)$$

$$= U(x)(c(zy) - c(z)) = U(x)U(z)c(y),$$

which, by linearity, shows that $U \colon G \to \mathsf{U}(\mathbb{H})$ is a homomorphism. We have the cocycle equation $c(xy) = c(x) + U(x)c(y)$ by definition, for all $x, y \in G$. □

Exercise 9.15 Let $c \colon G \to \mathbb{H}$ be as in the proof of Lemma 9.3.5. Let $V = \mathrm{cl}(\mathrm{span}(c(G)))$. Define $U(x)c(y) = c(xy) - c(x)$ for $y \in G$, and $U(x)v = v$ for $v \in V^{\perp}$.

Show that $U(x)$ is well defined and can be uniquely extended to a bounded linear operator on $\mathbb{H}$.

Show that $U(x)$ is a unitary operator. ▷ solution ◁

Exercise 9.16 Let G be a finitely generated group and μ a symmetric adapted probability measure on G.

Let $f \in \ell^2(G, \mu)$. Let $\check{f}(x) := f(x^{-1})$. Define $c \colon G \to \ell^2(G, \mu)$ by $c(x) = x\check{f} - \check{f}$.

Show that for the μ-random walk $(X_t)_t$,

$$\mathbb{E}\,||c(X_1)||^2 = 2\langle \Delta f, f \rangle.$$

Show that

$$\left\| \sum_x \mu(x)c(x) \right\| = ||\Delta f||. \qquad \rhd \text{solution} \lhd$$

We now prove the existence of a harmonic cocycle (Theorem 9.3.2).

Proof of Theorem 9.3.2 **Step I** Since G is amenable, there exists a sequence of Følner sets; that is, a sequence $(F_n)_n$ of finite subsets of G such that

$$\lim_{n\to\infty} \frac{1}{|F_n|} \sum_{x\in F_n} \mathbb{P}_x[X_1 \notin F_n] = 0.$$

Let $f_n = |F_n|^{-1/2} \cdot \mathbf{1}_{F_n}$. Note that $||f_n|| = 1$, and by Jensen's inequality,

$$||\Delta f_n||^2 = \sum_x \left| \sum_y P(x,y)(f_n(x) - f_n(y)) \right|^2 \leq \sum_{x,y} P(x,y)(f_n(x) - f_n(y))^2$$

$$= \frac{2}{|F_n|} \cdot \sum_{x,y} P(x,y)\mathbf{1}_{F_n}(x)\mathbf{1}_{G\setminus F_n}(y) = \frac{2}{|F_n|} \sum_{x\in F_n} \mathbb{P}_x[X_1 \notin F_n] \to 0.$$

Finally, note that for any $x, y \in G$ we have $\mathbb{P}_x[X_k = y] \to 0$ as $k \to \infty$ (because G is infinite, Exercise 5.5.2). Because F_n is finite, we also have that $\mathbb{P}_x[X_k \in F_n] \to 0$ as $k \to \infty$. Since

$$\langle P^k f_n, f_n \rangle = \frac{1}{|F_n|} \sum_{x\in F_n} \mathbb{P}_x[X_k \in F_n],$$

we get that

$$\frac{1}{m} \sum_{k=0}^{m-1} \langle P^k f_n, f_n \rangle = \frac{1}{m} \sum_{k=0}^{m-1} \frac{1}{|F_n|} \sum_{x\in F_n} \mathbb{P}_x[X_k \in F_n] \to 0,$$

as $m \to \infty$.

In conclusion, we may apply Lemma 9.3.4 to obtain a sequence $(\psi_n)_n \subset \ell^2(G, \mu)$ such that $\langle \Delta\psi_n, \psi_n \rangle = 1$ and $||\Delta\psi_n|| \to 0$.

Step II For every n define $c_n(x) = x.\check{\psi}_n - \check{\psi}_n$, where $\check{\psi}_n(y) = \psi_n(y^{-1})$. By Exercise 9.16 we have that

$$\left\| \sum_x \mu(x)c_n(x) \right\| = ||\Delta\psi_n|| \to 0.$$

Also,

$$\sum_x \mu(x)||c_n(x)||^2 = 2.$$

We choose some finite symmetric generating set $S \subset \text{supp}(\mu)$ (as μ is symmetric and adapted), and use the metric on G with respect to this S. Set $\mu_{\min} = \min_{s \in S} \mu(s) > 0$. Note that

$$\max_{s \in S} ||c_n(s)||^2 \le (\mu_{\min})^{-1} \cdot \mathbb{E}||c_n(X_1)||^2 = \frac{2}{\mu_{\min}}.$$

This implies that $||c_n(x)||^2 \le \frac{2}{\mu_{\min}}|x|^2$ for all $x \in G$. By Exercise 9.11, c_n is a cocycle. Thus, by Lemma 9.3.5 we obtain a homomorphism $U: G \to U(\ell^2(G, \mu))$ and a map $c: G \to \ell^2(G, \mu)$ such that $c(xy) = c(x) + U(x)c(y)$ for all $x, y \in G$, and such that c is a μ-harmonic map.

Since $\sum_x \mu(x)||c_n(x)||^2 = 2$ for all n, there must exist $x \in \text{supp}(\mu)$ such that $\limsup_{n \to \infty} ||c_n(x)|| > 0$. By Lemma 9.3.5, c may be chosen so that $||c(x)|| > 0$, and specifically c is nonzero.

Taking $\mathbb{H} = \ell^2(G, \mu)$ and letting G act on $\mathbb{H}$ via the homomorphism U (i.e. $x.v = U(x)v$), we have that $c: G \to \mathbb{H}$ is a nonzero μ-harmonic cocycle. $\qquad\square$

Corollary 9.3.6 Let G be a finitely generated infinite group and $\mu \in \text{SA}(G, 1)$. Then G admits a nonconstant Lipschitz μ-harmonic function. That is,

$$\dim \text{LHF}(G, \mu) > 1.$$

Proof If G is amenable, Theorem 9.3.2 shows that there exists a nontrivial harmonic cocycle $c: G \to \mathbb{H}$ for some Hilbert space $\mathbb{H}$ on which G acts by unitary operators. So there are $v \in \mathbb{H}, ||v|| = 1$, and $y \in G$ with $\langle c(y), v \rangle \ne 0$. The function $h(x) := \langle c(x), v \rangle$ is a Lipschitz harmonic function. If h is constant then $h(y) = h(1) = 0$, contradicting $h(y) = \langle c(y), v \rangle \ne 0$.

If G is non-amenable, we have already seen that (G, μ) is non-Liouville (Theorem 6.8.1) so by Theorem 2.7.1 we have $\infty = \dim \text{BHF}(G, \mu) \le \dim \text{LHF}(G, \mu)$. $\qquad\square$

9.4 Diffusivity

As an application of Theorem 9.3.2 we prove that for symmetric random walks on groups with finite second moment, the random walk does not exhibit "sub-diffusive behavior."

Theorem 9.4.1 *Let G be a finitely generated infinite group generated by a finite symmetric generating set S, and let $\mu \in SA(G, 2)$. Let $(X_t)_t$ be the μ-random walk.*

Then there exists a constant $c = c_{\mu,S} > 0$ such that for all $t \geq 1$ we have $\mathbb{E}|X_t|^2 \geq ct$.

Proof If G is non-amenable then we have already seen in Exercise 5.35 that $\mathbb{E}|X_t| \geq ct$ for some $c > 0$.

So assume that G is amenable. Let $c \colon G \to \mathbb{H}$ be a harmonic cocycle into a Hilbert space $\mathbb{H}$ on which G acts by unitary operators. Let $U_t = X_{t-1}^{-1} X_t$ be the "jumps" of the random walk. Then, for $t > 0$,

$$\mathbb{E}\|c(X_t) - c(X_{t-1})\|^2 = \mathbb{E}\|X_{t-1}.c(U_t)\|^2 = \mathbb{E}\|c(U_t)\|^2 = \mathbb{E}\|c(X_1)\|^2.$$

Also, since U_t is independent of $\mathcal{F}_{t-1}$,

$$\mathbb{E}\langle c(X_t) - c(X_{t-1}), c(X_{t-1})\rangle = \mathbb{E}\,\mathbb{E}[\langle c(U_t), X_{t-1}.c(X_{t-1})\rangle \mid \mathcal{F}_{t-1}]$$
$$= \mathbb{E}\left[\left\langle \sum_x \mu(x)c(x), X_{t-1}c(X_{t-1})\right\rangle\right] = 0.$$

So

$$\mathbb{E}\|c(X_t)\|^2 = \mathbb{E}\left[\|c(X_t) - c(X_{t-1})\|^2 + \|c(X_{t-1})\|^2\right.$$
$$\left. + 2\mathsf{Re}\,\langle c(X_t) - c(X_{t-1}), c(X_{t-1})\rangle\right]$$
$$= \mathbb{E}\|c(X_1)\|^2 + \mathbb{E}\|c(X_{t-1})\|^2 = \cdots = t \cdot \mathbb{E}\|c(X_1)\|^2.$$

By Exercise 9.8, we conclude that

$$t \cdot \mathbb{E}\|c(X_1)\|^2 \leq \mathbb{E}\|c(X_t)\|^2 \leq \left(\sup_{s \in S}\|c(s)\|\right)^2 \cdot \mathbb{E}|X_t|^2. \qquad \square$$

9.5 Ozawa's Theorem

In this section we will prove the following theorem (basically due to Ozawa, also based on works of Shalom), which is the main step in proving Gromov's theorem (Theorem 9.0.1).

Theorem 9.5.1 *Let G be a finitely generated infinite group and $\mu \in SA(G, 2)$. Let $(X_t)_t$ denote the μ-random walk. Assume that G is amenable.*

Then, at least one of the following holds:

- *There exists a normal subgroup $N \lhd G$ of infinite index $[G : N] = \infty$ such that G/N is isomorphic to a subgroup of $\mathsf{U}_d(\mathbb{C})$, or*
- *there exist normal subgroups $K \lhd N \lhd G$ with index $[G : N] < \infty$ and $[N : K] = \infty$ such that N/K is Abelian, or*
- *there exists $h \in \mathsf{LHF}(G, \mu)$ a nonconstant Lipschitz harmonic function on G such that*

$$\lim_{t \to \infty} \frac{1}{t} \sup_x \mathrm{Var}_x[h(X_t)] = 0.$$

The step from Ozawa's theorem to Gromov's theorem is not a difficult one. If G has polynomial growth then it is amenable. Using the polynomial growth together with Theorem 6.5.5, one rules out the third option in Ozawa's theorem. Either one of the first two options in Ozawa's theorem shows that G is virtually indicable. This will be spelt out in full detail at the end of the chapter, see Section 9.6.

For now, we build the necessary machinery to prove Ozawa's theorem.

Suppose that T is a bounded linear operator on a Hilbert space $\mathbb{H}$. We denote by $\mathbb{H}^T = \{v \in \mathbb{H} : Tv = v\} = \mathrm{Ker}(I - T)$, which we call the *subspace of T invariant vectors*. Since T is continuous, $\mathbb{H}^T$ is a closed subspace, so there exists an orthogonal projection onto $\mathbb{H}^T$.

The main tool used is the classical von Neumann ergodic theorem.

Theorem 9.5.2 (von Neumann ergodic theorem) *Let T be a bounded self-adjoint operator on a Hilbert space $\mathbb{H}$. Assume that $||T|| \leq 1$. Let $\mathbb{H}^T$ be the subspace of T invariant vectors. Let π be the orthogonal projection onto $\mathbb{H}^T$.*

Then, for any $v \in \mathbb{H}$ we have

$$\lim_{n \to \infty} \frac{1}{n} \sum_{k=0}^{n-1} T^k v = \pi v. \tag{9.3}$$

Proof Let V be the set of all $v \in \mathbb{H}$ such that (9.3) holds for v. We wish to show that $V = \mathbb{H}$.

It is an exercise to show that V is a closed subspace, and that $\mathbb{H}^T \leq V$.

Now, let $U = \{Tv - v : v \in \mathbb{H}\}$. Another exercise is to show that U is a subspace (but this time it is not guaranteed to be closed).

Now, if $u = Tv - v \in U$ then for any $w \in \mathbb{H}^T$ we have

$$\langle u, w \rangle = \langle v, Tw \rangle - \langle v, w \rangle = 0.$$

So $U \perp \mathbb{H}^T$, which implies that $\pi u = 0$ for all $u \in U$. Now, for any $u = Tv - v \in U$, we have

$$\left\| \frac{1}{n} \sum_{k=0}^{n-1} T^k u \right\| = \frac{1}{n} \|T^n v - v\| \le \frac{2\|v\|}{n} \to 0.$$

So we obtain that $U \le V$.

Since V is closed, we get that $W := \text{cl}\left(U + \mathbb{H}^T\right) \le V$. Now assume that $w \perp W$. Then $w \perp U$. Specifically, $w \perp Tw - w$. Since $\|Tw\| \le \|w\|$,

$$\|Tw - w\|^2 = \|Tw\|^2 + \|w\|^2 - 2\text{Re}\langle Tw, w \rangle \le -2\text{Re}\langle Tw - w, w \rangle = 0,$$

so $Tw = w$. But then $w \in \mathbb{H}^T$, and so $w \perp w$, implying that $w = 0$.

We arrive at the conclusion that $W^\perp = \{0\}$, which is to say that $\mathbb{H} = W \le V$, proving the theorem. $\square$

Exercise 9.17 Show that V in the proof of Theorem 9.5.2 is a closed subspace of $\mathbb{H}$.

Show that $\mathbb{H}^T \le V$.

Show that $U = \{Tv - v : v \in \mathbb{H}\}$ is a subspace of $\mathbb{H}$. ▷ solution ◁

9.5.1 Hilbert–Schmidt Operators

In the following, we develop the theory of the Hilbert–Schmidt spaces for general Hilbert spaces. We will, however, only require the theory in the separable case. Some readers may find it simpler to proceed with the assumption that all Hilbert spaces mentioned have a countable orthonormal basis (i.e. are separable). See Section A.3 for more details.

Let $\mathbb{H}$ be a Hilbert space. Consider the space of *Hilbert–Schmidt operators* on $\mathbb{H}$: for a bounded linear operator $T \colon \mathbb{H} \to \mathbb{H}$ define the **Hilbert–Schmidt norm**

$$\|T\|_{\text{HS}}^2 = \sum_{b \in B} \|Tb\|^2$$

(which may be infinite), where B is some orthonormal basis for $\mathbb{H}$. Note that this definition is independent of the specific choice of orthonormal basis.

Exercise 9.18 Show that if $B, \tilde{B}$ are two orthonormal bases for $\mathbb{H}$, then for any bounded linear operator $T \colon \mathbb{H} \to \mathbb{H}$,

$$\sum_{b \in B} \|Tb\|^2 = \sum_{b \in \tilde{B}} \|Tb\|^2 = \sum_{b,b' \in B} |\langle Tb, b' \rangle|^2.$$

Show that $\|T\|_{\text{op}} \le \|T\|_{\text{HS}}$. ▷ solution ◁

Definition 9.5.3 Let $\mathbb{H}$ be a Hilbert space. A bounded linear operator $A \colon \mathbb{H} \to \mathbb{H}$ is called a **Hilbert–Schmidt** operator if $\|A\|_{\mathrm{HS}} < \infty$. The space of all Hilbert–Schmidt operators is denoted by $\mathbb{H}^* \otimes \mathbb{H}$.

Exercise 9.19 Show that $\mathbb{H}^* \otimes \mathbb{H}$ is a Hilbert space under the inner product

$$\langle A, A' \rangle_{\mathrm{HS}} := \sum_{b \in B} \langle Ab, A'b \rangle,$$

where B is some orthonormal basis for $\mathbb{H}$.

Show that this defines an inner product and is independent of the choice of specific orthonormal basis.

Show that $\langle A^*, (A')^* \rangle_{\mathrm{HS}} = \langle A, A' \rangle_{\mathrm{HS}}$.   ▷ solution ◁

If $v, u \in \mathbb{H}$ then define the linear operator

$$v \otimes u \colon \mathbb{H} \to \mathbb{H} \qquad \text{by} \qquad (v \otimes u)w := \langle w, v \rangle\, u.$$

Exercise 9.20 Show that $v \otimes u \in \mathbb{H}^* \otimes \mathbb{H}$ for any $v, u \in \mathbb{H}$. Show that

$$\langle v \otimes u, v' \otimes u' \rangle_{\mathrm{HS}} = \langle v', v \rangle \cdot \langle u, u' \rangle,$$

so that

$$\|v \otimes u\|_{\mathrm{HS}} = \|v\| \cdot \|u\|.$$

Show that if B is an orthonormal basis for $\mathbb{H}$, then $\{b \otimes b' : b, b' \in B\}$ is an orthonormal basis for $\mathbb{H}^* \otimes \mathbb{H}$.   ▷ solution ◁

One way to think about Hilbert–Schmidt operators is as a generalization of matrices. One can think of $\langle Ab, b' \rangle$ as the coefficients of the operator A in the basis B, just like with matrices, where $M_{i,j} = \langle Me_j, e_i \rangle$ are the coefficients in the standard basis.

Exercise 9.21 Let $\mathbb{H}$ be a Hilbert space and let $A \in \mathbb{H}^* \otimes \mathbb{H}$ be a Hilbert–Schmidt operator. Let B be an orthonormal basis.

Show that there exists a countable subset $E = \{e_1, e_2, \ldots\} \subset B$ such that $Ab = 0$ for any $b \in B \backslash E$.

For any n, define $A_n \colon \mathbb{H} \to \mathbb{H}$ by

$$A_n v = \sum_{k=1}^{n} \langle v, e_k \rangle\, A e_k.$$

Show that for any $\varepsilon > 0$ there exists n_ε such that $\|A - A_n\|_{\mathrm{HS}} < \varepsilon$ for all $n \geq n_\varepsilon$.

       ▷ solution ◁

The natural topology on a Hilbert space is the one induced by the metric given by the norm. However, the unit ball is not necessarily compact in this topology.

Exercise 9.22 Let $\mathbb{H}$ be a Hilbert space. Show that the unit ball $\{u \in \mathbb{H} : ||u|| \leq 1\}$ is compact if and only if $\dim \mathbb{H} < \infty$.   ▷ solution ◁

Consider a different topology on $\mathbb{H}$.

Definition 9.5.4 We say that $(u_n)_n$ **converges weak*** to u (or, **in the weak*** **topology**), if for any $v \in \mathbb{H}$ we have $\langle u_n, v \rangle \to \langle u, v \rangle$.

Exercise 9.23 Show that if $(u_n)_n$ converges weak* to u then

$$||u|| \leq \liminf_{n \to \infty} ||u_n||.$$ ▷ solution ◁

Exercise 9.24 Let B be an orthonormal basis for $\mathbb{H}$.

Show that if $(u_n)_n$ is a sequence in the unit ball, then $(u_n)_n$ converges weak* to u if and only if for any $b \in B$ we have $\langle u_n - u, b \rangle \to 0$.   ▷ solution ◁

The weak* topology has the advantage that the unit ball is compact (see Lemma 9.5.5). However, unlike the norm topology, not every bounded operator is continuous with respect to the weak* topology. That is, if $(u_n)_n$ converges weak* to u, it does not necessarily hold that $||A(u_n - u)|| \to 0$ for a general bounded operator A. One feature of Hilbert–Schmidt operators is that they are still continuous with this weaker restriction (i.e. $||A(u_n - u)|| \to 0$ if $||A||_{HS} < \infty$).

Exercise 9.25 Let $\mathbb{H}$ be a Hilbert space and let $A \in \mathbb{H}^* \otimes \mathbb{H}$ be a Hilbert–Schmidt operator.

Show that $A : \mathbb{H} \to \mathbb{H}$ is a continuous function from $\mathbb{H}$ with the weak* topology to $\mathbb{H}$ with the norm topology.

That is, show that if $(u_n)_n$ is a sequence in $\mathbb{H}$ such that for some $u \in \mathbb{H}$ and any $v \in \mathbb{H}$, we have $\langle u_n, v \rangle \to \langle u, v \rangle$, then $||A(u_n - u)|| \to 0$.   ▷ solution ◁

Lemma 9.5.5 *Let $\mathbb{H}$ be a Hilbert space and let $A \in \mathbb{H}^* \otimes \mathbb{H}$ be a self-adjoint Hilbert–Schmidt operator.*

Consider $\varphi : \mathbb{H} \to \mathbb{R}$ given by $\varphi(u) = \langle Au, u \rangle$.

Then φ is maximized and minimized on the unit ball in $\mathbb{H}$. That is, there exist

$v, v' \in \mathbb{H}$ *with* $\|v\|, \|v'\| \leq 1$ *such that*

$$\varphi(v) = \max\{\varphi(u) : \|u\| \leq 1\} \quad \text{and} \quad \varphi(v') = \min\{\varphi(u) : \|u\| \leq 1\}.$$

Proof This is essentially due to the Banach–Alaoglu theorem, which tells us that the unit ball is compact in the weak* topology, and the fact that φ is continuous on this compact set.

However, we will provide a self-contained proof.

Let $K = \{u \in \mathbb{H} : \|u\| \leq 1\}$ denote the unit ball. Set $M = \sup_{u \in K} \varphi(u)$. Let $(u_n)_n$ be a sequence in K such that $\varphi(u_n) \to M$.

We claim that there exists a subsequence of $(u_n)_n$ that converges weak* in $\mathbb{H}$.

Let B be an orthonormal basis. Since $\sum_b |\langle u_n, b \rangle|^2 = \|u_n\|^2 < \infty$ for all n, we have that there exists a countable subset $E = \{e_1, e_2, \ldots\} \subset B$ such that $\langle u_n, b \rangle = 0$ for all n and any $b \in B \backslash E$.

The sequence $(\langle u_n, e_1 \rangle)_n$ is a bounded sequence. Thus there exists a sequence $(n_{1,j})_j$ such that

$$\lim_{j \to \infty} \left\langle u_{n_{1,j}}, e_1 \right\rangle = \lambda_1 \in \mathbb{C}.$$

Inductively, for $k > 1$, suppose we have a sequence $(n_{k-1,j})_j$ such that for all $1 \leq \ell \leq k - 1$ the limits

$$\lim_{j \to \infty} \left\langle u_{n_{k-1,j}}, e_\ell \right\rangle = \lambda_\ell$$

exist. Considering the bounded sequence $\left(\left\langle u_{n_{k-1,j}}, e_k \right\rangle \right)_j$, we may extract a further subsequence of $(n_{k-1,j})_j$, say $(n_{k,j})_j$ so that for all $1 \leq \ell \leq k$ we have the limits

$$\lim_{j \to \infty} \left\langle u_{n_{k-1,j}}, e_\ell \right\rangle = \lambda_\ell,$$

for some $\lambda_k \in \mathbb{C}$. Constructing this for every k, we now consider the diagonal subsequence $v_j := u_{n_{j,j}}$. We have that for any $k \geq 1$,

$$\lim_{j \to \infty} \left\langle v_j, e_k \right\rangle = \lim_{j \to \infty} \left\langle u_{n_{k,j}}, e_k \right\rangle = \lambda_k.$$

By Fatou's lemma,

$$\sum_k |\lambda_k|^2 \leq \liminf_{j \to \infty} \sum_k \left| \langle v_j, e_k \rangle \right|^2 = \liminf_{j \to \infty} \|v_j\|^2 \leq 1.$$

Thus, $u = \sum_k \lambda_k e_k$ is a well-defined vector in $\mathbb{H}$, with $\|u\|^2 \leq 1$. Note that for any $k \geq 1$ we have that

$$\langle u, e_k \rangle = \lambda_k = \lim_{j \to \infty} \left\langle v_j, e_k \right\rangle.$$

Also, $\langle u, b \rangle = 0 = \langle v_j, b \rangle$ for all $b \in B \backslash E$. By Exercise 9.24, $(v_j)_j$ is a subsequence of $(u_n)_n$ that converges weak* to u.

Now,

$$|\varphi(v_j) - \varphi(u)| = \left| \langle A(v_j - u), v_j \rangle + \langle Au, v_j - u \rangle \right| \le 2||A(v_j - u)|| \to 0,$$

by Exercise 9.25. Since $\varphi(v_j) \to M = \sup_{v \in K} \varphi(v)$ as $j \to \infty$, and since $u \in K$, we conclude that $\varphi(u) = M$.

This proves that φ is maximized on the unit ball in $\mathbb{H}$.

Replacing A by $-A$ proves that φ is also minimized. $\qquad \square$

The next lemma shows that, just like in the matrix case, a self-adjoint Hilbert–Schmidt operator always has a nontrivial eigenvector.

Lemma 9.5.6 *Let $0 \ne A \in \mathbb{H}^* \otimes \mathbb{H}$ be self-adjoint. Then there exists an eigenvector $0 \ne v \in \mathbb{H}$ with $Av = \lambda v$ for some $\lambda \ne 0$.*

Proof This is due to the fact that Hilbert–Schmidt operators are compact operators (on a Banach space). But we will give a self-contained proof.

Let $M = \max\{\langle Au, u \rangle : ||u|| \le 1\}$. Since $A \ne 0$, by perhaps replacing A by $-A$ we may assume without loss of generality that $M > 0$.

Fix $v, u \in \mathbb{H}$ and let $p : \mathbb{R} \to \mathbb{C}$ be

$$p(\xi) = \frac{\langle A(v + \xi u), v + \xi u \rangle}{\langle v + \xi u, v + \xi u \rangle} = \frac{\langle Av, v \rangle + \xi^2 \cdot \langle Au, u \rangle + \xi \cdot (\langle Av, u \rangle + \langle Au, v \rangle)}{||v||^2 + \xi^2 \cdot ||u||^2 + \xi \cdot (\langle v, u \rangle + \langle u, v \rangle)}.$$

Since the numerator and denominator are quadratic polynomials in ξ, we get that

$$p'(0) = \frac{(\langle Av, u \rangle + \langle Au, v \rangle) \cdot ||v||^2 - (\langle v, u \rangle + \langle u, v \rangle) \cdot \langle Av, v \rangle}{||v||^4}.$$

Consider $\varphi(u) = \langle Au, u \rangle$ and the *Rayleigh quotient* $\psi(u) = \frac{\varphi(u)}{||u||^2}$ defined for any $u \ne 0$. Note that $\psi(\alpha u) = \psi(u)$ for all $u \in \mathbb{H} \backslash \{0\}$ and all $\alpha \in \mathbb{C} \backslash \{0\}$.

By Lemma 9.5.5, we know that φ is maximized on the unit ball by some vector $u \in \mathbb{H}$ with $||u|| \le 1$. Since $\varphi(u) = M = \max\{\varphi(w) : ||w|| \le 1\} > 0$, we know that $u \ne 0$. Set $v = \frac{u}{||u||}$. Then, for any $w \in \mathbb{H}$ with $0 < ||w|| \le 1$,

$$\psi(v) = \psi(u) = \frac{\varphi(u)}{||u||} \ge \varphi\left(\frac{w}{||w||}\right) = \psi\left(\frac{w}{||w||}\right) = \psi(w).$$

So v maximizes ψ on the unit ball.

Then, with this v, the function $p(\xi) = \psi(v + \xi u)$ defined above is maximized at $\xi = 0$. By the above formula for $p'(0)$, we have for any $u \in \mathbb{H}$ that

$$\langle Av, u \rangle + \langle Au, v \rangle = (\langle v, u \rangle + \langle u, v \rangle) \cdot \langle Av, v \rangle.$$

Taking $\lambda = \langle Av, v \rangle$ we have that for any $u \in \mathbb{H}$,

$$\langle Av - \lambda v, u \rangle + \langle u, A^*v - \lambda v \rangle = 0.$$

Since $A = A^*$ we obtain $\mathrm{Re}\,\langle Av - \lambda v, u \rangle = 0$ for any $u \in \mathbb{H}$. So also

$$\mathrm{Im}\,\langle Av - \lambda v, u \rangle = \mathrm{Re}\,\langle Av - \lambda v, iu \rangle = 0$$

for any $u \in \mathbb{H}$. Thus, $Av = \lambda v$.   □

Now we add to the mix a group G acting on the Hilbert space $\mathbb{H}$ by unitary operators.

Exercise 9.26 Let G be a group acting on $\mathbb{H}$ by unitary operators.
Show that G acts on $\mathbb{H}^* \otimes \mathbb{H}$ by conjugation.
Show that this action is a unitary action on the Hilbert space $\mathbb{H}^* \otimes \mathbb{H}$.

▷ solution ◁

Exercise 9.27 Show that $x(v \otimes u)y^{-1} = yv \otimes xu$ (as operators on $\mathbb{H}$). Specifically, $x(v \otimes u)x^{-1} = xv \otimes xu$.

▷ solution ◁

Lemma 9.5.7 *Let G be a group. Let $\mathbb{H}$ be a Hilbert space on which G acts by unitary operators.*

Assume that $0 \neq L \in \mathbb{H}^* \otimes \mathbb{H}$ is a G-invariant vector; that is, $xLx^{-1} = L$ for all $x \in G$. Then, there exists a finite-dimensional subspace $V \leq \mathbb{H}$, $\dim V < \infty$, and some $v \in V$ such that $Lv \neq 0$ and such that $GV \leq V$ (i.e. V is a G-invariant subspace).

Proof Assume that $0 \neq L \in \mathbb{H}^* \otimes \mathbb{H}$ is a G-invariant vector. Since $xL^*x^{-1} = L^*$ as well (because the action is unitary), if we define $A = \frac{1}{2}(L + L^*)$, we have that A is a self-adjoint G-invariant vector in $\mathbb{H}^* \otimes \mathbb{H}$.

Since A is a self-adjoint Hilbert–Schimdt operator, there exists an eigenvector $0 \neq v \in \mathbb{H}$ with $Av = \lambda v$ for $\lambda \neq 0$, by Lemma 9.5.6. Let $V = \{u \in \mathbb{H} : Au = \lambda u\}$.

If $\dim V \geq n$ then we may find an orthonormal basis B for $\mathbb{H}$ such that $b_1, \ldots, b_n \in B \cap V$. However, in this case,

$$\|A\|_{\mathrm{HS}}^2 \geq \sum_{j=1}^{n} \|Ab_j\|^2 = n|\lambda|^2.$$

Hence $\dim V \leq |\lambda|^{-2} \cdot \|A\|_{\mathrm{HS}}^2 < \infty$.

We now claim that V is G-invariant. Indeed, for any $u \in V$ and any $x \in G$ we have that $Axu = xAu = \lambda xu$, so $xu \in V$ as well.

Finally, note that since $Av = \lambda v$, we have that

$$\operatorname{Re}\langle Lv, v\rangle = \frac{\langle Lv, v\rangle + \langle L^*v, v\rangle}{2} = \langle Av, v\rangle = \lambda||v||^2 \neq 0,$$

which implies that $Lv \neq 0$.   □

We now prove Ozawa's theorem.

Proof of Theorem 9.5.1   Let $\mathbb{H}$ be a Hilbert space on which G acts by unitary operators, with the guaranteed μ-harmonic cocycle $c: G \to \mathbb{H}$, from Theorem 9.3.2.

Let $(X_t)_t$ be the μ-random walk. For any map $f: G \to \mathbb{H}^* \otimes \mathbb{H}$ we will use the notation

$$\mathbb{E}[f(X_t)] = \sum_x \mathbb{P}[X_t = x] \cdot f(x).$$

Note that for this to be defined, it is required that the above sum converges in $\mathbb{H}^* \otimes \mathbb{H}$. If $\mathbb{E}[||f(X_t)||_{HS}] < \infty$ then $\mathbb{E}[f(X_t)]$ converges absolutely.

Note also that for any $x, y \in G$,

$$c(xy) \otimes c(xy)$$
$$= (c(x) + x.c(y)) \otimes (c(x) + x.c(y))$$
$$= c(x) \otimes c(x) + x(c(y) \otimes c(y))x^{-1} + c(x) \otimes x.c(y) + x.c(y) \otimes c(x).$$

Fix a finite symmetric generating set S for G. Let $\kappa = \max_{s \in S} ||c(s)||$. So $||c(x)|| \leq \kappa|x|$ for all $x \in G$.

Note that since $||x.c(X_t) \otimes c(x)||_{HS} = ||c(X_t)|| \cdot ||c(x)|| \leq \kappa^2|x||X_t|$, and since μ has finite first moment, the sum $\mathbb{E}[x.c(X_t) \otimes c(x)]$ converges absolutely. Similarly, since μ has finite second moment, for any $x \in G$ also $\mathbb{E}[c(xX_t) \otimes c(xX_t)]$ converges absolutely, as $||c(xX_t) \otimes c(xX_t)|| \leq \kappa^2(|x| + |X_t|)^2$.

Since c is μ-harmonic,

$$\mathbb{E}[x.c(X_t) \otimes c(x)] = \mathbb{E}[c(x) \otimes x.c(X_t)] = x.c(1) \otimes c(x) = 0,$$

so

$$\mathbb{E}[c(xX_t) \otimes c(xX_t)] = c(x) \otimes c(x) + x\,\mathbb{E}[c(X_t) \otimes c(X_t)]x^{-1}.$$

Let T be an operator on $\mathbb{H}^* \otimes \mathbb{H}$ given by $TA = \sum_x \mu(x)xAx^{-1}$. Define $A_t := \mathbb{E}[c(X_t) \otimes c(X_t)]$. Averaging the above over x according to μ we obtain that

$$A_t = A_1 + TA_{t-1} = A_1 + TA_1 + T^2A_{t-2} = \cdots = \sum_{k=0}^{t-1} T^k A_1.$$

It is an exercise to show that T is a bounded self-adjoint operator on $\mathbb{H}^* \otimes \mathbb{H}$ with $\|T\| \leq 1$. Thus, by the von Neumann ergodic theorem (Theorem 9.5.2),

$$\lim_{t \to \infty} \frac{1}{t} A_t = L,$$

where L is some T-invariant vector in $\mathbb{H}^* \otimes \mathbb{H}$ (in fact L is the orthogonal projection of A_1 onto the subspace of T-invariant vectors in $\mathbb{H}^* \otimes \mathbb{H}$).

We now have three cases.

Case I $L \neq 0$. By normalizing L, one may assume without loss of generality that $\|L\|_{\mathrm{HS}} = 1$. Since $TL = \sum_x \mu(x) x L x^{-1} = L$ is a convex combination of unit vectors, we obtain that L must be a G-invariant vector. See Exercise 9.30. By Lemma 9.5.7, there exists a subspace $V \leq \mathbb{H}$ such that $0 < \dim V < \infty$ and $GV \leq V$. Let $N = \{x \in G : \forall v \in V, \, x.v = v\}$. It is an easy exercise to show that N is a normal subgroup. Then G/N is isomorphic to a subgroup of $U_d(\mathbb{C}) \cong U(V)$, where $d = \dim V$.

Case Ia If $[G : N] = \infty$, then G/N is isomorphic to an infinite subgroup of $U_d(\mathbb{C})$.

Case Ib So assume that $[G : N] < \infty$. There exists $v \in V$ such that $Lv \neq 0$ (by Lemma 9.5.7). Consider the function $h(x) = h_v(x) := \langle c(x), v \rangle$,

$$0 \neq Lv = \lim_{t \to \infty} \frac{1}{t} \sum_{k=0}^{t-1} \sum_x \mu^k(x) \langle v, c(x) \rangle c(x),$$

which implies that there exists $x \in G$ such that $h(x) = \langle c(x), v \rangle \neq h(1) = 0$. So h is nonconstant. It is an exercise to show that since the action of G is by unitary operators, and since $x.v = v$ for all $x \in N$, we have that $h|_N$ is a homomorphism into $\mathbb{C}$. Consider the subgroup $\mathrm{Ker}(h)$. Note that $h|_N \equiv 0$ if and only if $N/\mathrm{Ker}(h) \cong h(N)$ is finite. Since we assumed that $[G : N] < \infty$, if $[N : \mathrm{Ker}(h)] < \infty$, then $h \equiv 0$. So it cannot be that $[N : \mathrm{Ker}(h)] < \infty$. Thus, $N/\mathrm{Ker}(h)$ is isomorphic to an infinite subgroup of $\mathbb{C}$, which must then be Abelian.

Case II $L = 0$. In this case, we find a Lipschitz harmonic function with sublinear variance. This is because c is a nontrivial cocycle, so there exists $y \in \mathrm{supp}(\mu)$ such that $c(y) \neq 0$. Define $h(x) := \langle c(x), c(y) \rangle$. It is immediate that $h(1) = 0 \neq h(y)$ and that $h \in \mathrm{LHF}(G, \mu)$.

Since c is μ-harmonic, we have that $\mathbb{E}[\langle x.c(X_t), c(y) \rangle] = 0$. Using

$$\langle v \otimes u, v' \otimes u' \rangle_{\mathrm{HS}} = \langle v', v \rangle \langle u, u' \rangle,$$

we arrive at

$$\mathbb{E}\,|h(xX_t)|^2$$
$$= \mathbb{E}\langle(c(x) + x.c(X_t)) \otimes (c(x) + x.c(X_t)), c(y) \otimes c(y)\rangle$$
$$= |\langle c(x), c(y)\rangle|^2 + \langle x(c(X_t) \otimes c(X_t))x^{-1}, c(y) \otimes c(y)\rangle$$
$$\quad + \langle c(y), c(x)\rangle \cdot \mathbb{E}[\langle x.c(X_t), c(y)\rangle] + \langle c(x), c(y)\rangle \cdot \mathbb{E}[\langle c(y), x.c(X_t)\rangle]$$
$$= |h(x)|^2 + \langle xA_t x^{-1}, c(y) \otimes c(y)\rangle,$$

which implies

$$\frac{1}{t}\mathrm{Var}_x[h(X_t)] = \frac{1}{t}\mathbb{E}\,|h(xX_t)|^2 - \frac{1}{t}|h(x)|^2$$
$$= \left\langle x\frac{1}{t}A_t x^{-1}, c(y) \otimes c(y)\right\rangle \le \left\|\tfrac{1}{t}A_t\right\|_{\mathrm{HS}} \cdot \|c(y)\|^2 \to 0. \qquad \square$$

Exercise 9.28 Let G be a group acting on a Hilbert space $\mathbb{H}$ by unitary operators. Let $V \le \mathbb{H}$ be a subspace. Let $N = \{x \in G : \forall\, v \in V,\ x.v = v\}$. Show that N is a normal subgroup of G.

Exercise 9.29 Let G be a group acting by unitary operators on a Hilbert space $\mathbb{H}$. Let $c\colon G \to \mathbb{H}$ be a cocycle. Let $v \in \mathbb{H}$ be such that $x.v = v$ for all $x \in G$. Show that $h(x) := \langle c(x), v\rangle$ is a homomorphism into $\mathbb{C}$.

Show that $h(G)$ is finite if and only if $h \equiv 0$.

Exercise 9.30 Let G act by unitary operators on a Hilbert space $\mathbb{H}$. Let μ be a symmetric and adapted probability measure on G. Let $T\colon \mathbb{H} \to \mathbb{H}$ be defined by $Tv = \sum_x \mu(x)x.v$.

Show that if $\|v\| = 1, Tv = v$ then $x.v = v$ for all $x \in \mathrm{supp}(\mu)$.

Conclude that $x.v = v$ for all $x \in G$.

(Hint: consider the random variable $\langle X.v, b\rangle$ for some fixed vector b, and X distributed according to μ. What is the variance of this random variable? What is the sum over b ranging over an orthonormal basis?)

▷ solution ◁

9.6 Proof of Gromov's Theorem

Finally we put all the pieces together to obtain Gromov's theorem (Theorem 9.0.1).

In light of the reduction in Proposition 9.1.1, we only need to show that any infinite polynomial growth group admits a finite index subgroup with a surjective homomorphism onto $\mathbb{Z}$. That is, it suffices to prove the following theorem.

Theorem 9.6.1 *Let G be an infinite finitely generated group of polynomial growth. Then there exists a finite index subgroup $[G : H] < \infty$ such that H admits a surjective homomorphism onto $\mathbb{Z}$.*

Proof Let G be a finitely generated group of polynomial growth. Choose any $\mu \in \mathsf{SA}(G, \infty)$. Since G is of sub-exponential growth it is amenable (Exercise 6.50). By Theorem 9.5.1 and Exercise 1.32, one of three cases holds:

(1) There exists a finite index subgroup $H \lhd G, [G : H] < \infty$ such that H admits a surjective homomorphism onto $\mathbb{Z}$.

(2) There exists a nonconstant $h \in \mathsf{LHF}(G, \mu)$ such that for any x,

$$\lim_{t \to \infty} \tfrac{1}{t} \operatorname{Var}_x[h(X_t)] = 0.$$

(3) There exists a homomorphism $\varphi \colon G \to \mathsf{U}_d(\mathbb{C})$ such that $|\varphi(G)| = \infty$.

If the first case holds, we are done.

If $h \in \mathsf{LHF}(G, \mu)$, then Theorem 6.5.5 tells us that for all t,

$$\left(\mathbb{E}_x |h(X_1) - h(x)| \right)^2 \le 4 \cdot \operatorname{Var}_x h(X_t) \cdot \left(H(X_t) - H(X_{t-1}) \right).$$

Using Exercises 6.24 and 6.30, since G has polynomial growth we have that $H(X_t) \le C \log(t + 1)$ for some $C > 0$ and all $t > 0$. So there are infinitely many t for which $H(X_t) - H(X_{t-1}) \le \frac{2C}{t+1}$. But then

$$\left(\mathbb{E}_x |h(X_1) - h(x)| \right)^2 \le 8C \cdot \liminf_{t \to \infty} \tfrac{1}{t} \operatorname{Var}_x h(X_t).$$

If $\frac{1}{t} \operatorname{Var}_x h(X_t) \to 0$ for all x then h is constant. This implies that the second case above cannot hold for groups of polynomial growth.

So we are left with the third case. There exists a homomorphism $\varphi \colon G \to \mathsf{U}_d(\mathbb{C})$ such that $|\varphi(G)| = \infty$. The group $\varphi(G)$ is a finitely generated infinite subgroup of $\mathsf{U}_d(\mathbb{C})$ of polynomial growth, so by Theorem 9.2.3 it is virtually Abelian and infinite. That is, there exists a finite index subgroup of $\varphi(G)$ with a surjective homomorphism onto $\mathbb{Z}$. Lifting this subgroup to a subgroup of G, we obtain a finite index subgroup of G with a surjective homomorphism onto $\mathbb{Z}$. $\qquad\square$

This completes the proof of Gromov's theorem, Theorem 9.0.1.

9.7 Classification of Recurrent Groups

Recall the notions of a recurrent and transient group from Section 4.12. We have seen that any group that is finite, virtually $\mathbb{Z}$, or virtually $\mathbb{Z}^2$ is recurrent. As an application of Gromov's theorem, we will now prove that these are the only recurrent groups.

Recall Exercise 8.19, which tells us that if G is a group of polynomial growth $\sim \left(r \mapsto r^d\right)$, and if $K \lhd G$ is such that $G/K \cong \mathbb{Z}$, then K is finitely generated and has polynomial growth $\preceq \left(r \mapsto r^{d-1}\right)$.

Also, Exercise 8.21 tells us that if $G/F \cong \mathbb{Z}$ for some finite subgroup $F \lhd G$, then G is virtually $\mathbb{Z}$.

Exercise 9.31 Let G be an infinite finitely generated group. Assume that G has growth $\preceq (r \mapsto r)$ (i.e. linear growth).

Show that G contains a finite index subgroup H such that $H \cong \mathbb{Z}$; that is, G is virtually $\mathbb{Z}$.   ▷ solution ◁

Exercise 9.32 Let G be an infinite finitely generated group. Assume that G has growth $\preceq \left(r \mapsto r^2\right)$.

Show that G contains a finite index subgroup H such that $H \cong \mathbb{Z}$ or $H \cong \mathbb{Z}^2$.

▷ solution ◁

Exercise 9.33 Let G be an infinite finitely generated group. Assume that G has growth $\preceq \left(r \mapsto r^{3-\varepsilon}\right)$ for some $\varepsilon > 0$.

Show that G is either virtually $\mathbb{Z}$ or virtually $\mathbb{Z}^2$. (Specifically, G actually has growth $\preceq (r \mapsto r^2)$.)   ▷ solution ◁

Theorem 9.7.1 (Varopolous' theorem) *Let G be a finitely generated recurrent group. Then G is virtually $\mathbb{Z}^2$ or G is virtually $\mathbb{Z}$ or G is finite.*

Proof We have seen in Exercise 9.33 that if G has growth $\preceq \left(r \mapsto r^{3-\varepsilon}\right)$ for some $\varepsilon > 0$ then G is either finite or virtually $\mathbb{Z}$ or virtually $\mathbb{Z}^2$. Thus we may assume that G has growth $\succeq \left(r \mapsto r^{3-\varepsilon}\right)$ for $\varepsilon > 0$ as small as we want. We then want to prove that under this condition G is transient. By Theorem 4.12.1, it suffices to find a probability measure $\mu \in \mathsf{SA}(G, 2)$ such that the μ-random walk is transient.

Fix some small $\varepsilon > 0$ $\left(\text{in fact } \varepsilon < \frac{1}{2} \text{ will suffice}\right)$.

Fix a Cayley graph of G, and let $B_r = B(1, r)$ denote the ball of radius r about the origin 1. Let $(\alpha_n)_{n \geq 1}$ be a sequence such that $\alpha_n > 0$ for all n and

$\sum_n \alpha_n n^2 < \infty$ and $\sum_n \alpha_n = 1$. Define a measure μ by

$$\mu(x) = \sum_{r=1}^{\infty} \frac{\alpha_r}{|B_r|} \mathbf{1}_{\{|x| \leq r\}}.$$

Then μ is fully supported (so obviously adapted) and easily seen to be symmetric $\left(\text{since } |x^{-1}| = |x|\right)$. Also,

$$\sum_x \mu(x)|x|^2 = \sum_x |x|^2 \sum_{r \geq |x|} \frac{\alpha_r}{|B_r|} = \sum_{r=1}^{\infty} \frac{\alpha_r}{|B_r|} \sum_{|x| \leq r} |x|^2$$

$$= \sum_{r=1}^{\infty} \frac{\alpha_r}{|B_r|} \sum_{k=1}^{r} k^2 \#\{|x| = k\} \leq \sum_{r=1}^{\infty} \alpha_r r^2 < \infty.$$

So μ has a finite second moment.

We now show that (G, μ) is transient.

Let $R \geq 1$, and define

$$\mu_R(x) = \sum_{r=1}^{R} \frac{\alpha_r}{|B_r|} \mathbf{1}_{\{|x| \leq r\}}$$

and $\nu_R = \mu - \mu_R$ (note that these measures are not probability measures).

We have that $\mu_R(G) \leq \sum_{r=1}^{R} \alpha_r$. Also, for any $x \in G$,

$$\nu_R(x) = \sum_{r > R} \frac{\alpha_r}{|B_r|} \mathbf{1}_{\{|x| \leq r\}} \leq \frac{1}{|B_{R+1}|} \cdot \sum_{r > R} \alpha_r.$$

It is an exercise to use these two estimates to prove by induction that

$$\mathbb{P}[X_t = 1] \leq (1 - \delta)^t + t \cdot \frac{\delta}{|B_{R+1}|},$$

where $\delta = \sum_{r > R} \alpha_r$ (which can be made as small as desired by choosing large enough R). Given this bound, we want to choose $R = R(t)$ so that $\sum_t \mathbb{P}[X_t = 1] < \infty$, which proves transience.

Let $k \geq 1, \beta > 0$ to be determined. Choose $\alpha_r = Cr^{-k}$ for some constant $C > 0$. So $\delta = O\left(R^{-(k-1)}\right)$. Then choose $R = R(t) = \left\lceil t^{\beta} \right\rceil$. It follows that

$$(1 - \delta)^t \leq e^{-\delta t} \leq \exp\left(-ct^{1 - \beta(k-1)}\right),$$

$$\frac{t\delta}{|B_{R+1}|} \leq O\left(t^{1 - \beta(2 + k - \varepsilon)}\right).$$

In order for the above to be summable, and to satisfy the requirements for the sequence $(\alpha_r)_r$, we require the conditions

$$\frac{2}{2 + k - \varepsilon} < \beta < \frac{1}{k-1} \qquad \text{and} \qquad k > 3.$$

If we take $k = 3 + \varepsilon$ and $\varepsilon < \frac{1}{2}$, this can be satisfied with $\beta = \frac{1}{2}$. $\qquad\square$

Exercise 9.34 Complete the details of the proof of Theorem 9.7.1. Show by induction: since μ_R, ν_R satisfy $\mu_R(G) \leq 1 - \delta$ and $\sup_{x \in G} \nu_R(x) \leq \frac{\delta}{|B_{R+1}|}$, that for all t and any $x \in G$ we may bound

$$\mathbb{P}[X_t = x] \leq (1 - \delta)^t + t \cdot \frac{\delta}{|B_{R+1}|},$$

where $\delta = \sum_{r > R} \alpha_r$.   ▷ solution ◁

9.8 Kleiner's Theorem

In light of the fact that $\dim \mathrm{BHF}(G, \mu) \in \{1, \infty\}$ (Theorem 2.7.1), we may naively think that the same holds for $\mathrm{HF}_k(G, \mu)$. In this section we will see that $\mathrm{HF}_k(G, \mu)$ can be finite dimensional (we have already seen in Corollary 9.3.6 that $\dim \mathrm{LHF}(G, \mu) > 1$). In fact Kleiner's theorem states that this is always the case when the growth of the group is polynomial.

Theorem 9.8.1 (Kleiner's theorem) *Let G be a finitely generated group of polynomial growth. Fix $k > 0$ and let $\mu \in \mathrm{SA}(G, 2(k + 1))$.*

Then the space $\mathrm{HF}_k(G, \mu)$ has finite dimension.

The proof requires two fundamental inequalities. The Saloff-Coste–Poincaré inequality and the reverse Poincaré inequality.

Let G be a finitely generated group with finite symmetric generating set S and let μ be a symmetric adapted probability measure on G. For a function $f : G \to \mathbb{C}$, define

$$A_r f := \frac{1}{|B(1, r)|} \sum_{x \in B(1, r)} f(x),$$

$$\|\nabla f\|_{\mu, r}^2 := \sum_{x, y \in B(1, r)} \mu\left(x^{-1} y\right) |f(x) - f(y)|^2.$$

Proposition 9.8.2 (Saloff-Coste–Poincaré inequality) *Let G be a finitely generated group, S a finite symmetric generating set, and μ an adapted probability measure on G such that $S \subset \mathrm{supp}(\mu)$. Consider the Cayley graph of G with respect to S.*

Then there exists a constant $C_\mu > 0$ such that for all $f : G \to \mathbb{C}$ and all $r > 0$,

$$\sum_{x \in B(1, r)} |f(x) - A_r f|^2 \leq C_\mu r^2 \cdot \frac{|B(1, 2r)|^2}{|B(1, r)|^2} \cdot \|\nabla f\|_{\mu, 3r}^2.$$

Proof Set $\alpha^{-1} := \min_{s \in S} \mu(s)$. If $|y| \leq 2r$, then for any $s \in S$,

$$\sum_{|x| \leq r} |f(xys) - f(xy)|^2 \leq \sum_{|z| \leq 3r} \sum_{s \in S} |f(zs) - f(z)|^2 \leq \alpha \cdot \|\nabla f\|^2_{\mu, 3r}.$$

Since any $y \in B(1, 2r)$ can be written as $y = s_1 \cdots s_n$ for $n = |y| \leq 2r$ and $s_1, \ldots, s_n \in S$, using the notation $y_0 = 1$ and $y_j = s_1 \cdots s_j$, we have by Cauchy–Schwarz,

$$\sum_{|x| \leq r} |f(xy) - f(x)|^2 \leq \sum_{|x| \leq r} n \sum_{j=1}^{n} |f(xy_j) - f(xy_{j-1})|^2$$

$$\leq 2r \sum_{j=1}^{n} \sum_{|x| \leq r} |f(xy_{j-1}s_j) - f(y_{j-1})|^2 \leq 4r^2 \alpha \|\nabla f\|^2_{\mu, 3r}.$$

This holds for any $|y| \leq 2r$. Summing over all such y, and using another Cauchy–Schwarz,

$$\sum_{|x| \leq r} \left(\sum_{|y| \leq 2r} |f(xy) - f(y)| \right)^2 \leq |B(1, 2r)| \sum_{|x| \leq r} \sum_{|y| \leq 2r} |f(xy) - f(y)|^2$$

$$\leq |B(1, 2r)|^2 4r^2 \alpha \|\nabla f\|^2_{\mu, 3r}.$$

Fix $|x| \leq r$. Any $|z| \leq r$ can be written as xy for some $|y| \leq 2r$. Thus,

$$|f(x) - A_r f| \leq \frac{1}{|B(1, r)|} \sum_{|z| \leq r} |f(x) - f(z)| \leq \frac{1}{|B(1, r)|} \sum_{|y| \leq 2r} |f(xy) - f(x)|.$$

Summing the squares, we have,

$$\sum_{|x| \leq r} |f(x) - A_r f|^2 \leq \frac{1}{|B(1, r)|^2} \sum_{|x| \leq r} \left(\sum_{|y| \leq 2r} |f(xy) - f(x)| \right)^2$$

$$\leq \frac{|B(1, 2r)|^2}{|B(1, r)|^2} 4r^2 \alpha \|\nabla f\|^2_{\mu, 3r}. \qquad \square$$

We also have the "reverse" inequality, but only for harmonic functions. This is also sometimes known as the *Cacciopolli inequality*.

Proposition 9.8.3 (Reverse Poincaré inequality) *Let G be a finitely generated group. Let $\mu \in \mathsf{SA}(G, 2(k+1))$ for some $k > 0$.*

Then there exist constants $C_\mu > 0$ such that for any $f \in \mathsf{HF}_k(G, \mu)$ and all $r > 0$,

$$\|\nabla f\|^2_{\mu, r} \leq C_\mu r^{-2} \cdot \sum_{|x| \leq 3r} |f(x)|^2.$$

Proof Since $|\zeta|^2 = |\text{Re}\zeta|^2 + |\text{Im}\zeta|^2$ for all $\zeta \in \mathbb{C}$, we have that

$$||\nabla f||_{\mu,r}^2 = ||\nabla \text{Re} f||_{\mu,r}^2 + ||\nabla \text{Im} f||_{\mu,r}^2$$

and

$$\sum_{|x| \le 3r} |f(x)|^2 = \sum_{|x| \le 3r} |\text{Re} f(x)|^2 + \sum_{|x| \le 3r} |\text{Im} f(x)|^2.$$

So, without loss of generality, it suffices to bound $||\nabla f||_{\mu,r}^2$ for real-valued functions $f \colon G \to \mathbb{R}$.

By Hölder's inequality with $p = \frac{k+1}{k}$ and $q = k+1$, since $\mu \in \text{SA}(G, 2(k+1))$ and since $f \in \text{HF}_k(G, \mu)$, there are constants $C, C_\mu > 0$ such that for all $r > 0$, we have

$$\mathbb{E}\left[1 + |f(X_1)|^2 \mathbf{1}_{\{|X_1| > r\}}\right] \le C \left(\mathbb{E}\left[(1 + |X_1|)^{2kp}\right]\right)^{1/p} \cdot (\mathbb{P}[|X_1| > r])^{1/q}$$

$$\le C \mathbb{E}\left[(1 + |X_1|)^{2(k+1)}\right] \cdot r^{-2} \le C_\mu r^{-2}.$$

This implies that

$$\sum_{|z| > r} \mu(z) \le \sum_{|z| > r} \mu(z) \left(1 + |f(z)|^2\right) \le C_\mu r^{-2}.$$

Define

$$\varphi(x) = \begin{cases} 1 & \text{if } |x| \le r, \\ 2 - \frac{|x|}{r} & \text{if } r < |x| \le 2r, \\ 0 & \text{if } |x| > 2r. \end{cases}$$

Note that $\varphi \colon G \to [0, 1]$ and $\text{supp}(\varphi) \subset B(1, 2r)$. Also, $|\varphi(xs) - \varphi(x)| \le \frac{1}{2r}$ for any $s \in S, x \in G$. Thus, for any $x, y \in G$ we have that $|\varphi(x) - \varphi(y)| \le \frac{|x^{-1}y|}{2r}$.

To ease the notation, we write $\varphi(x) = \varphi_x$, $f(x) = f_x$, and $P(x, y) = \mu\left(x^{-1}y\right)$. Note that $P(x, y) = P(y, x)$ because μ is symmetric.

For a finite subset $F \subset G$, define

$$J_F = \sum_{x,y \in F} P(x, y) \varphi_y^2 (f_x - f_y)^2,$$

$$A_F = \sum_{x,y \in F} P(x, y) \left(f_x \varphi_x^2 - f_y \varphi_y^2\right)(f_x - f_y),$$

$$B_F = 2 \sum_{x,y \in F} P(x, y) f_x \varphi_y (\varphi_x - \varphi_y)(f_y - f_x),$$

$$C_F = \sum_{x,y \in F} P(x, y) f_x (\varphi_x - \varphi_y)^2 (f_y - f_x),$$

$$D_F = \sum_{x,y \in F} P(x, y) \left(f_x^2 + f_y^2\right)(\varphi_x - \varphi_y)^2.$$

The identity

$$\varphi_y^2(f_x - f_y) = \left(f_x\varphi_x^2 - f_y\varphi_y^2\right) - 2f_x\varphi_y(\varphi_x - \varphi_y) - f_x(\varphi_x - \varphi_y)^2$$

implies that $J_F = A_F + B_F + C_F$.

We have

$$\|\nabla f\|_{\mu,r}^2 = \sum_{x,y \in B(1,r)} P(x,y)\tfrac{1}{2}\left(\varphi_x^2 + \varphi_y^2\right)(f_x - f_y)^2 \le J_F,$$

for any $F \supset B(1,r)$.

We bound A_F by utilizing the inequality $2|ab| \le a^2 + b^2$. Using the harmonicity of f, and symmetry $P(x,y) = P(y,x)$,

$$A_F = -2\sum_{x \in F}\sum_{y \notin F} P(x,y)\varphi_x^2 f_x(f_x - f_y)$$

$$\le \sum_{x \in F}\sum_{y \notin F} P(x,y)\varphi_x^2 \cdot 2f_x f_y$$

$$\le \sum_{x \in F} \varphi_x^2|f_x|^2 \cdot \sum_{y \notin F} P(x,y) + \sum_{x \in F} \varphi_x^2 \cdot \sum_{y \notin F} P(x,y)|f_y|^2.$$

Note that $|x| \le 2r$ and $|y| > 3r$ implies that $\left|x^{-1}y\right| > r$. Since $\varphi_x = 0$ for $|x| > 2r$, we may choose $F = B(1,3r)$ to obtain

$$A_F \le \sum_{|x| \le 2r}\left(1 + |f_x|^2\right) \cdot \sum_{|z| > r} \mu(z)\left(1 + |f_z|^2\right) \le C_\mu r^{-2} \cdot \sum_{|x| \le 2r}\left(1 + |f(x)|^2\right).$$

To bound B_F, we use the inequality $|ab| \le \tfrac{1}{4}a^2 + b^2$, so

$$|f_x\varphi_y(\varphi_x - \varphi_y)(f_y - f_x)| \le \tfrac{1}{4}(f_x - f_y)^2\varphi_y^2 + f_x^2(\varphi_x - \varphi_y)^2,$$

which implies

$$B_F \le \tfrac{1}{2}\sum_{x,y \in F} P(x,y)\varphi_y^2(f_x - f_y)^2 + 2\sum_{x,y \in F} P(x,y)f_x^2(\varphi_x - \varphi_y)^2 = \tfrac{1}{2}J_F + D_F.$$

To bound C_F we use the inequality $|a(a-b)| \le \tfrac{3}{2}a^2 + \tfrac{1}{2}b^2 < \tfrac{3}{2}\left(a^2 + b^2\right)$. We have that

$$C_F \le \frac{3}{2}\sum_{x,y \in F} P(x,y)\left(f_x^2 + f_y^2\right)(\varphi_x - \varphi_y)^2 = \frac{3}{2}D_F.$$

We conclude that

$$J_F \le A_F + \tfrac{1}{2}J_F + \tfrac{5}{2}D_F,$$

implying that $J_F \le 2A_F + 5D_F$.

Finally, to bound D_F, since $|\varphi_x - \varphi_y| \leq \frac{|x^{-1}y|}{2r}$, we have that

$$D_F \leq \frac{1}{2r^2} \sum_{x,y \in F} P(x,y) \left|x^{-1}y\right|^2 f_x^2$$

$$\leq \frac{1}{2r^2} \sum_{x \in F} f_x^2 \cdot \sum_z \mu(z)|z|^2.$$

Also, since f is a nonconstant harmonic function, $f \notin \ell^2(G, \mu)$, and specifically, $\sum_x |f(x)|^2 = \infty$.

By adjusting the constant C_μ, and combining all the above, we have that for all $r > 0$,

$$J_F \leq 2A_F + 5D_F \leq C_\mu r^{-2} \cdot \sum_{|x| \leq 3r} |f(x)|^2,$$

which completes the proof. $\square$

9.8.1 Proof of Kleiner's Theorem

In this section we prove Kleiner's theorem (Theorem 9.8.1). We will show that when G has polynomial growth, there exists some constant $K = K(G, \mu)$ such that any finite-dimensional subspace $V \leq \mathsf{HF}_k(G, \mu)$ has dimension bounded by K. Thus, any collection of more than K vectors in $\mathsf{HF}_k(G, \mu)$ is linearly dependent, implying that $\dim \mathsf{HF}_k(G, \mu) \leq K$.

It will be very helpful to consider the following canonical finite-dimensional Hilbert spaces. For $r > 0$ and $f, g \colon G \to \mathbb{C}$ define the semi-inner product

$$Q_r(f,g) = \sum_{|x| \leq r} f(x)\overline{g(x)}.$$

Implicitly, these depend on some fixed Cayley graph of the group G.

If $B = (b_1, \ldots, b_d)$ is a basis for a finite-dimensional vector space $V \leq \mathbb{C}^G$, then Q_r may be represented as matrices: the matrix of Q_r in the basis B is defined to be $[Q_r]_B \in \mathsf{M}_d(\mathbb{C})$ with entries given by $([Q_r]_B)_{i,j} = Q_r(b_i, b_j)$. Different bases give different matrices; however, all matrices of the same form are related via the following exercise.

Exercise 9.35 Show that if $B = (b_1, \ldots, b_d)$ and $C = (c_1, \ldots, c_d)$ are bases for a vector space V, then there exists an invertible matrix $M = M(B, C) \in \mathsf{M}_d(\mathbb{C})$ such that $[Q_r]_C = M[Q_r]_B M^*$. $\triangleright$ solution $\triangleleft$

Exercise 9.36 Show that for any finite dimensional vector space $V \leq \mathbb{C}^G$, there exists $r_0 = r_0(V)$ such that for all $r \geq r_0$ the space V with the inner product structure given by Q_r is a Hilbert space. ▷ solution ◁

A bilinear form Q on a vector space V is called **positive definite** if $Q(v, v) > 0$ for any $0 \neq v \in V$.

Exercise 9.37 Show that when Q_r is positive definite on a finite-dimensional $V \leq \mathbb{C}^G$, for any basis B of V, the matrix $[Q]_B$ is invertible. ▷ solution ◁

In light of this, for two positive definite bilinear forms Q_r, Q_R, the determinant

$$\det(Q_R/Q_r) := \frac{\det[Q_R]_B}{\det[Q_r]_B}$$

is well defined and independent of the choice of basis.

Exercise 9.38 Show that if Q_r, Q_R are both positive definite on a finite-dimensional $V \leq \mathbb{C}^G$, then one may choose a basis $B = (v_1, \ldots, v_d)$ for V for which $[Q_r]_B$ is diagonal and $[Q_R]_B$ is the identity matrix. ▷ solution ◁

Exercise 9.39 Assume that $R \geq r > 0$ and that Q_r is positive definite on a finite-dimensional $V \leq \mathbb{C}^G$. Show that for any basis B of V we have that $\det[Q_R]_B \geq \det[Q_r]_B$. ▷ solution ◁

We now move to prove Kleiner's theorem (Theorem 9.8.1).

Let G be a finitely generated group of polynomial growth. Fix some $k > 0$. Let $\mu \in \mathrm{SA}(G, 2(k + 1))$. Let $V \leq \mathrm{HF}_k(G, \mu)$ be a finite-dimensional subspace of dimension $\dim V = d$.

Set $\Lambda = 21$ (this just needs to be some large enough number). We will show that $\dim V \leq \Lambda^n$ for some large enough $n = n(G, \mu)$.

To this end, we fix a basis B of V such that $|b(x)| \leq (|x| + 1)^k$ for any $x \in G$ and all $v \in B$ (using the fact that $V \leq \mathrm{HF}_k(G, \mu)$). Let $r_0 = r_0(V)$ be large enough so that Q_r is positive definite on V for all $r \geq r_0$.

It will be useful to consider the function

$$h(r) = |B(1, r)|(\det[Q_r]_B)^{1/d}.$$

Note that for any $v \in B$ we have that $Q_r(v, v) \leq |B(1, r)| \cdot (r + 1)^k$, so by Hadamard's inequality, $(\det[Q_r]_B)^{1/d} \leq |B(1, r)| \cdot (r + 1)^k$. Thus, $h(r) \leq |B(1, r)|^2 (r + 1)^k$.

Set $g(r) = \log h(\Lambda^r)$. Since G has polynomial growth, there exists $D > 0$ such that

$$\liminf_{r \to \infty}(g(r) - Dr \log \Lambda) = 0.$$

Remark 9.8.4 Note that this only assumes that

$$\liminf_{r \to \infty} r^{-m} |B(1, r)| < \infty,$$

for some $m > 0$, which is an a priori weaker assumption than polynomial growth.

Claim 9.8.5 (Step I) There exists an integer $n_0 = n_0(G, \mu, V)$ such that the following holds for all $n \geq n_0$. First, Q_{Λ^n} is positive definite on V for all $n \geq n_0$. Additionally, there exist $b > a \geq n_0$ such that:

- $b - a \in (n, 3n)$,
- $g(b + 1) - g(a) < n \cdot 4D \log \Lambda$,
- $g(a + 1) - g(a) < 4D \log \Lambda$ and $g(b + 1) - g(b) < 4D \log \Lambda$.

Proof By Exercise 9.36, we can choose n_0 so that Q_{Λ^n} is positive definite on V for all $n \geq n_0$.

Fix any $n \geq n_0$. Let ℓ be such that $g(n_0 + 3n\ell) - g(n_0) < 4n\ell \cdot D \log \Lambda$. (This can be done when, for example, $2n_0 < n\ell$ and $2|g(n_0)| < n\ell \cdot D \log \Lambda$.) This implies (by a telescoping sum) that there exists $n_0 \leq j \leq n_0 + 3n(\ell - 1)$ such that $g(j + 3n) - g(j) < 4n \cdot D \log \Lambda$.

Telescoping again,

$$4n \cdot D \log \Lambda > g(j + 3n) - g(j)$$
$$= g(j + 3n) - g(j + 2n) + g(j + 2n) - g(j + n) + g(j + n) - g(j),$$

we see that since g is nondecreasing, there must exist $b \in [j + 2n, j + 3n - 1]$ and $a \in [j, j + n - 1]$ such that $g(b + 1) - g(b)$ and $g(a + 1) - g(a)$ are both less than $4D \log \Lambda$. Also, $g(b + 1) - g(a) \leq g(j + 3n) - g(j) < 4n \cdot D \log \Lambda$, by our choice of j. This proves Step I.   □

Step II Now choose some fixed $n \geq n_0$ so that

$$(C_\mu)^2 \Lambda^{-2n + 12D} < \tfrac{1}{2} \Lambda^{-8D},$$

where C_μ is the maximum of the constants from the Poincaré and reverse Poincaré inequalities (Propositions 9.8.2 and 9.8.3). Using a, b from Step I, we now choose $r = \Lambda^a$ and $R = \Lambda^b$. Thus,

$$(C_\mu)^2 \frac{|B(1, 7r)|^3}{|B(1, r)|^3} (r/R)^2 \leq (C_\mu)^2 \exp(3(g(a + 1) - g(a))) \Lambda^{-2n} < \tfrac{1}{2} \Lambda^{-8D}, \tag{9.4}$$

$$\frac{|B(1, \Lambda R)|}{|B(1, R)|} \leq e^{(g(b+1) - g(b))} < \Lambda^{4D},$$

$$\frac{|B(1, 2R)|}{|B(1, r)|} \leq e^{(g(b+1) - g(a))} < \Lambda^{4Dn}.$$

Choose a basis $\beta = (v_1, \ldots, v_d)$ for V so that $[Q_R]_\beta = I$ and $[Q_{\Lambda R}]_\beta$ is a diagonal matrix (that is, β orthonormal with respect to Q_R and orthogonal with respect to $Q_{\Lambda R}$).

Note that if $u = \sum_{j=1}^{d} \alpha_j v_j$, then

$$Q_{\Lambda R}(u, u) = \sum_{j=1}^{d} |\alpha_j|^2 Q_{\Lambda R}(v_j, v_j),$$

$$Q_R(u, u) = \sum_{j=1}^{d} |\alpha_j|^2 Q_R(v_j, v_j) = \sum_{j=1}^{d} |\alpha_j|^2.$$

Let ℓ be the number

$$\ell := \#\{1 \le j \le d : Q_{\Lambda R}(v_j, v_j) > \Lambda^{8D} \cdot Q_R(v_j, v_j)\}.$$

Then, since $\det(Q_{\Lambda R}/Q_R)$ is independent of the choice of basis,

$$\Lambda^{4D} \ge \exp(g(b+1) - g(b)) = \frac{h(\Lambda R)}{h(R)} \ge \det(Q_{\Lambda R}/Q_R)^{1/d}$$

$$= \left(\prod_{j=1}^{d} Q_{\Lambda R}(v_j, v_j) \right)^{\frac{1}{d}} > \left(\Lambda^{8D} \right)^{\frac{\ell}{d}}.$$

Thus $\ell < d/2$. We conclude that by taking

$$U_R = \text{span}\{v_j \mid Q_{\Lambda R}(v_j, v_j) \le \Lambda^{8D} \cdot Q_R(v_j, v_j)\},$$

then $U_R \le V$ has dimension $\dim U_R \ge \frac{1}{2} \dim V$, and also for any $u \in U_R$ we have

$$Q_{\Lambda R}(u, u) \le \Lambda^{8D} \cdot Q_R(u, u). \tag{9.5}$$

Exercise 9.40 (Step III) Fix $R > r > 0$. Show that one can find $m = m(r, R) \le \frac{|B(1, 2R)|}{|B(1, r)|}$ and points $x_1, \ldots, x_m \in B(1, R)$ such that

- $B(1, R) \subset \bigcup_{j=1}^{m} B(x_j, 2r)$ (the small $2r$-balls cover the big R-ball).
- The overlap is not too big: for any $x \in B(1, R)$, the number of $6r$-balls containing x, can be bounded:

$$\#\left\{1 \le j \le m : x \in B(x_j, 6r)\right\} \le \frac{|B(1, 7r)|}{|B(1, r)|}. \qquad \text{▷ solution ◁}$$

Step IV Now we combine the Poincaré and reverse Poincaré inequalities (Propositions 9.8.2 and 9.8.3).

Recall $m = m(r, R)$ from Step III, the number of balls in the covering of $B(1, R)$ by radius $2r$ balls. Define $\Psi \colon V \to \mathbb{C}^m$ by $\Psi(v) = (A_{x_1, 2r} v, \ldots, A_{x_m, 2r} v)$, where $A_{x,r}v = \frac{1}{|B(1,r)|}\sum_{y\in B(x,r)} v(y)$. Note that Ψ is a linear map, so $\dim \operatorname{Ker}\Psi = d - \dim \operatorname{Im}\Psi \geq d - m$. Let $K = \operatorname{Ker}\Psi$.

Because $B(1,R) \subset \bigcup_{j=1}^{m} B(x_j, 2r)$, instead of summing over $x \in B(1,R)$ we may sum over $j = 1, \ldots, m$ and then over $x \in B(x_j, 2r)$. Because of the properties of the overlaps in the covering,

$$
\sum_{j=1}^{m} \sum_{x,y\in B(x_j,6r)} \mu\left(x^{-1}y\right)|u(x) - u(y)|^2
$$

$$
= \sum_{x,y\in B(1,R+6r)} \mu\left(x^{-1}y\right)|u(x)-u(y)|^2 \cdot \#\{1 \leq j \leq m : x,y \in B(x_j,6r)\}
$$

$$
\leq \frac{|B(1,7r)|}{|B(1,r)|} \cdot \|\nabla u\|_{\mu,7R}^2.
$$

Let $u \in K$. So $A_{x_j,2r}u = 0$ for all $1 \leq j \leq m$. Specifically, by combining the Poincaré and reverse Poincaré inequalities (Propositions 9.8.2 and 9.8.3), and using (9.4),

$$
Q_R(u,u) \leq \sum_{j=1}^{m} \sum_{x\in B(x_j,2r)} (u(x) - A_{x_j,2r}u)^2
$$

$$
\leq \sum_{j=1}^{m} C_\mu 4r^2 \frac{|B(1,4r)|^2}{|B(1,2r)|^2} \sum_{x,y\in B(x_j,6r)} \mu\left(x^{-1}y\right)|u(x)-u(y)|^2
$$

$$
\leq \frac{|B(1,7r)|}{|B(1,r)|} \cdot 4C_\mu r^2 \frac{|B(1,4r)|^2}{|B(1,2r)|^2} \cdot \|\nabla u\|_{\mu,7R}^2
$$

$$
\leq (C_\mu)^2 \frac{|B(1,7r)|^3}{|B(1,r)|^3} \cdot (r/R)^2 Q_{\Lambda R}(u,u) \leq \tfrac{1}{2}\Lambda^{-8D} \cdot Q_{\Lambda R}(u,u),
$$

by our choices of r, R.

By (9.5), if $u \in U_R \cap K$ then we get

$$
Q_R(u,u) \leq \tfrac{1}{2}\Lambda^{-8D}Q_{\Lambda R}(u,u) \leq \tfrac{1}{2} \cdot Q_R(u,u).
$$

Thus $\Psi|_{U_R}$ is injective, which gives

$$
\dim V \leq 2\dim U_R \leq 2m \leq 2\frac{|B(1,2R)|}{|B(1,r)|} < 2\Lambda^{4Dn}.
$$

Since this bound depends only on G, μ, this completes the proof of Kleiner's theorem. $\square$

9.9 Additional Exercises

Exercise 9.41 Let G be a finitely generated group, and let $\mu \in \mathrm{SA}(G, 1)$.
Define

$$K = \{x \in G : \forall\, h \in \mathrm{LHF}(G, \mu),\ x.h - h \text{ is constant}\}.$$

Show that $K \lhd G$.

Show that for any $h \in \mathrm{LHF}(G, \mu)$, the function $f = h|_K - h(1)$ is a homomorphism from K into $\mathbb{C}$.

Show that if $h \in \mathrm{LHF}(G, \mu)$ satisfies $h|_K \equiv 0$ then $\bar{h}(Kx) = h(x)$ is a well-defined function on G/K, and that $\bar{h} \in \mathrm{LHF}(G/K, \bar{\mu})$, where $\bar{\mu}(Kx) = \mu(Kx) = \sum_{y \in K} \mu(yx)$ is the projection of μ onto G/K.

Conclude that

$$\dim \mathrm{LHF}(G, \mu) \le \dim \mathrm{Hom}(K, \mathbb{C}) + 1 + \dim \mathrm{LHF}(G/K, \bar{\mu}). \quad \triangleright \text{solution} \triangleleft$$

Exercise 9.42 Let G be a finitely generated group, and let $\mu \in \mathrm{SA}(G, 1)$ be such that (G, μ) is Liouville. Define

$$K = \{x \in G : \forall\, h \in \mathrm{LHF}(G, \mu),\ x.h - h \text{ is constant}\}.$$

Show that $Z_1(G) \lhd K$ (recall that $Z_1(G)$ is the center of G).

Similarly to Exercise 9.41, conclude that

$$\dim \mathrm{LHF}(G, \mu) \le \dim \mathrm{Hom}(Z_1(G), \mathbb{C}) + 1 + \dim \mathrm{LHF}(G/Z_1(G), \bar{\mu}),$$

where $\bar{\mu}(Z_1(G)x) = \mu(Z_1(G)x)$ is the projection of μ onto $G/Z_1(G)$. $\quad \triangleright \text{solution} \triangleleft$

Exercise 9.43 Show that if G is a finitely generated group and U is a finite generating set for G, then $\dim \mathrm{Hom}(G, \mathbb{C}) \le |U|$. $\quad \triangleright \text{solution} \triangleleft$

Exercise 9.44 Let G be a finitely generated nilpotent group, and let $\mu \in \mathrm{SA}(G, 1)$. Show (without using Kleiner's theorem, Theorem 9.8.1) that $\dim \mathrm{LHF}(G, \mu) < \infty$. $\quad \triangleright \text{solution} \triangleleft$

9.10 Remarks

Milnor was a major initiator of the study of growth of finitely generated groups. The Milnor–Wolf theorem (Theorem 8.2.4), published in Milnor (1968a) and Wolf (1968), was a special case of Gromov's theorem for the class of solvable groups. In the same year, Milnor (1968b) posed the following two questions:

(1) Do there exist groups of intermediate growth?
(2) What is the (algebraic) characterization of the groups of polynomial growth?

The first question was answered by Grigorchuk (1980, 1984), in which he constructed an intermediate growth group, now known as the *Grigorchuk group*. The second question was answered by Gromov (1981). This is Gromov's theorem (Theorem 9.0.1), stating that a finitely generated group has polynomial growth if and only if it is virtually nilpotent.

Kleiner (2010) gave a new proof of Gromov's theorem by showing that the space of Lipschitz harmonic functions has finite dimension (Theorem 9.8.1), and using the action on that space to provide a representation of the group as a linear group. The proof presented is from Kleiner (2010). The Poincaré inequality, Proposition 9.8.2, is attributed in Kleiner (2010) to Saloff-Coste. It is similar to an inequality appearing in Coulhon and Saloff-Coste (1993).

Ozawa (2018) provided yet another proof for Gromov's theorem; it is basically the one presented here.

Both Kleiner and Ozawa's proofs rely on the existence of harmonic coycles (Theorem 9.3.2). This follows from results of Mok (1995) or Korevaari and Schoen (1997). Kleiner provides a proof using *property (T)*. Ozawa provides a very short proof using the spectral theorem. The proof of Theorem 9.3.2 we present here is due to Lee and Peres (2013). This proof has the advantage of avoiding the use of *ultrafilters* and *ultraproducts*.

Theorem 9.4.1, stating that random walks on groups are never sub-diffusive, is due to Erschler, and appears in Lee and Peres (2013).

The proof of the von Neumann ergodic theorem (Theorem 9.5.2) presented is due to Riesz, and we have taken it from Tao's blog *What's New*.

Shalom and Tao (2010) gave a quantitative proof of Gromov's theorem, based on Kleiner's proof, which also shows that there is some $\varepsilon > 0$ such that any finitely generated group of growth $\leq r \mapsto r^{(\log\log r)^{\varepsilon}}$ is actually of polynomial growth. Thus, there is a "gap" for growth functions, where no groups exist. It is actually conjectured by Grigorchuk (1990) that there is a much more serious "gap."

Conjecture 9.10.1 (Grigorchuk's gap conjecture) Let G be a finitely generated group of growth $\leq r \mapsto \exp(r^{\alpha})$ for some $\alpha < \frac{1}{2}$. Then G is virtually nilpotent.

At the time of writing this remains a major open problem in the field.

As mentioned, Kleiner (2010) proved that polynomial growth implies that the space $\mathsf{HF}_k(G, \mu)$ is finite dimensional (and used this to prove Gromov's theorem). It was well known that $\dim \mathsf{LHF}(G, \mu) < \infty$ implies that G is virtually indicable (this is part of a broader phenomenon of *harmonic polynomials*, see

Meyerovitch et al., 2017). Together with the reduction in Proposition 9.1.1, this provides a proof for Gromov's theorem. However, polynomial growth is required for Proposition 9.1.1. So, a priori, it may be that some non-polynomial growth groups also have finite-dimensional spaces $\mathsf{HF}_k(G, \mu)$. It is conjectured in Meyerovitch and Yadin (2016) that this is not the case.

Conjecture 9.10.2 Let G be a finitely generated group and $\mu \in \mathsf{SA}(G, \infty)$. If $\dim \mathsf{LHF}(G, \mu) < \infty$ then G is virtually nilpotent.

9.11 Solutions to Exercises

Solution to Exercise 9.1 :(
The polarization identity tells us that

$$4\langle u, v \rangle = ||u + v||^2 - ||u - v||^2 + i||u + iv||^2 - i||u - iv||^2.$$

So (1) implies (2).
  (2) implies (1) by choosing $u = v$.
  If $a^* a = 1$, then $\langle av, av \rangle = \langle v, a^* av \rangle = \langle v, v \rangle$, so that (3) implies (1).
  Finally,

$$||a^* av - v||^2 = ||a^* av||^2 + ||v||^2 - 2\mathrm{Re}\,\langle a^* av, v \rangle = ||a^* av||^2 - ||v||^2.$$

If (1) holds, then since (2) also holds, for any $v \in \mathbb{H}$ we have that

$$||a^* av||^2 = \langle av, aa^* av \rangle = \langle v, a^* av \rangle = ||av||^2 = ||v||^2,$$

implying that (3) holds as well. :) ✓

Solution to Exercise 9.2 :(
For any $v \in \mathbb{H}$ we have $||xav|| = ||av||$ so $||xa||_{\mathrm{op}} = ||a||_{\mathrm{op}}$. Also, since $\left\| x^{-1}v \right\| = \left\| xx^{-1}v \right\| = ||v||$ for all $v \in V$, we have that

$$||a||_{\mathrm{op}} = \sup\left\{ ||av|| : \left\| x^{-1}v \right\| = 1 \right\} = \sup\{||axv|| : ||v|| = 1\} = ||ax||_{\mathrm{op}}. \qquad :) ✓$$

Solution to Exercise 9.3 :(
If $xv = \lambda v$ for $v \in \mathbb{C}^d$ with $||v|| = 1$, then because x is unitary

$$|\lambda| = ||\lambda v|| = ||xv|| = ||v|| = 1.$$

Let λ be the eigenvalue satisfying $\max_j |\lambda_j - 1| = |\lambda - 1|$. Let $v \in \mathbb{C}^d$ be the corresponding unit eigenvector, that is, $||v|| = 1$ and $xv = \lambda v$. Then $||(x - 1)v|| = |\lambda - 1|$.
  Moreover, if $v_1, \ldots, v_d$ is the corresponding orthonormal basis of eigenvectors corresponding to $\lambda_1, \ldots, \lambda_d$, then for any $u \neq 0$,

$$||(x - 1)u||^2 = \sum_{j=1}^{d} \left|\langle u, v_j \rangle\right|^2 \cdot ||(x - 1)v_j||^2 = \sum_{j=1}^{d} \left|\langle u, v_j \rangle\right|^2 \cdot |\lambda_j - 1|^2$$

$$\leq |\lambda - 1|^2 \cdot \sum_{j=1}^{d} \left|\langle u, v_j \rangle\right|^2 = |\lambda - 1|^2 \cdot ||u||^2,$$

by Parseval's identity. Thus, $||x - 1||_{\mathrm{op}} = |\lambda - 1|$. :) ✓

Solution to Exercise 9.4 :(

Just write

$$||x^n - 1||_{\text{op}} = \left\|x^n - x^{n-1} + x^{n-1} - 1\right\|_{\text{op}} \leq ||x^n(x-1)||_{\text{op}} + \left\|x^{n-1} - 1\right\|_{\text{op}}$$

$$= ||x - 1||_{\text{op}} + \left\|x^{n-1} - 1\right\|_{\text{op}} \leq \cdots \leq n||x - 1||_{\text{op}}.$$:)✓

Solution to Exercise 9.5 :(

On a finite-dimensional vector space, all norms are equivalent, and hence induce the same topology. Since $U_d(\mathbb{C})$ is a closed and bounded subset of the finite-dimensional normed vector space $\mathbb{C}^{d^2}$, it is compact by Heine–Borel.

For $x \in U_d(\mathbb{C})$ denote $B(x, \varepsilon) = \{y \in U_d(\mathbb{C}) : ||x - y||_{\text{op}} < \varepsilon\}$. The collection $(B(x, \varepsilon/2))_{x \in U_d(\mathbb{C})}$ is an open cover of $U_d(\mathbb{C})$, so there exists a finite sub-cover, say $U_d(\mathbb{C}) \subset \bigcup_{j=1}^{k} B(x_j, \varepsilon/2)$, for some $x_1, \ldots, x_k \in U_d(\mathbb{C})$ and some $k = k(\varepsilon/2)$. Thus, for any $y \in U_d(\mathbb{C})$ there exists $1 \leq j \leq k$ such that $||x_j - y||_{\text{op}} < \varepsilon/2$.

Let $G_\varepsilon \leq G \leq U_d(\mathbb{C})$ be as in the proof of Lemma 9.2.2. Let $T \subset G$ be a set of representatives for G_ε; that is, $G = \biguplus_{t \in T} G_\varepsilon t$. We would like to show that T is a finite set.

For any $t \in T$ we have that there exists $1 \leq j \leq k$ for which $||t - x_j||_{\text{op}} < \varepsilon/2$. We saw in the proof of Lemma 9.2.2 that for any $x, y \in G$, if $||x - y||_{\text{op}} \leq \varepsilon$ then $x^{-1}y \in G_\varepsilon$. If $t \neq s \in T$, then since these are representatives of different cosets of G_ε, it is impossible that $||t - s||_{\text{op}} \leq \varepsilon$. So the map from T to $\{x_1, \ldots, x_k\}$ mapping $t \mapsto x_j$ for which $||t - x_j||_{\text{op}} < \varepsilon/2$ is a well-defined injective map. Hence $|T| \leq k$.

:)✓

Solution to Exercise 9.6 :(

If $(N_1 x, \ldots, N_m x) = (N_1 y, \ldots, N_m y)$, then $x^{-1}y \in N_1 \cap \cdots \cap N_m = \{1\}$, so $x = y$. :)✓

Solution to Exercise 9.7 :(

The first assertion follows directly from

$$[G_1 \times G_2 : H_1 \times H_2] \leq [G_1 : H_1] \cdot [G_2 : H_2].$$

For the second assertion, assume that $H \leq G$ with $[G : H] < \infty$ and H Abelian. By possibly passing to a subgroup of H, using Theorem 8.4.2, we may assume that $H \triangleleft G$.

Then, for any $N \leq G$, we have that $[N : N \cap H] = [HN : H] \leq [G : H] < \infty$, with $N \cap H$ an Abelian group. :)✓

Solution to Exercise 9.8 :(

This follows from $||c(xy)|| = ||c(x) + x.c(y)|| \leq ||c(x)|| + ||c(y)||$. :)✓

Solution to Exercise 9.9 :(

Note that $c(1) = c(1 \cdot 1) = c(1) + 1.c(1) = 2c(1)$, implying that $c(1) = 0$.

Also, $0 = c(1) = c\left(x^{-1}x\right) = c\left(x^{-1}\right) + x^{-1}.c(x)$. :)✓

Solution to Exercise 9.10 :(

Note that if S is a finite symmetric generating set for G, setting $\kappa = \max_{s \in S} ||c(s)||$ we have that

$$\sum_y \mu(y) ||c(xy)|| \leq ||c(x)|| + \kappa \cdot \sum_y \mu(y)|y| < \infty,$$

so the corresponding series in $\mathbb{H}$ converges absolutely when μ has finite first moment.

If c is μ-harmonic, then $\sum_y \mu(y)c(y) = c(1) = 0$.

On the other hand, if $\sum_y \mu(y)c(y) = 0$ then for any $x \in G$,

$$\sum_y \mu(y)c(xy) = c(x) + x. \sum_y \mu(y)c(y) = c(x).$$:)✓

Solution to Exercise 9.11 :(

This is just $xy.v - v = x.v - v + x.(y.v - v)$. :)✓

Solution to Exercise 9.13 :(

If $f \in \ell^0(G)$ we have $f = \sum_{x \in F} f(x)\delta_x$ where $F = \text{supp}(f)$, we have that $\varphi(x) = \sum_y g(x, y)f(y)$ is a finite sum.

Also,

$$\Delta\varphi = (I - P) \sum_{k=0}^{\infty} P^k f = \lim_{n\to\infty} (I - P^n)f.$$

Since $P^n(x, y) \to 0$ as $n \to \infty$ (Exercise 5.5.2), and since f has finite support, we have that $P^n f \to 0$. Thus $\Delta\varphi = f$.

Finally, if $f \geq 0$, then $\langle P^k f, f \rangle \geq 0$ for all k, implying that

$$\langle \Delta\varphi, \varphi \rangle = \sum_{k=0}^{\infty} \langle f, P^k f \rangle \geq \sum_{k=0}^{n} \langle P^k f, f \rangle. \qquad \qquad :)\checkmark$$

Solution to Exercise 9.14 :(

Exercise 9.13 shows that $\Delta^{-1} = g$ satisfies the required properties, and we only need to show that for $f \in \ell^0(G)$ we have that $gf \in \ell^2(G, \mu)$.

Indeed, for any $y \in G$, we have that $(g\delta_y)(x) = g(x, y) = g\left(1, x^{-1}y\right)$. The bubble condition implies that $g\delta_y \in \ell^2(G, \mu)$ for all $y \in G$. Thus, for $f \in \ell^0(G)$ with $F = \text{supp}(f)$, we have $gf = \sum_{y\in F} f(y) \cdot g\delta_y \in \ell^2(G, \mu)$. $\qquad :)\checkmark$

Solution to Exercise 9.15 :(

Fix $x \in G$ and let $U = U(x)$. We use the notation from the proof of Lemma 9.3.5.

Let ℓ be such that $x = x_\ell$. Define U on all of $\text{span}(c(G))$ by

$$U \sum_{j=1}^{k} \alpha_j c(x_j) := \sum_{j=1}^{k} \alpha_j (c(xx_j) - c(x)).$$

Note that if

$$\sum_{j=1}^{k} \alpha_j c(x_j) = \sum_{j=1}^{k} \beta_j c(x_j),$$

then for any $m \geq \ell \vee k$,

$$\lim_{J_m \ni n\to\infty} \sum_{j=1}^{k} (\alpha_j - \beta_j)\tilde{c}_n(x_j) = 0.$$

Since U_n, x are unitary operators, one verifies that for $m \geq \ell \vee k$,

$$\left\| U \sum_{j=1}^{k} \alpha_j c(x_j) - U \sum_{j=1}^{k} \beta_j c(x_j) \right\| = \left\| \sum_{j=1}^{k} \alpha_j (c(xx_j) - c(x)) - \sum_{j=1}^{k} \beta_j (c(xx_j) - c(x)) \right\|$$

$$= \lim_{J_m \ni n\to\infty} \left\| \sum_{j=1}^{k} (\alpha_j - \beta_j) U_n x c_n(x_j) \right\|$$

$$= \lim_{J_m \ni n\to\infty} \left\| \sum_{j=1}^{k} (\alpha_j - \beta_j) \tilde{c}_n(x_j) \right\| = 0.$$

So U is a well-defined linear operator on $\text{span}(c(G))$.

It is very easy to show that if $v_n \to v$ in $\mathbb{H}$ and $v_n \in \text{span}(c(G))$ for all n, then defining $Uv = \lim_{n\to\infty} Uv_n$ extends U to a well-defined linear operator on all of V.

For a general $w \in \mathbb{H}$, writing $w = v + u$ for $v \in V$ and $u \in V^\perp$, we just define $Uw = Uv + u$, and this provides a linear operator on all of $\mathbb{H}$ that is the identity restricted to $V^\perp$.

The same argument as above shows that if U' is another bounded linear operator with $U'c(y) = c(xy) - c(x)$ for all $y \in G$, and $U'v = v$ for all $v \in V^\perp$, then $U' = U$.

Finally, to show that $U(x) \in \mathbb{U}(\mathbb{H})$ it suffices to show that $U(x)^* = U(x)^{-1}$.

We first claim that $U(x)^* = U\left(x^{-1}\right)$. Indeed, let $w \in H$ and write $w = v + u$ for $v \in V$ and $u \in V^\perp$. For any $y, z \in G$ we have that for some $k = k(x, y, z) \geq 1$ large enough,

$$\left\langle U\left(x^{-1}\right) c(y), c(z) \right\rangle = \left\langle c\left(x^{-1}y\right) - c\left(x^{-1}\right), c(z) \right\rangle = \lim_{J_k \ni n \to \infty} \left\langle U_n x^{-1} c_n(y), U_n c_n(z) \right\rangle$$

$$= \lim_{J_k \ni n \to \infty} \left\langle c_n(y), x c_n(z) \right\rangle = \lim_{J_k \ni n \to \infty} \left\langle U_n c_n(y), U_n x c_n(z) \right\rangle$$

$$= \left\langle c(y), c(xz) - c(x) \right\rangle = \left\langle c(y), U(x)c(z) \right\rangle.$$

Since $c(G)$ spans a dense subspace of V, and since $U\left(x^{-1}\right), U(x)$ are the identity restricted to $V^\perp$, this implies that $U\left(x^{-1}\right) = U(x)^*$ as claimed.

Finally, for any $y \in G$ we have that

$$U(x)^* U(x) c(y) = U\left(x^{-1}\right) \left(c(xy) - c(x)\right) = c(y) - c\left(x^{-1}\right) - c(1) + c\left(x^{-1}\right) = c(y),$$

which is sufficient to prove that $U(x)^* = U(x)^{-1}$.   :) ✓

Solution to Exercise 9.16   :(
Compute

$$||c(x)||^2 = \sum_y \left| \check{f}(x^{-1}y^{-1}) - \check{f}(y^{-1}) \right|^2 = \sum_y |f(yx) - f(y)|^2,$$

so

$$\mathbb{E} \, ||c(X_1)||^2 = \sum_{y,x} \mu(x)(f(yx) - f(y))^2 = 2 \langle \Delta f, f \rangle.$$

Also,

$$\left\| \sum_x \mu(x) c(x) \right\|^2 = \sum_y \left| \sum_x \mu(x) \left(\check{f}(x^{-1}y^{-1}) - \check{f}(y^{-1}) \right) \right|^2 = \sum_y |\Delta f(y)|^2 = ||\Delta f||^2.$$:) ✓

Solution to Exercise 9.17   :(
Let $A_n v = \frac{1}{n} \sum_{k=0}^{n-1} T^k v$. Then V is the set of $v \in \mathbb{H}$ such that $||A_n v - \pi v|| \to 0$. Since A_n, π are linear operators, V is easily shown to be a subspace by the triangle inequality.

Note that for any $w \in \mathbb{H}$ we have

$$||A_n w - \pi w|| \leq \frac{1}{n} \sum_{k=0}^{n-1} \left\| T^k w - \pi w \right\| \leq 2||w||.$$

(We have used that $||Tw|| \leq ||w||$ and that $||\pi w|| \leq ||w||$.)

So, if $(v_n)_n \subset V$ is a converging sequence $\lim v_n = v$, then for any $\varepsilon > 0$, let K be large enough so that $||v_K - v|| < \frac{\varepsilon}{4}$. Let N be large enough so that $||A_n v_K - \pi v_K|| < \frac{\varepsilon}{2}$ for all $n \geq N$. We then have that for all $n \geq N$,

$$||A_n v - \pi v|| \leq ||A_n v_K - \pi v_K|| + ||A_n(v - v_K) - \pi(v - v_K)|| < \frac{\varepsilon}{2} + 2||v - v_K|| < \varepsilon.$$

Thus, $||A_n v - \pi v|| \to 0$ as $n \to \infty$, implying that $v \in V$. Hence V is closed.

The other two assertions are very easy.   :) ✓

Solution to Exercise 9.18   :(
For any $b \in B$ we have

$$\sum_{b' \in B} |\langle Tb, b' \rangle|^2 = ||Tb||^2,$$

making the last identity immediate (by Tonelli's theorem, Exercise A.13).

Now, using this and Tonelli's theorem repeatedly,

$$\sum_{b\in B}||Tb||^2 = \sum_{b\in B}\sum_{a\in\tilde{B}}|\langle Tb,a\rangle|^2 = \sum_{a\in\tilde{B}}||T^*a||^2 = \sum_{a,a'\in\tilde{B}}|\langle T^*a,a'\rangle|^2 = \sum_{a'\in\tilde{B}}||Ta'||^2.$$

Also, if $v\in\mathbb{H}$ is such that $||v||=1$, then v is contained in some orthonormal basis for $\mathbb{H}$. Hence, $||Tv||^2 \le ||T||_{\text{HS}}^2$. This holds for all unit-length v, so $||T||_{\text{op}} \le ||T||_{\text{HS}}$. :) ✓

Solution to Exercise 9.19 :(

If $||A||_{\text{HS}} < \infty$ and $||A'||_{\text{HS}} < \infty$, then by Cauchy–Schwarz,

$$\sum_{b\in B}|\langle Ab, A'b\rangle| \le \sum_{b\in B}||Ab||\cdot||A'b|| \le \left(||A||_{\text{HS}}\cdot||A'||_{\text{HS}}\right)^{1/2},$$

so $\langle A, A'\rangle_{\text{HS}}$ converges absolutely. Thus, linearity of the Lebesgue integral will provide the properties of an inner product.

Also, if $\tilde{B}$ is another orthonormal basis for $\mathbb{H}$, then since $\sum_{\tilde{b}\in\tilde{B}}\langle A\tilde{b}, A'\tilde{b}\rangle$ is absolutely convergent, the set $\left\{\tilde{b}\in\tilde{B}:\langle A\tilde{b}, A'\tilde{b}\rangle\ne 0\right\}$ is countable. Also, since

$$||A||_{\text{HS}} = \sum_{\tilde{b}\in\tilde{B}}||A\tilde{b}||^2 < \infty,$$

and for any $\tilde{b}\in\tilde{B}$,

$$||A\tilde{b}||^2 = \sum_{b\in B}|\langle A\tilde{b}, b\rangle|^2 < \infty,$$

we may find countable subsets $\tilde{C}\subset\tilde{B}$ and $C\subset B$ such that

$$\sum_{\tilde{b}\in\tilde{B}}\langle A\tilde{b}, A'\tilde{b}\rangle = \sum_{\tilde{b}\in\tilde{C}}\sum_{b\in C}\langle A\tilde{b}, b\rangle\langle b, A'\tilde{b}\rangle$$

$$= \sum_{b\in C}\sum_{\tilde{b}\in\tilde{C}}\langle A\tilde{b}, b\rangle\langle b, A'\tilde{b}\rangle = \sum_{b\in B}\langle (A')^*b, A^*b\rangle$$

$$= \sum_{b,b'\in C}\langle (A')^*b, b'\rangle\langle b', A^*b\rangle = \sum_{b'\in B}\langle Ab', A'b'\rangle.$$

So $\langle A, A'\rangle_{\text{HS}}$ does not depend on the specific choice of orthonormal basis.

To show that $\mathbb{H}^*\otimes\mathbb{H}$ is a Hilbert space, we need to show that it is complete (with the Hilbert–Schmidt norm). The proof of this fact is basically the same as the proof that ℓ^2 is a complete space. Let $(A_n)_n$ be a Cauchy sequence in $\mathbb{H}^*\otimes\mathbb{H}$. Let B be an orthonormal basis of $\mathbb{H}$.

Let $v\in\mathbb{H}$. Note that

$$||A_{n+m}v - A_nv||^2 \le ||A_{n+m} - A_n||_{\text{HS}}^2\cdot||v||^2,$$

so $(A_nv)_n$ is a Cauchy sequence in $\mathbb{H}$. Thus the limit $Av := \lim_n A_nv$ exists in $\mathbb{H}$. This also immediately shows that A is a linear map.

Since

$$|||A_{n+m}||_{\text{HS}} - ||A_n||_{\text{HS}}| \le ||A_{n+m} - A_n||_{\text{HS}},$$

we get that $\lim_n ||A_n||_{\text{HS}}$ exists, and specifically, $M := \sup_n ||A_n||_{\text{HS}} < \infty$.

If $F\subset B$ is a finite subset then

$$\sum_{b\in F}||Ab||^2 \le 2\sum_{b\in F}||A_nb||^2 + 2\sum_{b\in F}||(A-A_n)b||^2$$

$$\le 2\sup_n ||A_n||_{\text{HS}}^2 + 2|F|\cdot||v||^2\cdot||A-A_n||_{\text{HS}}^2.$$

Taking $n\to\infty$ and the supremum over F we get that $||A||_{\text{HS}}^2 \le 2M^2 < \infty$. So $A\in\mathbb{H}^*\otimes\mathbb{H}$.

Now, since $||A_n||_{\text{HS}} < \infty$ and $||A||_{\text{HS}} < \infty$, there exists a countable subset $C\subset B$ such that for all $b\in B\setminus C$ we have $||A_nb|| = ||Ab|| = 0$ for all n. Write $C = \{c_1, c_2, \ldots\}$.

Let $\varepsilon > 0$. Let n_0 be large enough so that for all $n, m \geq n_0$, we have $||A_n - A_m||_{HS}^2 < \varepsilon$.

Fix some $r > 0$. Let $n_1 \geq n_0$ be large enough so that for any $m \geq n_1$, we have $||(A - A_m)c_j||^2 < \frac{\varepsilon}{r}$ for all $1 \leq j \leq r$.

We then have that for all $n \geq n_0$ and all $m \geq n_1$,

$$\sum_{j=1}^{r} ||(A - A_n)c_j||^2 \leq 2 \sum_{j=1}^{r} ||(A - A_m)c_j||^2 + 2 \sum_{j=1}^{r} ||(A_n - A_m)c_j||^2 < 2\varepsilon + 2||A_n - A_m||_{HS}^2 < 4\varepsilon.$$

Taking $r \to \infty$, we have that $||A - A_n||_{HS}^2 < 4\varepsilon$ for all $n \geq n_0$, implying that $A_n \to A$ in $\mathbb{H}^* \otimes \mathbb{H}$. :) ✓

Solution to Exercise 9.20 :(
Note that if B is an orthonormal basis then

$$\langle v \otimes u, v' \otimes u' \rangle = \sum_{b \in B} \langle b, v \rangle \langle v', b \rangle \langle u, u' \rangle = \langle v', v \rangle \cdot \langle u, u' \rangle,$$

$$||v \otimes u||_{HS}^2 = \sum_{b \in B} ||(v \otimes u)b||^2 = \sum_{b \in B} |\langle b, v \rangle|^2 \cdot ||u||^2 = ||v||^2 \cdot ||u||^2.$$

If $b, b', a, a' \in B$ then

$$\langle b \otimes b', a \otimes a' \rangle_{HS} = \langle a, b \rangle \cdot \langle b', a' \rangle = \mathbf{1}_{\{a=b\}} \cdot \mathbf{1}_{\{a'=b'\}}.$$

So $\{b \otimes b' : b, b' \in B\}$ forms an orthonormal system.

Also, for $A \in \mathbb{H}^* \otimes \mathbb{H}$ and $b, b' \in B$, we have

$$\langle A, b \otimes b' \rangle_{HS} = \sum_{b'' \in B} \langle Ab'', (b \otimes b')b'' \rangle = \sum_{b'' \in B} \langle Ab'', b' \rangle \cdot \langle b, b'' \rangle = \langle Ab, b' \rangle.$$

So if $\langle A, b \otimes b' \rangle_{HS} = 0$ for all $b, b' \in B$ then

$$||A||_{HS}^2 = \sum_{b \in B} |\langle Ab, b' \rangle|^2 = 0.$$

Thus, $\{b \otimes b' : b, b' \in B\}$ forms an orthonormal basis. :) ✓

Solution to Exercise 9.21 :(
Since

$$\sum_{b \in B} ||Ab||^2 = ||A||_{HS} < \infty,$$

there exists a countable subset $E = \{e_1, e_2, \dots\} \subset B$, as required.

For the second assertion, note that for any $b \in B$ we have that $A_n b = Ab$ if $b \in \{e_1, \dots, e_n\}$ and $A_n b = 0$ otherwise, so

$$||A - A_n||_{HS}^2 = \sum_{b \in B} ||Ab - A_n b||^2 = \sum_{k=n+1}^{\infty} ||Ab||^2.$$

As the tail of a converging series, this can be made as small as required by taking n large enough. :) ✓

Solution to Exercise 9.22 :(
If $\dim \mathbb{H} < \infty$ then $\mathbb{H}$ is isomorphic to $\mathbb{C}^n$ for some n so the unit ball is compact. If $\dim \mathbb{H} = \infty$ then let $(u_n)_n$ be an orthonormal sequence. So $||u_n|| = 1$ for all n and, specifically, the sequence is in the unit ball; however, for any $m \neq n$,

$$||u_n - u_m||^2 = ||u_n||^2 + ||u_m||^2 = 2,$$

by the Pythagorean theorem, so the sequence does not have a converging subsequence. :) ✓

Solution to Exercise 9.23 :(
For any subsequence $(u_{n_k})_k$ we have

$$||u||^2 = \langle u, u \rangle = \lim_{k \to \infty} \langle u_{n_k}, u \rangle \leq \limsup_{k \to \infty} ||u_{n_k}|| \cdot ||u||.$$

Taking the subsequence for which $\limsup_{k \to \infty} ||u_{n_k}|| = \liminf_{n \to \infty} ||u_n||$ completes the proof.   :) ✓

Solution to Exercise 9.24 :(
Let $v \in \mathbb{H}$. Fix $\varepsilon > 0$. Since $||v||^2 = \sum_{b \in B} |\langle v, b \rangle|^2 < \infty$, there is a finite subset $F \subset B$ such that for $v_F = \sum_{b \in F} \langle v, b \rangle b$, we have that $||v - v_F|| < \varepsilon$. Since v_F is a finite linear combination of basis elements, $\langle u_n - u, v_F \rangle \to 0$ as $n \to \infty$. Thus,

$$|\langle u_n - u, v \rangle| \leq |\langle u_n - u, v_F \rangle| + |\langle u_n - u, v - v_F \rangle| \leq |\langle u_n - u, v_F \rangle| + \sup_n ||u_n - u|| \cdot \varepsilon.$$

Taking $n \to \infty$ and then $\varepsilon \to 0$ completes the proof, utilizing the fact that $||u_n - u|| \leq 2$.   :) ✓

Solution to Exercise 9.25 :(
Let B be an orthonormal basis. For a finite subset $F \subset B$ define $A_F : \mathbb{H} \to \mathbb{H}$ by

$$A_F v = \sum_{b \in F} \langle v, b \rangle A b.$$

Let $\varepsilon > 0$. By Exercise 9.21, there exists a finite subset $F_\varepsilon \subset B$ such that $||A - A_{F_\varepsilon}||_{\mathrm{HS}} < \varepsilon$.
Let $(u_n)_n$ be a sequence converging weak* to u. For any finite $F \subset B$, we have $||A_F(u_n - u)|| \to 0$ as $n \to \infty$. Thus,

$$||A(u_n - u)|| \leq ||(A - A_F)(u_n - u)|| + ||A_F(u_n - u)|| \leq 2||A - A_F||_{\mathrm{HS}} + ||A_F(u_n - u)||.$$

This implies that for any $\varepsilon > 0$,

$$\limsup_{n \to \infty} ||A(u_n - u)|| \leq 2\varepsilon,$$

implying that $||A(u_n - u)|| \to 0$ as $n \to \infty$.   :) ✓

Solution to Exercise 9.26 :(
The action $x.A := xAx^{-1}$ is obviously a left action. It is unitary because for some orthonormal basis B of $\mathbb{H}$, since G acts unitarily on $\mathbb{H}$, the collection $xB = \{xb : b \in B\}$ also forms an orthonormal basis. Hence,

$$\left\langle xAx^{-1}, xA'x^{-1} \right\rangle_{\mathrm{HS}} = \sum_{b \in B} \left\langle xAx^{-1}b, xA'x^{-1}b \right\rangle = \sum_{b \in B} \left\langle Ax^{-1}b, A'x^{-1}b \right\rangle = \left\langle A, A' \right\rangle_{\mathrm{HS}}.$$:) ✓

Solution to Exercise 9.27 :(
For any $w \in \mathbb{H}$,

$$x(v \otimes u)y^{-1}w = \left\langle y^{-1}w, v \right\rangle xu = (yv \otimes xu)w.$$:) ✓

Solution to Exercise 9.30 :(
This is basically some form of Jensen's inequality.
Let $X = X_1$ have the distribution of μ. Fix any vector $b \in \mathbb{H}$. We have that $\mathbb{E}\langle X.v - v, b \rangle = \langle Tv - v, b \rangle = 0$. So using that $\mathbb{E}\langle X.v, b \rangle = \langle Tv, b \rangle = \langle v, b \rangle$, we arrive at

$$0 \leq \mathrm{Var}\langle X.v, b \rangle = \mathbb{E}|\langle X.v, b \rangle|^2 - |\langle v, b \rangle|^2.$$

Summing b over an orthonormal basis, and using the fact that $||X.v|| = ||v||$ because the action is unitary,

$$0 = \mathbb{E}||X.v||^2 - ||v||^2 = \sum_b \left(\mathbb{E}|\langle X.v, b \rangle|^2 - |\langle v, b \rangle|^2 \right) \geq 0.$$

So for any b we have $\mathrm{Var}\langle X.v, b \rangle = 0$, implying that $\langle X.v, b \rangle = \langle v, b \rangle$ a.s. That is, for any b we find that $\langle x.v, b \rangle = \langle v, b \rangle$ for any $x \in \mathrm{supp}(\mu)$. This implies that $x.v = v$ for any $x \in \mathrm{supp}(\mu)$. Finally, since μ is adapted, any $x \in G$ is the product of finitely many elements from $\mathrm{supp}(\mu)$, so $x.v = v$ for all $x \in G$.   :) ✓

Solution to Exercise 9.31 :(

By Theorem 9.6.1, we know that G has a finite index subgroup $[G : H] < \infty$ with a surjective homomorphism $\varphi : H \to \mathbb{Z}$. Set $K = \ker \varphi$.

Since H is finite index in G it has linear growth. Since $H/K \cong \mathbb{Z}$, by Exercise 8.19, it must be that K is finite. By Exercise 8.21, H is virtually $\mathbb{Z}$, implying that G is virtually $\mathbb{Z}$ as well. 　　　:) ✓

Solution to Exercise 9.32 :(

As before, by Theorem 9.6.1, G contains a finite index subgroup $[G : H] < \infty$ with a surjective homomorphism $\varphi : H \to \mathbb{Z}$. Set $K = \ker \varphi$. By Exercise 8.19, K must be finitely generated and has growth $\leq (r \mapsto r)$. Hence, by Exercise 9.31, K is either virtually $\mathbb{Z}$ or K is finite. In the latter case, we have seen in Exercise 8.21 that this implies that H is virtually $\mathbb{Z}$, so that G is virtually $\mathbb{Z}$ as well.

In the former case, when K is virtually $\mathbb{Z}$, we have that G contains a finite index subgroup H such that for some $N \lhd H$ we have $H/N \cong \mathbb{Z}$ and $N \cong \mathbb{Z}$.

Now, Exercise 8.19 tells us that $H = \mathbb{Z} \ltimes N$ for some action of $\mathbb{Z}$ on N. Since $N \cong \mathbb{Z}$, we only need to show that if $\mathbb{Z}$ acts on $\mathbb{Z}$, then $\mathbb{Z} \ltimes \mathbb{Z}$ is always virtually $\mathbb{Z}^2$.

Let $\mathbb{Z} \cong \langle a \rangle$ act on the additive group of integers $\mathbb{Z}$. That is, a is an automorphism of the additive group of integers. There are two possibilities for $a(1)$: either $a(1) = 1$ so the action is trivial, or $a(1) = -1$ and the action is a reflection around 0.

If the action is trivial, $a(1) = 1$, then it is easy to check that $(a^z, x) \mapsto (z, x)$ defines an isomorphism of $\langle a \rangle \ltimes \mathbb{Z}$ to $\mathbb{Z}^2$.

If the action is a reflection, $a(1) = -1$, then consider the subgroup $N = \langle a^2 \rangle \lhd \langle a \rangle$. We have that N acts trivially on the integers, and $N \ltimes \mathbb{Z}$ has index 2 inside $\langle a \rangle \ltimes \mathbb{Z}$. So $N \ltimes \mathbb{Z} \cong \mathbb{Z}^2$ is a finite index subgroup in $\langle a \rangle \ltimes \mathbb{Z}$. 　　　:) ✓

Solution to Exercise 9.33 :(

By Theorem 9.6.1, G contains a finite index subgroup $[G : H] < \infty$ with a surjective homomorphism $\varphi : H \to \mathbb{Z}$. Let $K = \ker \varphi$.

By Exercise 8.19, K must be finitely generated and have growth $\leq (r \mapsto r^{2-\varepsilon}) < (r \mapsto r^2)$. Thus, K is either finite or virtually $\mathbb{Z}$ or virtually $\mathbb{Z}^2$, by Exercise 9.32.

If K is virtually $\mathbb{Z}^2$ then K has growth $\geq (r \mapsto r^2) > (r \mapsto r^{2-\varepsilon})$, a contradiction. So K is either finite or virtually $\mathbb{Z}$.

If K is finite, then $H/K \cong \mathbb{Z}$ implies that G is virtually $\mathbb{Z}$, by Exercise 8.21.

If K is virtually $\mathbb{Z}$, then $H/K \cong \mathbb{Z}$ implies that G actually has growth $\leq (r \mapsto r^2)$, so G is virtually $\mathbb{Z}^2$ or virtually $\mathbb{Z}$ by Exercise 9.32. 　　　:) ✓

Solution to Exercise 9.34 :(

For $t = 0$ there is nothing to prove.

Assume for t, and we prove by induction for $t + 1$. Compute using the Markov property at time t,

$$\mathbb{P}[X_{t+1} = x] = \sum_y \mathbb{P}[X_1 = y] \cdot \mathbb{P}_y[X_t = x] = \sum_y \mu_R(y) \mathbb{P}\left[X_t = y^{-1}x\right] + \sum_y \nu_R(y) \mathbb{P}\left[X_t = y^{-1}x\right]$$

$$\text{by induction } \leq \sum_y \mu_R(y) \cdot \left((1-\delta)^t + t \cdot \tfrac{\delta}{|B_{R+1}|}\right) + \sup_{y \in G} \nu_R(y) \cdot \sum_z \mathbb{P}[X_t = z]$$

$$\leq (1-\delta) \cdot \left((1-\delta)^t + t \cdot \tfrac{\delta}{|B_{R+1}|}\right) + \tfrac{\delta}{|B_{R+1}|}$$

$$\leq (1-\delta)^{t+1} + (t+1) \cdot \tfrac{\delta}{|B_{R+1}|},$$

which completes the induction. 　　　:) ✓

Solution to Exercise 9.35 :(

Since B is a basis, for every j we can write

$$c_j = \sum_{k=1}^d M_{j,k} b_k,$$

for some $M_{j,k} \in \mathbb{C}$. The linearity of the form Q_r gives that

$$Q_r(c_i, c_j) = \sum_{k=1}^d \sum_{\ell=1}^d M_{i,k} Q_r(b_k, b_\ell) \overline{M_{j,\ell}} = (M[Q_r]_B M^*)_{i,j}. \qquad \text{:) ✓}$$

Solution to Exercise 9.36 :(

Let $B = (b_1, \ldots, b_d)$ be a basis of V. Since these are nonzero functions, there exists r_0 such that for every j there exists $|x| \leq r_0$ with $|b_j(x)| > 0$. Then, for any $r \geq r_0$, the restriction map $v \mapsto v|_{B(1,r)}$ is then a linear isomorphism from V into $\mathbb{C}^{B(1,r)}$. Thus, if $v \in V$ then $v(x) \neq 0$ for some $|x| \leq r$, implying that $Q_r(v, v) > 0$.

This turns Q_r into an inner product on V. Being finite dimensional, V is then obviously a Hilbert space with this inner product. :) ✓

Solution to Exercise 9.37 :(

Choose a basis E for V that is orthonormal with respect to the inner product. Then $[Q]_E$ is the identity matrix. So for any other basis $B = (b_1, \ldots, b_d)$ of V, we have that $[Q]_B = MM^*$ for some matrix M.

Since MM^* is self-adjoint, there exists a unitary matrix U and a diagonal matrix D such that $[Q]_B = MM^* = UDU^*$. So for any j,

$$D_{j,j} = (U^*[Q]_B U)_{j,j} = \sum_{k,\ell} U_{j,k}^* U_{\ell,j} Q(b_k, b_\ell) = Q(c_j, c_j),$$

where

$$c_j = \sum_{k=1}^d (U^*)_{j,k} b_k \in V.$$

Because U is invertible, $c_j \neq 0$ for all j, implying that $\det([Q]_B) = \det(D) > 0$. :) ✓

Solution to Exercise 9.38 :(

Let $E = (e_1, \ldots, e_d)$ be a basis that is orthonormal with respect to Q_R (this can be done since Q_R is positive definite, so defines a Hilbert space on V). So $[Q_R]_E = I$. Then, $[Q_r]_E$ is a self-adjoint matrix, so can be unitarily diagonalized. That is, $U[Q_r]_E U^* = D$ is a diagonal matrix for some unitary matrix U. Let $B = (b_1, \ldots, b_d)$ be the basis given by

$$b_j = \sum_{k=1}^d U_{j,k} e_k.$$

Then one may compute that

$$[Q_r]_B = U[Q_r]_E U^* = D \quad \text{and} \quad [Q_R]_B = U[Q_R]_E U^* = I. \qquad :) ✓$$

Solution to Exercise 9.39 :(

Since $\det(Q_R/Q_r) = \frac{\det[Q_R]_B}{\det[Q_r]_B}$ does not depend on B, we may assume that B is such that $[Q_r]_B$ and $[Q_R]_B$ are both diagonal matrices.

For $b \in B$ we have that

$$Q_R(b, b) = \sum_{|x| \leq R} |b(x)|^2 \geq Q_r(b, b).$$

Thus,

$$\det[Q_R]_B = \prod_{b \in B} Q_R(b, b) \geq \prod_{b \in B} Q_r(b, b) = \det[Q_r]_B. \qquad :) ✓$$

Solution to Exercise 9.40 :(

Inductively choose $x_1, \ldots, x_m$ as follows. Start with some $x_1 \in R$. For any $k \geq 1$, assume that we have chosen $x_1, \ldots, x_k$ such that the balls $(B(x_j, r))_{j=1}^k$ are pairwise disjoint. If $B(1, R) \subset \bigcup_{j=1}^k B(x_j, 2r)$, then set $m = k$ and the construction is complete. Otherwise, there exists $x_{k+1} \in B(1, R)$ such that $x_{k+1} \notin B(x_j, 2r)$ for all $1 \leq j \leq k$. Hence $B(x_{k+1}, r) \cap B(x_j, r) = \emptyset$ for all $1 \leq j \leq k$, and we can continue inductively.

With the above procedure, we have found $x_1, \ldots, x_m \in B(1, R)$ such that $B(1, R) \subset \bigcup_{j=1}^m B(x_j, 2r)$ and such that the balls $(B(x_j, r))_{j=1}^m$ are pairwise disjoint.

Since the balls $(B(x_j, r))_{j=1}^m$ are pairwise disjoint, we have that $m|B(1, r)| \leq |B(1, R+r)| \leq |B(1, 2R)|$.

We bound the overlap. Fix $x \in B(1, R)$. Let $J_x = \{1 \leq j \leq m : x \in B(x_j, 6r)\}$. Note that for any $j \in J_x$ we have that $x_j \in B(x, 6r)$. Because the balls $(B(x_j, r))_{j=1}^m$ are pairwise disjoint, we have that $|J_x| \cdot |B(1, r)| \leq |B(1, 6r + r)| = |B(1, 7r)|$. :) ✓

Solution to Exercise 9.41 :(

We have that

$$xy.h - h = x.(y.h - h) + x.h - h,$$
$$x^{-1}.h - h = -x^{-1}.(x.h - h),$$
$$x^{-1}yx.h - h = x^{-1}.(y.x.h - x.h),$$

and since the action on a constant function it trivial, this shows that $K \lhd G$.

Note that for any $y \in K$ and $x \in G$, we have that

$$h(yx) - h(1) = \left(y^{-1}.h - h\right)(x) + h(x) - h(1) = \left(y^{-1}.h - h\right)(1) + h(x) - h(1)$$
$$= h(y) - h(1) + h(x) - h(1).$$

This shows that $f = h|_K - h(1)$ is a homomorphism from K to $\mathbb{C}$.

Moreover, if $h|_K \equiv 0$, then for any $x \in G$ and $y \in K$, we have that $h(yx) = h(x)$, so that $\bar{h}(Kx) = h(x)$ is well defined. It is easy to see that $\bar{h} \in \mathrm{LHF}(G/K, \bar{\mu})$. Indeed,

$$|\bar{h}(Kxs) - \bar{h}(Kx)| = |h(xs) - h(x)| \le ||\nabla h||_\infty \cdot |s|,$$

$$\sum_{Kx \in G/K} \bar{\mu}(Kx)\bar{h}(KzKx) = \sum_{Kx \in G/K} \sum_{y \in K} \mu(yx)h(zx)$$
$$= \sum_{Kx \in G/K} \sum_{y \in K} \mu(yx)h\left(zy^{-1}z^{-1}zyx\right)$$
$$= \sum_{g \in G} \mu(g)h(zg) = h(z) = \bar{h}(Kz),$$

where we have used that cosets are pairwise disjoint, and that $h(zx) = h\left(zy^{-1}z^{-1}zyx\right) = h(zyx)$ for all $y \in K$ and $x, z \in G$.

Consider the liner map $\Psi(h) = h|_K$ on $\mathrm{LHF}(G, \mu)$. The image of Ψ is contained in the space V of homomorphisms from K to $\mathbb{C}$ plus a constant function. The kernel of Ψ, denoted $\mathrm{Ker}\Psi$, is the space of all $h \in \mathrm{LHF}(G, \mu)$ with $h|_K \equiv 0$. The map $\Phi(h) = \bar{h}$ from $\mathrm{Ker}\Psi$ into $\mathrm{LHF}(G/K, \bar{\mu})$ is easily seen to be an injective linear map. So we conclude that

$$\dim \mathrm{LHF}(G, \mu) \le \dim V + \dim \mathrm{Ker}\Psi \le \dim \mathrm{Hom}(K, \mathbb{C}) + 1 + \dim \mathrm{LHF}(G/K, \bar{\mu}). \qquad :) \checkmark$$

Solution to Exercise 9.42 :(

Let $Z = Z_1(G)$. If $z \in Z$ then for $h \in \mathrm{LHF}(G, \mu)$ and $x \in G$ we have that

$$|z.h(x) - h(x)| = \left|h\left(xz^{-1}\right) - h(x)\right| \le ||\nabla h||_\infty \cdot |z|,$$

so $z.h - h$ is a bounded μ-harmonic function. Since (G, μ) is assumed to be Liouville, $z.h - h$ is constant, implying that $z \in K$.

Proceeding exactly as in Exercise 9.41, we obtain

$$\dim \mathrm{LHF}(G, \mu) \le \dim \mathrm{Hom}(Z, \mu) + 1 + \dim \mathrm{LHF}(G/Z, \bar{\mu}). \qquad :) \checkmark$$

Solution to Exercise 9.43 :(

The map $\psi \mapsto (\psi(u))_{u \in U}$ is a linear map from $\mathrm{Hom}(G, \mathbb{C})$ into $\mathbb{C}^U$. Since U generates G, if a homomorphism ψ satisfies $\psi(u) = 0$ for all $u \in U$, then $\psi \equiv 0$. Thus, the above linear map is injective, so $\dim \mathrm{Hom}(G, \mathbb{C}) \le \dim \mathbb{C}^U$. $\qquad :) \checkmark$

Solution to Exercise 9.44 :(

The proof is by induction on the nilpotent step.

Base case: If G is 1-step nilpotent, then G is Abelian. So (G, μ) is Liouville (by the Choquet–Deny theorem (Corollary 7.1.2)), and $Z_1(G) = G$. By Exercise 9.43, we know that $\dim \mathrm{Hom}(G, \mathbb{C}) < \infty$, since G is finitely generated. Since $G/Z_1(G)$ is the trivial group, $\mathrm{LHF}(G/Z_1(G), \bar{\mu})$ is just the space of constants. Thus, by Exercise 9.42 we have that

$$\dim \mathrm{LHF}(G, \mu) \le \dim \mathrm{Hom}(G, \mathbb{C}) + 1 + 1 < \infty.$$

Induction step: Let $n > 1$. Assume the claim for nilpotent groups of step less than n, and let G be n-step nilpotent. We have that (G, μ) is Liouville because nilpotent groups are Choquet–Deny (Corollary 7.1.4). Also, $\dim \mathrm{LHF}(G/Z_1(G), \bar{\mu}) < \infty$ by induction, because $G/Z_1(G)$ is at most $(n-1)$-step nilpotent. Since G is a finitely generated nilpotent group, we know that $Z_1(G)$ is finitely generated, for example, by Rosset's theorem (Theorem 8.3.3) together with the fact that nilpotent groups have polynomial growth (Theorem 8.2.1). Thus, by Exercise 9.42,

$$\dim \mathrm{LHF}(G, \mu) \leq \dim \mathrm{Hom}(Z_1(G), \mathbb{C}) + 1 + \dim \mathrm{LHF}(G/Z_1(G), \bar{\mu}) < \infty. \qquad :) \checkmark$$

Appendices

Appendix A

Hilbert Space Background

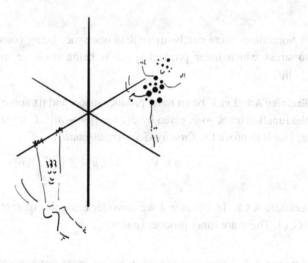

A.1 Inner Products and Hilbert Spaces

We record the necessary background regarding inner products and Hilbert spaces. Any basic book should contain the details. Most exercises and theorems are given without solutions or proofs in this section.

Definition A.1.1 A **complex inner product** space is a vector space V over the complex number field $\mathbb{C}$ with the additional structure of an **inner product**. This is a function $\langle \cdot, \cdot \rangle : V \times V \to \mathbb{C}$ such that for all vectors $u, v, w \in V$ and scalars $\alpha \in \mathbb{C}$,

- $\langle \alpha v + u, w \rangle = \alpha \langle v, w \rangle + \langle u, w \rangle$,
- $\langle v, u \rangle = \overline{\langle u, v \rangle}$ (complex conjugate),
- $\mathbb{R} \ni \langle v, v \rangle \geq 0$, and
- $\langle v, v \rangle = 0$ if and only if $v = 0$.

An inner product defines a **norm** on V given by $||v|| = \sqrt{\langle v, v \rangle}$ for all $v \in V$.

Sometimes there can be more than one space being considered. If we wish to stress which inner product space is being used we may write $\langle \cdot, \cdot \rangle_V$ or $|| \cdot ||_V$.

Exercise A.1 Let V be an inner product space, and fix some $v, u \in V$. Consider the function $\varphi \colon \mathbb{R} \to \mathbb{R}$ given by $\varphi(\xi) = ||\xi v - u||^2$. Compute the minimum of φ. Use it to prove the *Cauchy–Schwarz inequality*:

$$\forall\, v, u \in V, \qquad |\langle v, u \rangle| \leq ||v|| \cdot ||u||.$$

Example A.1.2 In Chapter 4 we consider examples of $\ell^2(G, c)$ for a network (G, c). These are inner product spaces. △▽△

Exercise A.2 Show that the norm on an inner product space V satisfies the following. For all vectors $v, u \in V$ and scalars $\alpha \in \mathbb{C}$,

- $||\alpha v|| = |\alpha| \cdot ||v||$,
- $||v + u|| \leq ||v|| + ||u||$, and
- $||v|| = 0$ if and only if $v = 0$.

Exercise A.3 Show that dist$(u, v) = ||v - u||$ defines a metric on an inner product space V.

Exercise A.4 Let V be an inner product space and fix some $u \in V$.

Show that the functions $v \mapsto \langle v, u \rangle$ and $v \mapsto \langle u, v \rangle$ and $v \mapsto ||v||$ are three uniformly continuous functions on V.

Definition A.1.3 An inner product space $\mathbb{H}$ is a **Hilbert space** if the metric induced by the norm on $\mathbb{H}$ is a complete metric; that is, every Cauchy sequence in $\mathbb{H}$ converges to a limit in $\mathbb{H}$.

Example A.1.4 We have that $\ell^2(G, c)$ is a Hilbert space, see Exercise 4.2 (Chapter 4). ▵▿▵

Definition A.1.5 Let $\mathbb{H}$ be a Hilbert space.

We write $v \perp u$ if $\langle u, v \rangle = 0$.

For subsets $S, T \subset \mathbb{H}$ we write $S \perp T$ if for all $s \in S$ and $t \in T$ we have $s \perp t$.

For a subset $S \subset \mathbb{H}$ we define $S^\perp = \{u \in \mathbb{H} : \forall v \in S, u \perp v\}$.

We also write $v^\perp = \{v\}^\perp$.

Exercise A.5 Show that $S^\perp$ is a subspace.

Exercise A.6 Show that $S^\perp$ is closed.

Definition A.1.6 Let $\mathbb{H}$ be a Hilbert space and $V, W \leq \mathbb{H}$ be closed subspaces. We write $\mathbb{H} = V \oplus W$ if $V \cap W = \{0\}$ and for every $u \in \mathbb{H}$ there exist $v \in V, w \in W$ such that $u = v + w$.

Theorem A.1.7 (Orthogonal projection) *Let* $V \leq \mathbb{H}$ *be a subspace. Then,* $(V^\perp)^\perp = \mathrm{cl}(V)$. *(So if V is closed then $(V^\perp)^\perp = V$.)*

Also, $\mathbb{H} = V \oplus V^\perp$.

Moreover, consider the maps $L \colon \mathbb{H} \to V$ *and* $R \colon \mathbb{H} \to V^\perp$ *given by* $Lu = v$ *and* $Ru = w$, *where* $u = v + w$ *is the unique decomposition for which* $v \in V$ *and* $w \in V^\perp$. *Then, these are both surjective linear maps, with*

$$Lv' = v', \qquad Lw' = 0, \qquad Rv' = 0, \qquad Rw' = w',$$

for all $v' \in V$ *and* $w' \in V^\perp$.

We call L the **orthogonal projection** *onto V (and similarly R is the orthogonal projection onto $V^\perp$).*

Finally, for any $u \in \mathbb{H}$ *we have that*

$$||u - Lu|| = \min\{||u - v|| : v \in V\}.$$

Exercise A.7 Show that if $V \leq \mathbb{H}$ is a closed subspace of a Hilbert space, and $\pi \colon \mathbb{H} \to V$ is the orthogonal projection, then $||\pi u|| \leq ||u||$ for all $u \in \mathbb{H}$.

Definition A.1.8 Let V be a complex vector space. A **linear functional** is a linear map from V to $\mathbb{C}$.

Example A.1.9 We have seen that in a Hilbert space $\mathbb{H}$, for $u \in \mathbb{H}$, the map $v \mapsto \langle v, u \rangle$ is a continuous linear functional. In fact, the next theorem states that these are the only possibilities.   △▽△

Theorem A.1.10 (Reisz representation theorem) *Let $\mathbb{H}$ be a Hilbert space. Let $\varphi \colon \mathbb{H} \to \mathbb{C}$ be a continuous linear functional. Then there exists $u \in \mathbb{H}$ such that $\varphi(v) = \langle v, u \rangle$ for all $v \in \mathbb{H}$.*

A.2 Normed Vector Spaces

Definition A.2.1 Let V be a vector space over $\mathbb{C}$. A **norm** on V is a function $V \ni v \mapsto ||v|| \in [0, \infty)$ such that for all vectors $u, v \in V$ and all scalars $\alpha \in \mathbb{C}$,

- $||\alpha v|| = |\alpha| \cdot ||v||$,
- $||v + u|| \leq ||v|| + ||u||$,
- $||v|| \geq 0$, and
- $||v|| = 0$ if and only if $v = 0$.

Such a space V is called a **normed space**. If the metric induced by the norm is complete, V is called a **Banach space**.

We have seen that any inner product induces a norm. But not every norm is induced by an inner product.

Exercise A.8 (Parallelogram law) Show that if V is a complex inner product space then for the norm induced by the inner product we have:

$$||u + v||^2 + ||u - v||^2 = 2 \left(||u||^2 + ||v||^2 \right)$$

for all $u, v \in V$.

Exercise A.9 (Polarization identity) Assume that V is a normed complex vector space such that for all $u, v \in V$, we have

$$||u + v||^2 + ||u - v||^2 = 2 \left(||u||^2 + ||v||^2 \right).$$

Show that

$$\langle u, v \rangle = \tfrac{1}{4} \left(||u + v||^2 - ||u - v||^2 + i||u + iv||^2 - i||u - iv||^2 \right)$$

is a well-defined inner product on V. Show that $\langle u, u \rangle = ||u||^2$.

So a norm is induced by an inner product if and only if it satisfies the parallelogram law.

Definition A.2.2 Let V, W be two normed spaces. For a linear operator $L : V \to W$, define

$$||L||_{V \to W} = \sup\{||Lv||_W \: : \: v \in V, \: ||v||_V \leq 1\}.$$

If $||L||_{V \to W} < \infty$, then L is called **bounded**.

Exercise A.10 Let V, W be two normed spaces. Let $L : V \to W$ be a linear operator.

Show that L is bounded if and only if L is continuous (as a function between two metric spaces).

Show that L is continuous if and only if L is continuous at $0 \in V$.

Definition A.2.3 Let V be a normed space, and let $T : V \to V$ be a bounded linear operator. Define the **subspace of T-invariant vectors** to be

$$V^T := \{v \in V \: : \: Tv = v\}.$$

Exercise A.11 Let V be a normed space, and let $T : V \to V$ be a bounded linear operator.

Show that V^T is always a closed subspace of V.

A.3 Orthonormal Systems

Lemma A.3.1 *Let X be a non-empty set (possibly uncountable). Consider the counting measure μ on X; that is, $\mu(A) = |A|$ for any $A \subset X$.*

For any $\varphi: X \to [0, \infty]$, the Lebesgue integral satisfies

$$\int \varphi d\mu = \sup_{\substack{F \subset X \\ |F| < \infty}} \sum_{x \in F} \varphi(x).$$

Definition A.3.2 Given a subset X of a Hilbert space $\mathbb{H}$, and a function $\varphi: X \to \mathbb{C}$, we define

$$\sum_{x \in X} \varphi(x) := \int \varphi d\mu,$$

where μ is the counting measure on X. This is only defined when the latter integral exists.

We say the sum $\sum_x \varphi(x)$ **converges** if this integral is defined and finite, and **converges absolutely** if $\sum_x |\varphi(x)|$ converges.

Definition A.3.3 Let $\mathbb{H}$ be a Hilbert space. A collection of vectors $N \subset \mathbb{H}$ is called **orthonormal** if for any $v, u \in N$ we have $\langle v, u \rangle = \mathbf{1}_{\{v=u\}}$.

Theorem A.3.4 *Let $\mathbb{H}$ be a Hilbert space. Let N be an orthonormal subset. The following are equivalent.*

- *N is a maximal orthonormal subset; that is, if M is an orthonormal subset with $N \subset M$, then $M = N$.*
- *The linear span of N (all finite linear combinations of elements of N) is a dense subspace of $\mathbb{H}$.*
- *For any $v \in \mathbb{H}$ we have $\|v\|^2 = \sum_{u \in N} |\langle v, u \rangle|^2$.*
- *For any $v, w \in \mathbb{H}$ we have $\langle v, w \rangle = \sum_{u \in N} \langle v, u \rangle \cdot \langle u, w \rangle$.*

Definition A.3.5 A maximal orthonormal subset N as above in a Hilbert space is called an **orthonormal basis**.

Theorem A.3.6 *Let $\mathbb{H}$ be a nontrivial Hilbert space. There exists an orthonormal basis for $\mathbb{H}$.*

In fact, for any orthonormal set $N \subset \mathbb{H}$ there exists an orthonormal basis B such that $N \subset B$.

Moreover, if B is an orthonormal basis for $\mathbb{H}$, then $\mathbb{H}$ is isomorphic as a Hilbert space to $\ell^2(B)$. Here, $\ell^2(B)$ is the space of all functions $\varphi: B \to \mathbb{C}$

such that $\langle \varphi, \varphi \rangle$ *converges, where the inner product is defined by* $\langle \varphi, \psi \rangle =$ $\sum_{b \in B} \varphi(b)\overline{\psi(b)}$. *In fact, the isomorphism is given by extending* $\mathbb{H} \ni b \mapsto \delta_b \in$ $\ell^2(B)$ *to finite linear combinations, and then extending to all of* $\mathbb{H}$ *continuously.*

Some remarks concerning infinite sums: When one deals with Hilbert spaces that have uncountable orthonormal bases, one is usually required to consider uncountable sums. A complication arises if we wish to take product spaces and double sums. One should be careful, as the counting measure on a non-countable space does not give a σ-finite space, and the Fubini–Tonelli theorem is not guaranteed. However, if $\sum_b |f(b)| < \infty$, then it is an exercise to prove that the set $\{b : f(b) \neq 0\}$ is countable. Hence, for absolutely convergent sums we can still interchange order of summation. When the summands are positive, we also can prove Tonelli's theorem.

Exercise A.12 Show that if $\sum_{b \in B} |f(b)| < \infty$ then $\{b \in B : f(b) \neq 0\}$ is countable.

Exercise A.13 (Tonelli's theorem for counting measures) Let A, B be non-empty sets. Let $f \colon A \times B \to [0, \infty]$.

For $a \in A$ define $f_a \colon B \to [0, \infty]$ by $f_a(b) = f(a, b)$ and for $b \in B$ define $f_b \colon A \to [0, \infty]$ by $f_b(a) = f(a, b)$.

Show that

$$\sum_{(a,b) \in A \times B} f(a, b) = \sum_{b \in B}\left(\sum_{a \in A} f_b(a) \right) = \sum_{a \in A}\left(\sum_{b \in B} f_a(b) \right). \qquad \triangleright \text{solution} \triangleleft$$

Exercise A.14 Let $\mathbb{H}$ be a Hilbert space with an orthonormal basis B. Show that if $f \in \ell^2(B)$ then there exists a unique vector $v \in \mathbb{H}$ such that for any $b \in B$ we have $\langle v, b \rangle = f(b)$.
$\triangleright \text{solution} \triangleleft$

For a Hilbert space $\mathbb{H}$ with orthonormal basis B, and for $f \in \ell^2(B)$, we use the notation $\sum_{b \in B} f(b)b$ to denote the vector $v \in \mathbb{H}$ for which $\langle v, b \rangle = f(b)$ for all $b \in B$.

A.4 Solutions to Exercises

Solution to Exercise A.13 :(
Define

$$S = \sum_{(a,b) \in A \times B} f(a, b) < \infty, \qquad S_A = \sum_{a \in A}\left(\sum_{b \in B} f_a(b) \right), \qquad S_B = \sum_{b \in B}\left(\sum_{a \in A} f_b(a) \right).$$

Because of the symmetry between the roles of A, B, we will only prove $S = S_A$.

Let $F \subset A \times B$ be any finite subset. Let $F_A = \{a \in A: \exists\, b \in B,\ (a, b) \in F\}$ and $F_B = \{b \in B: \exists\, a \in A,\ (a, b) \in F\}$. Then,

$$\sum_{(a,b) \in F} f(a, b) \leq \sum_{(a,b) \in F_A \times F_B} f(a, b) = \sum_{a \in F_A} \sum_{b \in F_B} f(a, b) \leq \sum_{a \in F_A} \left(\sum_{b \in B} f_a(b) \right) \leq S_A.$$

Taking the supremum over F shows that $S \leq S_A$.

Consider the sum $S_a := \sum_{b \in B} f_a(b)$.

If there exists $a \in A$ such that $S_a = \infty$, then $S_A \geq S_a = \infty$. Also, for any $R > 0$ we may choose a finite subset $F_{a,R} \subset B$ such that $\sum_{b \in F_{a,R}} f_a(b) > R$. By considering the finite subset $\{a\} \times F_{a,R} \subset A \times B$, we get that

$$S \geq \sum_{b \in F_{a,R}} f(a, b) > R,$$

so that $S = S_A = \infty$ by taking $R \to \infty$.

So assume that $S_a < \infty$ for all $a \in A$.

Fix $\varepsilon > 0$. For any $a \in A$ choose a finite subset $F_a = F_{a,\varepsilon} \subset B$ such that $\sum_{b \in F_a} f_a(b) > S_a - \varepsilon$. Let $F_A \subset A$ be any finite subset. Define $F_B \subset B$ by $F_B = \cup_{a \in F} F_a$. We then have that

$$S \geq \sum_{(a,b) \in F_A \times F_B} f(a, b) \geq \sum_{a \in F_A} \left(\sum_{b \in F_a} f_a(b) \right) \geq \sum_{a \in F_A} S_a - \varepsilon \cdot |F_A|.$$

Taking $\varepsilon \to 0$ and then a supremum over F_A we find that $S \geq S_A$. :) ✓

Solution to Exercise A.14 :(

Since $f \in \ell^2(B)$, there is a countable subset $C \subset B$ such that $f(b) = 0$ for all $b \in B \backslash C$. Enumerate $C = \{c_1, c_2, \ldots\}$. For every n, define $v_n = \sum_{k=1}^{n} f(c_k) c_k$. Note that by the Pythagorean theorem,

$$\|v_{n+m} - v_n\|^2 = \sum_{k=1}^{m} |f(c_{n+k})|^2 \leq \sum_{k>n} |f(c_k)|^2,$$

which shows that $(v_n)_n$ forms a Cauchy sequence in $\mathbb{H}$. Thus, $v = \lim_n v_n \in \mathbb{H}$ exists.

Also, if $b \in C$ then

$$\langle v, b \rangle = \lim_{n \to \infty} \langle v_n, b \rangle = f(b),$$

and if $b \notin C$ then, similarly, $\langle v, b \rangle = 0 = f(b)$.

If w is any vector with $\langle w, b \rangle = f(b)$ for all $b \in B$, then $\langle v - w, b \rangle = 0$ for all $b \in B \backslash C$, and $\langle v - w, b \rangle = f(b) - f(b) = 0$ for all $b \in C$, which implies that $v = w$. :) ✓

Appendix B

Entropy

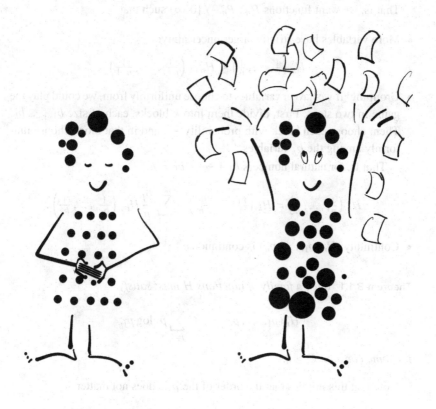

B.1 Shannon Entropy Axioms

A distribution on finitely many objects can be characterized by a vector $(p_1, \ldots, p_n)$ such that $p_j \in [0, 1]$ and $\sum_j p_j = 1$. We want to associate a "measure of uncertainty" for such distributions, say denoted by $H(p_1, \ldots, p_n)$. This is a number that only depends on the probabilities $p_1, \ldots, p_n$, not on the specific objects being sampled.

For an integer n, let $P_n = \left\{ (p_1, \ldots, p_n) : \sum p_j = 1 \right\}$ be the set of all probability distributions on n objects.

We want a measure of *uncertainty*. That is, we want to measure how hard it is to guess, or predict, or sample, a certain measure.

That is, we want functions $H_n : P_n \to [0, \infty)$ such that:

- More variables have strictly larger uncertainty:

$$H_n\left(\tfrac{1}{n}, \ldots, \tfrac{1}{n}\right) < H_{n+1}\left(\tfrac{1}{n+1}, \ldots, \tfrac{1}{n+1}\right).$$

- Grouping: If we have n variables to choose uniformly from, we could play the game in two steps. First, divide them into k blocks, each of size $b_1, \ldots, b_k$. Then, choose each block with probability $\tfrac{b_j}{n}$, and in that block choose uniformly among the b_j variables.

  That is, for natural numbers $b_1 + \cdots + b_k = n$,

$$H_n\left(\tfrac{1}{n}, \ldots, \tfrac{1}{n}\right) = H_k\left(\tfrac{b_1}{n}, \ldots, \tfrac{b_k}{n}\right) + \sum_{j=1}^{k} \frac{b_j}{n} H_{b_j}\left(\tfrac{1}{b_j}, \ldots, \tfrac{1}{b_j}\right).$$

- Continuity: $H_n(p_1, \ldots, p_n)$ is continuous.

Theorem B.1.1 *Such a family of functions H must satisfy*

$$H_n(p_1, \ldots, p_n) = -C \sum_{j=1}^{n} p_j \log p_j,$$

for some constant $C > 0$.

Note that this implies that the order of the p_j's does not matter.

Proof Set $u(n) = H_n\left(\tfrac{1}{n}, \ldots, \tfrac{1}{n}\right)$.

Step 1. $u(mn) = u(m) + u(n)$. Indeed, take $b_1 = \cdots b_m = n$ in the grouping axiom. Then,

$$u(nm) = u(m) + \sum_{j=1}^{m} \frac{1}{m} u(n) = u(m) + u(n).$$

So we get that

$$u\left(n^k\right) = ku(n).$$

Step 2. $u(n) = C \log n$ for $C = \frac{u(2)}{\log 2}$.

If $n = 1$ then grouping implies that

$$u(n) = u(n) + \sum_{j=1}^{n} \tfrac{1}{n} H_1(1),$$

so $u(1) = H_1(1) = 0$. So we can assume $n > 1$.

Let $a > 0$ be some integer, and let $k \in \mathbb{N}$ be such that $n^k \leq 2^a < n^{k+1}$. That is, $k \leq \frac{a \log 2}{\log n} < k + 1$ $\left(k = \lfloor \frac{a \log 2}{\log n} \rfloor\right)$.

Here, we use that u is strictly increasing: we get

$$ku(n) = u\left(n^k\right) \leq u(2^a) = au(2) < u\left(n^{k+1}\right) = (k+1)u(n).$$

So we have that for any $a > 0$,

$$\frac{k}{a} \leq \frac{\log 2}{\log n} < \frac{k+1}{a} \quad \text{and} \quad \frac{k}{a} \leq \frac{u(2)}{u(n)} < \frac{k+1}{a}.$$

Thus,

$$\left| \frac{u(2)}{u(n)} - \frac{\log 2}{\log n} \right| \leq \frac{1}{a}.$$

Taking $a \to \infty$ proves Step 2.

Step 3. Let us prove the theorem for $p_j \in \mathbb{Q}$. If $p_j = \frac{a_j}{b_j} \in \mathbb{Q}$ then define $\Pi_j = \prod_{i \neq j} b_i$ and $\Pi = \prod_{j=1}^{n} b_j$. So $p_j = \frac{a_j \Pi_j}{\Pi}$ and $\sum_j a_j \Pi_j = \Pi$. Thus, by grouping,

$$u(\Pi) = H_n\left(\frac{a_1 \Pi_1}{\Pi}, \ldots, \frac{a_n \Pi_n}{\Pi}\right) + \sum_{j=1}^{n} \frac{a_j \Pi_j}{\Pi} u(a_j \Pi_j).$$

Rearranging, we have

$$H_n(p_1, \ldots, p_n) = C \log \Pi - C \sum_{j=1}^{n} p_j \log(a_j \Pi_j) = -C \sum_{j=1}^{n} p_j \log\left(\frac{a_j \Pi_j}{\Pi}\right).$$

This is the theorem for rational values.

Step 4 The theorem follows for all values by continuity. $\qquad \square$

B.2 A Different Perspective on Entropy

Suppose that we are scientists, and we wish to predict the next state of some physical system. Of course, this is done using previous observations. Suppose that the distribution of the next state of the system is distributed over n possible states, say $\{1, 2, \ldots, n\}$. Assume that the (unknown) probability for the system to be in state j is q_j. Some scientist provides a prediction for the distribution: $p(1), \ldots, p(n)$. We want to "score" the possible prediction, or rather penalize the prediction based on "how far off" it is from the actual distribution $q_1, \ldots, q_n$.

If $B(p)$ is the penalization for predicting a probability $p \in [0, 1]$, then the expected penalization for a prediction $p(1), \ldots, p(n)$ is

$$\sum_{j=1}^{n} q_j B(p(j)).$$

How would this penalization work?

First, it should be continuous in changing the parameters.

Second, one can predict the distribution after two time steps. So this would be a prediction of a distribution on $\{(i, j) : 1 \leq i, j \leq n\}$; denote it by $p(i, j)$. Now, of course scientists all know logic and mathematics, so their prediction for two time steps ahead is consistent with their prediction for one time step ahead. Specifically, for all i, j we have $p(i, j) = p(j|i)p(i)$, where $p(j|i)$ denotes the predicted probability that state j will follow a state i.

Also, the penalization for predicting two time steps ahead should be the same as that for predicting one time step and then predicting the second step based on that. That is, $B(p(i, j)) = B(p(i)) + B(p(j|i))$.

Together these become: $B(xy) = B(x) + B(y)$ for all $x, y \in [0, 1]$.

It is a simple exercise to prove that $B(x) = \log_m x$ (where the base of the logarithm is a choice).

Thus, the expected penalization is

$$H(p(1), \ldots, p(n)) = -\sum_{j=1}^{n} q_j \log_m p(j).$$

(Negative sign is to show that this is a penalization.)

Proposition B.2.1 *Let $p_j, q_j \in (0, 1]$ be such that $\sum_{j=1}^{n} p_j = \sum_{j=1}^{n} q_j = 1$. Then, for any $m > 1$,*

$$-\sum_{j=1}^{n} p_j \log_m p_j \leq -\sum_{j=1}^{n} p_j \log_m q_j.$$

So $p_j = q_j$ minimizes the expected penalization.

Proof Note that $\log_m(x)$ has second derivative $-\frac{1}{x^2 \log m} < 0$ on $(0, \infty)$. So by Jensen's inequality,

$$\sum_{j=1}^{n} p_j \log_m \frac{q_j}{p_j} \le \log_m \sum_{j=1}^{n} p_j \frac{q_j}{p_j} = 0. \qquad \square$$

So if predicting an outcome distribution $p_1, \ldots, p_n$ "costs" the scientist

$$D(p|q) := \sum_{j=1}^{n} p_j \log \frac{p_j}{q_j},$$

where $q_1, \ldots, q_n$ is the real distribution, then the cost is always positive, and the cost is minimized by choosing the correct distribution.

Exercise B.1 Let $p_j \in (0, 1]$ be such that $\sum_{j=1}^{n} p_j = 1$. Define $r_j = \frac{1}{n}$ for all $1 \le j \le n$.

Show that for any $m > 1$,

$$-\sum_{j=1}^{n} p_j \log_m p_j \le -\sum_{j=1}^{n} r_j \log_m r_j = \log_m n.$$

Appendix C

Coupling and Total Variation

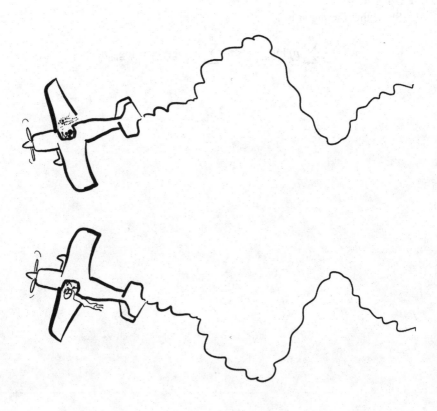

In this appendix we briefly review the connection between the *total variation distance* on probability measures, and *coupling*.

C.1 Total Variation Distance

Definition C.1.1 For two probability measures μ, v on a countable set Ω, define the **total variation distance** between μ and v to be

$$||\mu - v||_{\text{TV}} = \tfrac{1}{2} \sum_x |\mu(x) - v(x)|.$$

Exercise C.1 Let μ, v be two probability measures on a countable set Ω. Show that

$$\sup_{A \subset \Omega} |\mu(A) - v(A)| = \sum_{x:\mu(x)>v(x)} \mu(x) - v(x)$$

$$= \sum_{x:v(x)>\mu(x)} v(x) - \mu(x) = ||\mu - v||_{\text{TV}}. \qquad \triangleright \text{ solution } \triangleleft$$

Exercise C.2 Let μ, v be two probability measures on a countable set Ω. Show that

$$\sum_x (\mu(x) \wedge v(x)) = 1 - ||\mu - v||_{\text{TV}}. \qquad \triangleright \text{ solution } \triangleleft$$

C.2 Couplings

Given two probability measures μ, v on a countable set Ω, a **coupling** of (μ, v) is a probability measure λ on $\Omega \times \Omega$ such that the marginals of λ have the distributions μ, v respectively; that is,

$$\forall x, y \in \Omega, \qquad \sum_y \lambda(x, y) = \mu(x), \qquad \text{and} \qquad \sum_x \lambda(x, y) = v(y).$$

If X, Y are Ω-valued random variables with laws μ and v, respectively, a coupling of (X, Y) is defined to be a coupling of (μ, v). We say that (X, Y) is a coupling of (μ, v) if X has law μ and Y has law v, so that their joint distribution is a coupling of (μ, v).

Couplings enable us to put two probability measures into the same space, so we can compare them. This usually requires creativity in constructing the coupling, but finding a good coupling can be a very powerful tool.

A coupling of two probability measures always exists.

Exercise C.3 Show that the product measure $\lambda = \mu \otimes \nu$ given by $\lambda(x, y) = \mu(x)\nu(y)$ is a coupling of (μ, ν).

Exercise C.4 Let μ, ν be two probability measures on a countable set Ω. Show that if (X, Y) is a coupling of (μ, ν), then for any event A,

$$|\mu(A) - \nu(A)| \le \mathbb{P}[X \ne Y].$$

Conclude that

$$\sup_{A \subset \Omega} |\mu(A) - \nu(A)| \le \inf \mathbb{P}[X \ne Y],$$

where the infimum is over all possible couplings of (μ, ν).

Proposition C.2.1 *Let μ, ν be two probability measures on a countable set Ω. Then there exists a coupling (X, Y) of (μ, ν) such that $\|\mu - \nu\|_{\mathrm{TV}} = \mathbb{P}[X \ne Y]$. Thus, $\|\mu - \nu\|_{\mathrm{TV}} = \min \mathbb{P}[X \ne Y]$ where the minimum is over all possible couplings (X, Y) of (μ, ν).*

Proof Since

$$\|\mu - \nu\|_{\mathrm{TV}} = \sup_{A \subset \Omega} |\mu(A) - \nu(A)| \le \inf \mathbb{P}[X \ne Y],$$

where the infimum is over all couplings of (μ, ν), we only need to find a specific coupling (X, Y) of (μ, ν) satisfying $\mathbb{P}[X \ne Y] = \|\mu - \nu\|_{\mathrm{TV}}$.

For this, set $\varepsilon = \sum_x (\mu(x) \wedge \nu(x))$. We have seen in Exercise C.2 that $1 - \varepsilon = \|\mu - \nu\|_{\mathrm{TV}}$. Define a coupling (X, Y) as follows. Let ξ be a Bernoulli-ε random variable. Define

$$\lambda(x, y) = \begin{cases} \dfrac{\mu(x) \wedge \nu(x)}{\varepsilon} & x = y, \ \xi = 1, \\[2ex] \dfrac{(\mu(x) - \nu(x))_+}{\|\mu - \nu\|_{\mathrm{TV}}} \cdot \dfrac{(\nu(y) - \mu(y))_+}{\|\mu - \nu\|_{\mathrm{TV}}} & \xi = 0. \end{cases}$$

Note that

$$\sum_x \lambda(x, y) = (\mu(y) \wedge \nu(y)) + \frac{1}{\|\mu - \nu\|_{\mathrm{TV}}} \sum_x (\mu(x) - \nu(x))_+ \cdot (\nu(y) - \mu(y))_+$$

$$= \nu(y),$$

$$\sum_y \lambda(x, y) = (\mu(x) \wedge \nu(x)) + \frac{1}{\|\mu - \nu\|_{\mathrm{TV}}} \sum_y (\nu(y) - \mu(y))_+ \cdot (\mu(x) - \nu(x))_+$$

$$= \mu(x),$$

so that λ is indeed a coupling of (μ, ν). For (X, Y) with law λ we obtain that

$$\mathbb{P}[X \neq Y] = \mathbb{P}[\xi = 0] = 1 - \varepsilon = ||\mu - \nu||_{\mathsf{TV}},$$

which is what we wanted to prove. $\qquad\square$

As an example, Exercise 6.35 bounds the total variation distance between binomials. We recall that exercise here.

Exercise C.5 Show that the total variation distance between $\mathsf{Bin}(n, p)$ and $\mathsf{Bin}(n + 1, p)$ is bounded by $\frac{1}{\sqrt{2(n+1)}}$.

$\triangleright$ solution $\triangleleft$

C.3 Solutions to Exercises

Solution to Exercise C.1 :(
Let $E = \{x : \mu(x) > \nu(x)\}$ and let $F = \{x : \nu(x) > \mu(x)\}$. We have

$$\sum_{x \in E} \mu(x) - \nu(x) + \sum_{x \in F} \mu(x) - \nu(x) = \sum_x \mu(x) - \nu(x) = 0,$$

and we obtain that

$$2||\mu - \nu||_{\mathsf{TV}} = \sum_{x \in E} \mu(x) - \nu(x) + \sum_{x \in F} \nu(x) - \mu(x)$$
$$= 2 \sum_{x \in E} \mu(x) - \nu(x) = 2 \sum_{x \in F} \nu(x) - \mu(x).$$

For any $A \subset \Omega$, we have that

$$\mu(A) - \nu(A) = \sum_{x \in A} \mu(x) - \nu(x) \leq \sum_{x \in A \cap E} \mu(x) - \nu(x) = \mu(A \cap E) - \nu(A \cap E)$$
$$\leq \mu(A \cap E) - \nu(A \cap E) + \sum_{x \in E \setminus A} \mu(x) - \nu(x)$$
$$= \sum_{x \in E} \mu(x) - \nu(x) = ||\mu - \nu||_{\mathsf{TV}},$$

and similarly,

$$\nu(A) - \mu(A) \leq ||\mu - \nu||_{\mathsf{TV}},$$

so that

$$\sup_{A \subset \Omega} |\mu(A) - \nu(A)| \leq ||\mu - \nu||_{\mathsf{TV}}.$$

Since $||\mu - \nu||_{\mathsf{TV}} = |\mu(E) - \nu(E)| = |\nu(F) - \mu(F)|$, we are done. :)✓

Solution to Exercise C.2 :(
Let $E = \{x : \mu(x) > \nu(x)\}$. Compute

$$1 - ||\mu - \nu||_{\mathsf{TV}} = \tfrac{1}{2} \sum_{x \in E} (\mu(x) + \nu(x)) - (\mu(x) - \nu(x)) + \tfrac{1}{2} \sum_{x \notin E} (\mu(x) + \nu(x)) - (\nu(x) - \mu(x))$$
$$= \sum_{x \in E} \nu(x) + \sum_{x \notin E} \mu(x) = \sum_x (\mu(x) \wedge \nu(x)). \qquad :)✓$$

Solution to Exercise C.5 :(
See solution to Exercise 6.35.

References

Aldous, David, and Fill, James Allen. 2002. *Reversible Markov chains and random walks on graphs*. Unfinished monograph, recompiled 2014, available at www.stat.berkeley.edu/~aldous/RWG/book.html.

Amir, Gideon, and Kozma, Gady. 2017. Every exponential group supports a positive harmonic function. ArXiv:1711.00050.

Avez, André. 1972. Entropie des groupes de type fini. *Comptes rendus de l'Académie des Sciences Paris Series A–B*, **275**, A1363–A1366.

Avez, André. 1976. Croissance des groupes de type fini et fonctions harmoniques. Pages 35–49 of: *Théorie Ergodique*. Springer.

Bass, Hyman. 1972. The degree of polynomial growth of finitely generated nilpotent groups. *Proceedings of the London Mathematical Society*, **3**(4), 603–614.

Benjamini, Itai, Pemantle, Robin, and Peres, Yuval. 1998. Unpredictable paths and percolation. *The Annals of Probability*, **26**(3), 1198–1211.

Benjamini, Itai, Duminil-Copin, Hugo, Kozma, Gady, and Yadin, Ariel. 2015. Disorder, entropy and harmonic functions. *The Annals of Probability*, **43**(5), 2332–2373.

Benjamini, Itai, Duminil-Copin, Hugo, Kozma, Gady, and Yadin, Ariel. 2017. Minimal growth harmonic functions on lamplighter groups. *New York Journal of Mathematics*, **23**, 833–858.

Blackwell, David. 1955. On transient Markov processes with a countable number of states and stationary transition probabilities. *The Annals of Mathematical Statistics*, **26**, 654–658.

Carne, Thomas Keith, and Varopoulos, Nicholas. 1985. A transmutation formula for Markov chains. *Bulletin des Sciences Mathématiques*, **109**(4), 399–405.

Choquet, Gustave, and Deny, Jacques. 1960. Sur l'équation de convolution $\mu = \mu * \sigma$. *Comptes rendus de l'Académie des Sciences Paris*, **250**, 799–801.

Coulhon, Thierry, and Saloff-Coste, Laurent. 1993. Isopérimétrie pour les groupes et les variétés. *Revista Matemática Iberoamericana*, **9**(2), 293–314.

Cover, Thomas M, and Thomas, Joy A. 1991. *Elements of Information Theory*. John Wiley & Sons.

Derriennic, Yves, et al. 1980. Quelques applications du théoreme ergodique sous-additif. *Astérisque*, **74**, 183–201.

Druţu, Cornelia, and Kapovich, Michael. 2018. *Geometric Group Theory*. American Mathematical Society.

Durrett, Rick. 2019. *Probability: Theory and Examples*. Cambridge University Press.

Dynkin, Evgenii Borisovich, and Malyutov, Mikhail Borisovich. 1961. Random walk on groups with a finite number of generators. *Doklady Akademii Nauk*, **137**, 1042–1045.

Erschler, Anna, and Karlsson, Anders. 2010. Homomorphisms to $\mathbb{R}$ constructed from random walks. *Annales de l'Institut Fourier*, **60**(6), 2095–2113.

Frisch, Joshua, Hartman, Yair, Tamuz, Omer, and Vahidi-Ferdowsi, Pooya. 2019. Choquet–Deny groups and the infinite conjugacy class property. *Annals of Mathematics*, **190**(1), 307–320.

Furstenberg, Harry. 1963. A Poisson formula for semi-simple Lie groups. *Annals of Mathematics*, **77**(2), 335–386.

Furstenberg, Harry. 1971. Random walks and discrete subgroups of Lie groups. *Advances in Probability and Related Topics*, **1**, 1–63.

Furstenberg, Harry. 1973. Boundary theory and stochastic processes on homogeneous spaces. *Harmonic Analysis on Homogeneous Spaces*, **26**, 193–229.

Grigorchuk, Rostislav. 1980. Burnside problem on periodic groups. *Funktsional'nyi Analiz i ego Prilozheniya*, **14**(1), 53–54.

Grigorchuk, Rostislav. 1984. Degrees of growth of finitely generated groups, and the theory of invariant means. *Izvestiya Rossiiskoi Akademii Nauk. Seriya Matematicheskaya*, **48**(5), 939–985.

Grigorchuk, Rostislav. 1990. On growth in group theory. Pages 325–338 of: *Proceedings of the International Congress of Mathematicians*, vol. 1. Springer.

Gromov, Michael. 1981. Groups of polynomial growth and expanding maps (with an appendix by Jacques Tits). *Publications Mathématiques de l'IHÉS*, **53**, 53–78.

Guivarc'h, Yves. 1973. Croissance polynomiale et périodes des fonctions harmoniques. *Bulletin de la Société Mathématique de France*, **101**, 333–379.

Kaimanovich, Vadim A., and Vershik, Anatoly M. 1983. Random walks on discrete groups: boundary and entropy. *The Annals of Probability*, **11**(3), 457–490.

Karlsson, Anders, and Ledrappier, François. 2007. Linear drift and Poisson boundary for random walks. *Pure and Applied Mathematics Quarterly*, **3**(4), 1027–1036.

Kesten, Harry. 1959. Full Banach mean values on countable groups. *Mathematica Scandinavica*, 146–156.

Kleiner, Bruce. 2010. A new proof of Gromov's theorem on groups of polynomial growth. *Journal of the American Mathematical Society*, **23**(3), 815–829.

Korevaari, Nicholas J., and Schoen, Richard M. 1997. Global existence theorems for harmonic maps to non-locally compact spaces. *Communications in Analysis and Geometry*, **5**(2), 333–387.

Lee, James R., and Peres, Yuval. 2013. Harmonic maps on amenable groups and a diffusive lower bound for random walks. *The Annals of Probability*, **41**(5), 3392–3419.

Lyons, Russell. 1995. Random walks and the growth of groups. *Comptes rendus de l'Académie des sciences. Série 1, Mathématique*, **320**(11), 1361–1366.

Lyons, Russell, and Peres, Yuval. 2016. *Probability on Trees and Networks*. Cambridge University Press. Available at https://rdlyons.pages.iu.edu/.

Lyons, Russell, Peres, Yuval, Sun, Xin, and Zheng, Tianyi. 2017. Occupation measure of random walks and wired spanning forests in balls of Cayley graphs. ArXiv:1705.03576.

Meyerovitch, Tom, and Yadin, Ariel. 2016. Harmonic functions of linear growth on solvable groups. *Israel Journal of Mathematics*, **216**(1), 149–180.

Meyerovitch, Tom, Perl, Idan, Tointon, Matthew, and Yadin, Ariel. 2017. Polynomials and harmonic functions on discrete groups. *Transactions of the American Mathematical Society*, **369**(3), 2205–2229.

Milnor, John. 1968a. Growth of finitely generated solvable groups. *Journal of Differential Geometry*, **2**(4), 447–449.

Milnor, John. 1968b. Problem 5603. *American Mathematical Monthly*, **75**, 685–686.

Mok, Ngaiming. 1995. Harmonic forms with values in locally constant Hilbert bundles. Pages 433–453 of: *Journal of Fourier Analysis and Applications*. CRC Press.

Norris, James Robert. 1998. *Markov Chains*. Cambridge University Press.

Ozawa, Narutaka. 2018. A functional analysis proof of Gromov's polynomial growth theorem. *Annales scientifiques de l'École normale supérieure*, **51**(3), 549–556.

Perl, Idan, and Yadin, Ariel. 2023. Polynomially growing harmonic functions on connected groups. *Groups, Geometry, and Dynamics*, DOI 10.4171/GGD/660.

Peyre, Rémi. 2008. A probabilistic approach to Carne's bound. *Potential Analysis*, **29**(1), 17–36.

Pólya, Georg. 1921. Über eine Aufgabe der Wahrscheinlichkeitsrechnung betreffend die Irrfahrt im Straßennetz. *Mathematische Annalen*, **84**(1), 149–160.

Raoufi, Aran, and Yadin, Ariel. 2017. Indicable groups and $p_c < 1$. *Electronic Communications in Probability*, **22**, 1–10.

Rosenblatt, Joseph. 1981. Ergodic and mixing random walks on locally compact groups. *Mathematische Annalen*, **257**(1), 31–42.

Shalom, Yehuda, and Tao, Terence. 2010. A finitary version of Gromov's polynomial growth theorem. *Geometric and Functional Analysis*, **20**(6), 1502–1547.

Tao, Terence. *What's New* (blog). https://terrytao.wordpress.com.

Tits, Jacques. 1972. Free subgroups in linear groups. *Journal of Algebra*, **20**(2), 250–270.

Tointon, Matthew C.H. 2016. Characterisations of algebraic properties of groups in terms of harmonic functions. *Groups, Geometry, and Dynamics*, **10**(3), 1007–1049.

Varopoulos, Nicholas. 1985. Long range estimates for Markov chains. *Bulletin des Sciences Mathématiques*, **109**(3), 225–252.

Vershik, Anatolii Moiseevich, and Kaimanovich, Vadim Adol'fovich. 1979. Random walks on groups: Boundary, entropy, uniform distribution. *Doklady Akademii Nauk*, **249**, 15–18.

Woess, Wolfgang. 2000. *Random Walks on Infinite Graphs and Groups*. Cambridge University Press.

Wolf, Joseph A. 1968. Growth of finitely generated solvable groups and curvature of Riemannian manifolds. *Journal of Differential Geometry*, **2**(4), 421–446.

Index

378